Berndt Feuerbacher · Heinz Stoewer (Eds.)

Utilization of Space

Berndt Feuerbacher · Heinz Stoewer (Eds.)

Utilization of Space

Today and Tomorrow

With 315 Figures

 Springer

Professor Dr. Berndt Feuerbacher
Deutsches Zentrum für Luft- und Raumfahrt (DLR)
Institut für Raumsimulation
Linder Höhe
D-51147 Köln
Germany

Professor Heinz Stoewer
SAC Space Associates
Drachenfelsstraße 9
D-53757 Bonn
Germany

Cover photos from ESA/DLR/FU Berlin (G. Neukum) and Eumetsat

Library of Congress Control Number: 2005928158

ISBN-10   3-540-25200-2 Springer Berlin Heidelberg New York
ISBN-13   978-3-540-25200-9 Springer Berlin Heidelberg New York

Springer is a part of Springer Science+Business Media
springeronline.com

© Springer-Verlag Berlin Heidelberg 2006
Printed in Germany

Typesetting: Dataconversion by authors
Final processing by PTP-Berlin Protago-TEX-Production GmbH, Germany
Cover-Design: medionet AG, Berlin
Printed on acid-free paper   62/3141/Yu – 5 4 3 2 1 0

Preface

Utilization of space, what for? This book attempts to answer this question!

With this volume we intend to provide a single reference for the broad field of space utilization. Of all the books we know, this is the first to cover all aspects of scientific and application oriented activities in space, even though with limits.

We have attempted to document the current state of the art and open at the same time a perspective towards the future. We also want to bridge the gap between the many popular books dealing with space, and academic textbooks on specific research fields in space science, applications, or technology.

The book addresses a professional readership, while still offering much information to interested laymen. It should well serve students of physics, geodesy, informatics, mechanical, electrical or aerospace engineering. It should give scientists at universities and research institutions an overview of the extensive opportunities offered by space investigations, and industrial engineers and managers additional insights into the commercial potential of space. It should also help decision makers in agencies, governments, and industry to understand better the multidisciplinary interrelationships between utilization aspects and space infrastructure.

Our co-authors are amongst the world's most distinguished leaders in their fields. Their respective areas of research are presented by many illustrations and focus on the central messages rather than attempting to be exhaustive. Each chapter lists references for further reading, highlighting original publications, the most relevant textbooks, and major internet resources.

We are indebted to Susan Giegerich for ironing out language and style peculiarities, and to Sonja Gierse-Arsten for helping to format the book. We are in particular grateful to NASA, ESA and DLR for permitting to use their illustrations.

Cologne, May 2005

Berndt Feuerbacher and Heinz Stoewer

Contributors

Berndt Feuerbacher, Editor

Prof. Dr. Berndt Feuerbacher is Director of the Institute of Space Simulation at the German Aerospace Center (DLR) in Cologne, Germany, and holds the chair of Space Physics at the Ruhr-University in Bochum. He is active in several technical learned societies, amongst those he is full member of the International Academy of Astronautics, member of the Executive Board of the European Physical Society and Council Member the German Aerospace Society. He advises ESA and various national space agencies in a diversity of functions. He received his PhD in physics at the University of Munich. By joining the European Space Agency 1968 in the Space Science Department at ESTEC, Noordwijk (Holland), he started a career in space research with increasing responsibilities continuing until present. Berndt Feuerbacher was, among other functions, scientific investigator for lunar surface samples from the Apollo missions. He was active in the early developments of the orbital laboratory Spacelab and followed the First Spacelab Payload as the Project Scientist. He initiated and participated in numerous microgravity experiments in the field of materials sciences on platforms like Spacelab, Eureca, Foton, and Mir. He was active in scientific experiments on Mars and under his guidance the comet landing probe "Philae" for the Rosetta-Mission was designed and built.

In 1990 to 1993 he acted as a Founding Director for the institutes of Planetary Exploration and Space Sensor Technology in Berlin in their transition from the former East German Academy of Sciences. He holds 7 patents and has published 10 books, in addition to more than 170 articles in scientific journals.

Heinz Stoewer, Associate Editor

Prof. Dipl. – Ing. Heinz Stoewer, M. Sc. is President of the Space Associates GmbH, emeritus Professor of Delft University of Technology and member of several international scientific and industrial boards, such as the Board of Trustees of the International Academy of Astronautics (IAA), the Senate of the German Aerospace Society (DGLR), past Dean for Systems Engineering of the EADS/Astrium International Academy, chair of the Dutch Space Advisory Sub-Committee, Member of the Scientific Councils of the Dutch Aerospace Centers (NLR/NIVR) and the Dutch Space Research Organization (SRON), President of the International Council on Systems Engineering (INCOSE). He has authored numerous scientific/technical publications and holds prestigious national and international awards, such as the Medal of the German Parliament (Bundesrat) and exceptional NASA honours. Prof. Stoewer holds German and US degrees in technical physics, economics and systems management. He started his career in space industry in Germany (EADS) and the USA (Boeing), working on launch vehicles and human space systems. He became the first European Spacelab Programme Manager and founder of the Systems Engineering and Programmatics Department of the European Space Agency, ESA, at its Technical Centre ESTEC in The Netherlands. He served as Managing Director Utilization Programmes, in the German Space Agency, DARA, responsible for most of the national and international German space projects, until his retirement in 1995. He was Executive Chair of CEOS, the ESA Programme Board for Earth Observation and Meteorology, Member of the ESA Council and the EU Space Advisory Group.

Edward (Ed) Ashford

Ed Ashford has more than 40 years experience in the aerospace industry, during which he has worked in practically all areas of the field, in industry, the European Space Agency ESA, and academia. He started in the USA in 1962 working as a GN&C engineer. In 1970, he joined ESRO's then fledgling telecom satellite program. In ESRO and its successor ESA, he held a variety of increasingly senior posts, culminating in the early 1990 when he was appointed Head of the Communication Satellites Department. In 1997, Ed Ashford returned to the USA as Vice President (VP) for a subsidiary of Lockheed Martin Corporation. 1999, he moved to Luxembourg to work for SES ASTRA, helping to define its next generation space system. He later served as VP for Technology Development with SES GLOBAL. Since 2003 he has operated as an independent aerospace consultant. He is a Senior Member of the AIAA, past Chair of the International Astronautical Federation's Space Communications Committee, and a member of the International Academy of Astronautics. He is author of more than 50 technical papers and three books and serves on the Advisory Board and Curriculum Committee of the Delft University of Technology's Master for Space Systems Engineering program.

Stefan Dech

Prof. Dr. Stefan Dech, Ph.D. 1990 and habilitation 1997, at Würzburg University, is Director of the German Remote Sensing Data Center (DFD), an institute focused on earth observation services, methodological research and applications development. The institute is part of the German Aerospace Center (DLR) and is located in Oberpfaffenhofen near Munich (headquarter) and in Neustrelitz near Berlin. He has taught remote sensing at universities in Eichstätt, Munich and Heidelberg and holds an endowed chair in remote sensing at the Geographical Institute of the Bavarian Julius Maximilian University in Würzburg. He is a science consultant for diploma theses, dissertations, and habilitations in geography. Research emphases have been on operationalizing algorithms and systems processes for satellite remote sensing applications in environmental research; the conceptual design and realization of value-adding chains for generating geophysical satellite data products; desertification and GIS methodology in central Asia, and digital satellite image processing for visualization and computer animation. In 2004 he was visiting scientist at Scripps Institution of Oceanography in La Jolla, California, affiliated with the University of California in San Diego.

Ralf-Jürgen Dettmar

Prof. Dr. Ralf-Jürgen Dettmar is professor of astronomy at Ruhr-University Bochum since 1994. Born in 1955 he received his academic degrees from the University of Bonn including the doctorate in 1986 with a thesis project conducted at the Max-Planck-Institute for Radio Astronomy.

Between 1986 and 1994 he held a postdoctoral position at the University of Bonn, was a postdoctoral fellow at Lowell Observatory in Flagstaff, AZ (USA), and served as a staff member of the European Space Agency (ESA) at the Space Telescope Science Institute in Baltimore, MD (USA). His scientific interests are in the evolution of galaxies with a specialization in the interstellar medium in spiral galaxies. This particular topic requires to use observations in almost all parts of the electromagnetic spectrum and thus he has used optical and radio telescopes all over the world as well as astronomical satellites such as the Hubble Space Telescope or the XMM-Newton Observatory.

Hansjörg Dittus

Dr. Hansjörg Dittus studied Physics and Geophysics at the Ludwig-Maximilians University München, Germany, concluding in 1986 with a Ph.D. in Geophysics. In 1987, he moved to the Centre of Applied Space Technology (ZARM) of the University of Bremen, Germany, where he became Deputy Institute Director in 1990. Between 1987 and 1991 he was the Project Manager of the Bremen Drop Tower Project, a large microgravity laboratory installation. Since 1991, he also became Head of the Gravitational Physics Experimental Department (today Department of Gravitational Physics and Space Technology) at ZARM. His interests concern fundamental physics experiments (gravitation and quantum physics) under space conditions. He is member of numerous working groups for satellite and ISS Experiments (e.g. MICROSCOPE, STEP, LISA, OPTIS, HYPER). Since 2001, he is Chairman of the ESA Topical Team "Fundamental Physics on ISS".

Rupert Gerzer

Prof. Dr. Rupert Gerzer is Head of the Institute of Aerospace Medicine at the German Aerospace Center in Cologne since 1992 and Professor and Chairman of the Institute of Aerospace Medicine at Aachen University since 1993. He is a medical doctor with training in molecular and clinical medicine (University of Heidelberg, 1977-1980; Vanderbilt University, Nashville, USA, 1981-1983 and University of Munich, 1984-1992).

He has authored and coauthored over 200 publications and participated as a scientist and in his present function in many space missions. In 2003, he received the life science award of the International Academy of Astronautics. He served as President of the German Society for Aerospace Medicine from 1999 to 2001 and as Trustee of the International Academy of Astronautics since 2000.

Hartmut Graßl

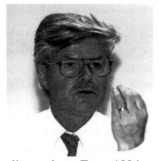

Prof. Dr. Hartmut Graßl is a physicist by training, but spent his scientific career in geosciences, in particular meteorology. Trained at the University of Munich in radiative transfer he started in his Ph.D. work (1967 to 1969) with remote sensing by deriving cloud droplet-size distributions from spectral solar radiances, when pointing a multi-channel spectrometer to the sun. Since he left Munich for the University of Mainz (1971) his main interest became optical properties of aerosol particles, both in the solar and terrestrial spectral domain, and their influence on clouds in a mix of experimental and radiative transfer model studies.

Directing institutes at the GKSS Research Centre in Geesthacht and Hamburg (Meteorological Institute of the University and Max Planck Institute for Meteorology) he also engaged in the transfer of knowledge from science to the public and to policy makers. From 1994 to 1999 he was Director of the World Climate Research Programme based at the World Meteorological Organization, WMO, in Geneva, Switzerland. He is a sought after expert on matters of climate and global change, advising many of the world's leading agencies and scientific organizations.

Günter Hein

Prof. Dr. Guenter W. Hein is Director of the Institute of Geodesy and Navigation of the University FAF Munich, Germany. He studied Surveying Engineering and Satellite Geodesy at the Universities in Mainz, Stuttgart and Darmstadt where he got also his Ph. D. and worked as a scientific research associate. After his habilitation in gravity gradiometry he became a "Privatdozent". At the age of 33 years he was appointed full professor at the University FAF Munich. His work in GPS had already started when he formed a research group that did pioneering R&D in the field of real-time kinematic GPS positioning and GPS/INS integration. He has contributed to more than 200 scientific publications in Geodesy and Satellite Navigation, has received over 100 research grants, was a Visiting Senior Scientist at the U.S. National Geodetic Survey, a visiting professor at the University of New South Wales in Sydney, at the University of Maine, and is a member of various national and international associations. He received Best Paper Awards in 1988, 1991 and 2000 from the U.S. Institute of Navigation. He is currently the European Technical Representative on the Institute of Navigation's Satellite Division Executive Committee, an instructor for Navtech Seminars, Inc. and Guest Professor in the Space Systems Engineering Master Program of Delft University. He is a member of the Advisory Boards of GPS World and Galileo's World. He is a member of the European Commission's Galileo Signal Task Force and joined as a German delegate and technical expert for the negotiations between the EU and the US on the interoperability of GPS and Galileo in the last years. In Sept. 2002 he received the prestigious Johannes Kepler Award of the US. Institute of Navigation, the the highest worldwide award in satellite navigation.

Ruth Hemmersbach

Dr. Ruth Hemmersbach is a zoologist and cell biologist in the Institute of Aerospace Medicine of the German Aerospace Center (DLR) in Cologne and teaches at the Rheinische Friedrich–Wilhelms University of Bonn, Germany. She has been active in gravity-related research on cellular systems for more than 20 years, e.g. as principal investigator for several biological experiments under varied gravitational stimulation in microgravity (parabolic flights and space shuttle), functional weightlessness (clinostats), and hypergravity (centrifuges). In 1991 she received the Junior Scientist Award of the DLR, in 1992 the Zeldovich Award of the COSPAR Life Sciences Group. She has published several papers and written a book related to "Gravity and the behavior of unicellular organisms".

Gerda Horneck

Dr. Gerda Horneck is former Deputy Director of the Institute of Aerospace Medicine at DLR and former Head of the Division of Radiation Biology at the Institute. She is trained in microbiology, radiation biology and genetics. Since Apollo 16, she has coordinated several national and European space projects in the fields of radiation biology, cell biology, and exobiology, and has recently led a study on human survivability in exploratory missions. She has been advisor to national and international space agencies in space-related issues and is author of several books related to astrobiology and search for life in space.

Ralf Jaumann

Dr. Ralf Jaumann, Ph.D. and habilitation at Munich University, is head of the Planetary Geology Section of the DLR Institute of Planetary Research in Berlin. He teaches Planetary Sciences at the University of Munich. His involvement in space missions comprises instrument development as well as scientific investigations for the exploration of Mars (ESA Mars Express), Saturn (NASA/ESA Cassini/Huygens), Venus (ESA Venus Express Mission), Comets (ESA Rosetta Mission) and Asteroids (NASA Dawn Mission). His many publications are devoted to the origin and evolution of planetary bodies, the composition of planetary surfaces and geologic processes in the solar system. He was visiting scientist at the University of Hawaii, the NASA Jet Propulsion Laboratory in Pasadena and the Brown University in Providence, USA.

Tillmann Mohr

Dr. Tillmann Mohr studied mathematics, physics and meteorology at the universities of Würzburg and Frankfurt. He graduated 1970 with a thesis on tropical meteorology. In 1965 Dr. Mohr joined the German Meteorological Organization. Serving in different functions in the Central Office he was appointed as its President in 1992. His involvement in satellite activities started in 1971 with ESRO, later ESA, as member and chairman of working groups and program boards of the METEOSAT program. Under his chairmanship the preparations for the establishment of the European Meteorological Satellite Organization, EUMETSAT in Darmstadt, Germany, were finalized. After its establishment in 1986 he headed the German delegation to the EUMETSAT Council and served 1989 – 1993 as chair of its Policy Advisory Committee. In 1995 he was appointed Director General of EUMETSAT, a position which he served until July 2004. In Nov. 2004 he accepted the position as Special Advisor on Satellite Matters to the Secretary General of the World Meteorological Organization.

Ron Nomen

Ir. Ron Noomen is Assistant Professor at Delft University of Technology from where he graduated in aerospace engineering in 1983. Since then he has been active in the area of Satellite Laser Ranging for a number of applications, such as precise orbit computation, crustal deformation studies (WEGENER-MEDLAS project), Earth's geocenter, quality control, and others. Main professional interests are orbital mechanics and space mission design. He is also Analysis Coordinator of the International Laser Ranging Service.

Lorenz Ratke

Prof. Dr. Lorenz Ratke holds a chair in Metal Physics at the University of Aachen (RWTH) and is head of the research group "Aerogels & Casting" at the Institute of Space Simulation at the German Aerospace Center DLR in Cologne. He received his diploma in Physics at the University of Muenster, Ph.D. in Metal Physics from the University of Aachen.

Prof. Ratke held positions at the Vereinigte Aluminium Werke Bonn, the Technical University in Clausthal, the Max-Planck-Institute for Metals Research in Stuttgart and visiting positions at the Technical University of Miskolc, Hungary and the University of Iowa. He has received the Georg-Sachs-Award of the German Society of Materials, Acta Metallurgica et Materialia Outstanding Paper Award, Senior Scientist Award of the DLR, the Silver Medal of the University of Miskolc and innovation award of the city of Cologne. He has published over 180 papers in the areas of corrosion, powder metallurgy, in-situ composites, polyphase solidification, coarsening and aerogels, is publisher of two books and has written a monograph on coarsening processes.

Johannes Schmetz

Dr. Johannes Schmetz is the Head of Meteorology, a Division in the Programme Development Department of EUMETSAT in Darmstadt, Germany. His Division develops meteorological satellite applications and covers the scientific and technical meteorological aspects of current and future satellite programs of EUMETSAT.

Dr. Schmetz received a Doctoral degree from the University of Cologne in 1981 on the topic of cloud-radiation interaction. He worked as research scientist at the Max-Planck-Institute for Meteorology in Hamburg and then as Senior Scientist with the European Space Agency (ESA). In 1995 he joined EUMETSAT in his current position.

Dr. Schmetz obtained the habilitation at the Johann-Wolfgang-von-Goethe-University of Frankfurt where he teaches courses in meteorology. He authored over 60 peer reviewed journal articles and serves on various international committees.

Ernst Schrama

Dr. Ir. Ernst J.O. Schrama received his Ph.D. in 1989 from Delft University of Technology. On the basis of a prestigious stipend from the Netherlands Academy of Sciences (KNAW) he spent 2 visits at the Space Geodesy branch of NASA's Goddard Space Flight Center. In 1994 he became Assistant Professor at Delft University of Technology and in 2001 was promoted Associated Professor. His main interests are in orbital dynamics and gravity field mapping, application of radar satellite altimetry missions for studying tidal physics, and related ocean model data assimilation studies. He acted as co- and principle investigator on several altimeter missions, and was involved in the development of ESA's GOCE gravity gradiometer mission.

Tilman Spohn

Prof. Dr. Tilman Spohn is Director of the Institute of Planetary Research of the German Aerospace Research Center in Berlin and Professor for Planetology at the Westfälische Wilhelms-University in Münster. He was trained in physics and geophysics at the University Frankfurt, where he received his Ph.D. in 1978. Tilman Spohn is member of the Space Science Advisory Committee of ESA and President of the Planetology Section of the American Geophysical Union. His scientific interest is focused on geophysics and thermodynamics of planetary bodies, comet physics, and dynamics and evolution of terrestrial planets. He participated scientifically in many space missions including Rosetta and Mars Express, and published more than 100 papers in scientific journals.

Bert Vermeersen

Dr. Bert L.A. Vermeersen graduated in astrophysics at Utrecht University (1987) and obtained his doctor's degree in geophysics at the same university in 1993. He spent post-doctoral terms with an European Space Agency (ESA) external fellowship in Bologna (1994-1996), with an Alexander von Humboldt fellowship in Stuttgart (1996-1998) and with an Italian Space Agency (ASI) fellowship in Milan (1998-1999).

In 1999 he became Assistant Professor at Delft University of Technology. His major areas of interest are on the interface between earth and planetary sciences and space geodesy, including post-glacial rebound, post-seismic deformation and rotational and gravitational variations. He is a member of ESA's GOCE satellite mission advisory group and president of the Geodynamics Division of the European Geosciences Union (EGU).

Pieter Visser

Dr. Ir. Pieter N.A.M. Visser graduated in aerospace engineering at Delft University of Technology (1988) where he later obtained his doctor's degree (1992). After spending a year at the Center for Space Research at the University of Texas at Austin, he returned to Delft, where he now has the position of Associate Professor. His major areas of interest are orbital mechanics and observation of the earth's gravity and magnetic fields by space-borne techniques. He is member of several European Space Agency mission and science advisory groups (GOCE, Swarm, CryoSat, METOP) and chair of the Technical Panel on Satellite Dynamics of the Committee on Space Research (COSPAR).

List of Contents

Space: Beyond the Horizon

Looking down: our Earth

Looking up: Stars and Planets

Any Limits?

Space: Beyond the Horizon

Overleaf image: The moon rising in the earth atmosphere (DLR)

1 Space Utilization

by Berndt Feuerbacher

The space age began on October 4, 1957 near the city of Leninsk (today: Baikonur), when a Russian Semjorka Rocket launched the first orbiting satellite "Sputnik." The characteristic "beep" received by many stations all over the world acoustically opened a new era, marked by the ability of mankind to access space. Starting from this event, outstanding progress has been made worldwide, including historical moments like the first step of a human on the moon. The technical challenge of space flight and astronautics is fascinating and innovating. Its final justification emerges from the benefit the utilization of these technical systems will bring to mankind. The present volume focuses on this point, discussing the utilization of space technology in all areas of application from scientific research to commercial applications on earth.

1.1 Space Has Changed Our Life

Space pervades our life in many aspects, sometimes quite obviously, in other cases largely unnoticed. Being a major source of innovation, it contributes to the improvement of welfare and quality of life. Space research as a source of knowledge expands the understanding of our origin and puts our position in the universe in perspective. From the viewing position in space, our home planet is seen in all its beauty and vulnerability. The fascination of space inspires our youth and opens their minds for science and technology. The space industry is an important factor in economic growth and international competitiveness. It is a driver for the development of technical progress. Space technology is also recognized as an important ingredient of national security and, even though this seems to be a contradiction, to international cooperation.

In the following, a few examples will be given to demonstrate the impact space activities have on our daily life.

Technological and Economic Impact

As we walk along the street and see the large number of satellite dishes on private homes and public buildings, the impact of *space communication* on society is apparent. Much less obvious is the role space plays in today's international broadcast systems. We expect video images from an event on the other side of the globe within hours in the newscast on our home TV screen. This is only possible using broadband space transmission channels.

In the 1970s space communication satellites started to revolutionize intercontinental communication by replacing transoceanic cables. We got used to a rapid response on the telephone, free of noise and echo effects. In the meantime, glass fiber cables can do the job in a similar way. Space communication now concentrates on other markets and improved technologies expanding the available bandwidth, for example for high-speed internet access. One important task is information dissemination in large-area developing countries, where surface cabling or wireless networks do not exist and the population is excluded from the benefits of the information society. A continuous increase in communication demand is expected globally, where space based techniques, particularly in the broadband regime, will contribute their share.

Satellite navigation is a common tool today, and modern cars are widely equipped with suitable devices to find the way on the road. In principle, we could use road signs or ask somebody, but on sea this becomes difficult and the impact of space technology is more apparent. Space navigation systems are used today in many applications to provide positioning down to the sub-centimeter scale and time

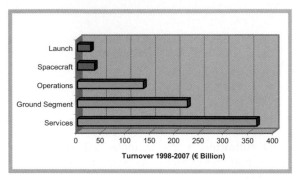

Figure 1-1: Value added projection for worldwide space communication and navigation, extrapolated over the time frame 1998 to 2007. Note the turnover on the ground (operations, ground segment, and services) is about ten times the space investment (European Commission 2001)

information on the nanosecond level. Apart from navigation on the ground, on the sea and in the air, it supports industries like agriculture and construction and helps to save lives. Most citizens are unaware of the fact that basic public services of our society, like telecommunication networks and energy provision, depend crucially on the global high precision time information provided by navigation satellite systems.

In both communication and navigation the most far reaching economic impact is made by the *value added*, not by building satellites and launching them. This is illustrated in figure 1-1, showing the value added chain of the global commercial market in an extrapolation from 1998 to 2007 (European Commission 2001). The expenditure in space, such as for satellite manufacturing and launch, is small compared to the turnover achieved on the ground. This demonstrates the lever action of space investment: a dollar spent in space leads to nearly 10 dollars turned over on the ground. For space navigation systems, this becomes quite apparent. The satellites generate a set of numbers relating to spatial coordinates and time. It is only the added value of the receiving system on the ground that converts these numbers into information useful for the customer, like road guidance or information for a search and rescue operation.

We all have become accustomed to the daily *weather report*, including a satellite map on TV, and benefit from its pretty high reliability. We know with some certainty the weather of tomorrow, which in past centuries used to be just a bit more than a matter of educated guessing. Obviously for agriculture and tourism weather predictions for a full week are invaluable. But the global view of space based weather services provides even more benefits. They warn of severe events like tropical storms, they contribute to the observation of our climate, including the influences of human activity, and they provide reliable high-altitude wind field measurements to safely guide air traffic on routes minimizing fuel consumption.

Figure 1-2: A three dimensional view of Mount Kilimanjaro taken from space. This image was obtained during the SRTM mission using radar interferometry. Apart from the spatial information it delivers height resolution of 6 m on a global scale. The color coding was introduced later to indicate heights (NASA/USGS)

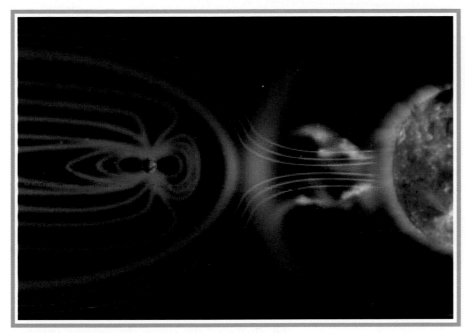

Figure 1-3: Schematic representation of Sun-Earth relations indicating solar wind flux emanating from the Sun. The magnetosphere protects the immediate Earth environment from hazardous radiation and particle impacts (DLR/ESA)

Many aspects of *earth observation* are of direct benefit to society. High resolution images are available for all kinds of purposes with only short delays from virtually any place in the world. Spectroscopic resolution gives additional value to agriculture and forestry applications. Radar techniques ease restrictions of cloud-free skies and solar illumination. In addition, radar interferometry is now able to map the earth in three dimensions with an accuracy hitherto unknown (figure 1-2). These new developments continue to integrate into our daily life in a way that is often not noticed on a short time scale, but with massive impact in the long run.

Space medicine is widely seen as a service for astronauts. This in fact was the origin of medical research in space. Today the focus is mainly on health care for everybody. The space microgravity environment allows study of a number of phenomena normally experienced with aging, but on a much shorter time scale and, in some instances, in a reversible way. Among these are cardiovascular diseases, bone demineralization, muscle atrophy, and problems with the vestibular system, the human organs for gravity sensing responsible for body equilibrium. Beyond this there is a major general impact of space medicine on human health care. In common life, the patient visits his physician, and medicine is based to a large extent on statistical experience with case studies including many patients. This is different in space medicine, where very few patients, the astronauts, require an individual rather than statistical approach, and they cannot visit their doctor from orbit. With recent advances in telemedicine, stimulated by space developments, a change of paradigm is entering health care, whereby the individual patient moves into the center of interest, with additional possibilities to consult specialist medical expertise remotely.

Scientific and Cultural Impact

Space research has led to a wealth of scientific discoveries. A number of these findings made by means of space techniques had a major impact on our present thinking, not only among scientists, as one would expect, but also on society at large.

Following the discovery of the earth radiation belts by an instrument of James A. van Allen in 1958 on the Explorer 1 satellite, the first major achievement in the scientific exploration of space was the detailed investigation of the *earth magnetosphere* (figure 1-3). Stimulated by the observation of polar light phenomena and theoretical predictions on their origin,

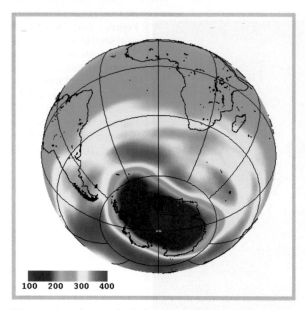

Figure 1-4: Ozone depletion over the Antarctic as observed by the Global Ozone Monitoring Experiment (GOME) on the Earth Resources Sensing (ERS) satellite of ESA. The image is composed of data obtained on 11 October 1996. The color scale on the lower left indicates the ozone column height in Dobson units (ESA/DLR)

the interaction of the earth magnetic field with that of the sun and the solar wind was studied by means of sounding rockets and satellites in near-earth or highly eccentric orbits. This resulted in a detailed understanding of structure, dimension and operation of the magnetosphere, which protects life on earth from high-energy radiation. Today continuous observation of the sun-earth interaction gives us real time information on "space weather", the dynamic processes in the earth magnetic environment.

In 1985 the British Antarctic Survey alarmed the world by demonstrating a depletion of the ozone layer in the upper atmosphere over the Antarctic region. Soon space instruments were used, in particular the Total Ozone Mapping Spectrometer (TOMS) on board various satellites, and later the Global Ozone Monitoring Experiment (GOME) on the Earth Resources Sensing (ERS) satellite, to provide a global view of ozone distribution, confirming the dramatic formation of an *ozone hole* (figure 1-4). This was the first evidence for the nowadays widely accepted fact that anthropogenic influences are in-

troducing a change in our environment on a global scale. It was particularly frightening as it led to the depletion of an essential constituent of the atmosphere which protects life on earth from hazardous ultraviolet radiation of the sun. These and other observations (like the greenhouse effect) indicate a change in world climate originating from human activity, which is a scary view and led to a change in mentality on a global scale. Major efforts by the governments of several nations to counteract such developments were initiated, including worldwide treaty initiatives. Today continuous observations monitor the ozone level in the atmosphere from space on a routine basis. Their results lead to a better understanding of the chemistry and dynamics involved, but do by no means reduce the alarming facts calling for global countermeasures.

In 1964 Arno Penzias and Robert Wilson, two engineers at the Bell Telephone Laboratories, searched for the source of noise in a microwave horn antenna and discovered the *cosmic background radiation*. This soon was identified as a fundamental breakthrough and was lauded with the Nobel Prize in Physics in 1978. Following earlier theoretical predictions, the discovery confirmed the Big Bang theory of the origin of our universe, but only rudimental information was possible from the ground due to disturbances arising from the "hot" earth. Space observations initiated by the Cosmic Background Explorer (COBE) confirmed the shape of the

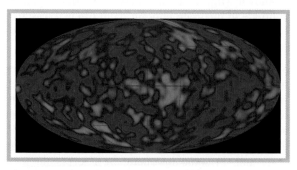

Figure 1-5: Cosmic background radiation as observed by the Cosmic Background Explorer COBE satellite in 1992. The structures represent tiny fluctuations in the sky brightness at a level of one part in one hundred thousand. The radiation is a remnant of the Big Bang, and the fluctuations are the imprint of density contrast in the early universe (NASA)

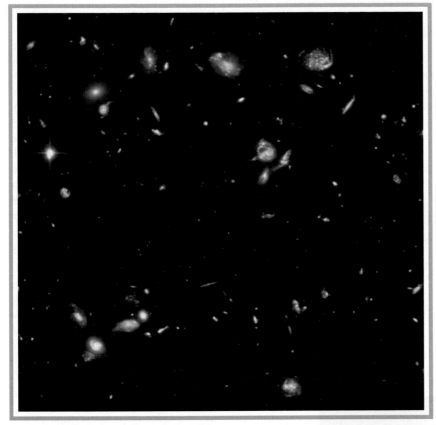

Figure 1-6: Ultra Deep Space image taken by the Hubble Space Telescope in 2004. It results from a total exposure time of more than 11 days over 400 orbits. It sees young galaxies just 800 million years after the Big Bang, in the so called "dark ages" when the first stars reheated the cold, dark universe. They reveal shapes far different from the older galaxies seen today (NASA)

spectrum as relating to a blackbody at 2.7 K, and allowed first measurements of the spatial structure of this radiation (figure 1-5). Our present understanding of the early development of the universe, the formation of stars and galaxies, relies to a large extent on these and later space measurements of the cosmic background radiation.

The Hubble Space Telescope had a bad start when it was launched in April 1990. Due to an avoidable manufacturing error in the optical system the images returned were disappointingly blurred. This situation changed after the repair action three years later, which revealed the incredible power of astronomical imaging from space. The beautiful pictures showing spectacular exotic worlds in our universe are now common property, and most people appreciate them simply for their decorative appearance. Astrophysical breakthroughs are reported at high frequency from the Hubble Telescope, but the most dramatic results arise from views in areas of the sky that are

virtually empty. Very long exposures in such dark fields reveal galaxies very far away and therefore very young. The recent Ultra Deep Field observation (figure 1-6), obtained with over a million seconds exposure time, sees very young galaxies, just 800 million years after the Big Bang, that look quite different from those we see today. Much has been learned about the evolution of our universe from such results, but most importantly detailed measurements lead to the conclusion of an *accelerated expansion of the universe*. This is in contradiction to current cosmological theories on the development of the universe and has led to the postulation of "Dark Energy" related to Einstein's cosmological constant. Science presently is not able to integrate these results in the existing theoretical framework, so we can expect a major overthrow of cosmology in the near future.

1.2
The Development of Space Utilization

Three phases follow each other in time, with substantial overlaps, in the development of space utilization. Those can be categorized in:

- The technology race
- Scientific exploration
- Early commercial utilization
- Exploitation in the service of society.

1.2.1
The Technology Race

The early driver of space exploration was the technology race between superpowers. Military interest in space technology, based on ballistic missile systems both for defense and deterrence purposes, and belief in global leadership through manned stations in space sustained an industrial machinery that found public resonance in a cold war environment. Enormous financial means were provided by govern-

Figure 1-7: Astronaut John W. Young, commander of the Apollo 16 lunar landing mission, jumps up saluting the U.S. flag at the Descartes landing site during the first Apollo 16 extravehicular activity (EVA-1). Astronaut Charles M. Duke, Jr., lunar module pilot, took this picture. The Lunar Module "Orion" is on the left. The Lunar Roving Vehicle is parked beside the module. Stone Mountain dominates the background in this lunar scene (NASA)

ments to celebrate spectacular pioneering ventures as achievements of their respective system, capitalist or communist. This spirit created the Luna program of the USSR and the Apollo program of the United States (figure 1-7). Some of the first flights to planets of our solar system also fall under this category. It is a characteristic of this phase that, while scientific objectives were promoted and pursued, they served mainly as fig leaves for the public. This is apparent from the fact that priorities during development and operations were clearly set for successful technological performance, while scientific requirements were subordinate. Outstanding scientific achievements were nevertheless possible.

1.2.2
Scientific Exploration

While Sputnik 1 was just a small radio transmitter in orbit, with no function except to demonstrate its existence, the first U.S. satellite Explorer-1 carried a scientific payload designed by J. A. van Allen in the form of a Geiger counter to measure cosmic rays. The observation of an unexpected over saturation of the detector in some orbital positions led to the discovery of the radiation belts and initiated fruitful research on the earth magnetosphere.

Science made efficient use of the early flights to the moon and to planets. The results of the Apollo program led to a better understanding of our planetary system; for example it clarified conflicting theories on the origin of our moon. Today we know that the moon was not an independent body caught by the earth, but rather originated from the earth in a violent collision with a smaller body.

After the early flyby missions of Mariner spacecraft to Mars and Venus, our perception of the solar system was influenced by the breathtaking views of planetary bodies provided by the images transmitted by the Pioneer-Voyager pair of deep space probes (figure 1-8). Photographs of all planets and many moons gave impressions of strange and unexpected worlds. Together with the results of remote sensing measurements, a wealth of scientific information was obtained. Insights from comparative planetology allowed conclusions on the development and destination of our Earth.

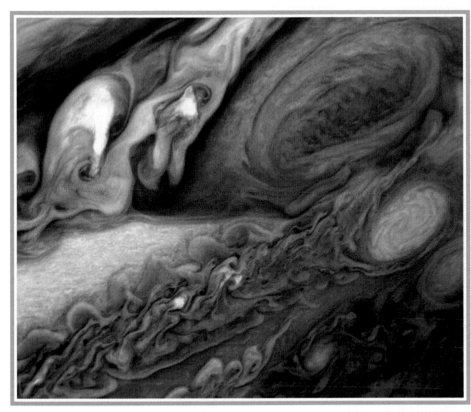

Figure 1-8: This view of Jupiter's Red Spot area was taken by Voyager 1 in 1999. The image was assembled from three black and white negatives taken through color filters and recombined to produce the color image. The structure relates to dynamic eddy currents in the Jovian atmosphere which appear stable over centuries (NASA)

Observation platforms in space, overcoming the limitations imposed by the earth atmosphere, opened new spectral windows for astronomical observations. Since the discovery of cosmic X-ray sources by Giacconi in 1962, which brought him the Nobel Prize in Physics in 2002, instruments making accessible the energetic spectral regions of X-rays and γ-rays revealed exotic objects like neutron stars or black holes and gave insight into the late stages of stellar evolution. The infrared wavelength region provided information on the birth of stars and planetary systems. With the Hubble Space Telescope, observation from space included, for the first time, also the visible wavelength region, which is accessible from ground. This gave rise to massive criticism, as many astronomers opposed this development, arguing that such results could be achieved better on Earth, provided comparable resources were invested. Their concerns seemed justified when the first defocused images were received on the ground. Since the correction of the optical system however, the results surprised not only professional astronomers but also the general public. With the unprecedented views into extreme deep space regions, this venture has justified itself in an overwhelming way.

The phenomenon of weightlessness has stimulated the imagination of scientists, engineers and space enthusiasts from the early days. New requirements and design criteria had to be taken into account in the development of spacecraft and instruments, for example depletion of fuel tanks without the assistance of gravity. It has been reported that Herrmann Oberth, one of the early space pioneers, studied the behavior of liquids in a bottle by jumping from a diving platform into a swimming pool in the early 1920s.

Early attempts to study the effects of weightlessness in space flight were of rather heuristic nature. The first scientifically prepared experiments took place during the Skylab mission. The results led to widespread enthusiasm and adventurous predictions about future commercial utilization, such as huge factories in space. The development of Spacelab (figure 1-9) in Europe and its first flight in 1983, jointly by NASA and ESA, introduced laboratory conditions for investigations in a space microgravity environment. This laid a scientific basis for extensive qualitative and quantitative measurements. A rapid development was initiated that delivered novel results and insights in specific areas of physics, materials science, human physiology and biology. Some of these early ideas were revealed as flops, others led to promising results. Predictions for mass production in space had soon been disproved. Results from space experimentation led to the clarification of scientific discrepancies and to applications on earth. Industrial methods on earth have been improved and new processes could be developed.

Research on Spacelab led to a wealth of scientific results in many disciplines. The absence of gravitational convection, sedimentation and hydrostatic pressure are the prominent features, so new results have been obtained in systems that contain at least one phase in liquid form, like solidification or phase separation phenomena. In biology, novel results on graviperception in plants and the role of the cytoskeleton in cells have been achieved. Physiologists investigated the function of the human organs perceiving gravity, the cardiovascular system, the skeleton, and the lungs.

Figure 1-9: Working atmosphere in the shirt-sleeve research environment of the Spacelab D2 science module during the 1993 STS-55 mission. Astronaut Jerry L Ross examines a sample tube at the "Werkstofflabor" rack, left, Bernard A. Harris, holding his arm, waits to have his blood drawn by Hans Schlegel (right). Wearing the baroreflex collar and waving is Ulrich Walter (DLR)

1.2.3
Early Commercial Utilization

Communication is the prominent field in commercial space utilization as it earns good money for the private investor. We will, however, include in the present discussion those application fields that rely, at least initially, on funding from the public sector, which extends the present considerations to navigation, earth observation and meteorology.

The potential of a global overview to observe the atmosphere for weather predictions was recognized very early in the space age. In 1960, NASA launched its first weather satellite under the name Television Infrared Observation Satellite (TIROS) into a polar sun synchronous orbit at 900 km altitude. The very first image of cloud formations, obtained by the on-board Vidicon camera, is shown in figure 1-10, allowing comparison with the progress made today as described in chapter 5. In the following years, dedicated satellite systems have been developed and operated in a continuous manner in the U.S. (GOES series, commencing in 1975), in Europe (Meteosat since 1977) and in Japan (GMS 1977), providing an increasing number of information channels beyond the daily weather chart on the TV screen.

The capability of land surface observation from space was first used routinely in the military sector for spy and reconnaissance purposes. High resolution imaging required low orbits but had the disadvantage of short spacecraft lifetimes. Photographic films were exposed and retrieved after traveling through the atmosphere to the ground for development and evaluation. These systems were later replaced by electronic cameras, which simplified operations and gave much more rapid access to the data. Major improvements in optical and electronic technology have been made for military or scientific purposes. Early attempts were made to enter the commercial markets with products of space imagery, like the commercialization of the French SPOT satellite in the SPOT IMAGE company. To date specialized products, tailored to the demands of the customer, can be obtained from several companies throughout the world. Commercial satellites are in orbit that deliver images with a resolution of well

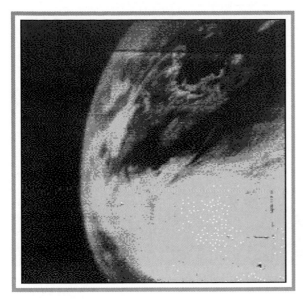

Figure 1-10: The very first television picture from space, taken by the TIROS 1 satellite on April 1, 1960 (NASA)

below 1 m, which are sold successfully to paying customers.

In communication technology, space exhibits attractive features. For any communication using electromagnetic waves, bandwidth and thus transmission capacity increases with frequency. On the other hand, diffraction effects also increase with frequency, so wave propagation is then more along straight lines like light. Therefore, higher and more closely spaced transmission stations are required. The geostationary orbital position is the highest conceivable transmission tower, as first pointed out by the ingenious author Arthur C. Clarke (1945). Following early communication demonstrations using the passive reflector satellites Echo 1 and 2 (30 m reflecting spheres visible to the unaided eye from the ground at night), the first commercial communication satellite was brought into orbit by the AT&T company. This active relay satellite enabled the first transatlantic television transmission in 1962. Due to a low orbit, transmission periods were limited by the brief visibility. This restraint was overcome by the first geostationary satellite Syncom 2 orbited by the Hughes Aircraft corporation, still with an inclination of 28°. Its successor Syncom 3, positioned in an equatorial geostationary orbit, was

used efficiently for television broadcasting of the 1994 Olympic Games in Tokyo.

Since ancient times, navigation techniques have made use of the sky to determine geographical latitude at sea. But only the development of long term accuracy in portable clocks enabled a full determination of position by including knowledge of the longitude. Precise navigation demands from the military side, especially for atomic submarines, led to the development of the "Transit" satellite navigation system in 1994, which was released for public use in 1997. This was succeeded by the Global Positioning System (GPS) in the U.S. and the Global Navigation Satellite System (GLONASS) in the USSR, both still in use today. Modern navigation satellite systems rely on accurate signal transit time measurements and are therefore technically very demanding. They require a constellation of several spacecraft on well defined orbits with high precision atomic clocks on board. The widespread use of these systems in a commercial environment stimulated a worldwide industry for receivers and application products.

1.2.4
Exploitation in the Service of Society

A fourth phase, characterized by a balance between operational and scientific utilization of space, has been triggered by the reduction of public funds for space activities, which is partially compensated by the much larger commercial investments made in the primary (producing) and secondary (supporting) industrial sectors. This phase is predominant today, so the bulk of this volume will be concerned with it.

With the end of the cold war the support for a technology race is fading away and space activities are increasingly driven by demand. Societies restrict spending to areas that hold promise of fulfilling their pressing demands, such as

- economic growth and employment

- industrial competitiveness

- sustainable development

- security and defense

- scientific progress.

National space policies evolve in many societies, giving various relative emphases to the points above.

Utilization Field	View to		Local Presence	Environment			Global/Cosmic Dimensions
	Earth	Space		Vacuum	Radiation	Microgravity	
Earth Surface	●●●						
Climate and Environment	●●●	●					●
Weather	●●●	●					●
Geodynamics	●●●						●
Astronomy and Astrophysics		●●●	●	●	●●		●●
Solar System Research		●●●	●●●	●	●●		●●
Communication	●●●						●
Navigation							●
Technology				●●	●	●●●	●
Fundamental Physics			●●		●	●●●	●●
Materials Science						●●●	
Life Sciences			●●		●	●●●	●

Table 1-1: The utilization fields discussed in this volume make use of specific qualities in space. The number of bullets schematically indicates their importance for the respective field

An important additional factor is international partnership, which helps to advance cooperation and peace worldwide. The International Space Station, presently in orbital assembly, is an excellent example. With more nations active in space, some of them even joining the small group of human space faring nations (like China), partnership becomes progressively more relevant.

1.3
Space Qualities for Utilization

Space offers a variety of special qualities that may be utilized for scientific, technical, or commercial objectives. From the orbital position, the view back to the earth surface is unique, and the view into the universe is unobstructed by atmospheric influences. Robotic presence in space allows the sampling of particles and fields or material from planetary bodies. Human presence adds flexibility and brainpower for complex tasks and research. Outside the earth atmosphere a high vacuum is encountered with a near infinite pumping speed. A spacecraft is subject to a complex radiation environment quantitatively and qualitatively different from that on earth. During space flight without propulsion, gravitational acceleration is compensated, leading to the phenomenon of "microgravity." Space also allows practically

unlimited use of spatial dimensions, for example for long baseline interferometry or gigantic detectors for gravitational waves. Table 1-1 shows how the various space utilization fields make use of these special qualities.

1.3.1
Unique Viewing Positions

View to Earth

One of the most fascinating experiences of astronauts in orbit is the view back to Earth. It shows our home planet in a global view, with its thin and vulnerable atmosphere, in an unusual perspective (figure 1-11). In fact the images we received from space contributed to a change of our present perception of our living environment from a local to a global point of view. Modern methods of earth observation continue to provide intriguing pictures of the earth surface revealing, with astonishing details, structures of anthropogenic or natural origin not perceivable from the ground. Weather forecasts provide us on a routine daily basis with information on cloud coverage, wind and rain expectations on scales sufficient to allow reliable predictions for several days. Investigations of the atmosphere in the top-down direction from space, looking from the thinner into the more dense regions of the height profile, measure con-

Figure 1-11: Clouds and sun glint over the Indian Ocean as seen during the STS-96 mission from the Space Shuttle Discovery (NASA)

stituents and physical state as a function of altitude not achievable from ground. Observation of the earth and sun from space provides insight into the development of the global climate and the impact of human influences. In addition to passive methods of earth observation that rely on natural radiation like solar illumination or thermal emission, active systems carry their own radiation source and therefore are independent of day and night cycles or even clouds, if the source is able to penetrate them. This is the case for radar satellites, as discussed in chapter 3, but also for lidar instruments that use the backscattered radiation of an emitted laser beam to study aerosols and atmospheric constituents.

A special case is the observation of geodynamics from orbit as described in chapter 6. Here the earth gravity field is measured with high precision, giving information on static and dynamic processes such as mantle dynamics (continental drift) and crustal deformation, both important inputs to the study of earthquakes. Other observed features include magnetic pole motion, water and ice storage or ocean topography

The unique position in space also creates new opportunities for communication. An elevated position like a transmission tower expands the surface area

reached by the transmitted signals, so it is obvious that the orbital location can be very attractive.

Observers of the earth surface prefer low earth orbits (LEO) if the objective is to look at details with high resolution. Here the limits will be given by the drag forces of the residual atmosphere and thus the lifetime of the spacecraft. Military satellites in 200 km orbits are reported to provide resolution of football-sized objects from space. Generally, if large surface coverage is a requirement, an orbit of high inclination will be chosen. As the orbital plane is inclined to the equatorial plane, the earth rotates under the satellite, exposing new areas to the nadir view of the payload. For a spacecraft on a typical LEO orbit of roughly 300 km altitude, the earth rotates 22.5 degrees or 2,500 km at the equator during a 90 minutes orbit, leading to a footprint coverage as shown in figure 1.12. If total coverage of the globe is a requirement, a polar orbit will be chosen. Special polar orbits make use of the perturbations due to the oblateness of the earth to introduce a precession such that the orbital plane rotates with the sun and therefore has a constant sun angle. These orbits, termed sun synchronous, see a particular site on the ground always under the same solar elevation, i.e. at the same time of day.

Some earth observation or communication objectives require rapidly repeated or continuous coverage of ground positions. Outside the geostationary orbit, this is obviously not possible using a single

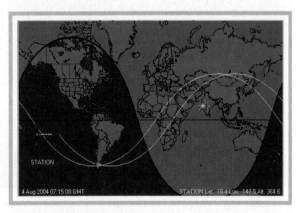

Figure 1-12: Ground track of the International Space Station in orbit at 350 km altitude and 56 degree inclination. As the Earth rotates under the station, the footprint shifts between orbits

spacecraft, but rather constellations of several or many satellites are required. An example is the navigation task, where several spacecraft are required within the viewing range of a ground receiver. In the case of GPS a satellite system of 24 spacecraft is located on 12 h orbits. Geostationary satellites remain in a fixed position in the sky only if they are in an equatorial orbit, otherwise they seem to move along a figure eight extending over their inclination angle. Therefore, the narrow band of equatorial geostationary positions is in high demand and international agreements regulate its use.

View into Space

For the naked eye our earth atmosphere looks completely transparent, at least if no clouds obscure the view. This observation holds true only in the narrow wavelength band of visible light. As seen in figure 1-13, wide regions of the total electromagnetic spectrum, extending from hard gamma rays to long wavelength radio waves, are more or less obscured by the atmosphere. Leaving the earth surface for an observation point above the atmosphere is therefore an attractive option for looking at the stars and our space environment, not only to avoid absorption, but also to improve the "seeing", without the blurring of high resolution images by the dynamic inhomogeneities of atmospheric transmission. Consequently, the view into deep space is the dominant motivation for astronomy and astrophysics to go outside the earth atmosphere, but it is also used by meteorologists and climatologists to study our earth's main source of energy, the sun.

Astronomy approaches not only collect photons in all regions of the electromagnetic spectrum to image objects and phenomena in deep space, but also use spectroscopic methods to characterize the radiation received. In the region of high energy gamma rays, detection takes place in solid crystals with capabilities of energy discrimination, but it is difficult to determine the spatial origin of the radiation, so imaging is not possible. Information is retrieved on the interaction between cosmic radiation and interstellar gas in galaxies, explosive events like gamma-ray bursts, and phenomena near supernova remnants or black holes. In the X-ray region imaging by grazing incidence telescopes become feasible. This wavelength range gives insight into energetic objects like pulsars and quasars, but also active galactic cores, which often contain black holes. White dwarfs, neutron stars and supernova remnants are other objects

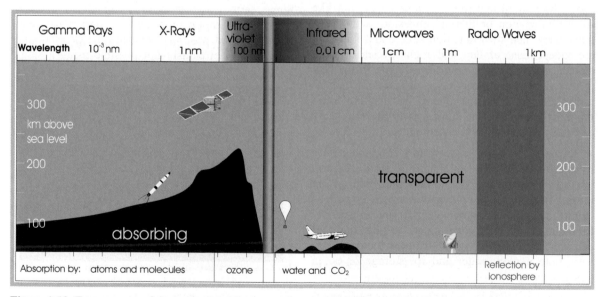

Figure 1-13: Transparency of the earth atmosphere to various wavelengths of the electromagnetic spectrum ranging from hard gamma rays to radio waves. The main absorption processes active in the respective wavelength band are indicated at the bottom

of observation.

The extreme ultraviolet range, strongly absorbed by the earth atmosphere, can be observed from space using adapted optical methods. Radiation in this region is emitted by very hot objects, young stars, and stars in late phases of development such as white dwarfs and neutron stars. Also supernova remnants can be studied in this wavelength range.

On the low energy side of the visible spectral range, in the infrared and far infrared, observations are possible in narrow transparent atmospheric windows from high mountain observatories, from aircraft or balloons. The full spectral range is accessible, however, only from space, where the required cooling of detectors complicates operations. Infrared radiation penetrates cosmic dust clouds and therefore reveals the regions of star formation. In addition, cool objects like planets and stars in their early stages of development can be observed in the infrared.

An important ingredient of our present knowledge in cosmology is derived from the observation of microwave background radiation from space. While the atmosphere is largely transparent in the microwave region, the demand of observing the low intensity

background for small variations, free of disturbances and for extended time periods, can only be fulfilled by space techniques.

Astronomical observation satellites in most cases do not impose heavy demands on the orbit. Frequently budgetary limits or procurement restraints confine the range of launchers available and thus the orbits that can be reached. Tradeoff factors are orbital lifetime, communications to ground stations, sky coverage, or protection from background radiation. Special requirements may arise for the observation of the sun, where uninterrupted observation is possible from a position at the L_1 sun-earth Lagrange point, which is one of five locations where the gravity fields of sun and earth combine to provide local equilibrium for a body in space (figure 1-14).

1.3.2
Local Presence in Space

Some investigations in the exploration of our space environment, especially within the solar system, call for presence on site in space. This may imply sending robotic systems to particular sites, or else sending astronauts. The latter is usually more demanding due to the requirements of life support and safe return to Earth, but it has the advantage of considerably increased flexibility. Robotic systems are far from achieving the intellectual performance of humans. A general preference, however, for either manned or unmanned space exploration does not make sense. Once the objectives of a particular mission are determined, the most efficient way to achieve those has to be chosen. This could be, depending on the task, human or robotic activity. To give an extreme example: for the investigation of the influence of low gravity on human physiology it does not make sense to send a robot.

A first step toward local presence is to approach objects near our earth for close observation by remote sensing techniques. All the planets of our solar system and many of their moons, as well as a few asteroids and comets have been visited this way. Images taken on these missions have opened to us unexpected, exotic worlds, and the results of remote measurements have extended our knowledge considerably. The next step is to land instruments or humans on these surfaces to make in-situ observations

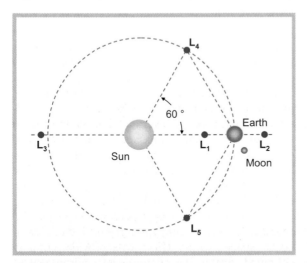

Figure 1-14: The five libration (or Lagrangian) points in the sun-earth system (not to scale). These represent the positions where the gravitational forces combine to keep a body in place relative to earth and sun. L_1 to L_3 are stable only in the plane perpendicular to the connection line; however, small forces are required to maintain a satellite in these positions

Figure 1-15: The Rover Sojourner during the Mars Pathfinder mission in 1997 using its Alpha Proton X-Ray Spectrometer to study composition of a rock dubbed Yogi. The image was taken by the camera of Mars Pathfinder over sols 8, 9, and 10, using the red, green and blue filters (NASA/JPL)

and measurements. Historically, these two steps have not followed consecutively but with strong time overlaps, and both are actively pursued today.

Once landed, the demand for mobility rises. Both on the moon and Mars the first mobile units have raised remarkable public interest (figure 1-15). Scientifically the demand extends to mobility under the surface, for example in the search for traces of life on Mars in subsurface soil samples, and possibly in the atmosphere (if present). Many other bodies will be subject to in-situ investigation in future, such as asteroids, comets, and the moons of Mars, Jupiter, and Saturn. Consecutively samples will be taken at remote locations, as done during the Apollo missions to the moon, and returned to Earth for detailed investigation. Finally, long term remote stations will be established, and humans will land on nearby bodies for extended presence. This will lead to requirements for extensive technology on site to ensure survival and protection, like using local resources for life support or construction of suitable habitats.

On site presence is a demand not only at the surface of planetary bodies. For the investigation of particles and fields, remote sensing methods have limited applicability, so a spacecraft has to be sent into the space region of interest in order to measure local fields and particle fluxes. Early observations of the earth magnetosphere have been achieved by satellites on highly eccentric orbits, covering extended regions of the near-earth environment and thus mapping the magnetosphere. The local properties are, however, highly dynamic both in time and space. This leads to the demand of multiple local platforms. Satellite systems, such as the European Cluster mission consisting of four spacecraft flying in formation, give detailed information on the dynamic interaction of the solar wind with the earth magnetic field, resolved in space and time.

Similar requirements arise in the search for micrometeorites and interplanetary or interstellar dust. Samples of such material can give insight into the evolution of our planetary system and the origin of life. They have to be retrieved and investigated either in situ or returned to Earth without changing their properties. In the study of the conditions of survival of primitive forms of life in space, samples of living matter are placed in space for exposure to radiation and other environmental influences (see chapter 13.2). The resulting effects are observed either in situ or after retrieval in the laboratory.

A particular case is the presence of humans in low earth orbit. The first human orbital flights by Yuri Gagarin and John Glenn were pioneering achievements, demonstrating the capability of human survival in the hostile environment of space. Early space stations like Salyut or Mir were designed for military purposes mainly, even though Mir was used for science investigations later. Skylab, the first U.S. space station, used a converted Saturn V third stage and had three crewed mission between 1973 and 1974. It operated with remarkable success but had a limited lifetime and reentered the earth atmosphere in 1979. Carried by the space shuttle, Spacelab, with its first flight in 1983, introduced laboratory working conditions in space for up to two weeks of in-orbit operations. Once assembly is completed, the International Space Station (ISS) will provide virtually unlimited time in the microgravity environment of space (Messerschmid 1999).

The requirements concerning orbits and transport to achieve presence in space are widespread. Reaching planets or other bodies in the planetary system requires sophisticated orbital mechanics to minimize

The earth atmosphere is a complex system. This is recognized from the change of temperature with altitude, as illustrated in the figure. The temperature parameter is used in the most common classification of atmospheric regions (Prölss 2004). Close to the earth surface, temperature decreases with altitude, a common experience as we climb a mountain. This is because the land and ocean surface is heated due to the combined effect of direct radiation absorption and the greenhouse effect. With increasing altitude radiative cooling leads to a decrease of temperature in the atmosphere. This region, where clouds may form and other weather phenomena take place, is termed the troposphere and extends to the tropopause at an altitude of 10 to 12 km. Above this limit, temperatures rise due to increasing absorption in the ultraviolet, mainly by ozone, in a zone called the stratosphere. At an altitude of about 50 km a temperature maximum close to the ground level values marks the stratopause.

A turn in temperature behavior arises in the mesosphere. Here absorption processes are overcompensated by heat radiation from trace

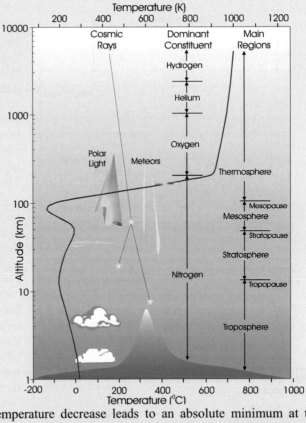

gases, mainly carbon dioxide. The resulting temperature decrease leads to an absolute minimum at the mesopause, 80 to 90 km above ground. Up to this level the atmosphere is well mixed, leading to a constant average composition. At higher altitudes gravitational de-mixing of the constituents is observed, with heavier gases decreasing in abundance with height. This is accompanied by a dramatic temperature rise, leading to the name thermosphere for this region, arising from absorption in the solar extreme ultraviolet (below 242 nm) with no efficient heat loss processes active. Above 200 km the temperature asymptotically approaches the thermopause limit, which varies with solar wind activity and earth magnetic field effects between 1000 and 2000 K.

fuel mass and transfer times (Messerschmid 2000). In many cases, gravity assist or swing-by maneuvers are helpful to adjust the trajectory of a spacecraft. Here momentum transfer between a spacecraft and the orbital motion of a planetary body around the sun is achieved during a close flyby.

1.3.3
The Space Environment

In space utilization, the environment is an important parameter during the complete lifetime cycle of a spacecraft and its payload. As an integral part of the manufacturing, assembly, and integration process

the environment is well defined, carefully controlled and documented for both the spacecraft and payload. Severe mechanical or thermal environmental conditions may prevail during the launch and transport phase in space. Those are important design criteria and frequently dominate the test specifications. In the present context we will concentrate on the environmental conditions during space operations. In many cases these are the very reason for the space mission and decisive from the utilization aspect. Both the spacecraft and the payload have to be designed and tested for this environment. Sensors frequently make use of specific local environmental

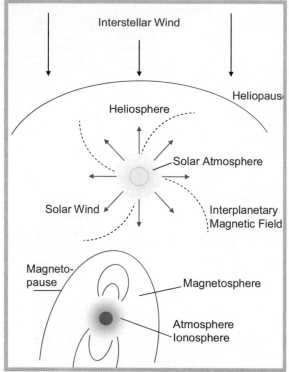

Figure 1-16: Structure of the Earth's space environment (Prölss 2004)

conditions like the space vacuum. It also has to be ensured that the spacecraft will not change its own environment in orbit, for example by excessive outgassing or loose particles that may influence scientific measurements or operational objectives.

The characteristics of the space environment of our earth are illustrated in figure 1-16. The surface density of 3×10^{19} molecules per cubic centimeter decreases exponentially in the atmosphere and ionosphere. Above the atmospheric range, in the magnetosphere, the local properties are governed by the earth magnetic field. Beyond the magnetopause, interplanetary space is dominated by the solar wind streaming away from the sun, with the solar magnetic field frozen in and forming a spiral structure. The area of solar influence, the heliosphere, is bordered by the heliopause to the interstellar medium. The nearest star can be found at a distance of 3.5 light years from the sun.

Vacuum

In the present context we do not consider the influence of the atmosphere during launch or reentry. From the utilization point of view, the main influence of the atmosphere is in terms of aerodynamic drag or torque and the corrosive effects of residual gases, mainly atomic oxygen. In addition, the atmosphere is of interest as a subject of investigation through the payload, with the unique position of an upside-down view, where constituents can be observed from the lower density to the thicker medium. These aspects will be discussed in chapters 4 and 5.

Aerodynamic drag and torque on a satellite depend on its cross sectional area, shape and attitude, the velocity relative to local winds, and the atmospheric density. The latter is strongly variable with day and night cycle, solar activity, and earth magnetic field. For practical purposes empirical tables, so called "Standard Atmospheres", are used. At perigees below 120 km, drag leads to unpractical short lifetimes; orbits for operational satellites use altitudes above 200 km. Above 600 km drag is usually not the limiting factor of orbital lifetimes, which range beyond 10 years. At distances more than 1000 km from the earth surface, radiation pressure from the sun exceeds the atmospheric drag. This is usually taken as the limit of the earth atmosphere.

Atmospheric drag not only influences the orbital lifetime but also induces a force that has an influence on the low gravity environment. Satellites dedicated to measurements of the gravitational field or of fundamental quantities related to gravity are therefore designed to actively compensate for drag (see chapters 6 and 11). Payloads making use of orbital low gravity have to take into account the disturbance introduced by aerodynamic drag.

Electromagnetic and Particle Radiation

The atmosphere protects the earth surface from many kinds of dangerous radiation, in fact it enables the very existence of life. On the other hand, this protection impedes investigations of space features from the ground in various wavelength regions of the electromagnetic spectrum, as shown in figure 1-13. Electromagnetic radiation outside our atmosphere extends over wavelength ranges from hard X-

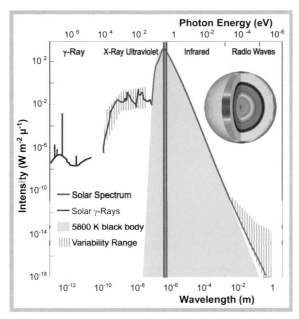

Figure 1-17: Solar radiation spectrum outside the earth atmosphere. The insert shows solar rotation and flows as observed during the SOHO mission (ESA/SOHO)

rays to long radio waves. The by far strongest source is the central star of our planetary system, the sun.

At altitudes above 200 km a spacecraft is exposed to a wide spectrum of electromagnetic radiation with various intensities, as shown in figure 1-17. Here the observed solar spectrum is compared to that of blackbody radiation at 5,800 K. The spectral maximum is at the ultraviolet end of the visible spectral region. A nearly exponential decay occurs to longer wavelengths, roughly along the blackbody spectrum, with deviations at very long wavelengths. On the high energy side, radiation intensity is enhanced compared to a blackbody in the extreme ultraviolet, arising from high temperature regions of the solar corona. Variability is strong at the extreme sides of the spectrum. However, the total integrated irradiation intensity near earth, the solar constant of 1.37 kW/m^2, is hardly affected by these variations, which appear exaggerated in the double logarithmic plot of figure 1-17.

Corpuscular radiation is a flow of elementary particles, ions, and heavy nuclei which share with hard electromagnetic radiation the feature of penetrating inorganic or organic matter. They therefore are a potential hazard not only to humans in space, but also to electronic components and detectors. On the other hand, high energy radiation and particles are an important source of scientific information on stars, galaxies and other objects in the universe.

The sun emits a continuous stream of matter, the solar wind, consisting to 99% of protons, helium nuclei (alpha particles) and electrons. Close to the earth the velocity averages 500 km/s, corresponding to a kinetic temperature near 10,000 K and a particle energy of roughly 1 keV. Much higher energies originate from solar events like solar storms or solar flares, which are short time (a few days) statistical events and produce protons in the range of 1 MeV to several hundred MeV, respectively. Cosmic radiation particles may reach the inner solar system from our own galaxy or from other objects in the universe. Their flux rate is very low compared to solar particles, in the order of 10 events per square centimeter and second. Consisting of protons (85%), alpha particles (14%) and heavy charged particles (termed HZE for High Z-number and Energy) with masses up to iron nuclei, cosmic radiation has energies above 10 GeV.

Charged particles like electrons and protons may enter the earth magnetosphere to get trapped in the Van Allen radiation belts with energies greater than 30 keV. Reflected to and back within the earth magnetic field, particles can reach life times of several years in those belts. A summary of particle fluxes in the vicinity of Earth is depicted in figure 1-18 (Messerschmid 1999). The influence on humans and instrumentation in space depends both on energy and flux. Biological effects are most severe in the 0.1 to 10 MeV range. Very high energy particles tend to penetrate without serious consequences, but secondary effects like bremsstrahlung and cascading in shielding materials have to be accounted for. Precautionary measures in the orbit chosen and suitable shielding for humans and sensitive instruments remain necessary.

Gravity Environment

Gravity can neither be switched off nor screened. The influence of gravity is ubiquitous in space and originates from a superposition of fields of the

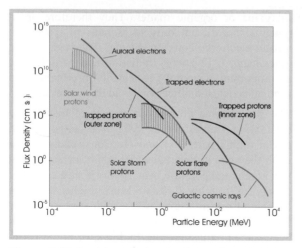

Figure 1-18: Particle radiation in the near earth environment, originating from the solar wind (yellow/orange), particles trapped in the Van Allen belts (purple), and cosmic rays (green). Auroral electrons are shown in blue

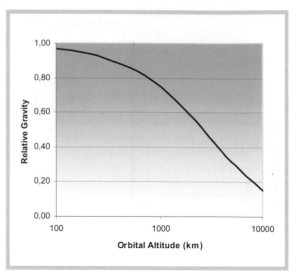

Figure 1-19: Gravitational attraction to Earth as a function of orbital altitude

masses distributed in our universe. The force acting on a reference mass decreases with the square of the distance to a massive body; therefore only the nearest heavy masses will be of influence in any place in space. Within our planetary system, the sun contains about 750 times more mass than the planets and moons, so it dominates its environment.

In the vicinity of the earth, the gravitational attraction towards the earth surface decreases with altitude, as shown in figure 1-19. It is apparent that there is no means to reach a gravity free environment by going into orbit. The term "zero gravity" or "microgravity", used to describe the fact that astronauts float around in spaceships, is thus misleading. A better term is "weightlessness" or "free fall conditions." The observed phenomenon relies on a compensation of the gravitational attraction to Earth by the centrifugal force in a circular orbit. But this is only a special case of a free fall condition applying to orbital motion. In fact, as soon as a spacecraft switches off its engine outside an atmosphere, free fall conditions prevail and weightlessness is perceived.

Weightlessness is attractive for research applications, because objects under this condition do not fall, bubbles do not rise and convection does not occur in fluids. This fact has stimulated a whole

range of investigations. Here the term microgravity has become accepted to describe a nearly weightless condition, as in reality disturbing forces prevent pure free fall conditions. On the ground, microgravity conditions can be obtained only for short periods of time like in a drop tower for a few seconds, in a parabolic aircraft flight for about 20 seconds, or by suborbital rocket flight for a few minutes.

In the field of physics, the lack of gravitational convection, sedimentation or hydrostatic pressure stimulated scientific investigations of effects, including at least one fluid phase. Novel results could be achieved in solidification of metals or semiconductors with interesting applications. In biology, new insight into the graviperception of plants, the function of the cytoskeleton of cells, and into the processes of signal transduction was achieved. Research in human physiology led to a number of discoveries including the function of the human gravity sensing organ, the lung function, the performance of the cardiovascular system or bone resorption.

Based on the results of Spacelab, which flew on numerous shuttle mission between 1983 and 1997, a solid scientific foundation was laid for the utilization of the International Space Station (ISS). The fundamental effects are well investigated, processes are understood and evaluated, and the design of instrumentation is based on knowledge and experience. This allows for efficient utilization, once assembly is complete, in terms of well defined demands from industry and science.

1.3.4
Global and Cosmic Dimensions

Out in space the inspiration of a scientist or engineer is not confined by spatial limitations. In a boundless environment with no gravity constraints, structures or systems of gigantic dimensions are conceivable. This has stimulated early ideas on huge orbital stations, free floating space colonies to harbor the overflow of population on Earth. Other thoughts projected large orbital energy farms to collect solar radiation, convert it into transportable energy and send it to Earth, or big mirrors to illuminate certain areas on the earth surface during night. Many scientific ideas make use of the available dimensions, like very large baseline interferometry, which has the potential of observations with spatial resolution inconceivable by other means, or using the long mean free paths in the space vacuum to investigate lifetimes of excited states of molecules and atoms.

Two projects making use of space dimensions have taken shape recently. A development jointly agreed between ESA and NASA for launch in 2011 is the Laser Interferometer Space Antenna (LISA). This instrument, designed to open a novel window to unravel the secrets of our universe by observing gravity waves, will be described in more detail in chapter 11. It consists of a triangular interferometer with a side length of 5 million kilometers, orbiting around the sun by trailing the earth in its orbit. A view of the dimensions involved is given in figure 1-20. A mission of more moderate dimensions in this category is the X-Ray Evolving Universe Spectrometer (XEUS). This advanced X-ray spectroscopy mission is designed to improve the observation of massive black holes or early galaxy evolution by two orders of magnitude over existing instruments. It consists of two spacecraft, one carrying the photon collecting mirrors, the other the detectors, flying in formation about 50 m apart. Tentatively scheduled for launch in 2014 it is planned to make use of the International Space Station for assembly and operation in orbit.

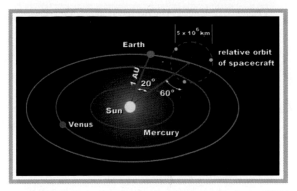

Figure 1-20: A schematic diagram of the three LISA spacecraft flying in formation as they orbit around the sun. The spacecraft are separated by 5 million km and trail behind the earth at a distance of 50 million km (equivalent to 20 degrees) (ESA)

1.4
Concluding Remarks

Space utilization is an indispensable part of our present society. It permeates our daily life in many technical and commercial aspects, expands our knowledge and continues to influence the self conception of humans. Today it is a mature field. It has developed since the early days of space flight and has undergone several phases of varying emphasis. As a multidisciplinary theme it makes use of the full variety of utilization qualities our capabilities of access to space open. The exploitation of space becomes successively more demand oriented, with the requirements of commercial markets, military use, and society needs shaping the developments into the future.

References and Further Reading

Clarke AC (1945) Extra-Terrestrial Relays, Wireless World, October 1945, p 305

Davies JK (1997) Astronomy from Space: The Design and Operation of Orbiting Observatories Praxis Publishing Ltd, Chichester

Hallmann W and Ley W (1988) ed: Handbuch der Raumfahrttechnik Carl Hanser Verlag München

European Commission (2001): "Space Industry

Developments in 2000", Enterprise Directorate-General (web:http://europa.eu.int/comm/enterprise/aeros *pace/)*

Fortescu P and Stark J (1995) Spacecraft Systems Engineering, John Wiley and Sons, Chichester

Larson WJ and Wertz JR (1992) ed: Space Mission Analysis and Design Microcosm, Inc, Torrance and Kluwer Academic Publishers, Dordrecht

Messerschmid E and Bertrand R (1999) Space Stations, Systems and Utilization, Springer Verlag, Heidelberg

Messerschmid E and Fasoulas S (2000) Raumfahrtsysteme, Springer Verlag, Heidelberg

Prölss GW (2004) Physik des erdnahen Weltraums, Springer Verlag, Heidelberg

Rycroft M (1990) ed: The Cambridge Encyclopedia of Space Cambridge University Press, New York

2 Access to Space – the Prerequisites for Space Utilization

by Heinz Stoewer

This chapter describes the prerequisites and means needed to make space utilization possible. The means to enable utilization are the launch vehicles, the spacecraft, satellites, or space stations, and the ground infrastructure necessary to support launch and operation of space vehicles. The chapter also provides an overview of some of the enabling competences which play an important role in the realization of space vehicles. These are project management, systems engineering and product assurance. All three have significant implications for how utilization aspects are integrated into the conception, development, manufacturing, testing and operation of space vehicles. The chapter also addresses the role of technology research and development. Many of the building blocks needed to design rockets and spacecraft, including their sensors and instruments needed for space utilization, originate from advanced technology developments.

2.1
Launch Vehicles

Launch vehicles are the foundation of mankind's access to space. Here a concise overview is presented on past, present and future developments.

2.1.1
The Pioneers – from Goddard and the V2 to Saturn and Apollo

Theoretical foundations for space rockets were notably created by three space pioneers, namely Konstantin Ziolkowsky from Russia, Robert Goddard from the USA, and Hermann Oberth from Germany. They created the mathematical formulations and theoretical underpinnings, postulating the possibility to reach velocities that would allow rockets to orbit Earth and fly beyond its gravity. Earth bound rockets, using solid propellants, had been developed

millennia ago by the Chinese for fireworks and military purposes. The feasibility of space rockets received a major boost, however, when it was proven that liquid propellants could be harnessed in a controlled explosion, expelling exhaust gases through specially adapted Lavalle nozzles.

Robert Goddard, USA, was the first to successfully test a small liquid propellant rocket (figure 2-1 left). In the late 1920s, an experimental rocket group in Berlin, Germany, which was eventually joined by the young Wernher von Braun, also demonstrated successful applications of liquid propulsion. The first large scale developments were undertaken during the 1930s and early 40s by the Peenemünde team led by Wernher von Braun. Their missile developments, up to the V2 and A4 rocket, for the first time demonstrated the interplay of the many technologies needed for controlled long distance flight, i.e. propulsion, guidance and control, and structures (figure 2-1 right). Most early post World War II rocket developments were based on the Peenemünde technical breakthroughs.

Soviet rocket developments around the famous chief designer Sergej Koroljow led to a number of missiles and space rockets eventually launching the world's first satellite, Sputnik (4 Oct. 1957), then Laika, the first animal in space, followed by Lunar and Venus probes, and eventually the launch of the first cosmonaut, Juri Gagarin (12 April, 1961).

In the USA a number of parallel competing missile developments were led by the different fractions of the armed forces. Soviet competition and U.S. President Kennedy's challenge to launch humans to the moon initiated the establishment of the famous NASA centers in Houston, Huntsville and Cape Canaveral. Wernher von Braun directed the development of the Saturn launch vehicles, which propelled the Apollo spacecraft, Lunar Excursion Mod-

Figure 2-1: (Left) In 1926 Robert Goddard realized the first successful liquid rocket flight, using gasoline and liquid oxygen as propellants. (Right) the Peenemünde developments created the first large scale missile, commonly known as the V2, which for the first time demonstrated the complex interplay of technologies needed for the later development of space launchers

ules and astronauts on their journeys to explore our moon.

2.1.2
Expendable Launch Vehicles – Today's Fleet

The fleet of proven space launch vehicles at the beginning of the 21st century shows a broad spectrum of small and large rockets.

European launcher developments were originally based on the British Blue Streak Intercontinental Ballistic Missile. The unsuccessful attempts by the European Launcher Development Organization (ELDO) in the 1960s and early 70s to produce a

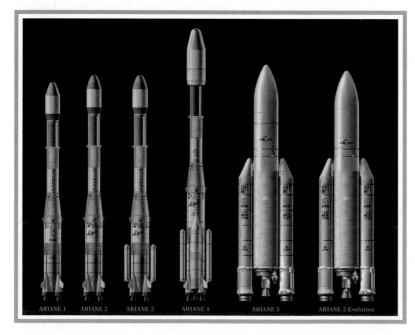

Figure 2-2: Ariane 4 and today's Ariane 5 families have become reliable space launchers for Europe's access to space and serious competitors in the worldwide commercial space launch market (ESA)

Figure 2-3: The predominantly Ukrainian-built Zenith, the Soyuz, and Proton launchers are the showpieces and money makers in the Russian space rocket stable. They are launched mostly from Baikonour in Kazachstan, Pletzesk in Central Russia, from a swimming platform in the Pacific (in a joint venture with Boeing), and the Soyuz soon also from Kourou (Rosaviakosmos)

viable space launcher eventually led to Europe's workhorse in space, Ariane (figure 2-2). Ariane is launched due East and North from the near equatorial space center in Kourou, French Guyana, on South America's Atlantic coast.

The Soviet and Russian space developments have generated a large and reliable family of mostly missile derived space launchers, the best known being the Zenith, Soyuz and Proton (figure 2-3). They are marketed worldwide by Russian companies and through joint ventures with Western enterprises. In addition Russia is producing a number of smaller

Figure 2-4: The U.S. expendable space launch vehicle stable serves a wide array of scientific, applications and defense needs. They are launched rather frequently from either East or West Coast bases. (Right) Titan launcher at the Kennedy Space Center or Cape Canaveral ready for launch; (center) Atlas en route to a geostationary transfer orbit with a commercial telecom satellite; (left) Delta launching the Mars Opportunity mission (NASA)

missile derived launch vehicles, such as the Cosmos and Rockot, serving the small satellites market worldwide.

U.S. developments have also produced a series of missile derived space launchers, the most famous being the Delta, Atlas and Titan families (figure 2-4). Each has a number of models, which allow the use of tailored versions for different mission needs. The principle U.S. launch sites are Kennedy Space Center on Cape Canaveral, Florida, and Vandenburg, California. Amongst the smaller commercial launch vehicle developments, Pegasus is the only current operational rocket deployed from a high flying aircraft.

Other nations with successful space launch vehicles are Japan, Israel, India and China (figure 2-5). These launchers underline a determination to be independent of other nations for the launch of national satellites and the realization that launch vehicles can be profitably sold worldwide for commercial launches.

All of the launchers noted above, and others not included here, undergo a continuous evolution and capabilities upgrade process to achieve better performance, lower cost, or both. Their development today is mostly governed by commercial and national security or national independence policies. We can expect continuous development of the existing fleet and the entry of new players in an attempt to secure a part of the commercial and institutional market. From a utilization point of view this stable of expendables will continue to provide good and diverse, but relatively expensive access to space for many years to come.

2.1.3
Reusable Launch Vehicles – Space Shuttle and Beyond

Reusable launchers in the long-term have the potential of substantial cost reductions. Reusable launchers have been the dream of rocket designers since the beginning. The German rocket pioneer Eugen Sänger first developed such concepts in the 1930s. The ideal long-term goal for reusable launcher developments remains to create a vehicle that would allow airline type operations at relatively low cost compared to the expendables.

Some of these projections accompanied the early proposals for the development of the Space Shuttle. But in this respect, the Shuttle could not live up to expectations. According to original plans, the Shuttle was to fly comparatively inexpensively roughly once per week and replace all but the smallest of the

Figure 2-5: Japan has developed the H series of launch vehicles, originally derived from the U.S. Delta rocket. China has evolved its Long March rocket series and has proven with its Shenzhou missions the capability of launching humans into space (JAXA, CNSA)

U.S. launchers. Together with another rocket, a propulsive stage to be carried in the orbiter's payload bay, the Shuttle was designed to also launch commercial and other satellites to higher altitude orbits, including the geosynchronous, 24 hour orbits. History proved otherwise! The Shuttle fleet of five could not be developed and operated as easily and cheaply as was anticipated by its proponents. The idea of an all purpose space launcher had to be abolished.

The Shuttle's principal mission then evolved predominantly into launching humans and their cargo into low Earth orbit. Nonetheless, the Shuttle became the most inspiring space transport system of the past decades. Launches in particular with the space laboratory, Spacelab, and those for the International Space Station, ISS, also enabled much of the microgravity based research of the 1980s and 90s through today.

The Space Shuttle (figure 2-6) moreover performed ground breaking servicing missions in orbit, extending, e.g., the life and capabilities of the Hubble Space Telescope. It has launched huge radar antennas into orbit, mapping the planet in ways otherwise not possible. The Shuttle has enabled new forms of space utilization and opened perspectives for continued human space exploration.

The Space Shuttle is a highly capable, albeit expensive and complex launch system. Its two catastrophic failures have had dramatic impacts on the international space agendas. But it has proven many technologies needed for future space utilization. The

Figure 2-6: The Space Shuttle, since the 1970s, has become the workhorse for heavy lift launches, such as the Hubble Space Telescope, twenty four Spacelab missions, Mir resupplies and many International Space Station deliveries (NASA)

Shuttle has also demonstrated that technology is not yet sufficiently mature to allow single stage to orbit reusable launch vehicles to be built. It has proven that an aircraft can fly safely up to 25,000 kilometers per hour, i.e., that a single aerodynamic shape can function in the hypersonic, supersonic and subsonic regimes. Finally, the Shuttle made the point that, based on today's design knowledge and technology, reusable vehicles are extremely complex and not

Figure 2-7: Typical past and present concepts for reusable launch systems need to await technology advances prior to realization. Left, the Sänger two stage reusable hypersonic aircraft with a space plane concept (EADS); center, Istar, a NASA and U.S. industry research concept (NASA); right, SpaceShipOne (Scaled Composites) and other ballistic concepts could stimulate technologies and know-how for future reusable launchers

necessarily fail-safe.

Many studies during the past decades have yielded new single stage, one and a half stage, and two stage reusable launcher concepts on paper. None of them has to date yielded a viable system design which could be produced and operated economically and safely to the point that inexpensive transport of goods into low earth orbit could open new dimensions of utilization. The principle constraints still lie in the unavailability of suitable technologies, such as propulsion techniques and materials at favorable cost, yielding sufficient performance and robust operations. However, it is only a matter of time until such constraints will fade away.

Examples of concept studies of reusable or partially reusable launchers of the past decades are the British Hotol, the German Sänger and the U.S. NASP concepts, or more recently, various ESA and commercial ideas, such as Lockheed's Venture Star and more exotic NASA or USAF ideas (figure 2-7). The X-price competitors for vertical ascent to 100 km, e.g., Space ShipOne (also figure 2-7), or others could further advance concepts and technologies towards this end.

2.1.4
Rockets for Space Utilization – Summary

Today's international spectrum of expendable launch vehicles has proven to be very capable of launching a large variety of spacecraft and humans into space. Principle constraints remain the high cost per kilogram payload to orbit. Further refinements of the present fleet and a number of current attempts by commercial companies to produce lower cost launch vehicles may reduce cost to some extent.

New concepts could in the meantime provide impulses for ballistic suborbital flights and maybe long distance aircraft. A further step in reducing the cost of space transportation and hence a major advance in space utilization will only be possible when the necessary technologies have sufficiently matured to yield viable partially or fully reusable launch systems, capable of advancing substantially beyond the Space Shuttle heritage. Figure 2-8 shows historic cost per kg to Low Earth Orbit (LEO) for past and current expendable launchers and for projected future fully re-useable concepts. Potential cost gains to be achieved depend however also strongly on the payload capability envisaged for such vehicles. Sub-

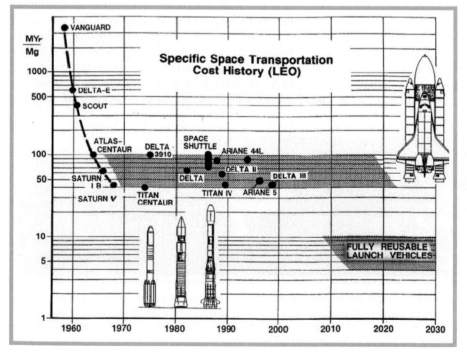

Figure 2-8: History of cost per kg to orbit for expendable and reusable launch vehicles. (Note: one man year (MYr) corresponds to approximately 210,000 € or 232,000 $ at mid-2004 economic conditions) (D. E. Koelle 2003; web: astrobooks). Economy of scale factors tends to favour re-usable launcher concepts with a relatively large cargo capability when compared to today's expendable launchers

stantial advances can only be expected for very large vehicles; cost per kg advantages for vehicles with payloads below 10 tons could be marginal or negative.

Since the Space Shuttle is supposed to be phased out by the end of the decade, it remains unclear what launcher(s) can take its place. None of the various presently studied system concepts for new, reusable launchers seems sufficiently mature to be realized in the short term. More work will be needed to generate a viable technology base from which an economic, reliable and robust concept can be built. The next 10 or more years are likely to yield the necessary breakthrough and could then open new vistas for space utilization.

2.2
Spacecraft and Satellites

The terms spacecraft and satellite are not well differentiated and are in practice used interchangeably. In this chapter spacecraft is used in its generic meaning representing satellites and space vehicles, including crewed laboratories. Space probes are generally referred to as spacecraft destined beyond Earth's gravity field, i.e., planetary and deep space explorers. Spacecraft have two principle functions: first the so-called housekeeping functions, which encompass everything necessary to maintain the spacecraft operational in orbit. They provide the needed support to the payload. Second the payload function, which assures the scientific measurements, data formatting, sample storage or similar functions, i.e., its primary space utilization capability.

2.2.1
Spacecraft Classes – a Variety Store

Spacecraft can function as observatories, looking out into space, at a moon, comet, asteroid, planet, sun, or the universe at large, or down towards Earth. They can function as communication nodes or relay stations for terrestrial information or localization networks. Once landed on a moon or planet, they can function as in situ laboratories with multiple analysis and observation functions. Defense dedicated spacecraft sometimes have additional functions regarding signal analysis, information management, or the supply of positioning signals to a variety of users.

Early satellites, such as Sputnik (4 October 1957) or Explorer (31 January 1958) consisted principally of housekeeping functions with a very small payload. Their objective was to prove the technological feasibility of launching and operating spacecraft in orbit. Since then by the year 2000 some 4000 to 5000 satellites had been launched from Earth into various orbits in space.

Today's spacecraft are highly complex multifunctional designs, often with a multitude of instruments. Depending on their mission objectives, they come in

Figure 2-9: Typical science observatories in various earth orbits are complex assemblies of telescopes, cameras, spectrometers, antennas and other instruments. Left, the Infrared Space Observatory (ISO), in the ESA - ESTEC Test Center, right, Hubble Space Telescope during in-orbit servicing (ESA, NASA)

Figure 2-10: The Solar and Heliospheric Observatory (SOHO) continuously monitors and analyzes the sun and its environment. It is stationed in a Lagrangian orbit, a "point" between the Earth and the sun where the forces of gravity exerted by these two bodies are equal and the spacecraft can thus maintain a "fixed" position (ESA)

Figure 2-11: Telecommunications and navigation spacecraft carry a variety of parabolic or specially shaped antennas connecting users in different regions. Shown is a "Hotbird" telecom satellite" during final integration (Eutelsat)

many different sizes and shapes. We distinguish the following principle classes of (unmanned or automated) spacecraft:

- LEO (earth or space oriented)
- LEO and MEO (various functions)
- GEO (mostly communications and meteorology)
- Lagrange point observatories
- Deep space probes and orbiters
- Lunar, planetary, and other landers

Spacecraft design is strongly governed by the launch vehicle chosen to propel it into the desired orbit. The launcher determines mass, volume and many more specific spacecraft design parameters, such as static and dynamic loads, acoustic environment, power and communication interfaces, packaging for launch and in-orbit deployment, and the like. The launcher selection also strongly influences mission cost. The most important constraint on a spacecraft design results from the environment in which it has to operate (see also chapter 1). Different orbits have different radiation, thermal, solar flux, atmospheric den-

Figure 2-12: The design of typical earth observing spacecraft is governed by the need to accommodate large and sophisticated instruments, such as antennas for altimeters and radars, radiometers to analyze atmospheric or meteorological phenomena, or multispectral cameras simultaneously observing in different wavelengths. Shown are the NASA Landsat 4 and the Eumetsat Metop 4 configurations (left to right) (NASA, Eumetsat)

sity, and other parameters which affect spacecraft subsystems, such as attitude and orbit control, power generation, or energy and thermal management characteristics.

Spacecraft are adapted or optimized for their specific mission purposes. Figures 2-9 to 2-14 show examples of today's many different types of spacecraft, selected to show representative designs from the spectrum of the many different configurations and applications.

Common design criteria for all spacecraft include mass (and power) minimization since the largest proportion of today's space mission cost remains launch cost. Reliability and availability of service are also driving design criteria. The assurance that users can depend on a spacecraft or a constellation of spacecraft for the intended functions during the projected lifetime has become an essential competitivity differentiator in commercial space ventures.

Recent developments have emphasized satellite constellations, i.e., a number of spacecraft operating in close proximity or as an orbital network, providing integrated services, or interferometer capabilities.

The 1990s saw a strong emergence of so-called micro-and mini-satellites in the 10 to 50 and 50 to 500 kg ranges. They were driven by the emergence of new technologies, the need to reduce cost, or the push from developing countries, research centers and universities to participate in the realization of spacecraft that could be built and launched with low budgets and managed by small groups of engineers and users. It is expected that this trend will continue in view of further progress in technology miniaturization. In constellations with multiple satellites the utilization potential of small satellites is further enhanced. Advanced concept studies currently explore the potential of swarms of tens or hundreds of micro- or nano-satellites, weighing only a few kilograms and less, for future mission applications.

Future spacecraft challenges are summarized as follows:

- Continued minimization of mass, power, energy
- Optimization of instruments and payload performance
- Autonomy, robustness and security of operations
- In-orbit reconfigurability and adaptability
- Cost, especially launch cost reductions

Figure 2-13: Deep space exploration spacecraft are highly complex and densely packed robots with a variety of sensors and the ability of increasingly autonomous operations in regions far in our solar system. Shown are artist renderings of the Rosetta Lander (left), projected as landed on the comet Churyumov-Gerasimenko in 2014, and the Huygens spacecraft (right) as projected during its descent upon the Saturn moon Titan in January 2005 (ESA)

Figure 2-14: Typical defense satellites acquire or relay data and information around the globe and supply defense and security forces with up-to-date geographic, intelligence, communications, command and other information. Shown is "SAR Lupe," a radar satellite constellation of five small satellites, with its ground segments (OHB Systems)

- Reliability, availability, dependability
- Extended mission duration
- Maximization of utilization potential
- Shortened time to destination for deep space missions
- Precision constellation capabilities

2.2.2
Spacecraft Housekeeping Functions

Housekeeping functions enable a spacecraft's operation and provide the necessary support to the payload. Typical spacecraft housekeeping functions include

- structure to which everything else is attached,
- mechanical assemblies such as booms, motors, gyros, antennas,
- thermal control, with passive and active components,
- environment and life support for crewed spacecraft and laboratories,
- electrical power generation, energy storage and management,
- telemetry to monitor a spacecraft's health,
- communications to receive and transmit data (signals), including commands
- data management to interact or deal with the housekeeping and payload information streams,
- attitude and orbit control to point and stabilize the spacecraft, maintain the desired orbit, and sometimes
- propulsion to change orbits or flight paths, initiate atmospheric re-entry, or boost into a "graveyard" orbit

These housekeeping functions together compose what is often referred to as a spacecraft's platform or bus. Buses or platforms come in a large variety of different designs. Standardization of platforms is common practice for commercially sold telecommunications satellites, but is also applied more and more in other mission applications. Cost savings and the reuse of know-how are the decisive criteria. A number of standard platforms are offered off the shelf or are available at short notice, with limited

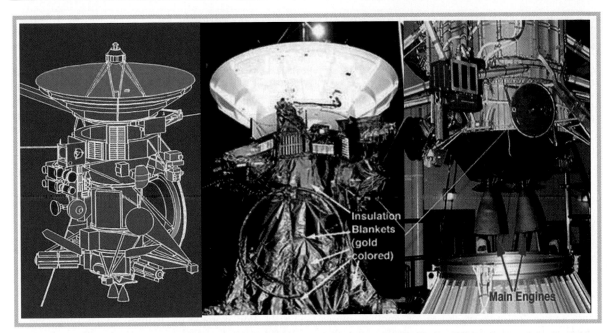

Figure 2-15: Typical spacecraft housekeeping subsystems and instruments are shown for the Cassini spacecraft orbiting Saturn. The large High Gain Antenna dominates the upper spacecraft and during cruise shades the probe from direct solar illumination. During the descent of the Huygens probe to Titan's surface, the antenna is pointed to the predicted landing site to collect Huygens' data for onboard storage. Cassini is then re-oriented to its normal attitude and communications re-established with Earth to transmit the recorded data. The various mechanical assemblies, thermal shielding and propulsion subsystems are shown in the different views (NASA)

modifications, from a number of suppliers around the world.

Some typical housekeeping subsystems and instruments for the Cassini spacecraft are shown in figure 2-15, e.g. mechanical assemblies prior to coverage with multilayer insulation, and its thermal and propulsion subsystems.

Trade-offs between subsystems involve complex analyses to identify, for example, the most propellant saving attitude and orbit control strategy which at the same time does not impose undue penalties on the thermal and energy management subsystems design. Such trade-offs often involve payload requirements and utilization options and eventually result in detailed designs and operations profiles for a mission. Trade-offs concerning for example the degree of autonomy of a spacecraft in orbit, or the interdependence of instruments and spacecraft subsystems, involve in depth analyses and design studies to find the best compromise between conflicting requirements and constraints.

2.2.3
Payloads and Instruments

Instruments, sensors, and antennas, together with their supporting electronics, software, thermal control and mechanical support constitute a spacecraft's payload. The payload is the key to and enabler of utilization. Payloads acquire, process, and transfer the user information from scientific exploration or commercial exploitation projects. They are the interface to the users and markets and the "raison d'être" for going into space. Spacecraft are built to accommodate the needs of these instruments and of the user community which specifies or proposes and, depending on the situation and field, often builds and finances the instruments.

Payloads of today's spacecraft are usually made up of a multitude of different sensors or instruments. These could be designed to measure solar flux or magnetic fields, look into deep space, image and

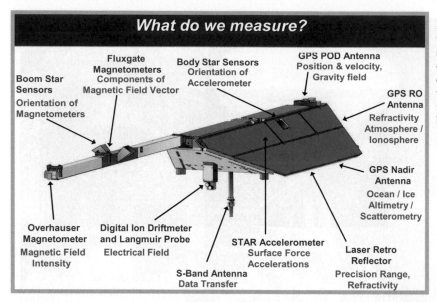

What do we measure?

Boom Star Sensors
Orientation of Magnetometers

Fluxgate Magnetometers
Components of Magnetic Field Vector

Body Star Sensors
Orientation of Accelerometer

GPS POD Antenna
Position & velocity, Gravity field

GPS RO Antenna
Refractivity Atmosphere / Ionosphere

GPS Nadir Antenna
Ocean / Ice Altimetry / Scatterometry

Overhauser Magnetometer
Magnetic Field Intensity

Digital Ion Driftmeter and Langmuir Probe
Electrical Field

STAR Accelerometer
Surface Force Accelerations

Laser Retro Reflector
Precision Range, Refractivity

S-Band Antenna
Data Transfer

Figure 2-16: A typical relatively small science satellite, such as CHAMP, has a number of measurement objectives and a diverse set of instruments which together deliver the needed information to the users (Reigber)

characterize features on Earth or another planet, reflect or transmit information, or analyze the characteristics of materials, or the responses of humans to the space environment. Instruments could be flux meters, pressure sensors, telescopes, cameras, antennas, biosensors, small laboratories or ovens with gas chromatographs. Often they are highly complex and costly. For example the Hubble Space Telescope or XMM optical and sensor payload packages exceed several hundred kilograms in mass and hundreds of millions of dollars in cost. The same can apply to some of the complex medical or biological laboratories in a space station. In contrast, the instruments of the relatively small gravity field spacecraft CHAMP

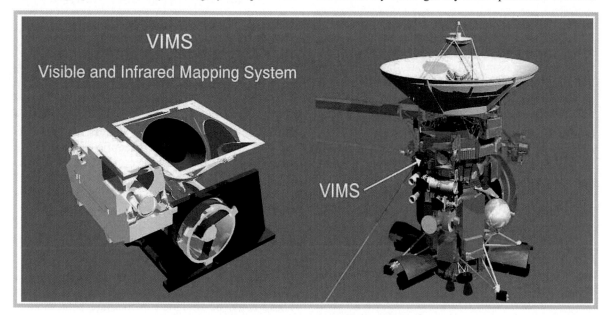

Figure 2-17: One of several of Cassini's complex instruments, the Visible and Infrared Mapping system, together with its electronic box, is shown separately and as mounted on the spacecraft (NASA)

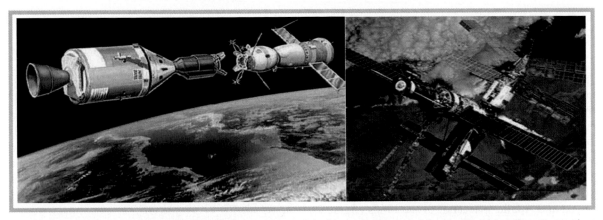

Figure 2-18: Shows, left, the Apollo-Soyuz rendezvous and docking mission and right, the Russian Mir space station, which was the only permanently crewed space laboratory of the 1980's and 1990's (NASA, RSA)

and GRACE are comparably inexpensive. CHAMP measurement objectives and instruments are shown in figure 2-16, indicating the diversity typical for many of today's spacecraft. CHAMP measures, for example, gravity and magnetic fields, atmospheric phenomena and distances to the Earth surface with extremely high accuracies.

Cassini's Visible and Infrared Mapping System (VIMS), a planetary observing instrument, is one of several complex sensors on board a large and expensive scientific planetary exploration spacecraft (figure 2-17). It combines optical assemblies with extremely sensitive detectors and electronics to measure planetary phenomena in different wavelengths.

The very faint signals are processed, compressed, stored, and eventually transmitted to Earth via the spacecraft communication subsystem.

Radiometers for weather satellites or multispectral cameras for Earth or planetary observation are instruments of comparable complexity and cost.

Cost drivers for many instruments are often the space qualification of terrestrial (laboratory) instruments. Space qualification entails minimization of mass and power consumption, reliability and safety upgrades, space hardening of components, and the like. Space qualification is expensive and time consuming. It includes considerable efforts for testing and documentation.

Figure 2-19: Left, the Apollo Lunar Excursion Module and right, the Skylab Space Station represent the early crewed U.S. space habitats (NASA)

2.3
Space Stations (Human Space Laboratories)

Human presence in space has multiple dimensions, utilization for scientific knowledge being one of them. The first mini-stations for humans were the early capsules built in the Soviet Union and USA to launch the first cosmonauts and astronauts into ballistic trajectories and low earth orbits.

The best known early spacecraft accommodating humans in their quest for space exploration are the Soyuz and Apollo spacecraft of the 1960s (figure 2-18, left) illustrated during the joint docking mission in 1975. Soviet developments then focused on the Salyut and Mir space stations, with continuous presence in space from 1986 to 2001. The Mir space

station accommodated many astronauts or cosmonauts from around the world during its many years of operation and played an important role in the know-how buildup for the ISS.

Early U.S. developments culminated with the Apollo program and the first human space habitat and laboratory on the moon, the Lunar Exploration Module (LEM) (figure 2-19, left). The only (Apollo derived) U.S. space station was Skylab, launched in 1973 and operated for some two years before descent into the Pacific Ocean (figure 2-19, right). Further U.S. space station plans were shelved in favor of the Space Shuttle development.

The need for extending the Space Shuttle's capability for research and experimentation led to development of the European Spacelab. Spacelab was a

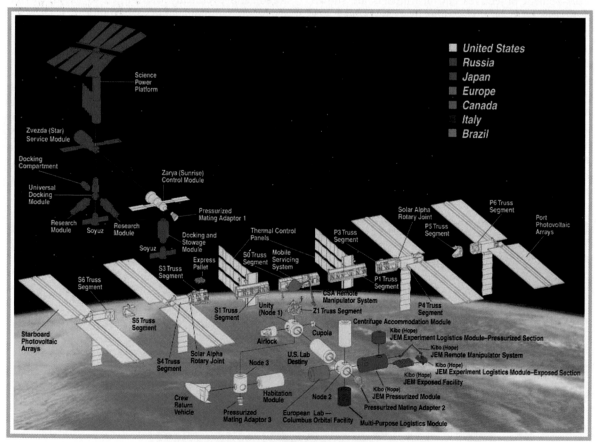

Figure 2-20: Shows the original ISS development partners sharing in one of mankind's most ambitious and complex-cooperation projects ever (NASA)

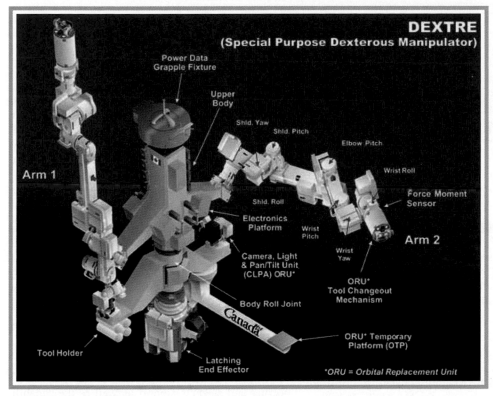

Figure 2-21: The multi-arm Canadian Dextre robot is designed to conduct a variety of tasks, supporting or replacing astronauts in complex and expensive EVA excursions (MacDonald Dettwiler)

laboratory for multidisciplinary research in a microgravity environment. It was used on 24 Shuttle flights by engineers or scientists from around the world, notably from Europe and the USA. An add-on platform allowed the accommodation of experiments needing direct exposure to space.

In the mid 1980s U.S. President Ronald Reagan invited his Western partners to jointly develop an international space station, then called Space Station Freedom. This station was intended to realize the goal of permanent presence of humans in space, which until then had only been demonstrated by the Soviet Union. The European, Japanese, and Canadian space agencies eventually joined in and agreed to develop their own visible contributions, the Columbus laboratory, the Japan Experiment Module, as well as the Canadian Canadarm 2 as part of this project. After the end of the cold war, Russia, in 1993, was invited to become a partner in this venture as well.

The ISS development, shown in an exploded view with the original partners in figure 2-20, is the largest international scientific and engineering venture ever. There has been a substantial evolution of this configuration since the originally proposed Freedom concept, which was even more elaborate than the current ISS configuration.

In January 2004 President Bush announced a new U.S. Space Exploration Initiative (SEI). It foresees limiting the use and lifetime of the ISS to 2016 and phasing out the Space Shuttle by 2010, replacing it with a yet to be defined new space transportation system. Future U.S. efforts should hence focus on combined robotic and human exploration, notably of the Moon and Mars. Many studies have yet to be completed before the programmatic and engineering content of this venture will become clearly visible.

Along the path of defining the new space exploration strategy, a robotic servicing mission to the Hubble Space Telescope could become a key milestone for future human space flight missions. This mission, if decided, could re-enforce the application of modern highly capable multi-armed dexterous human assistants, such as the Canadian Dextre robot

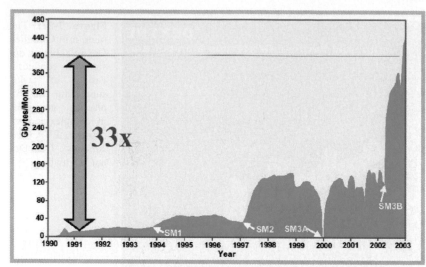

Figure 2-22: The monthly science data production of the Hubble Space Telescope as a function of operation year shows characteristic step increases relating to the service missions (SM). These resulted in a 33-fold increase in scientific data flow which generally equates to a similar increase in scientific performance

(figure 2-21) designed for operation on the ISS and capable of complementing and replacing some human operations in space.

Since its existence the Hubble Space Telescope had been serviced and upgraded in orbit on average every 4 years. Figure 2-22 shows the 33-fold increase in dataflow, which has resulted from these upgrades. This can be roughly equated to a 33-fold scientific performance gain for the HST. It brings into evidence how rapidly instrument technology has advanced during the orbital lifetime of HST, i.e. in just 15 years. And it underlines the enormous inherent potential for product improvement associated with a strategy of planned in-orbit servicing and technology or instrumentation upgrades for today's complex in orbit observing systems and other spacecraft. The success of this HST servicing approach resides in a synergistic interplay between robotic means and astronauts. It further demonstrates the enormous recent advances in robotic technologies and its future potential for space exploration.

In summary, human presence in space has become a matter of routine. Many important utilization results have emerged from these missions, particularly during the past two decades. More are to be expected from the ISS. Human missions serve many objectives. The scientific and commercial utilization potential is relatively limited as compared to automated spacecraft, especially when the budgets needed to sustain these programs are considered in

such a trade-off. The strong impact of human spaceflight and exploration upon mankind's imagination and advancement is however unquestionable.

2.4
Ground Support Centers

Ground facilities, or support centers, provide engineering and science support as well as the capabilities of command and communications with space vehicles and their payloads. They support the integration and test phases of space vehicles, space launches, and operations of spacecraft during their lifetimes.

There are characteristic differences between a Launch Control Center (LCC), a Mission Control Center (MCC) and a User Support and Operation Center (USOC). LCCs support and control launchers in their preparatory phases at the launch site and during launch until the final orbit destination has been reached. This includes propelling the upper stage into a non-conflicting graveyard orbit, or into an Earth re-entry trajectory. The MCC usually takes over control of the mission from the point of separation between launcher and payload.

Ground control centers consist of work stations, data, tools, communications provisions and human experts who have the capability to direct operations of launchers and spacecraft and solve problems during launch or in orbit as they occur.

Figure 2-23: Space vehicles are tested for structural integrity on dynamic shakers simulating the highest anticipated dynamic and static loads expected during the launch phase. Right, Mercury spacecraft "Messenger" during Vibration test at John Hopkins Lab in Baltimore. Thermal vacuum chambers, left, verify the anticipated thermal behavior of a spacecraft during its journey through space. Shown is Canada's Radarsat inserted into the thermal vacuum test chamber at the CSA David Florida Laboratory in Ottawa (CSA, NASA)

USOCs supplement the MCCs by providing the necessary interaction with the instruments and experiments or scientist astronauts for crewed missions. They receive the data flow from payloads or instruments and relay information to the users on the ground. Ground stations enable up- and down-link communications for telecommunications satellites in orbit and receive, process, and distribute data for others.

Spacecraft integration and test centers support the assembly of test articles and spacecraft to simulate or verify integrity, functionality, performance and workmanship of space vehicles. By means of mechanical shakers, thermal vacuum chambers (figure 2-23), acoustic, electromagnetic or antenna test ranges (figure 2-24) and dedicated test rigs, various pieces of equipment, subsystems, instruments or spacecraft are subjected to tests which simulate the severe environment to be encountered during their life cycle. This includes hardware and software, and as many functions as can be reasonably tested on the ground. To protect and avoid contamination of the sensitive equipment, much of this is done in clean rooms.

Checkout stations with dedicated equipment and emulators serve to verify functionality and performance or enable spacecraft communications and command functions during the development, test, and integration process. Often checkout stations contain the same functions, hardware, and software, as the ground control centers. Sometimes checkout stations are adapted into such centers or can, in the extreme, serve as a ground control center for a spacecraft or its payload in orbit.

System tests generally involve the complete spacecraft or significant elements thereof. Tests of subsystems and equipment are designed to verify function, performance or workmanship prior to system integration. Examples are tests of thrust chambers, antennas, attitude control components, mechanical assemblies, batteries or electronics, including associated software.

2.5
Enabling Competences

Design, development, manufacturing and operation of space vehicles rely on many different technical, administrative and scientific competences. Space projects are highly multidisciplinary. Margins of error are very small and the cost of failure is high. To cope with these risks and to ensure success, space developments place particular emphasis on three special competences. These are project management, systems engineering and product assurance. All three are directed to cope with the high risks of space developments and to generate check and decision points during a project's development cycle.

Space agencies and industry have over decades developed processes, standards and practices directed to identify and successively reduce risks for space missions. Some of these practices are specific to space, some have been derived from defense projects and others originate from industrial standards and practice.

A fourth competence essential for the conception and realization of space projects, namely development or selection of the needed technologies for new projects, is also addressed in this chapter.

2.5.1
Project Management

Project management ensures that the design, development, manufacturing, operation and utilization of space projects are executed within pre-defined cost, schedule and performance limits. Project management also encompasses the actions necessary to meet or arbitrate stakeholder requirements, including user requirements, within the constraints of a project.

Project management teams are usually established early in the life cycle of space projects and are generally composed of a combination of experts associated with the early studies of a new mission and of those with experience in the implementation of previous projects. They encompass experts from important technological fields of a project, system engineers, payload specialists, user representatives, subsystem experts, project controllers, contracts and financial specialists, as well as product assurance, manufacturing and operations staff. Most frequently project teams work in matrix interaction with other enterprise experts.

Generally the project manager has full empowerment for day to day matters and reports on a regular basis to the company or agency management. For public financed projects there are project teams at the space agencies and at the (prime) contractors responsible for the industrial management and realization of space missions. For privately funded space

Figure 2-24: Left, electromagnetic or electrostatic radiated emission testing at the ESA/ESTEC test center verify signal characteristics and compatibility of emitters and electronic equipment. Right, final antenna alignments and performance are verified at a Loral facility (ESA, Loral)

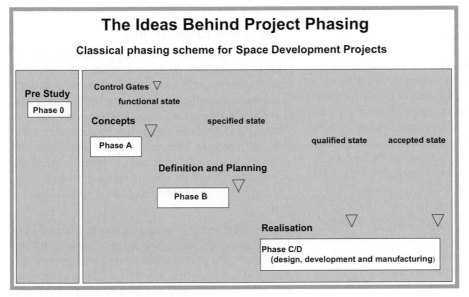

Figure 2-25: Space projects go through various phases during their development. This classical chart shows the ideas behind this phasing along with the different stages of accomplishments and associated control gates

projects there is usually a single project team, and additional organizational elements or people representing various stakeholder requirements, such as market demands, technical and operations requirements, return on investment criteria and the like.

Space projects go through different study, development and operations or utilization phases (figure 2-25). Initially pre-studies, often referred to as phases 0 and A, address alternative market, mission and system concepts. They are followed by system definition, design and development phases, which include test and verification efforts and the manufac-

turing of the prototype and flight vehicle(s) (phases B and C/D). Operational or utilization phases up to and including the disposal of a spacecraft are referred to as phases E to G.

There are different terms, number of phases and decision points in use by different agencies and enterprises. In principle the objectives are similar, namely to enable successive and iterative study, development and operational risk reductions for a project. Each phase is designed to provide successively better insight into design and programmatic parameters, such as performance, cost, return on

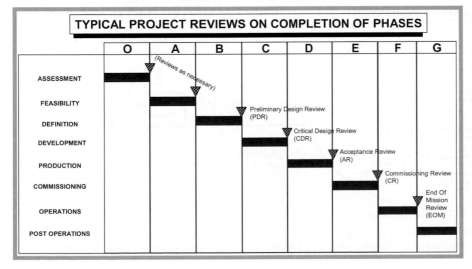

Figure 2-26: Project reviews are key milestones during the life of a spacecraft. They bring together a number of experts from all fields concerned to verify the technical and programmatic status of the project concerned. Shown is the ESA standard process which is typical for the space developments (ESA)

41

investment, schedule, and associated risks.

Another important characteristic of space projects management is the organization of project reviews. These are important checkpoints during the study and implementation phases, whereby progress of a project is assessed. The process consists of experts reviewing all aspects of a space project, i.e. its mission, user and other requirements, system design, test plans and results, manufacturing drawings, materials lists, etc. with the objective of finding flaws in the technical, programmatic or quality state of a project. These are checkpoints, or control gates, which, if exercised properly and with the help of independent, experienced experts, attest a project good health, or otherwise.

For management they provide feedback on progress and a check of confidence. Problems identified are generally documented in Review Item Dispositions, actioned and followed through until closure by the authorized person or board. Reviews are sometimes conducted during early mission phases and with increasing depth during the definition, design and manufacturing phases. Launch readiness reviews and in-orbit operations reviews are part of this process.

Figure 2-26 shows the ESA standard project review process as documented in the European Cooperation for Space Standardization (ECSS) document series.

2.5.2
Systems Engineering

Systems engineering and project management go hand in hand. Both have been practiced since millennia, albeit mostly ad hoc. Architects and engineers who built the Egyptian pyramids, the Eiffel tower or the Hover dam showed exceptional skills in planning and executing high risk projects with complex technologies and many specialist disciplines involved. Modern defense and aerospace projects are all executed with the help of project management, systems engineering, and integration techniques and skills.

Systems engineering focuses upon the integration of the technical and user aspects of a project. It conducts the trade-offs needed to ensure a proper balance between scientific, market, stakeholder or other user requirements with the technical and programmatic implementation alternatives, including cost and schedule variables. The definition of systems engineering by the International Council on Systems Engineering, (INCOSE), is shown in a separate box.

Modern systems engineering is conducted end to end in interaction with other disciplines concerned. For commercially oriented projects there is a significant focus on non-technical factors (figure 2-27).

Systems engineering concentrates upon the systems architecture and design aspects as well as implementation processes. Examples for the latter are the tasks

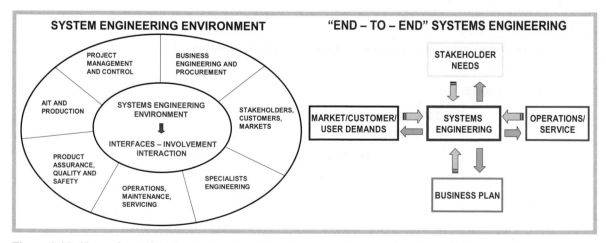

Figure 2-27: Shows the environment and process for end-to-end systems engineering, which emphasize the contextual relationships of systems engineering within an enterprise and with the non-technical factors of a project or system

Systems Engineering is an interdisciplinary approach and means to enable the realization of successful systems. It focuses on defining customer needs and required functionality early in the development cycle, documenting requirements, then proceeding with design synthesis and system validation while considering the complete problem:

Operations	Cost & Schedule
Performance	Training and support
Test	Disposal
Manufacturing	

Systems Engineering integrates all the disciplines and specialty groups into a team effort forming a structured development process that proceeds from concept to production to operation. Systems Engineering considers both the business and technical needs of all customers with the goal of providing a quality product the meets the user needs.

(Definition of the International Council on Systems Engineering INCOSE, see web: *Incose*).

logic flow during individual project phases and the technical and programmatic planning. The goal is to ensure that the end product, e.g. a ground segment, spacecraft, or mission, is realized in a most efficient manner with identified and manageable risks. The process generally follows a progressive top-down definition and a bottom-up integration flow, as shown in the "Vee-chart", figure 2-28.

Mission design and definition processes are increasingly conducted with the help of computer based tools and simulations. Concurrent systems design and engineering and model based systems engineering have made enormous progress in the past decade. NASA JPL metrics (figure 2-29) show the kind of progress and efficiency gain possible with the introduction of integrated conceptual mission design centers. Cost reductions by a factor of 8 and schedule gains by factors of 10 to 20, at the same or improved output quality, have been achieved for early study phases since the inception of the process. European integrated design centers, e.g. at EADS or ESA, point to similar results.

Systems (and concurrent) engineering plays a particularly strong role in controlling cost during the early phases of a project. During these phases only a relatively small amount of a project's total cost is accrued, on average between 5 and 15%. However, 70 to 90% of a project's technical and programmatic parameters are fixed during these phases (figure 2-30). Issues which are not resolved then, i.e. prior to entering the development and manufacturing phases of a project, are successively more costly to resolve in later phases. The extent to which such cost multipliers can grow depends on the individual project characteristics and life cycle.

In summary, from a project utilization and operations point of view, systems engineering plays an important role in a project. Most frequently the trade-offs between user requirements and technical solutions or operational profiles are conducted by the system engineers. The associated processes result in the technical and operational baselines for a project, which determine amongst others its utilization or operations characteristics and user access and interfaces.

2.5.3
Product Assurance

Product assurance encompasses a number of specific disciplines oriented towards ensuring the quality of a product in the broadest sense. The European Space Agency defines product assurance as to "assure that the space products accomplish their defined mission objectives and more specifically that they are safe, available, and reliable."

It defines the objective of product assurance management as "to ensure and achieve an adequate, effective and efficient coordination and implementation of the product assurance activities through a proper integration of the product assurance discipline as well as the integration of product assurance with all management and engineering activities." Other space agencies and industries generally have similar objectives and comparable approaches.

Product assurance encompasses quite a number of different but strongly interrelated fields. Examples are: quality, reliability, dependability, safety, and selection of components, materials, mechanical parts, manufacturing processes, and software standards. In day to day project implementation this

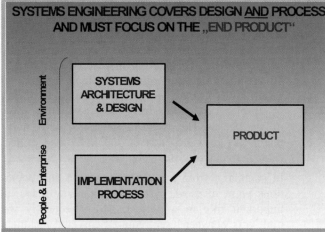

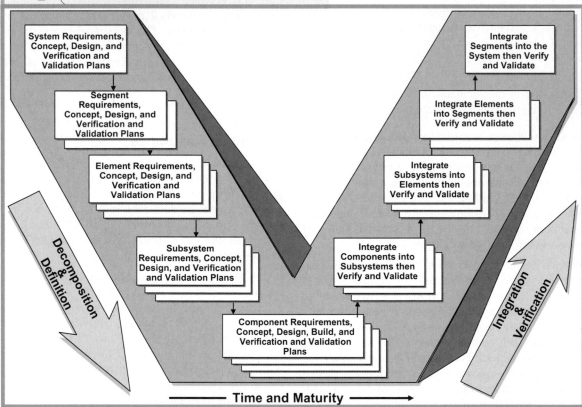

Figure 2-28: Systems Engineering is most effective when it is conducted in an environment which values design and process, both focused upon the final product (left). Lower chart, systems are defined top-down and integrated bottom-up. The famous Vee chart shows the relationships from mission to components in a hierarchical and time sequence (Center for System Management, CSM)

means product assurance gives attention to all factors that could compromise the quality and dependability of space vehicles throughout their lifetime. To ensure this, it focuses on all related factors, in particular those with a potential for failure. Product assurance addresses relevant activities in a project at different levels of execution. In some ways it consists of common sense principles for good management, engineering, manufacturing and operation as well as of a series of accumulated good practices and detailed processes such as soldering, outgassing, components selection, etc. Product assurance ana-

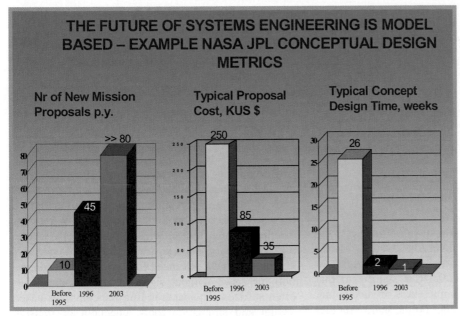

Figure 2-29: Demonstrates the advances made since the introduction of concurrent systems and mission engineering processes at NASA JPL (Steve Wall, JPL).

lyzes the potential influence of different parameters and conducts tradeoffs to identify best solutions. Examples are: different redundancy provisions for a particular design solution for improved reliability, the application of different software standards, or test approaches to be used during subsystems and integrated systems test.

For human space systems safety plays an overriding role. Compromises with human safety are unthink-

able when it comes to the design and operation of crewed vehicles. Extra quality, redundancies, and fail safe modes are built in from the outset. On the other hand cost associated with the design assurance, extended test and qualification processes, etc. for crewed systems are substantial.

Product assurance plans are established early in a project's life, consistent with project objectives, requirements, and constraints. Part of the process is

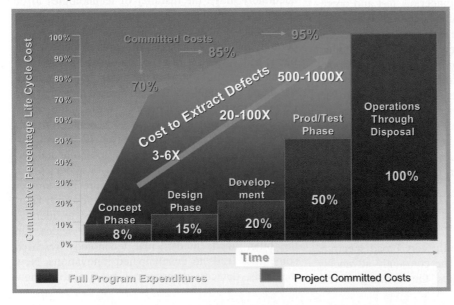

Figure 2-30: Shows that mastering cost at the earliest phases of a project has strong advantages and avoids solving problems at a high price later (Defense Acquisition University, DAU)

Why do projects fail?	
Incomplete requirements	13.1%
Lack of user involvement	12.4%
Lack of resources	10.6%
Unrealistic expectations	9.9%
Lack of executive support	9.3%
Changing requirements or specifications	8.7%
Lack of planning	8.1%
Didn't need it any longer	7.5%

The answer to the question why (space) projects fail are manifold. The table gives one set of answers which are fairly representative, even though not derived from space projects. It is significant that misunderstandings or incompleteness of stakeholder requirements range fairly high on the scale. Another frequent reason for failures, or for cost and schedule impäacts, not evident from this particular source, results from bad choices of technologies.

to ensure allocation and availability of adequate resources, personnel and facilities to do the job. These are different for different projects, e.g., small or large, high or low risk. Whether this function is carried out by specially trained product assurance staff or by the nominal engineering, test and operations staff is a secondary consideration. When seen from a broader perspective one can note that different approaches can be successful and that product assurance is handled differently in different companies, user institutes, or countries. Experience shows, however, that a combination of independent product assurance staff and a project team trained to pay attention to quality, reliability and the like works most effectively. This applies to a spacecraft's bus or a launch vehicle as much as to a user's instrument. A mission can be lost if either of them fails!

2.6
The Technology Base

Space developments rely on a very broad spectrum of different technologies. The constantly advancing technology frontier supplies the building blocks from which new space vehicle designs and capabilities are derived in the quest for lighter weight, lower power and energy consumption, better detection capabilities, communications throughput, mission flexibility, etc.

During the early decades, space equipment relied mostly on technologies developed specifically for space purposes. Often those were derived from defense applications. Recent developments in terrestrial industrial technologies have created in the past two decades or so new technology capabilities very useful for space applications. They may be less expensive and more readily available compared to dedicated space technologies. The most significant examples come from information technology where, e.g., processors, storage devices, and software developments have advanced to a point where adaptation for space without significant additional effort often suffices. The same is true for computer based design and simulation tools, checkout or mission control software and the like.

Technology advances as measured in mass and power reductions have significantly progressed during the last decades (figure 2-31). At the same time performance capabilities, especially in the information processing and sensor technologies, have grown substantially.

A major challenge for space vehicle designers is to accurately judge the maturity of technologies to be adopted in a project. This entails difficult decisions on whether to adapt or develop new technologies for certain functions, or to rely on proven ones. Engineers and scientists must continuously assess new technologies "bubbling" in the technology and innovation pipeline and evaluate associated technical and programmatic risk parameters during the early phases of new projects. The decision points for technology choices are generally prior to project go-ahead (Phases C/D). Technology research and development management in industry or agencies is a complex process of anticipating future mission requirements and feeding these into the research laboratories. This top-down process helps to identify technology research priorities and time scales. On the other hand, fundamental technology research sometimes leads to unanticipated innovations. New materials, devices, or processes have enabled new mission ideas and concepts that would not have been thought of without the bottom-up process. The top-down and bottom-up processes together are conducted in a complex interplay between mission planners, including user representatives, and technology specialists and scientists across many different fields. The challenge remains to decide for each new project upon a proper mix of proven and new technologies, independent of whether they come from the space, military or commercial research and development pipelines.

The spectrum of technologies used for space developments is huge. The risk of delay caused by a retarded development or qualification of any single new technology is high and can seriously impact an entire project. On the other hand a new sensor or propulsive device can substantially advance a system's capability and utilization potential and enable new ideas for the execution of missions.

Typical examples for new high potential technologies for space applications are plasma propulsion, nuclear power generation, new materials, microminiature equipment, electronically steerable antennas, nano components, or high precision atomic clocks. Figure 2-32 attempts to present a cross section of space vehicle and payload technologies from the many different sectors which play essential roles in the realization of space projects.

In summary, advancements in space utilization are strongly interrelated with advancements in technology research and development. The technological building blocks are the basis for the design and development of space projects. Together with new science, other user ideas or market opportunities they fuel innovation and progress in space systems and their utilization.

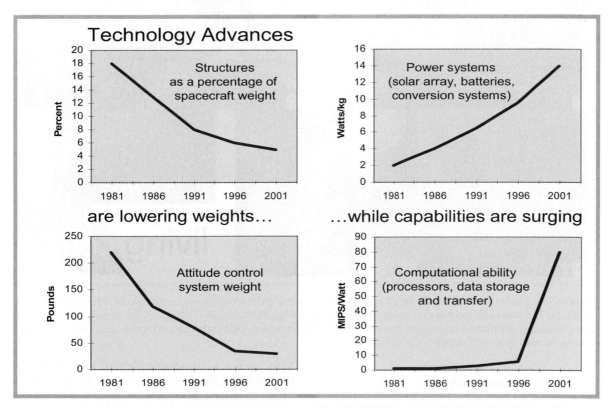

Figure 2-31: Shows the accelerating technology advances in relationship with surging performance capabilities for some typical space equipments (adapted from Pedro Rustan, Scaled Composites)

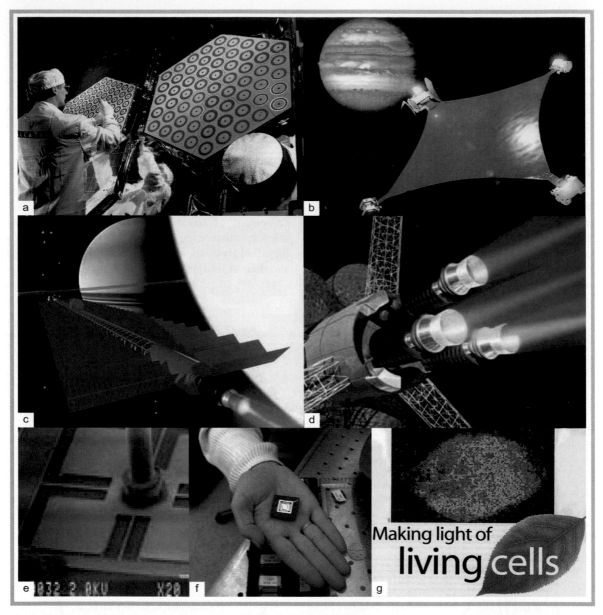

Figure 2-32: Shows examples of the broad spectrum of the many different technologies that are used for space developments. (a) Electronically steerable antennae for a telecom satellite, (b) solar sail concept, (c) fusion propulsion for deep space missions, (d) plasma propulsion concept, (e) microminiaturized solid state gyro, (f) active pixel sensor, (g) biosystems technologies with potential for space applications

2.7 Conclusions

Launcher, spacecraft and ground segments are the enabling means for space utilization. These elements have reached a remarkable state of maturity during the past decades. Progress has been substantial. Many down-to-earth practical, ambitious, and far reaching utilization objectives have thus been enabled. New utilization objectives and market opportunities will amongst others continue to depend upon technology advances of these elements also in the future. The need for lower cost solutions has probably become the most challenging and paramount driver for the space business in recent years.

Early space systems consisted primarily of light-weight mechanical structures with thermal protection, guidance and attitude control, propulsion, some electrical, electro-mechanical and electronic components, rudimentary computing capabilities and simple instruments. Modern space systems employ distributed architectures with extensive embedded information processing capabilities. They are driven by ever more sophisticated instruments and payloads, which account for the majority of advances in space science and practical space uses. Software has become a dominating element in the design and operation of spacecraft and ground systems.

Satellite platforms today can be bought off the shelf for a number of mission applications. Reuse of components and equipment, standardization of design and engineering practices have become commonplace. A good proportion of ground segments are procured as commercial terrestrial products. Only some two decades ago most of the computing and data processing equipment was custom made with enormous research and development investments in dedicated software and hardware.

Modern spacecraft and launch systems have grown enormously in complexity as measured in terms of number of functions and components when compared to their predecessors. This is analogous to a similar trend in terrestrial products. Today's telephones, household machines, cameras, or children's toys often include more computing power and control loops than did early spacecraft.

On account of technology miniaturization, especially in the fields of electronic and electromechanical components, satellites have grown smaller for comparable performance. Today's observation or communications satellites of some 500 kg mass sometimes outperform their predecessors of some 2,000 kg of only 10 years ago. The trend towards smaller satellites will continue to some extent. The advent of so called "micro" and "nano" spacecraft in the class of 50 kg of mass and below has enabled relatively low cost missions for special applications. Such spacecraft have become quite popular for university research, technology tests, specialized missions and with newcomers in the space field. When used in constellations of two or more and eventually as swarms of tens or even hundreds, their mission applications potential and impact on current practices could grow substantially in the long term.

On the other hand some spacecraft classes have been driven by economy of scale factors, for example geostationary commercial telecommunication satellites. They have grown in size, mass, power, and number of transponders, adapting to the increasing launcher capabilities available on the market. Similar considerations apply to some scientific spacecraft developments where the design of instruments is driven by the ambition of more complex or comprehensive measurements.

The utilization phase for the ISS has barely started. Compared to the broader goals of demonstrating mankind's ability to live and work in space, scientific and commercial utilization will remain a secondary objective. The Space Exploration Initiative has created a need for new infrastructure developments. It remains to be seen how and when it will advance into realization and to what extend it can stimulate utilization.

The biggest constraint for space utilization remains the high cost of launching spacecraft into their final destinations. New technologies may eventually allow the construction of economic, safe and easy to operate, reusable space launchers with designs substantially beyond the heritage of the Space Shuttle. This could, some time in the future, open a new era of space utilization. In the meantime evolutionary advances in expendable launcher designs will dominate the scene. This will mean that transportation cost per kilogram mass into orbit will remain high.

The challenge of lowering the cost for space missions to enable ambitious space utilization objectives to be realized is likely to remain the key driver for years to come.

References and Further Reading

Boden DG, Larson WJ (1996) eds: Cost-effective Space Mission Operations, McGraw-Hill Publishers, New York

European Space Agency (1996) European Cooperation for Space Standardisation, ECSS, Space Engineering and Space Project Management Series, ESA Publications (http://www.ecss.nl)

Forsberg K, Mooz H, Cotterman H (2000) Visualizing Project Management, Wiley Publishers, New York

Fortescue P, Stark J and Swindon J (2003) Spacecraft Systems Engineering, John Wiley and Sons Publishers, Chichester

Hallmann W, Ley W (1999) ed: Handbuch der Raumfahrttechnik, Carl Hanser Verlag, München

Koelle DE (2000) Handbook of Cost Engineering for Space Transportation Systems TCS-TR-168 TCS-TransCostSystems, Ottobrunn

Larson WJ, Pranke LK (1999) eds: Human Space Flight Mission Analysis and Design, McGraw Hill, New York

Messerschmid E, Bertrand R (1999) Space Stations – Systems and Utilization, Springer Verlag, Berlin Heidelberg

Messerschmid E, Fasoulas S (2004) Raumfahrtsysteme, Second Edition, Springer Publishers, Berlin Heidelberg

Mooz H, Forsberg K, Cotterman H (2003) Communicating Project Management, John Wiley and Sons Publishers, Hoboken

Sellers JJ (2000) Understanding Space, An Introduction to Astronautics, McGraw Hill, New York

Shishko R (1995) NASA Systems Engineering Handbook, NASA SP 6105

Wertz JR, Larson WJ (1999) ed: Space Mission Analysis and Design, Microcosm Press and Kluwer Academic Publishers, Torrance and Dordrecht

Wertz JR, Larson WJ (1996) eds: Reducing Space Mission Cost, Space Technology Library, Microcosm Press and Kluwer Academic Publishers, Torrance and Dordrecht

Web References:

astrobooks: http://www.astrobooks.com, see Launch Systems, TransCost-7.1

incose: http://www.incose.org

Looking down: our Earth

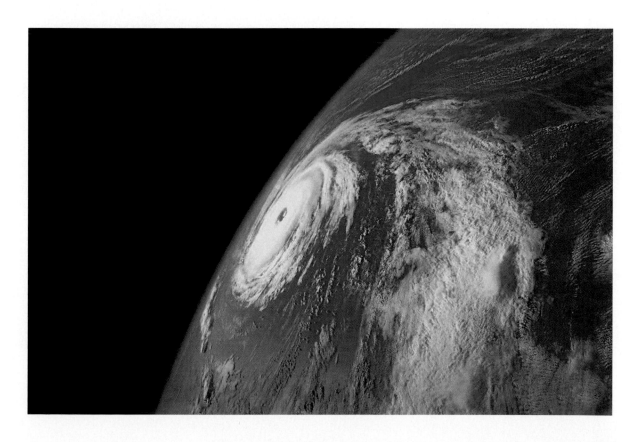

Overleaf image: Weather phenomena are easily observed from space (NASA)

3 The Earth Surface

by Stefan Dech

Looking down at Planet Earth from the vantage point of space has always exerted a great fascination, whether on astronauts orbiting the earth or on earthbound beholders of spectacular satellite images. We have become aware of the fragility and vulnerability of the earth, learned to comprehend it as a unit irrespective of political borders, seen that gossamer veil of an atmosphere which protects living organisms from aggressive cosmic rays and, not least, been struck by the beauty of our planet. Space activities at the inception of the 21st century are giving us new views of our "spaceship earth," leading us to new insights.

3.1
Sensing the Earth Surface

Remote sensing of the earth from space was stimulated by the space race of the 1960s between the USA and the USSR. The view from above offered numerous possibilities for observing inaccessible territory and monitoring activities there, particularly from military considerations. But rather soon the great potential for scientific investigation of the earth and its complex ecosystem became evident. The Earth Resources Technology Satellite (ERTS), launched in 1972 and originally designed for military applications, marked the beginning of civil remote sensing from space, at least as far as the land surface was concerned. Soon after launch ERTS was rechristened Landsat 1, and satellites of the Landsat series have been setting milestones in earth observation ever since in the development of successor systems, a multitude of analytical techniques, geoscientific insights and commercial utilization. This chapter provides an overview of current and future possibilities in earth observation and the associated systems with respect to the earth's land surface, oceans and cryosphere. Following chapters will elucidate other application areas, such as atmospheric research (in chapters 4 and 5) and geophysics (in chapter 6).

Geoinformation from Space

Remote sensing is first of all a measuring technique for obtaining data about the earth's surface and the atmosphere without coming into physical contact with the objects to be measured. The procedure can be used to study any celestial body (as has ESA's Mars Express since January 2004). We speak of earth observation if the remote sensing is carried out from space with the earth as target. The goal is to obtain terrestrial measurements of high temporal and spatial resolution for large areas by utilizing the spectral characteristics of surfaces and objects on earth. The data obtained in this way are converted into various geophysical or biophysical information layers with the help of physical-mathematical algorithms. These can serve as the basis for subsequent analyses or be integrated into mathematical models describing many earth processes. The derivation and mapping of environmentally relevant variables and object identification also play an important role in many applications. All told, objective and independent geoinformation can be obtained with the help of modern earth observation technologies for use in the geosciences, marine and life sciences and atmospheric research, as well as in many practical applications concerned with local, regional and environmental planning, and with disaster management and humanitarian aid activities that fall under the category of civil security.

Only earth observation from space allows spatially integrated and continuous recording of measurements covering large areas and at various scales ranging from the global views needed for climate research to the local perspectives required for mapping urban areas. In the 1980s, earth observation started to make serious progress. Geoscientists from the most varied disciplines were quick to appreciate its enormous research potential for environmental studies and monitoring. Thanks to the previously unimaginable availability of high quality, repeatable measurements for vast areas, almost all of the important ecosystems on earth negatively affected by human activity—from tropical rain forests (land clearance) to semiarid and arid regions (desertification) to temperate zone forests (forest dieback) to oceans and coastlines (material transport, algal blooms, El Niño)—could be regularly monitored and analyzed in context.

Promising perspectives for the commercialization of this technology were also soon on the horizon. For example, the vast realms of agriculture, forestry, public planning and environmental concerns turned out to offer a variety of interesting points of departure for usefully integrating the methodology and products of earth observation. Its imaging measurement principle was particularly advantageous for the rapid growth in applications for earth observation data. Thanks to the continuous recording capability of remote sensing instruments, the sum of all the individual measurements (pixels) and the information products derived from them can be represented in the form of digital satellite maps. Seen together, all the individual measurements form a three-dimensional data set in raster format, which results in a continuous, spatial pattern allowing mapping of the area under observation. Such results can easily be integrated in models or put into geographical information systems where they can be combined with other types of data, which can be either in the form of data points or area representations, before further analysis. Another weighty advantage is the option of regularly repeating measurement of the same area without undue logistic effort, making it possible for the first time to extensively monitor landscape changes caused by natural processes or human activity. Finally, rapid access to data from earth observation satellites is encouraging new areas of application. The first step was meteorological remote sensing from a geostationary orbit, which now makes near-real-time weather reports routine. With today's land and ocean observing satellites, images of an affected area can be ideally supplied within hours of a catastrophe (forest fires, floods, earthquakes) to crisis situation centers. Earth observation data are increasingly becoming an important

resource for investigating and minimizing damage, supporting emergency humanitarian efforts, and planning measures which improve civilian defense. Thus the information technology known as earth observation is increasingly finding a widely recognized role in a broad range of scientific and public applications.

3.2
Remote Sensing Basics

Data from earth observation instruments have been received for some 35 years. They are usually operated from unmanned satellites which, depending on the application, may be placed in any of a number of orbits. In order to obtain data of high temporal resolution (such as required for meteorology), equatorial, geostationary orbits at 36,000 km altitude are preferred. At that location, one orbital circuit is identical to the length of one day (24 hours) so the earth is viewed by the sensor always from the same viewing angle and as if it were standing still. The disadvantage is that only about 25% of the earth's surface is visible; the polar regions cannot be observed at all from this vantage point. For surface monitoring satellite orbits which cross the polar regions and are inclined some 98 degrees with respect to the equator are preferred. Satellite altitudes may vary between about 300 and 850 km. These almost-polar orbits take about 100 minutes for one circuit and offer sun synchronicity. In other words any point on earth is always observed at the same local solar time of day, a considerable advantage when comparing images. Another advantage of this orbit is that almost any point on the earth's surface can be viewed, the only exceptions being regions very near the poles in the case of sensors with limited side viewing capabilities. Optical earth observation sensors usually detect the irradiance reflected from the earth's surface along the flight path of the satellite and a certain distance to the right and left depending on the defined viewing angle of the sensor in question (so-called nadir viewing). Radar instruments always view to one side of the satellite track.

Earth observation from satellites uses electromagnetic radiation from the optical to the microwave part of the spectrum (figure 3-1). Optical sensors operate in the wavelength region from 0.3 to 14 μm, while microwave sensors cover the mm to dm wavelength range. A distinction is made between passive instruments like digital cameras, radiometers and spectrometers that measure reflected solar or earth-emitted radiance, and active instruments like laser or radar systems that send out signals from their own source of power and measure the returned ("back-scattered") signal, which is a small fraction of the originally transmitted power. There are a number of physical mechanisms involved in the propagation of electromagnetic waves through the earth-atmosphere system. Absorption and scattering take place both in the atmosphere and at the earth's surface. In addition, there is reflection at the surface and self-emission of radiation according to Planck's law. The first 20 years of satellite earth observation were dominated by passive optical sensors which detected primarily irradiance in the visible and near, short-wave or thermal infrared parts of the electromagnetic spectrum, depending on the application. Using such multispectral data (ca. 1-10 spectral channels) the most important methodologies for today's appli-

cations were developed. The optical systems are usually optoelectronic or electromechanical scanners which split the incoming irradiance spectrally with the help of a grid or prism and then pass it on via elaborate mirror constructions to individual detectors or CCD (charged-coupled device) arrays, whereby each detector in the array senses one pixel in the image field. Optical earth observation data which includes different looks of one and the same object from slightly different viewing angles were successfully processed to yield so-called digital elevation models of the earth's surface. The optical region contains many diagnostic features in the shape of the reflectance and emissivity spectra. In addition, the surface temperature which can be derived from the data is an important biophysical and climatological parameter. However, there are drawbacks. Optical imagery frequently suffers from haze and cloud contamination, and here radar has the advantage. Since microwaves penetrate clouds and atmospheric trace gases to a large extent, the quality of microwave images is almost independent of cloud cover-

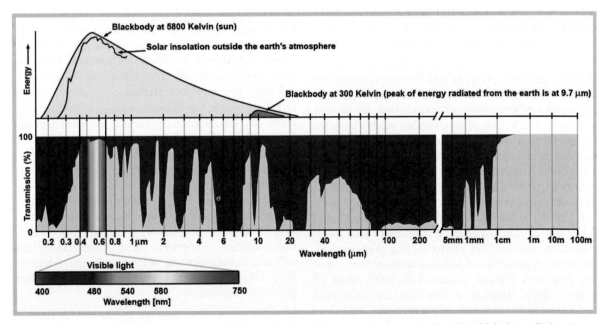

Figure 3-1: The electromagnetic spectrum. Below: the dark blue areas are those portions in which the radiation is entirely absorbed by the atmosphere and thus unavailable. The "windows" (light blue) which let through radiation in the visible (c. 400-700 nm wavelength), infrared (0.7-300 μm) and microwave (1 mm-1 m) portions are used in earth observation. Above: Planck radiation curves for blackbodies at 5800 K (sun) and 300 K (earth). The radiation maximum is in the visible range

age, so data can be collected regardless of local weather conditions at the time of recording. Since radar instruments send out their own signals, they are not dependent on solar illumination as are the optical systems, so they can be operated independently of the time of day. Since the early 1990s elevation models can also be derived using data from radar systems. By collecting radar echoes coming from one and the same object from slightly different positions in space, phase differences can be analyzed not only to derive these models but also to detect small, centimeter-range horizontal movements, subsidence and uplifting of the earth's surface. Moreover, multifrequency and multipolarization active microwave systems can be used to derive information on soil roughness and moisture, and they enable good differentiation of diverse vegetation types and urban surfaces. This technique permits a multitude of additional applications, some of which are described below. Each type of sensor and spectral region has its own specific advantages and drawbacks, depending on the target to be viewed. Optical and radar are on principle complementary recording systems, and in the ideal case multisensor optical and microwave imagery are always combined for a detailed retrieval of earth surface information.

Whether passive optical or active radar systems, the signals detected by the sensor are converted into electronic information and transmitted to ground stations via telemetry. The telemetry signal includes the actual payload data, information about the state of the sensor (housekeeping data), and the orbit data, all of which together make it possible to create a calibrated and referenced image. Satellite lifetimes in the inimical environment of space can vary from one year up to ten years under favorable conditions. As many as ten different instruments have been placed on one satellite platform, and they may be operated separately or in various combinations, as is the case with Envisat, launched in 2002. Most of today's earth observation satellites are developed and operated by national or international organizations (NASA, ESA, CNES, ISRO and others) with public funding and in the context of science programs (NASA's Mission to Planet Earth, ESA's Living Planet Program, etc.). Currently, long-term and overall continuity of earth observation is not assured, in contrast with the situation for meteorological satellites. Since the end of the 1990s an intensification of commercial activity could be observed, particularly regarding high spatial resolution systems (like Ikonos, QuickBird). Lastly, there are a number of small earth observation systems (on so-called micro- or mini-satellites) being designed, built and operated by research institutions and/or universities worldwide to meet specific teaching and research goals (BIRD, SUNSat, etc.).

It cannot be attempted here to give even a superficial survey of former, current and planned earth observations missions. There are simply too many, and the field is very dynamic. Table 1 (at the end of this chapter) lists only the most important current and planned missions for sensing the earth's surface and supplies some technical information about them. For the most thorough summary available see Kramer (2002).

3.2.1
Optical Remote Sensing

In the optical spectral region, for wavelengths $\lambda = 0.3\text{-}14$ μm, a radiative transfer code is used to remove the atmospheric influence on the measured sensor signal and to obtain information from the earth's surface. In the solar-dominated region ($\lambda < 2.5$ μm) the surface reflectance signature can be retrieved, while the thermal region ($\lambda > 8$ μm) enables the calculation of surface temperature and emissivity spectra. For minerals, rocks, and soils, electronic and vibrational transitions are induced by the electromagnetic interaction, and these cause diagnostic absorption features in reflectance and emissivity spectra. The electronic transitions require higher energy, and thus take place at shorter wavelengths than do vibrational transitions. For vegetation canopies, important factors determining the reflectance behavior are the leaf pigments, particularly chlorophyll-a and chlorophyll-b, leaf water, and canopy geometry.

Depending on the number of available spectral bands (multispectral instruments: typically 5-15 bands; hyperspectral: 50-300 bands), the reflectance signature allows a detailed quantitative extraction of geophysical, chemical and biophysical parameters. For example, different vegetation canopies have different amounts of chlorophyll-a and -b pigments and plant water. The pigments are primarily responsible for the reflectance behavior in the visible (0.4-0.7 µm) region. Plants appear green because of the higher pigment absorption in the blue and red parts of the spectrum. The reflectance in the near infrared plateau (0.7-1.2 µm) is due to the cell structure, and also influenced by the water content ($\lambda > 0.96$ µm). In the 1.2-2.5 µm region the plant's liquid water is the major factor determining the reflectance curve. In the framework of multitemporal monitoring of agricultural areas with satellite sensors, pigment concentration and leaf water content can be derived. This information can be used to assess crop health status and to predict crop yield, in combination with other parameters. Further examples from different application fields are discussed in subsequent sections of this chapter.

3.2.2
Radar Remote Sensing

Synthetic aperture radar (SAR) sensors use microwave pulses of 3 to 25 cm wavelength to image the earth surface. As said above, these waves penetrate clouds and since the radar carries its own illumination, SARs work independently of weather and sunlight. Figure 3-2 depicts the SAR imaging geometry. A radar sensor transmits short microwave pulses (typically 1,000-5,000 per second) to the earth in a side-looking fashion and receives the waves reflected back from the ground. The two coordinates of the image are formed by two different mechanisms:

In the range direction (perpendicular to the flight path) the swath is scanned by the wave pulse sweeping over it at the speed of light. Different objects on the ground are distinguished by their distance to the SAR and, hence, by the arrival times of their wave echoes. The resolution in range is given by the length of the transmitted pulse. The width of a pulse and its bandwidth are inversely proportional. Given a bandwidth of W [in Hz] and a local incidence angle θ_i the ground range resolution in meters is

$$\rho_{g/r} = \frac{c}{2\,W\,\sin\theta_i}.$$

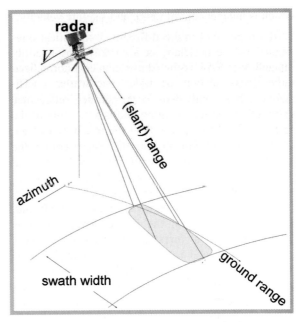

Figure 3-2: Imaging geometry of a synthetic aperture radar

The European Remote Sensing Satellites' (ERS-1, ERS-2) family of SAR instruments uses a bandwidth of 15 MHz and an incidence angle of about 23 degrees, resulting in a ground range resolution of 25 m. The German TerraSAR-X satellite will boast a 300 MHz bandwidth.

In the azimuth direction (along the flight track) the swath is scanned by the flight motion of the radar. The resolution in azimuth is given by the extent of the illuminated patch on the ground. Focusing it to a few meters, as required for remote sensing, would call for an antenna of several kilometers length, which is not feasible. Instead, the trick of aperture synthesis is employed: the radar returns are recorded coherently, i.e., their amplitudes and phases are measured. The resulting raw data set resembles a hologram. In a second step this is focused to a high resolution image by dedicated digital signal processing software packages, so-called SAR processors (Curlander 1991). The finally achievable azimuth resolution of a SAR with an antenna of physical length L in the flight direction is

$$\rho_{az} = \frac{L}{2},$$

which is independent of range and satellite velocity.

SAR images look quite different from optical ones (figure 3-7, left) (Ulaby et al. 1986). The quantity imaged by a SAR is the (dimensionless) normalized radar cross section or radar backscatter coefficient σ^0. It is equivalent to the surface reflectance known from optical imaging and is a measure of the ability of the ground to reflect the microwaves back to the radar. σ^0 depends on the roughness of the surface, its dielectric constant, the wavelength, and the incidence angle. Surfaces smoother than a wavelength (asphalt, calm water, slick rock) act as mirrors reflecting the waves away from the SAR and, hence, appear dark in the image. Rough surfaces (agricultural fields, forests) scatter the microwaves in all directions and a part of the wave energy can be received by the SAR, leading to brighter areas in the image. Besides the roughness the dielectric properties of the surface play an important role: the higher the dielectric constant the higher σ^0. The dielectric constant is in turn a function of the water content of soil or plants. SAR images are not only maps of the surface. Unlike in the optical regime, microwaves can penetrate vegetation, soil, ice and snow. While X-band ($\lambda = 3$ cm) gets reflected in the layers close to the surface, L-band ($\lambda = 25$ cm) partially passes through canopy and gets double-bounce reflected by the soil and trunks. C-band ($\lambda = 6$ cm) provides measurements from the middle of plant bodies. Finally, SAR allows the polarization of transmitted and received waves to be controlled. Fully polarimetric SARs can transmit and receive horizontally (H) and vertically (V) polarized waves quasi simultaneously. Each pixel of such an image consists of a set of numbers representing the different polarization states, e.g., HV for H on transmit and V on receive, or HH, VV. Each scattering mechanism, like surface, volume or double-bounce, has its own polarimetric signature that can be used to infer the type of surface or land cover.

Interferometric SAR (InSAR)

Since the phases of the received echoes are recorded and preserved throughout all processing steps, each pixel of a so-called single-look complex SAR image is a vector of two numbers, the in-phase (or real) part and the quadrature (or imaginary) part. The length of the vector represents the pixel brightness and its angle the phase. The phase is an extremely sensitive measure for range. A change in range as small as a fraction of the wavelength (λ), i.e., centimeters or millimeters and, hence, much smaller than the range resolution, will lead to a detectable phase variation. The phase information from a single image does not carry useful information, since the scattering at the earth surface may introduce a random phase. However, in the phase difference of two images taken of the same area on the ground, this random phase cancels out to a large degree. InSAR exploits phase differences of at least two SAR images to derive more information about the imaged objects compared to using a single image (Bamler and Hartl 1998). The images are first co-registered accurately and then the phase differences are computed on a pixel-by-pixel basis. The resulting "image" of phase differences is the interferometric phase map or—in short—the interferogram. If the range of an object differs by ΔR between the two images, the interferogram pixel representing the position of the object will exhibit an interferometric phase of

$$\phi = \frac{4\pi}{\lambda} \Delta R .$$

In case the images are taken from the (ideally) same orbit but at different times, we talk about differential InSAR. Only objects and areas that have moved between the two acquisitions will give rise to an interferometric phase. Two-dimensional maps of ground motion can be generated by this method. Landslides, glaciers, volcanoes, land subsidence and earthquakes are typical targets of investigation using differential InSAR.

If the images are taken from different positions but at the (ideally) same time, this is referred to as across-track InSAR (figure 3-3). It allows us to recover the three-dimensional coordinates of every pixel and to generate a digital elevation model of the imaged area. As shown in figure 3-2 a SAR system maps the three-dimensional world into the two cylindrical coordinates range and azimuth. The third dimension, the angle θ, is accessible by across-track InSAR. It is directly related to the range parallax ΔR (figure 3-3) and, hence, to the interferometric phase. The center of figure 3-7 (see page 67) shows the

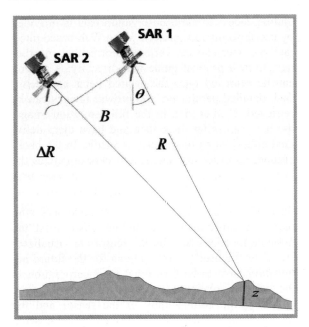

Figure 3-3: Imaging geometry of an interferometic synthetic aperture radar

across-track interferogram formed from the image of figure 3-7 (left) and an interferometric companion. The iso-phase lines, the so-called fringes, resemble height contours. The finally derived digital elevation model is shown in figure 3-7 (right) in a perspective view.

3.2.3
Processing Data to Information

With the growing complexity of available recording technologies for multispectral optical, polarimetric and interferometric radar systems there is ever more demand for standardized, operational (regular and routine in near-real-time) processing of the raw data into information products. This is accomplished with the help of suitable software which uses algorithms to derive the sensor-specific, viewing, physical irradiance and geometric characteristics of a satellite image and then to generate from them products of varying levels of sophistication (value adding levels). The specific requirements depend on whether attention is being directed at the atmosphere, the solid earth, the oceans or the land surface. For many scientific studies, basic georeferenced and calibrated products are required, for example, surface reflec-

tance or radiance values for each recorded channel (or range within the electromagnetic spectrum), along with the associated geographic coordinates, upon which subsequent calculations and analyses can be based. However, critical to the dissemination of earth observation data for a wide range of applications to people who are not remote sensing experts is being able to generate products which provide information in the form of well-known and scientifically accepted geophysical or biophysical variables like surface temperature, land use, biomass content or evaporation, in familiar cartographic projections. Such earth observation information products can be directly integrated in planning, environmental or climatological analyses. They can be immediately understood by people knowledgeable in the interpretation of the variables presented, but without specific remote sensing expertise. Of course, generally accepted algorithms and margins of error must be identified. As in many sciences, the development of new procedures and the derivation of appropriate algorithms under defined, well known boundary conditions form the basis for methodological innovation. On the other hand, applicability under other conditions (in the case of remote sensing this might involve the use of different spectral channels, resolutions, recording intervals, atmospheric conditions or test sites) is particularly difficult to achieve, since the algorithms often have to be modified for each new case. Standardization and agreement on generally valid and applicable algorithms and procedures is a challenge which must be faced by anyone using earth observation as a technology to gain geoinformation, if broad acceptance and robust applications are to be assured.

3.2.4
Managing Data and Products

Collecting and archiving earth observation data on the atmosphere, solid earth, oceans and land surface and then processing them into information products of various levels of sophistication is the task of so-called payload data ground segments. They also distribute data and products to users who may include the science community, government authorities, commercial companies, the media, or private individuals. For optimal exploitation of sensors on board the satellite, payload ground segments also enable the user to initiate acquisitions of desired locations at desired times. Comprehensive functionality has to be provided in order to establish a bridge between satellite sensors and data users, including:

- Online access via the Internet to existing data, as well as to initiate new acquisitions,
- Order management and delivery of the requested earth observation products to the user,
- Reception of new acquisitions downlinked from the satellites,
- Archiving and cataloging the acquired raw data and derived information,
- Processing the raw data to higher information levels.

Information and communication technology plays a key role in establishing these functions. Depending on the amount of data and number of missions and sensors to be supported, clusters of high performance servers are needed to process the data. Cataloging and order management require database management systems, whereas online access and delivery rely on Web and ftp servers. Local area networks (LANs) which connect the systems on a 1 gigabit/sec basis and high-speed wide area networks (155 megabit/sec) are needed for remotely distributed facilities. In the case of near-real-time services such as providing satellite data for weather forecasts, the possible failure of hardware must be anticipated so that remedial procedures are automatically implemented.

A central part of a payload ground segment is the archiving and cataloging system. All received raw data and derived higher level products have to be stored in a digital data library for comfortable access and long-term preservation. Catalog information describing each product is managed in appropriate databases to enable the search for a desired product among millions of archived items. The image data themselves are kept in automated archives on foreground disks and background tapes which provide online access to recently used data and near-line access to older products. Initiated by a user request, a robot is activated to fetch the tape where the older product is located and put it into a tape drive; finally the data is accessible online on disk without operator interaction.

Online access to earth observation data is provided by user information services. Via a Web-based interface the user can get various information on products from a product guide and directory; a catalog can be searched once the desired items are identified; detailed parameters and browse images can be retrieved. If interested in the full resolution image, the user can order these data and have them delivered either online or shipped on media. In the background, an order management system organizes the ordering and delivery process, handles the user profiles and considers data policy and license issues.

Specific features are provided for customers who request acquisitions by the satellite sensor. First, the possible locations seen by the sensor are visualized, and then the user's specifications for the future acquisition have to be forwarded for incorporation in the mission planning process. Mission planning, including the commanding of the sensor and the uplink to the satellite, are performed by the mission operations segment supported by the instrument operations segment. Once the sensor has taken the measurement and the satellite is in the visibility range of a receiving station, the acquisition data is downlinked and archived and the raw data can be processed to higher information levels.

Although missions are planned and operated by space agencies, the use of data is not limited by political boundaries. The data management systems implementing payload ground segments are linked with the help of networks and international data representation and communication standards (see for example www.ceos.org). This enables the combining of data from various sources to even higher level earth observation information, for example multidimensional georeferenced coverage for use in geographic information systems.

Building and operating payload ground segments is a complex task necessary for systematic and successful exploitation of the acquired data. Considerable savings can be achieved by employing multimission ground segment infrastructures capable of integrating the new data formats and processing facilities of future missions into existing operational structures.

3.3
State-of-the-Art Applications

What follows is a look at a representative selection of current earth observation applications, chosen from a sizeable list. The organization is by field of investigation rather than by the particular technology applied, in keeping with the motivation behind earth observation activities, to obtain information which describes and facilitates understanding of processes taking place on earth.

3.3.1
The Land Surface

Earth's land surface is mankind's primary living space and economic area. Pressures caused by an increasing world population and the accompanying demand for energy and food have led to human activities which have decisively disrupted nature's balance, sometimes irreversibly, on large portions of the land surface. These include agriculture, forestry, pasturage, mining, expansion of urban areas, ground sealing, and a multitude of planning measures too numerous to mention. The consequences of all these influences taken together are formally documented as contributing to climate change (IPCC 2001). Looking beyond the land surface, they are also attributed to changes that can be readily sensed in the atmosphere (key words being the ozone hole, high levels of air pollution, increases in the occurrence of extreme events like major storms), as well as in the oceans and polar regions (pollution, overfishing, melting of inland ice and reduction in the thickness of sea ice). The following section is an introduction to the contributions earth observation can make toward acquiring parameters of the land surface relevant for environmental and planning activities, with the ultimate goal of contributing to sustainable ecological and economic management.

Land Use and Land Cover Mapping

Up-to-date information on land cover (classifying the surface in principal categories) and land use (discrimination within these categories, e.g., corn, rice or wheat crops within the category "agriculture") is a fundamental requirement for a variety of applications in sustainable land management and ecological mapping and can be regarded as one of the most important multipurpose products based on remote sensing technology. As examples, an analysis of current land use and its changes over time can provide a basis for spatial planning, land, water and coastal zone management, sectoral analyses in agriculture and forestry, monitoring soil sealing and urban sprawl, nature conservation, and to a lesser extent for decisions regarding transport, telecommunications and navigation. It can also be used to assess the impact of agricultural policies or of the environmental effects of trans-European transport networks, to support the implementation of biodiversity conventions and similar programs, and to assess air emissions and air quality.

Land cover and land use mapping based on remote sensing is being performed on various scales. On the global scale a land cover map with a coarse resolution of 1 km was generated in the framework of the International Geosphere-Biosphere Programme (IGBP-DIS) (Belward et al. 1999). Recent data sets at 1 km spatial resolution were produced, for example, in the Global Land Cover 2000 (GLC2000) project (Bartholome et al. 2002) and on the European level in the Pan-European Land Cover Monitoring (PELCOM) project (Mucher et al. 2000). At a medium resolution mapping scale of 1:100,000 the most important activity in Europe is currently the CORINE Land Cover program.

The objective of this European-wide program is the generation and regular updating of consistent and comparable information on land cover in Europe for environmental assessment and planning purposes. The mapping is done at a scale 1:100,000 according to a European-wide harmonized nomenclature of land cover and land use classes. The CORINE land cover database, first generated in the 1990s and updated in the CORINE Land Cover 2000 (CLC2000) project, provides up-to-date information for the reference year 2000 and for land cover changes in Europe during the previous decade. The mapping is based on computer-assisted analysis of Landsat 7 ETM+ satellite images with a spatial resolution of 15 m (panchromatic) and 30 m (multispectral). The CLC2000 project is led by the European Environment Agency in cooperation with the member states of the European Union. The database covers an area of about 4.5 million square kilometers.

Approximately 300 ETM+ satellite scenes are necessary for complete coverage of the European Union countries. The data are taken from the vegetation period of the year 2000. In case of cloud cover, scenes from 1999 or 2001 are used instead. Based on these satellite data the mapping is performed in each of the participating countries. The nomenclature of CLC2000 distinguishes 44 land cover and land use categories in three hierarchical levels. The main categories are urban and artificial surfaces, agricultural areas, forests and semi-natural areas, wetlands, and water bodies. In figure 3-4 the land cover map for Germany is shown as an example, where 36 of these 44 European land cover classes are relevant. The CORINE land cover database is being used as a multipurpose product for a wide range of applications at European and national levels.

Agriculture and Forestry

Regularly updated space-based land-use information is one of the key resources for the management of rural and forested areas, including agriculture. Whereas in third world countries agriculture is essential for the daily basic local food supply, in industrialized countries agro-business is gaining relevance. Remote sensing technology is assisting both objectives. On the one hand, the identification of

drought and the potential shortages likely to occur in developing countries as a result are fundamental to government and international response programs and relief efforts. At the other extreme, crop assessment as a basis for granting subsidies in industrial agriculture in order to control production has necessitated the development of sophisticated new technologies. Earth observation techniques for crop and yield forecasting have therefore different meanings in a global context. While farmers strive for profitable, efficient and sustainable production from renewable resources, it is political decision makers who have to address and respond to issues of over- and/or underproduction, imports, exports, quotas, conservation, protection, food security, subsidy allocation and administration.

The manifold applications of remote sensing in agriculture make use of radar, standard optical, or new hyperspectral sensors, depending on scale and specific research field. Innovative concepts such as precision farming combine satellite navigation (GPS or the future Galileo) with satellite-derived information (current crop condition) and geographical information systems (soil characteristics, topography, past yield history) to allow the farmer to optimize agricultural production and the distribution of fertilizers, herbicides and pesticides for each individual

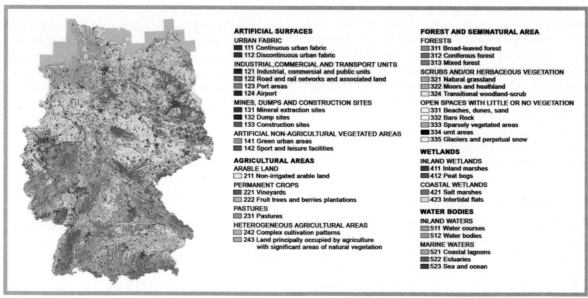

Figure 3-4: The 36 land cover classes for Germany resulting from Corine Land Cover 1990 activities

field, while minimizing environmental impact and encouraging sustainable land management. Numerous factors such as soil variability, field size, and economic and technological possibilities influence the decision for or against precision agriculture, so its adoption is unlikely to be uniform across farm types and sizes. However, traditional management techniques and structures will be—at least partially—replaced as essential data on soil and vegetation can be provided by satellite remote sensing. Vegetation information such as biomass estimation, plant phenology assessment, dryness or wetness of crops, early crop disease detection, yield estimation and land use classification data can be provided by remote sensing, among other sources (Clevers 1999). The synoptic view and the repetitive cover afforded by satellite data allow multitemporal observations of seasonal changes, which is crucial for detecting dynamic vegetative phenomena. Optical multispectral or hyperspectral data can be used to assess crop and soil conditions and to measure quantitatively important crop parameters such as plant canopy water content, nitrogen, chlorophyll, and leaf area index values. In addition, information on soil temperature, soil sealing and land degradation can be provided to the farmer to enhance his production methods (figure 3-5). Crucial information on physical soil properties such as the amount of organic matter has been correlated to specific spectral responses to show the potential for automated classification. Nevertheless, such direct applications of remote sensing for soil mapping are limited because several other variables can impact soil reflectance, such as tillage practices and moisture content.

Forests cover about one third of Earth's land surface. They produce the majority of land dwelling species; they are the land ecosystems with the highest biomass content and primary production (biomass); their economic use is manifold. They constitute a shelter and living space for many prominent as well as many precious species in economic terms, such as tropical timber (mahogany) or flowering plants (orchids). The world's forests play a vital role as a global climate regulator, as storage for water, energy and carbon, and as a storehouse for genetic resources. As producers of timber they create a livelihood for millions of people who extract wood products for various markets ranging in scale from indus-

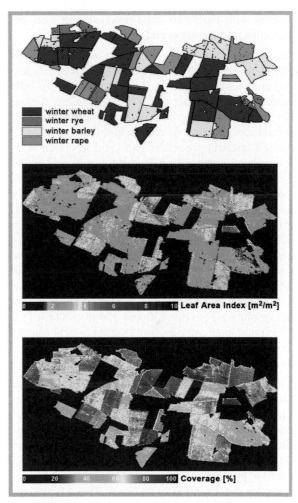

Figure 3-5: This image of farmland in the county of Demmin (Mecklenburg-Western Pomerania, Germany) was recorded by the Landsat 7 ETM+ sensor on May 1, 2000. It shows (above) which crops were planted on these fields, (center) the leaf area per unit ground surface as an indication of vegetation status and (below) how much of the ground was covered with crops. Such information is being used to analyze how much rainwater can be intercepted by vegetation for protection against erosion in endangered areas

trial production to the heating and cooking needs of individual households, with all the positive and negative consequences. There is a broad interest in forests on the part of the scientific community, public authorities, and organizations and companies dealing with their harvested products and their exploitation, preservation and management.

Earth observation offers the means to collect data on global forest resources in a cost-effective, efficient and comparable way, sometimes serving as the only means of data gathering over inaccessible regions like mountainous terrain without infrastructure or population centers. A comprehensive picture of a forest's spectral properties can yield information on forest stand composition, density, health and evolution, biomass content and sub-canopy flooding. This technology also presents a window in time by documenting past changes in Earth's forests as data for the same region are collected and assessed year after year. Early applications focused on the use of optical mid-resolution (30-60 m) to low-resolution (over 1000 m) sensors like Landsat's MSS, TM and NOAA's AVHRR. In recent years medium resolution sensors like MODIS or MERIS with their spatial resolution between 250 and 500 m, ASTER (15-30 m) and even very high-resolution optical sensors like IKONOS or QuickBird (0.6-1 m) are also being used. Synthetic Aperture Radar (SAR) has been used to detect deforestation hot spots.

Standardized forest data and information products can satisfy a range of scientific, public and commercial interests. Remote sensing data has already been successfully used for accurate quantification of forest extension and distinction and quantification of relative proportions of natural forests and plantations, for determining forest health, forest types, age classes, tree density, biomass quantity, and adjacent vegetation types. Commercial and noncommercial forestry uses remote sensing to manage diverse forests worldwide. Due to the versatility and scale of remote sensing, it is invaluable in all stages of forest management (clear-cutting, reforesting, afforestation, harvesting, and damage assessment). Additionally, building on the ideas behind the Kyoto Protocol, the trade of emission certificates will require an up-to-date, timely, global and accurate assessment of the carbon balance in forested areas, for which earth observation can be an essential tool.

Modeling Fluxes

Numerical models are utilized to simulate and predict future developments in the realms of climate and environmental research, as well as for many practical planning tasks. Models are used as a way to link earth's different spheres (the atmosphere, land surface, biosphere, hydrosphere, cryosphere) for the purpose of describing quantitatively the ongoing exchanges of energy, momentum and matter (for example, carbon and water fluxes from the atmosphere to the vegetation on land and ocean and vice versa). Starting from known, measurable conditions, the models extrapolate into the future, simulating the effects of various (human) influences, such as the emission of carbon dioxide or chlorofluorocarbons, the regulation of river beds, deforestation, overgrazing, etc. Reliable models are accordingly an important resource both for scientific research and decision makers. The closer the starting parameters of a model match reality at the beginning of a simulation, the preciser the prognosis. This is where remote sensing is playing an increasingly important role. It provides a number of initial parameters used to drive the model; the process is called parametrization. But remote sensing data can also be used to check the results of models which are already running, by making it possible to compare their prediction of the current situation with actual data. If this succeeds, then it means that the complex interrelationships within the model have been adequately described and it is robust enough to be used for prognoses. Possibilities for modeling material flows which involve vegetation are briefly described below.

Net Primary Productivity (NPP), the uptake of atmospheric carbon dioxide by plants, essentially defines the amount of carbon removed from the atmosphere and fixed as vegetative carbon. Direct observations of NPP are not available globally, but computer models derived from local observations have been developed to represent global NPP (Cramer and Field 1999). On the other hand, monitoring the vegetative carbon sinks is uniquely facilitated through the use of remote sensing, since the imagery can provide timely information on vegetation state over a truly global extent.

The emission of biogenic compounds has implications for both air quality and climate stability either through direct effects, or through changes they cause in atmospheric chemistry. Some important compounds currently being studied are the so-called biogenic volatile organic compounds (BVOC) that include isoprene, monoterpene and methane. Until recently, the role of remote sensing in BVOC estimation was limited to the classification of land-cover types. However, remote sensing data provide much more information and great strides have been made to derive from satellite data related to vegetation information which can be coupled with models of biogenic compound emission in order to make more accurate flux estimations. Through the use of modeling techniques, many scientists are involved in the effort to understand how carbon exchange and BVOC emission can best be estimated. The resulting NPP and BVOC models are of varying complexity and require, as inputs, information about soil, plants, and weather. Since much of the information required for soils is static (texture, soil type, porosity), it is readily available for many areas of the world in well-populated databases. Additionally, meteorological information such as rainfall and temperature is also relatively easy to obtain for much of the world through established data collecting networks. However, obtaining information on other variable inputs, such as soil moisture, photosynthetic active radiation (PAR) and vegetation state, which do not have well established data collection networks, requires the use of remotely sensed data.

When remote sensing products showing current land use or land cover, leaf area index (LAI), land surface temperature (LST) and PAR are coupled with NPP or BVOC models, these data will allow for the characterization of ground pixels in terms of net primary productivity or emissions of BVOCs to improved estimates of these parameters on a continental scale and to increased accuracy of flux predictions by providing timely assessments of the NPP or BVOC sources.

The potential for gaining insight into the way vegetation changes over time and space is based on the knowledge of how reflected solar radiation is altered by vegetation. Sensors that acquire data in the visible and near-infrared portions of the spectrum are usually employed; for estimating biomass radar data are also utilized.

Desertification

Desertification is a process of landscape change in the context of landscape degradation. Mapping of desertification processes involves two steps, measuring parameters for desertification, and detecting changes for assessment of transformation between two or more dates. Remote sensing is exceptionally valuable for monitoring because it provides contiguous and calibrated spatial data at specific temporal intervals and is unbiased and quantifiable for statistical analysis. Satellite earth observation enables analysts to draw conclusions about the speed, stage, and spatial variability of landscape change, and make comparisons to other regions in the world.

Here, desertification describes the process by which landscapes arrive at ecological characteristics truly related to deserts. It involves the biosphere, pedosphere, morphosphere, and hydrosphere and is the final stage of landscape degradation, resulting in a growth of deserts into areas which have not been desert before (Mensching and Seuffert 2001). It requires an inherent natural potential of the region, such as high variability of precipitation, thin soils susceptible to erosion, and negative human impacts like unsustainable land use. Environmental indicators for desertification monitoring can be arranged into four major groups: vegetation, soil, morphology, and hydrology.

Typically, vegetative processes are estimated by time series of indices such as the Normalized Difference Vegetation Index (NDVI) or derivates. The NDVI standardizes the difference between the near infrared and red portions of the electromagnetic spectrum, whereas chlorophyll reflection is maximal in the near infrared and minimal in the visible red. Therefore, vegetation indices provide a measurement of healthy green vegetation density convertible to biomass. Biophysical parameters including Leaf Area Index (LAI) and Fraction of Absorbed Photosynthetic Active Radiation (FAPAR) indicate vegetation coverage and phenological state. Airborne hyperspectral remote sensing reveals non-green vegetation, using specific wavelengths in the shortwave infrared. While sensors with high temporal global coverage facilitate time series analysis, satellites with very high spatial resolution or hyperspectral sensors allow conclusions about ground vegetation distribution, i.e., dispersed or contracted vegetation—an important criterion for deserts.

Pedological indicators encompass soil properties showing changes in soil carrying capacity and top surface layers. Detailed surface characteristics such as grain size, litter, salt, carbon, or iron can be sensed with airborne hyperspectral instruments. High spatial resolution images might depict erosion features such as gullies or arroyos. Remote sensing in the microwave portion of the spectrum penetrates vegetation and the uppermost surface layers, yielding information about soil texture, soil moisture, and salinity.

Morphological processes in semiarid and arid environments are characterized by catastrophic events, usually triggered through intensive rainstorms or long-lasting landfall. Wind is the other agent active in sparsely vegetated areas. Earth's morphology is mainly related to topography, which is best sensed with radar interferometry techniques since they can detect slight changes such as surface creeping, denudation processes, and dune movement.

Remote sensing of hydrological parameters for dryland detection is accomplished by a variety of instruments. Geostationary meteorological satellites record patterns of cloud cover and high water-vapor content in very high temporal resolution, commonly every 30 minutes, and even every 15 minutes with Meteosat Second Generation (MSG) satellites. Precipitation is monitored using weather satellites with sensors which can detect cold temperatures at cloud top, or passive microwave measuring techniques. The measurement of precipitation, hence water availability, allows estimates of variations in regional climate and productivity. Another important issue is surface hydrology in arid areas prone to desertification. Commonly inflow and outflow data are provided by in situ measurements. However, moderate to high spatial resolution earth observation sensors with thermal detectors facilitate calculating surface energy balances in semiarid regions while actual evapotranspiration remains as residual.

A prominent example for a desert building process caused by human activity and documented with satellite remote sensing is the Aral Sea in central Asia (figure 3-6).

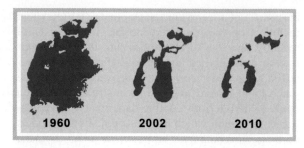

Figure 3-6: Once the world's fourth largest freshwater-fed lake, the Aral Sea has been continuously shrinking since the 1960s primarily due to intensified irrigation using water from its feeder rivers for crop cultivation. GIS bathymetry data combined with multisensor remote sensing data used to monitor the desiccation support the prediction that the present southern lake will split into an eastern and western part by 2010. The lake's salt content is increasing due to this tremendous volumetric shrinking, causing fish extinction and an end to commercial fishery. Toxic salts and dust blown from the now dried lake bottom are being deposited onto the surrounding farmland. These are just two of the many consequences of this water management disaster. The images show so-called "sea masks" which are calculated using historic map information (left), different ratios in the visible and near-infrared portion of Terra-MODIS satellite data (middle), and GIS model output (right)

Topography

Information on the earth's relief is a key parameter for almost any geoscientific analysis and for all precise land-oriented applications and planning purposes. Moreover, precise global topographic data in conjunction with GPS data can play a key role in enhancing civil flight security during bad sight conditions. To better understand what earth observation techniques can provide, we first have to discriminate between three different terms used for the digital representation of the surface topography. The Digital Elevation Model (DEM) is the least specific term and usually refers to a raster or a regular grid of spot heights. The Digital Terrain Model (DTM) describes the true shape of the bare earth terrain, while the Digital Surface Model (DSM) adds to the DTM man-made objects and the height of the vegetation canopy. Today, digital elevation models are one of the most important data applications used in geospatial analysis. Remote sensing from space has evolved into an important supplement to ground observations and aerial photogrammetry. From a global point of view it is the only practicable solution for mapping large areas.

Both optical visible and near-infrared as well as SAR data provide DEM generation capability. Several methods for elevation extraction have been developed, one of which is SAR interferometry. As depicted in figure 3-3 two SAR antennas fly on (ideally) parallel tracks viewing the earth's surface from two slightly different positions. Given the sensor locations and the two range distances R and R + ΔR, the three-dimensional position of every point imaged on the surface can be determined by triangulation. The range difference or parallax ΔR is proportional to the terrain height and inversely proportional to the wavelength of the radar. The separation of the flight paths—the so called baseline B—is small (less then 1 km) compared to the ranges R and R + ΔR (up to 800 km). Therefore, ΔR must be determined very precisely. For this purpose interferometry uses the phase information of every pixel. The interferogram provides the phase difference. It is defined by the complex multiplication of both SAR images. The evaluation of the phase difference allows measurement of the parallax to some few millimeters accuracy. The intersection point of the two range circles of the radiuses R and R + ΔR around the antennas then provides the three-dimensional position of the ground target. The Shuttle Radar Topography Mission (SRTM) of 2000 is the most prominent interferometry mission to date. The baseline B was realized by mounting the secondary antenna on top of a mast that was deployed in space to a total length of 60 m from the shuttle's cargo bay, where the primary antenna was located.

Interferometric data can also be acquired by revisit flights where one or two satellites observe the area on the ground from approximately the same orbit on two different dates. This is the ERS-1/ERS-2 situation. However, changes in the atmosphere, and in particular changes on the ground, reduce the coherence of the signals. Loss of coherence means that the signals cannot be evaluated. Additionally, the baseline needs to be known very precisely. During SRTM it could be directly measured, while for repeat pass interferometry it has to be determined from the two independent and less accurately known orbits. New concepts propose simultaneous acquisitions by two very closely positioned spacecraft or the use of passive small satellites operated in parallel to an active SAR.

Figure 3-7 shows the radar image acquired by the onboard antenna, the interferogram of both radar images and the final DEM of the Cotopaxi volcano in Ecuador (left to right). As the viewing angle of the two radar antennas is almost the same, the radar images look very much alike. Therefore, no parallaxes could be measured. However, looking at the color coded interferogram the relief is already visible. The same phase differences and range parallaxes appear along one colored fringe. It can also be seen that the distance between the fringes varies depending on the terrain slope so that they already look like elevation isolines. Within this interfero-

Figure 3-7: Radar image, color-coded interferogram and color shaded DEM of the Cotopaxi volcano in Ecuador (left to right)

gram the phase ambiguity has not been removed. This means that the phase difference, e.g., from yellow to yellow, is 2π or $360°$. The phase unwrapping process determines the absolute phase values that will finally be transformed into the individual heights in the triangulation step mentioned above.

DEMs are also generated from optical images in the visible and near-infrared range with stereo-matching techniques. The major disadvantage is the dependency on cloud-free observation conditions, which places limits on this technology for global mapping.

Earthquakes and Volcanic Eruptions

Earthquakes occur all over the world due to plate tectonics. The Global Seismic Hazard Assessment Program has noted that a tenth of the world's population lives in areas classified as having a medium-to-high seismic hazard. Therefore, monitoring these seismic areas on a global basis is important.

For a long time seismology provided the only way to study continental earthquakes. The development of new earth observation techniques has expanded the capability of scientists worldwide to monitor earthquakes using satellites. Besides spaceborne high resolution optical sensors (Ikonos, QuickBird), a variety of sensors measure wavelengths of energy that are beyond the range of human vision, for example ultraviolet, infrared, and microwave. A great contribution is made by Synthetic Aperture Radar interferometry (InSAR), which was first applied to map co-seismic displacements during the 1992 Landers earthquake in California. An interferogram was constructed by combining two SAR images acquired by the ERS-1 satellite before and after the earthquake. The interferogram of this region shows at least 20 fringes (lines of the same phase differences) representing a displacement of up to 560 mm. The range from the ground surface to the satellite agrees well with field measurements and dislocation models. These remote sensing observations initiated a new era of hazard monitoring. Since then, InSAR has been used to map displacements resulting from some 30 earthquakes (figure 3-8). For comparison, conventional surveying techniques had detected the deformation of less than 15 earthquakes before 1992. Conventional observation techniques have some limitations. A network of digital creep meters

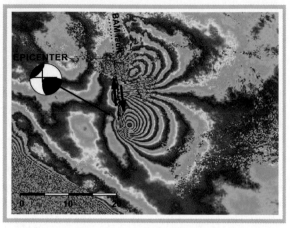

Figure 3-8: Differential radar interferometry can be used to map very small displacements of the earth's surface. This displacement map, derived from Envisat ASAR data provided by ESA, shows co-seismic surface displacements during the December 26, 2003 earthquake in Bam, Iran. Each color cycle (fringe) corresponds to 28 mm relative motion in the line-of-sight direction of the satellite. Such displacement maps help to further understand fault mechanics and the earthquake cycle and may in the future help to better predict the probability of an earthquake

has to be installed for reliable monitoring. Differential interferometry in comparison has high spatial resolution, millimeter measurement precision (Colesanti et al. 2003) and covers a large area (typically 100 km x 100 km for a single image). These advantages of the InSAR methodology have the possibility to overcome the limitations of conventional measurements. Furthermore, many of the last decade's significant earthquakes occurred on faults that had not previously been known to produce large earthquakes (e.g., Northridge, California, 1994; Kobe, Japan, 1995; Athens, Greece, 1999). InSAR helps to improve understanding of earthquake mechanisms and has a better predictive capability with respect to the seismic shock and occurrences of earthquakes. Remote sensing techniques can thereby support better hazard mitigation.

Another advantage is the possibility to observe areas that are dangerous or difficult to access, like volcanoes, without putting field crews at significant risk. There are about 1,000 potentially active volcanoes in the world and it is not feasible to monitor all of them with ground-based methods. For the purpose of studying volcanoes, remote sensing is the detection by a satellite's sensors of electromagnetic energy that is reflected, radiated, or scattered from the surface of a volcano or from its erupted material in an eruption cloud. Different sensors provide different types of useful information. Hyperspectral optical sensors can detect plant dieback on the volcano's surface due to gas effusions, for example, permitting conclusions about volcanic activity. Thermal infrared sensors can detect hot spots in the summit crater or along the flanks of the volcano's edifice, indicating new volcanic activity. SAR interferometry can provide coverage over the entire deforming region around an active volcano. Detailed spatial coverage often gives clues to magma migration and other underground processes, leading to improved understanding of the volcanic processes concerned. The most promising application of radar analysis to volcano hazards might be the detection of pre-eruptive ground deformation due to magma migration.

Mount Etna, the largest volcano in Europe and one of the most active in the world was often the target of remote sensing research. Lundgren et al. (2003) observed the volcano from quiescence in 1993 through the initiation of renewed eruptive activity in late 1995 until 1996 by analyzing interferograms. He evaluated the vertical and horizontal components of the displacement field and detected an onset of sliding coincident with a new cycle of volcanic activity. Field measurements confirmed that the displacements started simultaneously with the volcanic activity.

Soil and Mineral Exploration

Terrestrial life depends to a large extent on the fragile crust of soil that coats the land surface. Just a single inch of soil can take centuries to build up but, if mistreated, it can be blown and washed away in a few seasons. Remote sensing techniques can be used to estimate important factors characterizing soil type and functioning: mineral composition, soil moisture, organic matter content, and soil texture.

The penetration depth of radar waves in the soil crust is related to its moisture content and thus SAR systems can be used to derive this parameter. On the soil reflectance spectra recorded from optical systems, soil moisture differences cause different brightness levels. The mineral composition of soils also affects the reflectance spectrum. Depending on the amount of important soil minerals such as iron oxides, clay minerals, carbonates or gypsum, diagnostic absorption features show up in the reflectance signature of soils that can be measured with spectroscopic methods.

Soil mineral composition, organic components, and moisture content determine to a large extent the possible land use forms and consequently the productivity of arable lands in all climates. Particularly semiarid ecosystems providing important land resources for adapted agricultural production and grazing systems are often at risk due to climatic change and land degradation dynamics. With increasing aridity good irrigation practices become especially important, otherwise free carbonates and salts will ascend with the water table due to capillary action. The consequences are salinization of the soil crust, reduced productivity and ultimately abandonment of previously arable lands. All of the key parameters characterizing (semi-)arid soils such as high concentrations of carbonates, soluble salts and, depending on the parent substrate, SiO_2-rich components in the upper soil horizons can be identified with remote sensing based imaging spectrometry. Thus, based on information derived from remote sensing, land use practices can be adapted leading to sustainable use of the soil crust.

Remote sensing techniques are well-established tools in mineral exploration. The methodology focuses on the specific information necessary to discover mineral or hydrocarbon deposits: knowledge about the geological rock formations and the structure surrounding a deposit, and the abundance of key indicator minerals. Generally, geologic structures and mineral alteration patterns can be quite vivid on satellite imagery. Subtle "color" variations that would go unnoticed on the ground can be made quite bold using the different spectral measurements modern satellite data provide. SAR is especially useful for obtaining topographical information and drainage patterns, particularly in tropical forests or other densely vegetated areas. For humid climates often under heavy cloud cover, spaceborne SAR systems provide excellent additional capabilities for base mapping and for structural and tectonic analysis of many newly attractive exploration regions. Geological mapping based on the use of multispectral optical sensors such as the Landsat (Enhanced) Thematic Mapper, (E)TM, has strongly increased the potential for finding new mineral resources and simultaneously helped reduce the investments necessary for classical mineral exploration based on field geology, core logging and ground based or airborne geophysical methods.

Hyperspectral optical remote sensing makes use of the unique spectral characteristics of many alteration and rock-forming minerals that can be recorded remotely. Information extraction algorithms have been developed that make specific use of diagnostic absorption features of minerals to allow their remote identification and even an estimate of their relative abundance (Kemper and Sommer 2003), see figure 3-9 (top). The interpretation of such mineral abundance maps can make a significant contribution to the field of exploration geology. Especially altera-

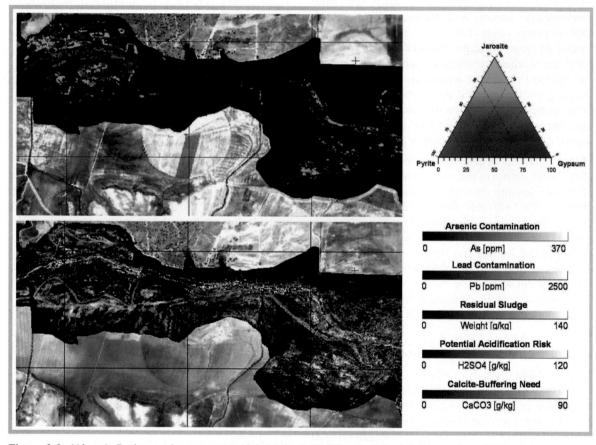

Figure 3-9: (Above): Pyrite, a mine waste material, contaminated the Guadalquivir River after a 1998 mining accident. When exposed to the atmosphere it transforms to jarosite, producing considerable amounts of acidity in the process. Lime was used during the clean-up activities to neutralize the acids, producing gypsum. Pyrite, jarosite, gypsum and other minerals produced during the oxidation process can be identified and quantified using satellite or airborne imaging spectroscopy techniques and presented in a form that can be understood by environmental officers and used directly in the remediation process. (Below): Poisonous contaminants such as arsenic and lead usually leach out of the surroundings under acid conditions. It is possible to determine the levels of these contaminants from the minerals mapped (top) since they are an indicator of acidity. The data were acquired with the HyMap™ sensor in 1999. The affected area is the black part of the image. It was superimposed on a false-color image of the surroundings for better orientation. (Source: Thomas Kemper, used by permission of JRC)

tion minerals such as clays are of interest to the exploration and mining industry because of their potential association with occurrences of precious metals such as gold or silver. In arid or semiarid regions, where the vegetation cover is sparse, airborne hyperspectral sensors have been used for detailed mineral mapping and have led to the discovery of gold or diamond deposits. Currently, efforts are being made to bring the methodology to an operational level for use by the exploration industry. Initiatives towards hyperspectral capabilities in space are being taken in USA, Germany, Canada, Australia and Japan.

Compositional mapping has long been the dream of geologists since it focuses on fundamental geological attributes, not just pictures of the terrain. Data collected by airborne hyperspectral spectrometers have already been used to demonstrate that it is possible to identify certain types of exposed minerals, to automatically label them, in some cases to determine their chemistry, and to ascertain the fractions of the minerals occurring in small, sub-pixel units. Thus a new type of map, a mineral map, is helping explorers home-in on zones of mineral alteration around mineral deposits, detect previously unrecognized mineral patterns across whole mineralized districts, document mineralogical components of the weathered regolith, and locate waste products such as the sulphate minerals which cause acid mine run-off from mine tailings.

Pollution

Ecosystems and their components have been considerably affected, and in some cases destroyed, as a result of many decades of human economic activity and increased consumption of natural resources, particularly in the industrialized countries. Causes are the introduction of harmful matter into the ambient air or water bodies, where they can be further distributed by such forces as wind or currents, or the direct deposition of harmful substances onto the land surface (garbage, industrial waste, sewage). Common catchwords characterizing the consequences are forest death, acid rain, land contaminated with inherited toxic waste from now-defunct refineries, factories or military installations, heavy metals in rivers or garbage dumps, the eutrophication of lakes or the pollution of agricultural land by overfertilization

(primarily nitrates and phosphates). We could add the sometimes very considerable poisoning of the environment caused by accidents in the chemical industry, such as the Sandoz catastrophe of 1986 when water used by firefighters flushed huge amounts of pesticides into Europe's busiest waterway, or in mining (the 1998 contamination of the Doñana wildlife reserve in southern Spain after the collapse of a retention dam for mine tailing slurry loaded with poisonous heavy metals from the Aznalcóllar iron mine upstream, see figure 3-9.

The measuring approaches described at the beginning of this chapter and the unique advantages of remote sensing predestine earth observation as a supporting tool for locating many cases of environmental pollution, analyzing the processes at work, and monitoring the consequences. One cannot usually detect the contamination itself with remote sensing of the surface from space. Exceptions are perhaps open ground with surface contamination by highly concentrated materials that can be identified in hyperspectral optical data on the basis of their unambiguous spectral behavior, like hazardous military waste (Dech and Glaser 1993). In general, indirect evidence is used, for example, observing stress symptoms in vegetation (leaf yellowing or dieback) or algal blooms in inland waters. Besides the spectral analysis of surfaces under environmental pressure, change-detection procedures are especially suitable. They provide for every pixel in an image a measurement of changes occurring between observations. Thus, regular repeated mapping of potentially endangered areas can reveal changes in ecological constraints, or suspicious deviations from normal spectral behavior, which may hint at pollution. Ground samples can then be collected on location and the cause for the spectral changes investigated.

In addition to these quantitative and statistical approaches, earth observation data can be consulted at short notice for information about the affected surfaces, both during an environmental catastrophe for purposes of limiting damage, or later for ecological mapping and monitoring of contaminated areas.

3.3.2
The Oceans and the Cryosphere

Over two thirds of the earth's surface is covered with water, in the polar regions some of it in the frozen form of sea ice. In addition, an entire continent, the Antarctic, as well as Greenland are covered with huge ice shields containing about half of Earth's fresh water inventory. The oceans and the cryosphere profoundly influence the climate, on the global level (both as a sink and a source of carbon) as well as on the regional level (consider the effects of warm ocean currents like the Gulf Stream). They are also important tracers indicating global climate change, admittedly with long reaction times, similar to the situation in the atmosphere. Sea surface temperature, ocean height, sea ice thickness and extent and amount of fresh water entering the oceans as a result of glacier or shelf ice calving are important evidence of global climate change. The oceans are also one of mankind's economic areas; whether for the food supply, as transport medium, or as a renewable energy source. Large-scale, continuous, objective measurement of parameters relevant for the processes occurring in the oceans and cryosphere is one of the pillars of earth observation, making monitoring and scientific analyses possible. The data are also used to optimize the economic exploitation of the oceans (fishery), to monitor ecological changes (algal blooms), to increase the safety of ship traffic (wave heights and direction) and to plan off-shore wind energy converters (wind speeds). In both realms we are talking about a two-sided coin; the basic processes can be used for good or ill, as with all technologies. Being able to locate fish can lead to aggressive overfishing, for example. And indeed, most of the sea surface temperature maps generated primarily for climate research are popular with the fishing industry because of their usefulness in optimizing fishing fleet harvests.

Sea Surface Temperatures

Sea surface temperature (SST) is an important geophysical parameter because it provides the boundary condition used to estimate heat flux at the air-sea interface. As the most widely observed variable in oceanography, SST is used in many different studies of the ocean and its coupling with the atmosphere. Mapping SST makes it possible to find out, for example, where water is cold or warm enough for certain species, where currents are carrying water, where storms gain energy from the ocean or where ocean mixing is happening. On the global scale this is important for climate modeling and study of the earth's heat balance, and gives insight into atmospheric and oceanic circulation patterns and anomalies. On a local scale, SST can be used operationally to assess eddies, fronts and upwelling, and to track biological productivity. Over the long term, SST data can be used to study annual shifts in ocean currents and temperatures. SST data are also crucial to understanding periodic phenomena (like El Niño) and long-term climate shifts attributable to global warming.

Although a key parameter in many scientific fields, SST is difficult to define exactly because of the highly variable and complex vertical structure of the upper ocean (to about 10 m in depth) (Donlon et al. 2002). The depth at which measurements are made and the method of measurement can have a significant effect on the SST that is reported. The SST at the surface, within the top few micrometers, is termed the "skin" SST, and the SST immediately below, down to 1 m depth, the "bulk" SST. The skin SST is usually cooler relative to the bulk SST immediately below it because heat transfer is in general from the ocean to the atmosphere and the ocean loses heat to the atmosphere by molecular conductance. But during daytime, solar insolation and low wind speeds can cause increasing skin SST, while the bulk SST is relatively free of these effects.

There are two main sources of SST measurements: in situ and satellite. Traditional in situ observations are direct temperature measurements made from ships and buoys, whose ranges are limited. Satellite remote sensing by contrast makes possible a synoptic view of the ocean allowing the examination of basin-wide upper ocean dynamics as well as measurements of a given area several times per day. Methods for determining SST from satellite remote sensing include thermal infrared and passive microwave radiometry; both methods have their strengths and weaknesses. Thermal techniques are the most accurate and have the best resolution, but require cloud-free conditions (figure 3-10). Microwave approaches have the great advantage of being able to

0°C   12°C   24°C

Figure 3-10: Since 1993, multitemporal SST maps (composites) have been compiled at DLR based on thermal infrared AVHRR measurements. Products are generated in a daily, weekly and monthly fashion, which allows enough time to capture cloud-free pixels for the entire region. This unique image averages SST data for the European seas over the ten years from 1993 to 2003 and was derived from monthly AVHRR composites. In total, roughly 18,000 AVHRR images were analyzed to calculate this data set which serves as a climatological reference from which temperature anomalies can be easily detected

measure under all weather conditions excepting rain, because the radiation at these longer wavelengths is largely unaffected by clouds. In either case the radiometer measures radiation over a number of finite channels. For any radiance measured within a specified wavelength window, there is an associated temperature (called the brightness temperature) which can be determined by ascertaining at which temperature a blackbody would emit the same radiation. The emissivity of an object is the ratio of the amount of radiation emitted by the object to that of a blackbody at the same temperature and wavelength. Therefore, knowing the brightness temperature and the emissivity of the ocean surface allows determination of SST. Various algorithms exist for converting radiance to SST (McClain et al. 1985, Wentz et al. 2000). Because satellite sensors can only measure the skin SST, the retrieval algorithms are tuned

against in situ bulk temperatures from ships and buoys in an attempt to determine a "pseudo-bulk" SST. The best satellite-derived SST absolute accuracy possible today is on the order of 0.3 to 0.6 K.

Topography and Currents

Before the first oceanographic civil radar satellite (Seasat, operating in L band in the 30 cm wavelength range) was launched in 1978, many scientists, including radar experts, doubted that its images would show ocean waves. A new quality of sea state measurement capabilities was demonstrated when the Seasat images not only showed ocean-wave-like structures in general, but very distinct features of ocean waves, internal waves and bathymetry. Further, it could be shown by their direction and wavelength that these waves could be traced back to the storm centers that had generated them.

When ERS-1 was launched 13 years later, SAR images of ocean waves started to become available on a regular basis. They yield high resolution two-dimensional images of the radar backscatter properties of the sea surface and can thus be used to measure wind fields and sea state from space. Sea surface features, spatial variation of wind speed, rain, current and motion of the sea surface all have imaging effects which can be used to derive directional ocean wave spectra, revealing wave height, wavelength (the distance from wave crest to wave crest) and the direction the wave is propagating. This information is important input for improving wave model predictions. Other information can be gathered from the radar images directly, like individual wave height, crest length and grouping of ocean waves. Thus the distribution of maximum wave height can be investigated globally, or wave refraction and diffraction in coastal areas can be studied in detail (figure 3-11).

These results of basic research efforts are starting to bear fruit in practical applications. There is now 12 years' worth of global ocean wave measurements that can be analyzed for decadal variations. ERS image mode data (100 x 100 m coverage at 30 m resolution) show wave refraction and wave reflection at coastlines and are being drawn on for coastal applications like offshore wind farming or oil well siting, coastline detection, current feature determina-

Figure 3-11: The potential of SAR data for providing directional ocean wave information and topography is illustrated in this ERS-2 SAR image of an area about 10 x 7 km in size. It was acquired over the north tip of the island of Sylt in the North Sea on 15 Oct. 1998, 10:06 UTC. Ocean waves of about 100 m wavelength are approaching the coast. As they get closer to the shoreline their wavelength becomes shorter

tion, and ship and harbor safety considerations. ERS wave mode data (5 x 10 km coverage at 30 m resolution acquired once every 200 km along the satellite track) are used for detecting sea surface features, sea ice, storm tracks, the propagation direction of ocean waves and wave groups, and for measuring high individual ocean waves, as well as to improve sea state forecasts at weather centers by assimilation of satellite SAR spectra. An improved wave mode became available with the Advanced SAR (ASAR) on Envisat, making it possible to acquire images every 100 km along the satellite track; some 2,000 are in the meantime available per day as a standard data product.

Primary Production

Primary production is the process of building up plant tissue by photosynthesis. For the oceans and other waters phytoplankton species are the dominant primary producers, converting inorganic material (nutrients) into new organic compounds through photosynthesis. The production of phytoplankton starts the marine food chain and is thus one of the most important and basic processes for marine life in general (Lalli and Parsons 1993).

Another aspect is the consumption of carbon dioxide (CO_2) by photosynthesis. Recent estimates equate the total amount of CO_2 fixation by phytoplankton in the global ocean to the amount fixed by the entire rain forest. Although the mechanisms of marine primary production, drivers and limitations differ from those for terrestrial primary production, and although the CO_2 fixation is compensated to some extent by respiration, it is still a significant factor for the balance in the global CO_2 cycle.

Primary production by phytoplankton is a process which cannot be accessed by remote sensing techniques alone. A description and determination is only possible using biophysiological models involving a complexity of different parameters, such as: species, abundance and vertical distribution of phytoplankton in the water column, availability of nutrients (e.g., nitrate, phosphate) as well as physical parameters and processes (temperature, available light, currents, turbidity in the water, vertical mixing) and more. Due to this complexity there is no generic, globally applicable model to derive primary production on a global scale, ranging from various coastal ecosystems to large basins and the global open ocean. Hence, up to now assessment of global marine primary production remains an estimate with a significant level of uncertainty.

Earth observation is the ideal technique to provide a number of essential parameters in time and space for model computations on different scales (Platt and Sathyendranath 1988):

- Surface chlorophyll concentration, representing the biological state with respect to phytoplankton;
- Turbidity of the water column;
- Sea surface temperature (SST);
- Photosynthetically available radiation (PAR);
- Wind, waves and currents as driving forces for physical processes in the water column.

The focus here is on chlorophyll and turbidity since the other parameters have already been discussed above. After passing through the atmosphere, sunlight penetrates the water body, where some portions of the light are absorbed and some are scattered by water molecules or particles in the water. These absorption and scattering processes determine the wavelength composition (spectrum) of the light remitted from the water body or, in other words the "color" of the water. Pure water itself almost totally absorbs at wavelengths below about 400 nm and above 750 nm. For this reason only the wavelength range between these two borders may be used to look into the water body (the visible and near infrared spectral range). Three main water constituent groups additionally influence the spectrum: the phytoplankton, showing scattering due to its cell structure and absorption due to pigments, unpigmented suspended or particulate matter showing scattering, and dissolved organic matter showing typical absorption behavior. Each component may be characterized by its specific spectral optical properties.

For the open oceans phytoplankton and covarying components are the dominating constituents in the water. As its major part contains chlorophyll-a as pigment, the concentration of chlorophyll-a is used to characterize phytoplankton abundance in the water. Since chlorophyll is a direct measure for the ability to perform photosynthesis, it is an essential parameter for the calculation of primary production. Pigments other than chlorophyll (e.g., carotenoid) which also play an important role are accounted for by special correction procedures.

In coastal zones or closed basins the situation is more complicated than in the open ocean: in addition to phytoplankton the water here mostly contains other suspended material (e.g., sediment) and dissolved organic matter (e.g., humic acids). These components may vary independently of the phytoplankton and therefore need to be treated separately.

With respect to water constituents remote sensing uses the measurement of the light spectrum remitted from the water body (ocean color) in the visible and near-infrared spectral range by imaging radiometers or spectrometers.

Knowledge about the specific optical properties of the different water constituents is used to calculate maps of concentrations for chlorophyll, suspended matter and dissolved organics (figure 3-12). Knowing the concentrations, one is able to calculate the extinction of light in the water body, and therewith turbidity, which is a measure for the penetration depth. Both parameters are essential inputs for models estimating the primary production.

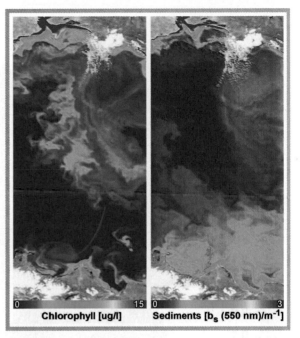

Figure 3-12: Algorithms making use of different channels of the MOS sensor can reveal in a single image two different kinds of phytoplankton blooms in the Black Sea: the presence of chlorophyll (left) or sediment produced by lime-depositing or calcareous algae (right)

Sea Ice and Polar Ice Sheets

As polar regions are particularly inaccessible to regular monitoring from the ground, satellite images provide the most important spatial data source for scientific evaluation at different scales. Due to the orbit characteristics of polar orbiting satellites, scenes of neighboring orbits widely overlap, which leads to excellent scene coverage for most polar regions. Important information about the cryosphere is derived from instruments acquiring images in visible, infrared, and passive microwave wavelengths, but also from active radar sensors.

Large portions of the polar and subpolar oceans are covered with ice year-round. For all of Earth's oceans taken together, about 5% is covered with ice in March; in September the figure is some 8%. Temperature and salinity determine when seawater freezes; usually not above -2 °C.

Depending on age and compression, sea ice can become remarkably thick, up to about 12 m. In the

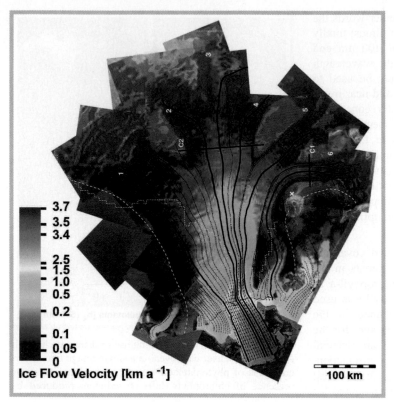

Ice Flow Velocity [km a⁻¹]

3.7
3.5
3.4

2.5
1.5
1.0
0.5

0.2

0.1
0.05
0

100 km

Figure 3-13: Interferometrically derived two-dimensional surface velocity map of the Antarctic Thwaites Glacier. The glacier was mapped with 45 interferograms based on data collected between 1995 and 2000 by the ERS-1 and -2 satellites. Topographic phase contributions were eliminated with a coarse external DEM. The velocity map covers almost 180,000 square km and comprises more than 80 percent of the glacier's catchment. Six individual tributaries were identified; their center-line velocities increase from 0 at the catchment boundary to some 0.3 km per year when they join the main glacier trunk. Velocity increases up to 3.6 km per year on the floating tongue. This value indicates that Thwaites Glacier is one of the fastest moving ice streams in the world

Arctic Ocean the average is 3 m, near Antarctica only about 1 m. Depending on how dirty the surface is, sea ice has a very high albedo (between 0.6 and 0.9). Up to 90% of incoming solar illumination is accordingly reflected back to space from this light surface. As a consequence, sea ice has the effect of an energy sink and therefore plays a significant role in the climate system. Sea ice also acts as a layer of insulation preventing heat exchange between the relatively warm ocean (-1°C) and the colder atmosphere. (-30°C). Over sea ice the atmosphere is much colder than it is over the open ocean. Sea ice also influences the oceans by facilitating the production of deep water. Ocean water has an average salinity of 34‰, sea ice only about 5‰. When sea ice freezes it releases to the ocean a considerable amount of salt, which makes the surface water heavier, causing it to sink down to deeper ocean levels. This dense, heavy, deep seawater generated in winter in polar regions drives oceanic thermohaline deep circulation. For such reasons monitoring the formation and extent of sea ice is an important aspect of climate research, and remote sensing technology has been contributing to this effort for the past three decades by recording the dynamic process of sea ice expansion and contraction. Besides being used for sea ice research, satellite data on sub-polar waters is in high demand as reliable information for vessels at sea (ice floes and iceberg warning).

Optical sensors, passive microwave radiometers and radar systems of a variety of designs are utilized, during polar night especially radar or thermal infrared sensors such as AVHRR on NOAA satellites. The reliable data on sea ice extent based on earth observation data available since the 1970s reveal that the ice cover increased in the mid-1970s, followed by a noticeable decrease between 1978 and 1990. Since that date the situation has remained fairly constant. In the Antarctic surface ice dramatically decreased in the 1970s and has been slowly increasing since 1980 (Eicken and Lemke 2001). The significant global warming of air temperatures observed since the 1990s has not yet had an effect on the extent of sea ice in either polar region.

Whether it has had an effect on sea ice thickness is one of the burning issues of climate research. Remote sensing techniques are being developed to estimate the age of sea ice by measuring its thickness with optical and passive microwave instruments.

Satellite images are used to investigate changes in large Arctic and Antarctic ice shelves as well as to track the drifting of icebergs and monitor glaciers (figure 3-13, Lang et al. 2004). In January 1995, 4,200 square kilometers of the northern Larsen Ice Shelf on the Antarctic Peninsula broke away. Satellite images, complemented by field observations, showed that the two northernmost sections of the ice shelf fractured and disintegrated almost completely within a few days. This break-up followed a period of steady retreat that coincided with a regional trend of atmospheric warming. The observations imply that after an ice shelf retreats beyond a critical limit, it may collapse rapidly as a result of perturbed mass balance. Calving of large portions of a polar ice sheet is a regular process. The broken parts sometimes reach the size of small countries and drift for many years in the polar sea. In order to prevent collisions with ships, icebergs are permanently monitored using satellite images supplied by various sensors. Images and information are regularly updated and are accessible via the Internet (Web: *National Ice Center*).

Off-shore Wind Farming

Because of a shortage of suitable sites on land, wind farms (arrays of turbines rotated by wind-catching blades and thereby converting wind energy to electrical energy) are increasingly moving offshore. So far about 280 MW_e of power has been harvested by offshore installations in the North and Baltic Seas, advantageous locations because their shallow waters and high mean wind speed promise a vast potential for wind farms. In the near future wind farms with over 100 wind turbines and an output of over 5,000 MW covering areas over 200 square km are planned or already under construction (Web: *North and Baltic Seas*). The largest wind park in operation is Horns Rev situated in the North Sea on the west coast of Denmark, where 80 wind energy converters are producing 160 MW_e. Looking at the global picture, wind energy is about to replace hydropower as

the most important commercial source of clean, renewable energy, with Europe as the main market and leading operator.

An innovative application for earth observation technologies for all those interested in tapping this energy source is based on the fact that active microwave radar instruments transmit and receive radar signals with wavelengths in the range of a few centimeters to one meter, making them suitable for measuring the roughness of the sea surface, and thus determining ocean wind and wave fields. It has been demonstrated that SAR systems can provide information about the wind across areas up to 500 x 500 km in size (nearly the entire North Sea, for example) at resolutions down to 100 m (Horstmann et al. 2002). Techniques developed to measure wind fields in coastal regions are relevant for offshore wind farming, since local wind speed is the key parameter for estimating the generating power of a wind farm (power output is in a first approximation proportional to the cube of the wind speed). Wind turbines are usually operated at wind speeds between 4 and 25 m per sec. In the range of interest, even small uncertainties about what is the average wind speed result in big differences in the estimated energy output. For this reason, offshore wind farming requires high resolution regional operational forecasting of meteorological and ocean conditions, with earth observation technologies playing a significant role.

Determining wind fields with SAR is typically a two-step process. In the first step, wind direction is determined, a necessary input for the second step, which is to obtain wind speeds from the intensity values recorded as the wind-roughened ocean surface was imaged by the SAR. Wind direction is determined by identifying wind-induced phenomena aligned in the wind direction, which are usually visible in SAR images.

Wind, hail, strong rain or surface slicks change the roughness of the ocean surface and thus the intensity of the SAR image, and accordingly what speed is calculated. But wind directions can also change, as for instance within an atmospheric front. In addition, areas with ocean current shear can show a pattern in the scale of wind streaks that can be misinterpreted as the wind direction. To avoid the influence of such features not due to the local wind, complementary wind measurements have to be considered. Depending on the specific application (optimal siting of the wind farm or optimization of the wind farm design at a given site) these are either obtained from atmospheric models or in situ measurements.

Since SAR images have been acquired over the oceans on a continuous basis by radar systems on the ERS and Envisat satellites for the past 12 years, it is possible to compare current and historic data. This exercise yields valuable information for site planning. Relevant geophysical parameters are mean wind speed and direction, wind variability (changes over time in wind direction, speed and intensity), and turbulences induced by individual wind turbines and by the entire wind park taken as a whole (which affect output and can damage equipment).

Optimal siting means not only optimizing the power output but also minimizing the impact of wind parks on the environment. SAR data gathered from ERS, Envisat, and Radarsat sensors enable investigation of changes in wind field due to the presence of wind turbines, such as turbulent wakes and blockage effects in front of the wind farm, as well as ocean surface wave fields, all of which are of environmental relevance. Although some studies have been made on the effect on birds and fishery, no joint studies have been undertaken on how this new technology can be assisted by operational wind forecasting. The situation is similar with respect to the offshore oil industry.

Pollution

The generic term pollution refers to a broad variety of substances and the mechanisms that bring them into marine environments, including all influences which change marine ecosystems (open oceans, coastal waters, sea bed) in a manner causing harm or threat to flora and fauna or people, as well as pollution from waste material in general. Most sources of pollution are caused by human activity. However, there are also some natural processes causing marine pollution, such as oil seepage from the ocean floor. Earth observation offers a variety of ways to detect and monitor all the main categories of marine pollution:

- *Air pollution* due to fuel burning and industrial combustion places significant amounts of problematic substances (hydrocarbons, nutrients, sulphur oxides) into the oceans, either through rainwater or by runoff from land.
- *Drainage and runoff from land and rivers* can be polluted by agriculture and farming activities (nitrates, phosphates) and industrial waste. These substances are transported into coastal waters through surface water runoff, groundwater or rivers.
- *Dumping, leakage and accidents* on ships or marine platforms put large amounts of oil and other chemicals as well as bacteria and other invasive species into ocean waters; marine oil disposal and bilge water flushing are common.
- *Aquacultures* may cause overloads of nutrients or organic waste material on local or regional scales.

Pollutants can be detected by remote sensing technologies if they change either the optical properties (color), surface roughness, or temperature of the water. Since chemicals (nitrates, phosphate, toxic acids) are dissolved in the water and most bacteria do not show specific optical properties, these components may only be detected by in situ biochemical analyses. However, the spread of substances discharged by rivers can be traced and monitored by the suspended matter to which they are bound. Suspended matter is a water constituent which can accurately be accessed by optical remote sensing instruments.

Oil slicks (from tanker leaks or dumping) and natural surface slicks (extreme algal blooms) flatten the roughness of the sea surface by changing the surface tension of the water and thereby the properties and occurrence of capillary waves, which can be precisely detected by radar imaging. However, it is only possible to detect the occurrence and areal distribution of such spills, not the abundance of oil in total. This can only be done by specially equipped airplanes.

Biological "pollution" in the form of invasive species or harmful algal blooms is a phenomenon of growing importance. Worldwide shipping introduces through bilge water dumping alien species of bacteria, algae, macroalgae and macrophytes into basins far from their natural occurrence. Changing ecosystem conditions (nutrient enrichment, warming) may support the invasion of these species into new habitats, which may cause damage to native species or fish schools and disrupt entire ecosystems.

Instruments primarily designed for scientific research open up the possibility of operational detection and monitoring of pollution in the oceans from space. Before this can be achieved, however, optical and radar systems have to be merged to obtain the most useful data, and if long-term monitoring capability is the goal, dedicated operational systems will have to be put in place.

3.3.3
Security, Disaster Management and Humanitarian Aid

One can readily come up with a long list of disasters for which the vantage point of a satellite might be desirable, even critical: pollution on land or sea, floods, tsunami, hurricanes, storms, droughts, failed harvests, desertification, erosion, fires, earthquakes, volcanic eruptions, landslides, wars, terrorist attacks, refugee migrations, and technological disasters, not even to mention the sometimes horrendous results of incompetence, ignorance or criminal negligence leading to bad construction, engineering and siting, or errors of judgment. All too often daily news broadcasts confront us with the depressing consequences, whether they be to persons, property, infrastructure, the environment, the economy, or even to the very social order. This is not the place to speculate on the causes, and as a matter of fact earth observation technologies cannot themselves hinder the natural processes, looming crises or military conflicts that undergird such catastrophes. What they can do, in many different ways, is help increase safety and security by contributing to the early recognition of potential problems, in some cases even predict approaching catastrophes, monitor an ongoing crisis, assess damage, and contribute to the organization of relief measures during and after the event.

Earth observation can be used to record and keep track of those environmental conditions which exert a significant negative influence on human living conditions by leading to drought, floods, erosion,

and scarcities. These kinds of environmental changes can quickly give rise to destabilizing economic and social problems (water shortages or failed harvests leading to poverty, hunger, health hazards and migration, for example). Interpreters of remote sensing imagery are also able to contribute to effectively monitoring international treaties, with the aim of supporting the prevention or early warning of disasters, be they environmental, political, economic or social. Satellite data can be used to generate better estimates of static (informal settlements) and dynamic (refugee migration) populations on a global scale. They can also contribute to better surveillance of borders and sensitive infrastructure like nuclear power plants, and in some cases to monitoring the proliferation of weapons of mass destruction. A new application since September 11th 2001—at least in the USA—is the use of remote sensing techniques for homeland security.

Floods, Forest Fires

River valleys have always been desirable locations for human settlements. All over the world, as populations increased rivers were modified to facilitate shipping, their beds were confined, their natural catchment areas were drained and their courses were redirected. Such human intervention often exacerbates the consequences of natural flooding by permitting the flood surge to spread faster over wider areas than it would naturally do. Large scale flooding of settlements, towns, transportation routes, industrial areas, forests and crops are the damaging consequence. The 1988 Ganges-Brahmaputra flood in Bangladesh, the 2000 flooding of the Mekong in southeast Asia and of the Zambesi in Mozambique, the German Elbe floods of 2002, and the 2003 flooding of the Rhone River in France are recent examples.

Because of the temporal and spatial dimension of catastrophic floods and tsunami, it is often possible to record their effects and document their progress with earth observation technology. Both optical and radar sensors capable of recording the spatial distribution of the floodwater are utilized. Depending on the time and location of the area flooded, data from satellite or airborne sensors with their different spatial resolutions can be consulted. There are several methods to generate a water mask indicating pre-

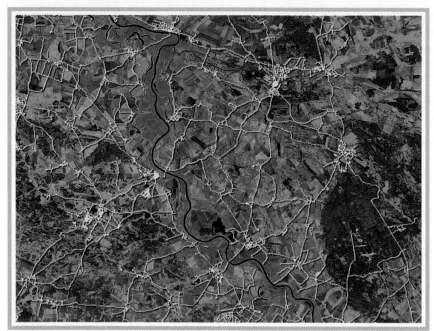

Figure 3-14: Flooding of the Elbe River near Torgau, Germany in August 2002 as seen in a Landsat image overlaid with information about the surrounding roads and settlements. The flooded areas (in lilac) were extracted using a change detection method based on two different ETM data sets, the first representing regular river level conditions, the second the situation during the flooding. As a result, a "water mask" was calculated showing the flooded areas. This mask was superimposed on the reference, permitting a view of the land structures below the water

sumably flooded areas from optical data based on calculations of the relationships and indices of different spectral channels. For example, the surface of clear water appears dark in the spectral range between 0.8 and 2.3 µm, since reflection off clear water is very low in this part of the spectrum. Flooded areas can accordingly be classified on the basis of their spectral and textural characteristics, although precisely what can be discriminated depends on the spatial and spectral capabilities of the sensor. Thermal data can be additionally consulted wherever there is a significant difference in the surface temperatures of land and water.

Computed satellite maps of flooded areas can be underlaid with traditional maps showing topography, land use, or the transportation network to produce a valuable tool for use by decision makers in crisis teams and situation centers when assessing threats or damage (figure 3-14). The ambitious precondition is timely and routine access to reliable and up-to-date data on location, ideally as soon as possible after the image has been recorded. In the context of the "International Charter on Space and Major Disasters" founded in 1999 it is possible for a limited community of users to obtain satellite data free of charge for rapid assessment of natural catastrophes (see Web: *Disasters*).

In addition to spatially detailed damage assessment, early warning methodologies based on satellite data are also being investigated. Changes in ground moisture over a defined period can be analyzed with data from scatterometers for the purpose of identifying trends and developments. Since floods are not only influenced by intensive and complexly structured rainfall but also especially by the condition of the soil in the area of interest (whether the soil is dried out or soaked, for example), predictions about flood disposition can be derived from ground moisture indicators.

An intensifying problem common to semiarid regions, boreal evergreen forests and tropical rain forests alike are catastrophic fires, often resulting from human activity. One underlying cause is clearance by fire of land to be made available for agriculture. Another is the pressure to find housing for growing populations and the associated land speculation. On average, some 22 million hectares of boreal evergreen forests (Cofer et al. 1996) and 600,000 hectares of brushland and forests in the Mediterranean area are consumed annually by fire. Both the number of fires and their extent has multiplied in recent years, whereas the share of fires in-

Figure 3-15: During the disastrous forest fires in Portugal of 2003, some 300,000 hectares burned, about 30 percent of the country's forests. This image of actively burning fires was recorded by the experimental DLR satellite BIRD, which detects not only hot spots, but also information about the physical properties of the fires, such as fire temperatures over a wide range above 200 degrees C (this example) and released energy. Mapping useful for setting fire fighting priorities in the field thus becomes possible. For this image the sensor's near infrared channel was used for the background grey values; the fire scars are dark. The mid- and thermal-infrared channels were used for the hot spots and reveal the fire temperatures

duced by natural causes has remained roughly the same at 1-5% of the total.

This increase, along with improvements in earth observation technology over the past few decades, led in the 1980s to fire monitoring by satellite. Near-real-time recording of actively burning fires is accomplished using satellites equipped to record channels in the mid- and thermal-infrared ranges (3-6 μm and 8-14 μm), including experimental satellites such as the DLR micro-satellite BIRD (figure 3-15). Primarily optical satellite data are used for the subsequent spatial damage assessment, whereby burned areas are classified using traditional procedures which make it possible to estimate the extent of affected area. Depending on requirements, either the extent of large fires or small-scale damage, particularly to infrastructure, can be determined.

Improving Safety at Sea

Individual ocean waves of exceptional height and shape, popularly known as rogue waves, are believed to have been the cause of significant damage to offshore constructions and the reason for the loss of many large ships at sea. For this reason, studies are being undertaken to detect, investigate and explain rogue wave phenomena (Sand et al. 1990). Plans are being made to publish in the form of an atlas the statistics which are being gathered on these extreme wave events for different areas of the ocean, together with historical information wherever the relevant data sets are available for analysis. This resource will be available to the science community, the shipping industry, marine designers and engineers, port authorities, certifying institutions, insurers and various international organizations.

For a variety of periods and geographical areas with different sea state conditions, data gathered by spaceborne SAR sensors are being combined with more traditional data from wave-riding buoys, from stationary marine radar installations on oil platforms, and from marine radar used on board container vessels for ship traffic control and navigation at sea. Whereas buoy records provide reliable information about the temporal variability of wave phenomena at a fixed ocean position, ship and platform radar systems provide spatial information on the nearby sea surface at a given time, and additionally a temporal sequence of consecutive radar sea surface images. Satellites complement the picture by observing the ocean surface continuously on a global scale, providing a synoptic picture of waves over large areas, including the less-studied vast southern oceans, and extreme events such as hurricanes. Traditional methods of analyzing radar images are being extended by new analytical approaches which exploit the information contained in satellite images for better description and proper identification of individual waves and wave groups.

Depending on the receiving mode, SAR images of the sea surface can be obtained for areas extending anywhere from 10 x 5 km ("imagettes") up to 500 x 500 km. Beyond providing a synoptic overview, they have been successfully used to derive mean sea state parameters in the open ocean. From a collection of radar imagettes for areas in the North and South Atlantic Ocean for which complementary buoy and marine radar data were also available, algorithms have been developed to detect and identify individual waves, as well as wave groups.

The sea surface images which radar systems provide are a function of many electromagnetic scattering mechanisms at the sea surface, influenced by currents, wave tilting, velocity of water particles, local wind, etc. Taken together, all these phenomena, known as radar imaging effects, yield a single radar measurement of intensity. Hence radar images contain information about how the sea surface backscatters the radar fields, rather than the wave elevation itself. Therefore, to detect individual waves, it is necessary to reverse the relationship or invert the radar imaging effects in order to obtain an estimation of the original sea surface scanned by the radar sensor, typically in the form of elevation maps. Once the wave elevation map is obtained, detection of single waves is possible, for example, by determining wave height in the spatial domain. It is also possible to trace a maximum in the elevation map through all the available radar images for a comparison of its spatial and temporal evolution.

Wave groups also play an important role in the design and assessment of offshore platforms, breakwaters or ships, because successive large single wave crests or deep troughs can cause severe damage due to their impact, or they can excite the resonant frequencies of the structures. For ships, an encounter with wave groups can sometimes cause capsize or severe damage. An extreme wave can develop from a large wave group due to interference of its harmonic components. Therefore the detection of wave groups in space and time is of high interest for ocean engineers and scientists.

Swell tracking is another type of study which can be carried out on a global scale with satellite data. All such earth observation techniques help to find empirical relationships between mean sea state characteristics and probabilities of extreme wave events. Furthermore, they help to identify hot spots and thus to improve risk maps (figure 3-16). At the time of writing, only 4,000 images for 27 days, corresponding to three weeks of data, have been processed. This data set is too small for firm conclusions, but ten years of SAR raw data are ready and waiting for processing, subsequent analysis, and comparison with complementary data of all types, thanks to the archiving around the world of an untapped treasury of satellite images.

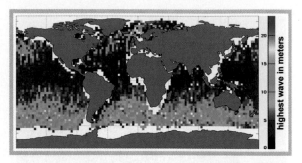

Figure 3-16: Map showing maximum single wave heights derived from three weeks of ERS-2 SAR data acquired in August-September 1996. The rough areas in the southern hemisphere and the path of Hurricane Fran in the northern Atlantic are visible

Political Conflicts, Effects of War

Politically and strategically, empires and nations have always needed means to keep watch over their territories, allies, and enemies. In the past, ground based reconnaissance and surveillance were used to penetrate enemy defenses, risking human lives and armed conflicts. From the 1960s on, space technology was also employed for these tasks and insured that more information of better quality could be collected without the risks, embarrassment and consequences associated with conventional international espionage. Remote sensing, imagery intelligence, reconnaissance, and surveillance all became terms used in billion-dollar space programs designed at first by the United States and Russia to spy on each other's abilities, tactics, hardware, and military activities. Since these beginnings, nonintrusive reconnaissance using satellites to obtain information about other nations has gained international acceptance as a legitimate procedure.

Conclusions about whether activities might be security relevant must be based on long-term observation of areas suspected to be breeding grounds for international threats. In the West, primarily the USA has access to powerful satellite surveillance networks; France also operates spy satellites and other European nations like Germany will soon follow with their own systems. The technical capabilities of these military systems are not made public in full detail and the imagery obtained is only made available to the satellite operators themselves. However,

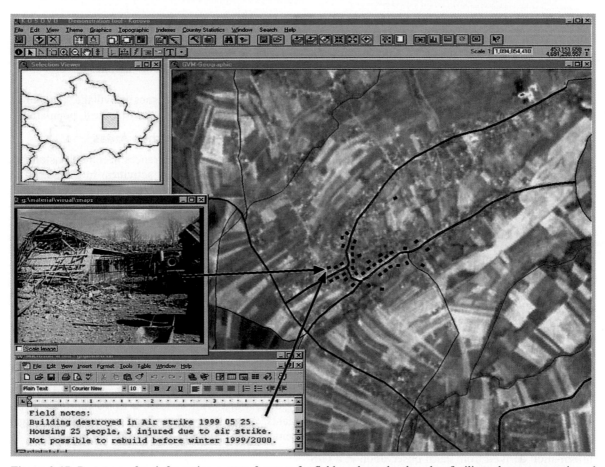

Figure 3-17: Prototype of an information system foreseen for field workers, developed to facilitate the reconstruction of Kosovo after the 1999 Balkan war. The information layers include a land use classification, the network of streets, rivers and railways, settlements, a map and a satellite image, all in the same map projection (UTM zone 34, WGS 84) and draped over a digital elevation model. This screenshot also shows photographs and field notes. A notebook contains part of the database, which is the reference for further data acquisition. A GPS for positioning, a digital camera for snapshots of infrastructure and a mobile phone for data transmission to and from the central database are connected as external elements

the recording instruments on board military reconnaissance satellites closely resemble those used in civilian remote sensing systems. The former have higher resolution capabilities, but the borders between civilian and military systems are disappearing and this trend will continue. Already today, imagery from civilian satellites is also being used for surveillance purposes. The highest optical imagery resolution from civilian satellites is currently about 60 cm (QuickBird) and will improve in the future (Ikonos Block II: 50 cm and better). Optical systems, however, always depend on good weather and visibility

conditions. They need to be complemented by radar sensor systems for utilization in any kind of weather, day and night. Civilian radar satellites so far offer up to 10 m resolution (Canada's Radarsat); systems providing 1-3 m resolution (Germany's TerraSAR-X) are under development.

Due to the large number of images, the evaluation of any kind of data set for purposes of detecting suspicious changes or identifying security-relevant relationships can only be accomplished with the help of highly automated image processing procedures.

Techniques like data fusion, knowledge-based image interpretation and data mining are required. It is also necessary to make use of geographic information systems (GIS) so that remote sensing data can be combined with other data sources providing, for example, information on population statistics, border locations and transportation infrastructure in crisis situations. What is being undertaken in postwar Kosovo (Ehrlich et al. 2000) can serve as an example:

The 1999 Balkan war has had dramatic economic and humanitarian consequences for this province of the former Yugoslavia. It has been estimated that one third of the dwellings have been damaged or destroyed, a half million people displaced, half of the farm land damaged, and telecommunication infrastructure severely disabled (European Commission and World Bank 1999). The economic and social restoration process requires up-to-date, precise and readily available spatial information, and a decision was made to set up a GIS suitably customized for Kosovo which can be consulted in the offices of government authorities or relief agencies, or on location in the field (figure 3-17). It benefits from the ongoing addition from many sources of maps, data from earth observation sensors and field campaigns, and ancillary information like cadastral boundaries and demographic statistics. Embedding information derived from earth observation in a GIS is an ideal way to facilitate its interpretation, exploit its strengths and temper its weaknesses. The following is typical of the kind of information the Kosovo GIS is providing:

- Perspective landscape views (generated by combining digital elevation models with satellite data);
- A variety of maps, including land use maps based on satellite data (from which availability of timber for reconstruction can be calculated, or crop yields predicted);
- Visualization of data at different scales (building clusters, settlements, regions), depending on the area of interest;
- Location of any of Kosovo's almost 2,000 villages;
- Highlighting of villages where hospitals, schools or utilities are available;

- Display of roads with visualization of the distance and elevation profiles, also based on satellite imagery and digital elevation models (for planning activities involving transport and calculating distances);
- Visualization of statistics by village, municipality or province (damage assessment, population).

Management of Urban Areas

The reason natural occurrences become disasters has much to do with human activities. In the course of industrialization, the creation of agroindustrial structures, and modification of formerly natural ecosystems into forms that are more economically efficient (wetland drainage, riverbed regulation, consolidation of arable land), one of the most significant developments of the last century has taken place, namely urbanization. In 2007 for the first time in human history more people will be living in towns and cities than in rural areas. Almost two thirds of the world's population of currently 4.9 billion people will live in cities by 2030. Most of this growth will be concentrated in the developing countries (United Nations 1999). This trend will be accompanied by an increasing emergence of megacities—urban agglomerations with at least 8 to 10 million inhabitants. Accordingly, the current number of 40 megacities is projected to rise to 60 within the next 15 years.

One of the consequences of this migration is that people are settling on land that was previously avoided because of natural processes (periodically flooded coastal areas, slopes prone to mudslides); another is the socioeconomic and ecological problems accompanying the formation of megacities. Their existence increases the likelihood of a large number of victims in any natural disaster. One reason is the concentration of many people in a relatively small area, complicating any systems for their early warning and mass evacuation. Another is buildings whose construction is inadequate to withstand stresses from earthquakes and storms, and whose high density almost assures the propagation of damage.

Rapid urbanization comes along with numerous ecological, social and economic challenges that hold risks of profound ecological or socio-economic

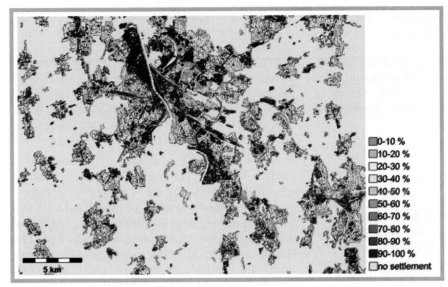

Figure 3-18: The color coding in this Landsat 7 ETM+ image of Germany's Rhein-Neckar region indicates the percentage of land surface sealed within built-up areas on 12 August 2001 (the disregarded countryside is white). The highly sealed area at top center represents the cities of Ludwigshafen and Mannheim, including a large industrial chemical facility

crises. The development of effective policies and plans for sustainable management of urban areas significantly relies on a proper understanding of both the dynamics of the urban system and the interaction between human activity, city structure, and the surrounding environment. Earth observation satellites offer promising and cost-effective opportunities to provide some of the required spatial and socio-economic information on a regular basis.

On a global scale remote sensing is a valuable technology for investigating the interactions between urban agglomerations or megacities and environmental phenomena like air or water pollution or ozone depletion, and their consequences. In this context satellite data play a significant role in monitoring relevant atmospheric parameters and improving atmospheric and hydrological models.

In a regional urban context remote sensing is mainly employed to monitor urban sprawl, population growth patterns and ground sealing (figure 3-18) and is usually based on optical satellite imagery with ground resolutions of ca. 5-30 m. Especially with regard to megacities the detection and mapping of informal settlements is of particular interest. Analysis of time series allows a detailed review of the spatial behavior of urban sprawl, the tracing of emerging environmental or socio-economic risks and a rough estimation of the population development. Urban heat islands have also been investigated

successfully in order to verify the climatic effects and impacts of urban agglomerations on both the urban system and the surrounding environment.

On a local scale the spatial characteristics of typical urban structures demand satellite data having a ground resolution of at least 5 m, down to 1 m. Such data can be used to map urban land use and existing infrastructure, analyze housing characteristics (a socio-economic feature), risk assessment in view of man-made or natural hazards (urban vulnerability, disaster management) and the position or distribution of selected public utilities. A highly differentiated analysis of urban surface types as well as the direct measurement of water or ground pollution on a local scale can be achieved using hyperspectral optical data, currently only available on airborne platforms, but expected to be available on satellites in the near future.

As far as standardized information extraction over urban areas is concerned, the lack of a classification approach sophisticated enough for automated data analysis has been keenly felt. Procedures are being developed to locate and group pixels into objects meaningful for urban analyses (buildings, streets, parks), to distinguish between different structures in close proximity by considering their likely function, based on their immediate surroundings (whether a green patch is recognized as a lawn or an agricultural crop, for example), and to display different objects in a single image at the various scales appropriate for the applications (a park might be simply identified as such without further elaboration, whereas a city block might be resolved into individual buildings and passageways). If successful, such procedures will make possible a wide variety of applications, from long-term urban planning to short-term disaster relief.

3.4
Future Developments

It can be predicted with some confidence that earth observation data for research in the geosciences will continue to gain in importance. In combination with other types of geodata from nonsatellite sources, or when integrated into models, remote sensing data will lead to new insights and improved understanding of ecological and climate processes. At the same time, scientific methodology and procedures will continue to be operationalized in order to make their output available to a community of users far more numerous than the original remote sensing science teams. The gap between 30 meter Landsat-like resolution and meteorological satellites is being closed by special and daily-revisit medium resolution imagers. Global daily vegetation status is being provided by the MODIS sensor currently operated on two U.S. satellites. This kind of environmental observation with multiple spectral bands is becoming a standard for global weather and climate observation. Hyperspectral imagers are driving the demand for high spectral resolution and will become operational assets in the coming years.

As of today, the use of earth observation data for environmental monitoring and mapping is primarily driven by government funded research projects and a few operational mapping programs. The vast majority of the medium resolution optical and SAR data is being used for such purposes. Commercial satellite operators also rely heavily on governmental users. In the United States as well as in Europe government agencies form the largest single class of customers. With the increased demand for security and military applications and a 6-14% prospected growth of the satellite image based geoinformation market (according to various forecasts), this dependency on governmental contracts will even increase. The total volume of the earth observation market in 2001 was estimated for the United States at $2.4 billion (ASPRS 2004) and in Europe at $1.2 billion (Frost & Sullivan 2001). These numbers do not include the in-orbit and ground segment systems, but might include the entire aerial survey market, which still forms the bulk of the geoinformation market. In the U.S. aerial platforms are used about twice as often as space platforms to collect remotely sensed data.

Fundamental to the establishment of commercial markets is on the one hand a clear market for the generated products (for use in agriculture or cartography, for example), and on the other hand a reliable guarantee of long-term availability of the basic data. And commercial value adding, that step from raw data to information products, will only be possible when the available instruments and systems are designed to meet market demands, provide good value for money, and are reliably available long term. The failure of a single component on Landsat 7 has vividly demonstrated the vast implications for the many studies worldwide that have been based for decades on data provided by the Landsat series.

Since the late 1990s former military optical reconnaissance technology has emerged on the public and commercial satellite imaging scene. The availability of very high resolution imaging has provided the largest stimulation to the earth observation market during the last few years. However, the new commercial providers—some of whom originated in defense companies—were forced to learn many lessons about the needs of a diversified commercial and governmental market for high quality services at affordable prices. Due to the new threats to security and the associated reorganization of national military capabilities, the highest demand for these kinds of data is coming from military agencies.

On the nongovernmental customer side, the largest revenue in earth observation services comes from mapping services (cartographic, thematic including environmental, and elevation), followed by various kinds of agricultural monitoring. While the majority of applications is still based on medium to high resolution satellites (above 15 m), the demand for very high resolution data is increasing. Interestingly, it is observed (ASPRS 2004) that very high resolution space data and aerial information are not competing but rather augmenting each other. This is also because traditional aerial analog imaging technologies are now becoming digital. Also, laser based imaging (lidar) can only be operated from aircraft.

Currently, the establishment of sturdy and commercial structures making use of remote sensing for deriving market-tested information layers is significantly dependent on public investment. And thus decisions made in the political arena will shape the future of commercial applications.

Political Framework

The monitoring of the earth from space is considered to be the most important tool for assessing the environmental state of the earth's ecosystems and delivering important information to allow forecasts of global ecosystems development and climate change (table 3-1). Triggered by findings about the depletion of the ozone layer, the Kyoto Protocol fostered many activities using a multitude of satellite based earth observation systems to monitor environmental change. Using existing global science entities and the framework of meteorological organizations, space agencies are contributing their data to global observing systems established from 1991 onwards in response to climate, ocean and terrestrial concerns. To focus on specific environmental themes, integrated global observing systems for atmospheric chemistry, global carbon, the global water cycle and the oceans were created between 1999 and 2002. All these programs are meant to be a framework to harmonize international policy, space programs and science topics. Satellite operation and data supply are left to the space agencies and meteorological satellite operators. Meanwhile, the European Space Agency is evaluating first results of studies in order to define the needs for next generation operational environmental monitoring satellites.

In July 2003, the U.S. government invited the international community to an "Earth Observation Summit" held in Washington, DC. Over 30 countries and 20 international organizations agreed at this summit on a "Global Earth Observation" (GEO) initiative, aimed at better integrating the existing earth observation systems and closing identified gaps in the observation of environmental parameters.

Under the umbrella of the European Union and ESA, the European countries are contributing to this process with the "Global Monitoring for Environment and Security" (GMES) initiative. After having agreed on a European satellite based navigation system (Galileo), the ESA and European Commission councils approved in 2001 a "political ... initiative to secure Europe with an autonomous and operational information production system in support of environment and security policies" (ESA and European Commission 2003, European Commission 2003). GMES concluded its initial period in 2003 and is aiming for a fully operational system in 2008.

GMES also addresses the security of European citizens from the viewpoint of civil protection. This somehow mirrors the U.S. homeland security program, where satellite based information is used to monitor critical infrastructures. Driven by the new geopolitical situation and manifested in their "Common Foreign and Security Policy" (CFSP), European governments wrote the strategic advantage of satellite imaging into their long term plans. Following the French Helios optical reconnaissance system, Germany and Italy started to build high resolution satellite radar reconnaissance systems (SARLupe and COSMO SkyMed). Meanwhile and in addition, commercial high resolution systems are used as sources to fill the increased need for data for humanitarian aid, peacekeeping missions and military intelligence.

In the early half of the first decade of the new millennium, the general political environment for earth observation is characterized by tighter governmental budgets, efforts to establish viable markets for earth observation data, and the implications of new U.S. space priorities. Opportunities can be seen in the increasing awareness of European and other countries of the importance of earth observation, the establishment of public funding schemes to launch operational missions, and the development of innovative technologies for both space and ground segments, all helping to make space-based observation an affordable tool for earthbound problems.

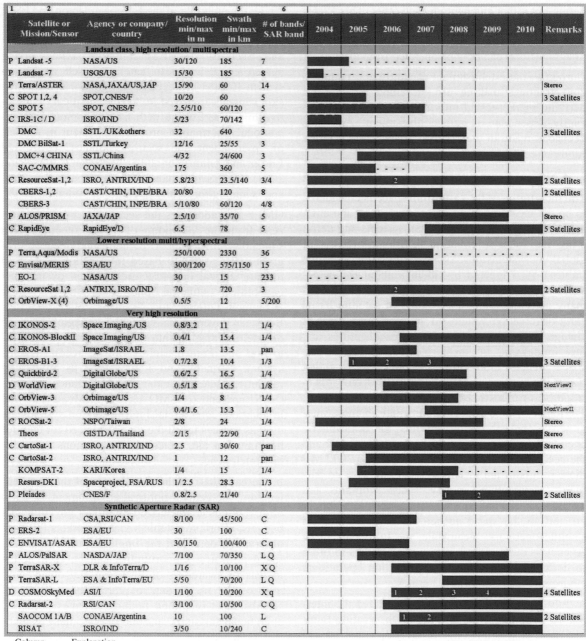

Satellite or Mission/Sensor	Agency or company/country	Resolution min/max in m	Swath min/max in km	# of bands/SAR band	2004	2005	2006	2007	2008	2009	2010	Remarks
Landsat class, high resolution/ multispectral												
P Landsat -5	NASA/US	30/120	185	7								
P Landsat -7	USGS/US	15/30	185	8								
P Terra/ASTER	NASA, JAXA/US, JAP	15/90	60	14								Stereo
C SPOT 1,2, 4	SPOT, CNES/F	10/20	60	5								3 Satellites
C SPOT 5	SPOT, CNES/F	2.5/5/10	60/120	5								
C IRS-1C / D	ISRO/IND	5/23	70/142	5								
DMC	SSTL/UK&others	32	640	3								3 Satellites
DMC BilSat-1	SSTL/Turkey	12/16	25/55	3								
DMC+4 CHINA	SSTL/China	4/32	24/600	3								
SAC-C/MMRS	CONAE/Argentina	175	360	5								
C ResourceSat-1,2	ISRO, ANTRIX/IND	5.8/23	23.5/140	3/4			2					2 Satellites
CBERS-1,2	CAST/CHIN, INPE/BRA	20/80	120	8								2 Satellites
CBERS-3	CAST/CHIN, INPE/BRA	5/10/80	60/120	4/8								
P ALOS/PRISM	JAXA/JAP	2.5/10	35/70	5								Stereo
C RapidEye	RapidEye/D	6.5	78	5								5 Satellites
Lower resolution multi/hyperspectral												
P Terra, Aqua/Modis	NASA/US	250/1000	2330	36								
C Envisat/MERIS	ESA/EU	300/1200	575/1150	15								
EO-1	NASA/US	30	15	233								
C ResourceSat 1,2	ANTRIX, ISRO/IND	70	720	3			2					2 Satellites
C OrbView-X (4)	Orbimage/US	0.5/5	12	5/200								
Very high resolution												
C IKONOS-2	Space Imaging/US	0.8/3.2	11	1/4								
C IKONOS-BlockII	Space Imaging/US	0.4/1	15.4	1/4								
C EROS-A1	ImageSat/ISRAEL	1.8	13.5	pan								
C EROS-B1-3	ImageSat/ISRAEL	0.7/2.8	10.4	1/3		1	2	3				3 Satellites
C Quickbird-2	DigitalGlobe/US	0.6/2.5	16.5	1/4								
D WorldView	DigitalGlobe/US	0.5/1.8	16.5	1/8								NextViewI
C OrbView-3	Orbimage/US	1/4	8	1/4								
C OrbView-5	Orbimage/US	0.4/1.6	15.3	1/4								NextViewII
C ROCSat-2	NSPO/Taiwan	2/8	24	1/4								Stereo
Theos	GISTDA/Thailand	2/15	22/90	1/4								Stereo
C CartoSat-1	ISRO, ANTRIX/IND	2.5	30/60	pan								Stereo
C CartoSat-2	ISRO, ANTRIX/IND	1	12	pan								
KOMPSAT-2	KARI/Korea	1/4	15	1/4								
Resurs-DK1	Spaceproject, FSA/RUS	1/ 2.5	28.3	1/3								
D Pleiades	CNES/F	0.8/2.5	21/40	1/4					1	2		2 Satellites
Synthetic Aperture Radar (SAR)												
P Radarsat-1	CSA,RSI/CAN	8/100	45/500	C								
C ERS-2	ESA/EU	30	100	C								
C ENVISAT/ASAR	ESA/EU	30/150	100/400	C q								
P ALOS/PalSAR	NASDA/JAP	7/100	70/350	L Q								
P TerraSAR-X	DLR & InfoTerra/D	1/16	10/100	X Q								
P TerraSAR-L	ESA & InfoTerra/EU	5/50	70/200	L Q								
D COSMOSkyMed	ASI/I	1/100	10/200	X q			1	2	3	4		4 Satellites
C Radarsat-2	RSI/CAN	3/100	10/500	C Q								
SAOCOM 1A/B	CONAE/Argentina	10	100	L				1	2			2 Satellites
RISAT	ISRO/IND	3/50	10/240	C								

Column	Explanation
1	C = commercial mission or commercialized outside country of origin; P = public/private partnership between space agency and commercial partner; D = dual use: military and commercial; no sign indicates limited regional or science availability
6	Capital letters C, L, X denote radar frequency bands; P = fully polarized; p = partially polarized; pan = optical panchromatic
7	Numbers denote launch of satellite in a series; mission end dates are as planned or estimated; "- -" denotes that satellite might operate longer
8	Numbers of satellites in the mission; special capabilities are abbreviated

Table 3-1: Overview of current and anticipated land surface earth observation satellites

References

ASPRS (2004) 10-Year Industry Forecast, Phases I – III documentation, Photogrammetric Engineering and Remote Sensing 70 pp. 1-10

Bamler R, and Hartl P (1998) Synthetic aperture radar interferometry, Inverse Problems 14, pp. R1-R54

Bartholome E, Belward AS, Achard F, Baratalev S, Carmona-Moreno C, Eva H, Fritz S, Gregoire JM, Mayaux P, Stibig HJ (2002) GLC 2000 – Global Mapping for the year 2000. European Commission Joint Research Centre

Belward AS, Estes JE, Kline KD (1999) The IGBP-DIS Global 1-km Land Cover Data Set DIS-Cover: A Project Overview, Photogrammetric Engineering and Remote Sensing 65 (9), pp. 1013-1020

Clevers J (1999) The use of imaging spectrometry for agricultural applications, Journal of Photogrammetry and Remote Sensing 54, pp. 299-304

Cofer WR, Winstead EL, Stocks BJ, Overbay LW, Goldammer JG, Cahoon DR, Levine JS (1996) Emissions from boreal forest: are the atmospheric impacts underestimated? In: Levine JS ed: Biomass Burning. Cambridge, MA: MIT Press, pp. 834–839

Colesanti C, Ferretti A, Prati C, Rocca F (2003) Monitoring landslides and tectonic motions with the Permanent Scatterer Technique, Engineering Geology 68, pp. 3-14

Cramer W, and Field CB (1999) The Potsdam NPP Model Intercomparison, Global Change Biology 5, pp. 1-15

Curlander J, and McDonough RN (1991) Synthetic Aperture Radar: Systems and Signal Processing. John Wiley and Sons, New York

Dech SW, and Glaser RJ (ed.) (1993) Fernerkundung von Umweltbelastungen auf dem militärischen Übungsgelände in der Colbitz-Letzlinger Heide. Forschungsbericht der Deutschen Forschungsanstalt für Luft- und Raumfahrt (DLR-FB 93-46), Köln, p. 106

Donlon CJ, Minnet PJ, Gentemann C, Nightingale TJ, Barton IJ, Ward B, and Murray MJ (2002) Toward Improved Validation of Satellite Sea Surface Temperature Measurements for Climate Research, Journal of Climate 15, pp. 353-369

Ehrlich D, Hansen C, Louvrier C, Hubbard N, Richards T, Reinartz P, Mehl H (2000) Use of Satellite Imagery in the Set-Up of a GIS to support Reconstruction of Kosovo. GIS (GeoInformationsSysteme) 13 (5), pp. 25-28

Eicken H, and Lemke P (2001) The response of polar sea ice to climate variability and change. In: Lozán et al. ed: Climate of the 21st century: changes and risks, GEO, Hamburg, Germany, pp. 206-211

ESA and European Commission (2003) Global Monitoring for Environment and Security, Final report for the GMES initial period (2001-2003), version 3.5

European Commission (2003) White Paper Space: a new European Frontier for an Expanding Union, COM (2003) 673, Brussels

European Commission and World Bank (1999) Towards Stability and Prosperity - A Program for Reconstruction and Recovery in Kosovo, Report, http://www.seerecon.org/Kosovo/documents/kosovo_toward_stability_and_prosperity_1999.pdf

Frost & Sullivan (2001) European Commercial Remote Sensing Data End-User Markets, Report B020 (www.frost.com)

Horstmann J, Koch W, Lehner S, Tonboe R (2002) Ocean Winds from RADARSAT-1 ScanSAR, Canadian Journal of Remote Sensing 28 (3), pp. 524-533

IPCC (2001) Third Assessment Report: Climate Change, Geneva

Kemper T, and Sommer S (2003) Mapping and monitoring of residual heavy metal contamination and acidification risk after the Aznalcóllar mining accident (Andalusia, Spain) using field and airborne hyperspectral data, in Proceedings, 3rd EARSeL Workshop on Imaging Spectroscopy EARSeL Secretariat, Paris, pp. 333-343

Kramer HJ (2002) Observation of the Earth and Its Environment. Springer Verlag, Berlin

Lalli CM, and Parsons TR (1993) Biological Oceanography: An Introduction. Pergamon Press, Oxford

Lang O, Rabus BT, Dech SW (2004) Velocity map of the Thwaites Glacier catchment, West Antarctica. Journal of Glaciology 50 pp. 46-56

Lundgren P, Berardino P, Fornaro G, Lanari R (2003) Coupled magma chamber inflation and sector collapse slip observed with synthetic aperture radar interferometry on Mt. Etna volcano, Journal of Geophysical Research 108 (B5), doi: 10.1029/2001JB000657

McClain EP, Pichel WG, Walton CC (1985) Comparative performance of AVHRR-based multichannel sea surface temperatures. Journal of Geophysical Research 90, pp. 11587-11601

Mensching HG, and Seuffert O (2001) (Landschafts) Degradation – Desertifikation: Erscheinungsformen, Entwicklung und Bekämpfung eines globalen Umweltsyndroms, Petermanns Geographische Mitteilungen 145 (4), pp. 6-15

Mucher CA, Steinnocher KT, Kressler FP, Heunks C (2000) Land Cover Characterisation and Change Detection for Environmental Monitoring of Pan-Europe, International Journal of Remote Sensing 21 (6/7), pp. 1159-1182

Platt T, and Sathyendranath S (1988) Oceanic Primary Production: Estimation by Remote Sensing of Local and Regional Scales, Science 241, pp. 1613-1620

Sand SE, Hansen NEO, Klinting P, Gudmestad OT, Sterndorf MJ (1990) Freak wave kinematics. In: Torum O and Gudmestad OT eds: Water Wave Kinematics, Kluwer Academic Publ., pp. 535 – 549

Ulaby FT, Moore RK, Fung AK (1986) Microwave Remote Sensing, III. Artech House, Norwood, MA

United Nations (1999) "Key Findings" World Urbanization Prospects: The 1999 Revision. United Nations, New York

Wentz FJ, Gentemann CL, Smith DK, Chelton DB (2000) Satellite measurements of sea surface temperature through clouds. Science 288, pp. 847-850

Web References

Disasters: http://www.disasterscharter.org

National Ice Center: www.natice.noaa. gov/icebergs.htm

North and Baltic Seas: www.prokonnord.de/ pages/projekte/index_projekte.html (select Offshore Windpark Borkum-West) or www.bsh.de/ de/ Meeresnutzung/Wirtschaft/Windparks/index.jsp

4 Climate and Environment

by Hartmut Graßl

The earth can be described as a system composed of several major components which interact on very different space and time scales. The result is fantastic complexity, high variability on time scales of minutes up to billions of years, however, also a remarkable stability that has sustained life for more than three billion years. Mankind needed a long time to be able to correctly give the astronomical causes of an earth day and an earth year: The earth is close to a sphere which rotates on an ellipse around the sun in a year and around its own axis, oblique to the orbital plane, within a day. This constellation creates daily and annual cycles in those parts of the system that react rapidly to insolation changes.

4.1
The Earth System Components

A first simple subdivision into system components orients itself on the thermodynamic phases gaseous, liquid and solid. The atmosphere is the gaseous envelope of our planet in which we live close to the solid surface; the ocean is the liquid main reservoir within the global water cycle that also dominates the energy cycle of the Earth system; the lithosphere, i.e., the solid upper-most part of the earth's mantle, is often also called the earth's crust. A next step in a finer subdivision of Earth system components includes the cryosphere, i.e., all parts containing frozen water at or below the earth's surface, and the biosphere, composed of living and dead organic matter. Together with the earth's core all these components can be observed from space either directly via electromagnetic radiation scattered or emitted by the object of interest, like the atmosphere and the surface, or indirectly via the measurement of gravity and magnetic field by low earth orbiting satellites.

A full understanding of the earth's climate and of our environment does need the monitoring of all the above system components with all types of satellites. At present we are still far from a full monitoring. However, progress in our understanding has been accelerated most by earth observation satellites, high tech sensors always challenging material sciences. It is fair to say that the broad political consensus to view development and environment as inseparable twins has been strongly promoted by satellite remote sensing allowing a truly global view.

4.1.1
Component Interactions

All Earth system components interact on time scales ranging over orders of magnitude. Therefore, reactions to disturbances cannot only be delayed by up to thousands or even millions of years, but can often also not be unraveled if no global long-term observation system exists.

An illustrative, but typical example of human action on a system component is: higher economic return now, costly consequences later. In practical terms, let us look at nitrogen fertilization. A farmer boosts agricultural yield by fertilizing with nitrogen compounds, e.g., ammonium-nitrate ($NH_4 NO_3$). A small part of it, up to a few percent, ends up in the atmosphere as dinitrogen oxide (N_2O, nitrous oxide), a potent greenhouse gas with a lifetime of about 120 years (e-folding time). Photolysis of N_2O in the stratosphere, the only major sink mechanism, leads in parts to nitrogen monoxide (NO) that is the key substance for one of several catalytic ozone destruction processes. Nitrogen fertilization will thus contribute significantly both to the enhanced greenhouse effect and later to ozone depletion in the stratosphere, for more than a hundred years.

Can we monitor it by remote sensing from space? Yes, in large part. Besides several vegetation parameters we would need to observe at least vertical profiles of N_2O, O_3 and NO as well as other trace substances. Many more examples of this kind could be named, suffice it to mention the global methane budget, the carbon dioxide sink and source assignment, climatologies of contrails, and noctilucent clouds.

4.1.2
Changes in Atmospheric Composition since Industrialization began

The atmosphere of our planet is characterized by the dominance of trace gases with respect to their relevance for the energy budget. Less than three permille of its mass, including water vapor, cloud water and cloud ice, determine how much solar radiation reaches the surface and how the energy is radiated back into space. All three long-lived greenhouse gases, namely carbon dioxide (CO_2), nitrous oxide (N_2O) and methane CH_4 (numbers 2, 4 and 5 in a ranking of natural greenhouse gases, see also table 1), are increasing because of human activities: from 280 to 376 parts per million by volume (ppmv) for CO_2, from 0.7 to 1.75 ppmv for CH_4 and from 0.28 to 0.32 ppmv for N_2O since massive industrialization began. We are just now beginning to observe column contents of these gases from space in order to better understand their budgets, and we hope to get dedicated missions for their monitoring on a global scale.

This enhanced greenhouse effect no doubt will change global climate; however, much higher precision in measurements is needed to monitor changes of parameters like temperature and precipitation than for a strong concentration change of greenhouse gases. Only very recently did the evaluation of satellite series give first trend analyses of temperature for thick tropospheric layers (see also section 3.2.4 on studies of change) or of cloud reflectance as a consequence of changed air pollution.

4.1.3
Variability versus Change

Observation of an environmental parameter over a certain time period will often show strong variability on all time scales pertinent to the entire observation period. Whether the trend sometimes observed is due to a real change, e.g., caused by a variation of external parameters like solar radiation, or is due to a low frequency component interaction process, cannot be decided without additional knowledge. In other words: trends might be part of internal long-term oscillations or real changes can be masked by component interaction processes. Using a combination of satellite remote sensing with in situ data and global model studies is the best way to decide on variability versus change. It became tradition to speak of internal component interaction and external forcing, whereby the latter often includes influence on climate by volcanoes and human activities. Here the use of the term external forcing is restricted to changes in solar output, the earth's orbit, as well as human activities.

4.1.4
Advantages of Earth Observation from Space

When laymen hear about the costs of earth observation satellite missions they often might not believe that 0.5 billion euro per mission is not only less than the yearly expenditure for existing in situ systems in meteorology, but that observations from space are now also of comparable or in parts superior quality and can deliver many parameters for which no in situ networks exist. An obvious and well accepted advantage of satellite remote sensing is good and very frequent coverage by geostationary missions and global as well as frequent coverage by polar orbiters. In addition, satellite observations are con-

No.	Gas		Characteristic life time	Relative contribution to the greenhouse effect
1	Water vapor	H_2O	~ 9 days	~ 65%
2	Carbon dioxide	CO_2	~ 5-7 years	~ 15%
3	Ozone	O_3	Days to months	~ 10%
4	Dinitrogen oxide	N_2O	~ 120 years	Several percent
5	Methane	CH_4	~ 10 years	Several percent

Table 4-1: Gases in air ordered according to their contribution to the natural greenhouse effect of the atmosphere

sistent, as the same sensor observes the entire globe, albeit this is only true for a single satellite but not a series. Homogenization of time series from a satellite series is, however, as demanding as homogenization of time series from in situ stations or networks, but it is valid on a global scale at once. On top, new hyperspectral sensors, be it spectrometers like the Atmospheric Infrared Sounder (AIRS) or interferometers like the Infrared Atmospheric Sounding Interferometer (IASI), can deliver a three dimensional view of the entire atmosphere that approaches the vertical resolution of planned so-called active sensors, like lidars and radars (Wulfmeyer et al. 2005).

In view of these major advantages the drawbacks in comparison to in situ networks should not be neglected. These are:

- physical barriers: no penetration of electromagnetic radiation into the earth's interior or into thick clouds in the visible and thermal infrared wavelengths;
- short lifetime of satellites, although it has been increased systematically in recent years;
- lack of or insufficient intercalibration for satellite series;
- partial lack of in-flight calibration;
- inadequate ground segment for experimental satellites;
- launch delay or failure interferes with the demand for global time series, i.e., high risk may delay commitments of national services needed to support pre-operational missions.

Operational meteorological satellite series demonstrate, however, that continuous global monitoring is possible and risks can be significantly reduced by cooperation.

4.1.5
Detection of Changes in Global Biogeochemical Cycles

Mankind interferes strongly with the two most important global cycles: the carbon cycle (see figure 4-1) on which life is founded and the water cycle which is fundamental for the planetary energy budget and for life as well. Before human activities influenced nature on a global level, the carbon cycle showed an average net flux of below 10^9 tons of

carbon per year into the atmosphere or the biosphere and the ocean. The imbalance at present has reached in a multiyear average about $3 \cdot 10^9$ tons of carbon per year. The water cycle on continents is severely changed by mankind as formerly major rivers are reduced to very small ones; some even do not reach the sea for longer periods (e.g., Colorado, Indus, Yellow River). In order to monitor carbon fluxes, something required for compliance with international treaties like the Kyoto Protocol of the United Nations Framework Convention on Climate Change, we need to know the regional sources and sinks. While we have started to observe column contents of CO_2 from space (see section 4.4.1) the monitoring of stream flow from space is still a wish. However, the observation of lake and river levels using altimeters on satellites has already shown its great potential. Satellite remote sensing is the only way to monitor the net freshwater flux at the ocean surface, and can answer whether or not the global water cycle intensifies with mean global warming at the surface in a world with further enhanced turbidity caused by air pollution from industry and traffic and by vegetation fires in emerging and developing countries.

4.1.6
Earth System Analysis and Sustainability

The ultimate goal of mankind, as set out in international treaties, is sustainable development. The absence of sustainability is apparent from the major anthropogenic disturbances of biogeochemical cycles that are the basis of life on earth and are strongly influenced by life itself. In order to find the corridor to sustainability one has to analyze, to the extent possible, the Earth system (Schellnhuber, 2001) under its present pressures. This analysis needs global monitoring and global scenario calculations, whereby as many feedback processes as possible should not be prescribed, but modeled as part of the system. This task calls for many data sets from satellite sensors to be evaluated synergetically. We are far from such a global observing and data assimilation system. But we see recent attempts to reach it. One can mention the Global Earth Observation System of Systems (GEOSS) urged by the Group on Earth Observation (GEO) to which most

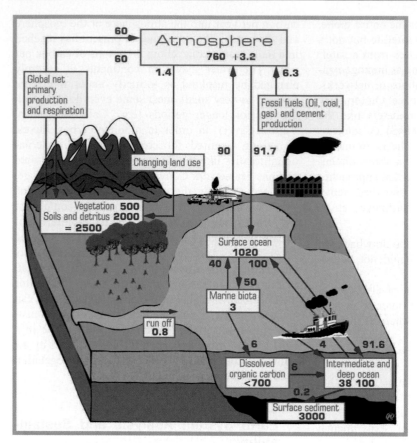

Figure 4-1: Schematic illustration of the global carbon cycle. Reservoirs indicated in billion tons, fluxes in billion tons per year; values taken from Prentice et al. (2001)

developed and some emerging and developing countries committed themselves to contribute. It is clear that the corridor can only be found by intensified global change research and a continuous dialog with decision makers. Earth observation results will play a decisive role in this dialog.

4.2
Challenges for Climate and Environmental Monitoring

In the debate about global change issues it became clear that *the* dominating issue of global dimensions is anthropogenic climate change; it is a concern for all, the highest impacts will very likely be in areas where people did not contribute considerably to changed atmospheric composition. New weather extremes will accompany this change, loss of lives and damage costs can be so high as to throw back countries in their (economic) development. If envi-

ronmental monitoring from space is to help in understanding interaction processes to analyze trends and show consequences of action, it must not only reach a high level of sophistication that often still does not exist, but it must also be continuous and re-evaluate all former satellite series in the light of algorithms improved with new, more sophisticated sensors, in order to establish series long enough for trend analyses.

4.2.1
First Challenge: Calibrated Global Data Sets

A satellite data set can be useful to establish a regional or global climatology of a certain geophysical parameter only if the different types of calibration relate counts to physical quantities, mostly radiances. Besides the pre-launch calibrations, in-flight calibrations and vicarious calibrations have to be combined for a single satellite's time dependent calibration coefficient covering all spectral channels. In a next step overlap periods of two satellites looking at nearly the same locations with only a short time shift have to be used to find potential systematic drifts of sensors and their detectors. In an attempt to establish a time series of the Advanced Very High Resolution Radiometer (AVHRR) on board the NOAA operational meteorological satellite series to be used for a trend analysis of the coldest cloud top temperatures, we found, for example, that two different types of calibrations were used for different time periods, causing erroneously low cloud top temperatures (< -95°C) in one multisatellite series. A recalibration of the data set using new calibration proposals of NOAA for the entire period has removed the obvious bias, but fortunately did not remove the downward trend found earlier.

4.2.2
Second Challenge: Assimilation of Remote Sensing Data into Earth System Models

The goal of most geoscientists is to extract the predictable portion of the Earth system and to use starting fields for forecasting or projections (scenario calculations). This needs small error bars in observations and sophisticated models using the observations as starting fields. The interface between models and observations is called data assimilation. In simplified terms it is the ingestion of any useful observation at a certain place and time into the model prior to the start of the numerical prognosis. This involves a test whether the information is useful in a physically consistent model. Very often it is advisable to ingest satellite radiances instead of geophysical parameters as the models have to contain a radiative transfer sub-model anyway. As is often the case, meteorologists are the most advanced also in this technique, and from the European Centre for Medium-range Weather Forecasts (ECMWF) in Reading, UK, we learn that through assimilation of data from many satellite sensors weather forecasting skill in the Southern Hemisphere has now reached or exceeded that for the Northern Hemisphere. This is as it should be, because orography makes forecasting more difficult in the Northern Hemisphere. The breakthrough is due to four-dimensional variational

data assimilation that allows any information at any time from any location to be considered for the consistent starting field. We therefore have the advantage of being able to plan the coming week with improved weather forecasting, due in large part to satellite remote sensing financed by the global meteorological service community. Similar assimilation techniques have already been used in oceanography and will soon enter first Earth system models, which include for example air chemistry processes, assimilating data on trace gas columns (NO_2, CO, etc.). But also data of water cycle components like cloud water are candidates for assimilation into global or regional circulation models.

4.2.3
Third Challenge: Evaluation of Satellite Series for Environmental and Climate Monitoring

The basis for many decisions by citizens, companies, and governments is reliable climate information, especially now in an era with climate change. Thus information will come with growing share from satellites. It is in this context that the constitution of the European Institution for the Exploitation of Meteorological Satellites (EUMETSAT) has been amended to contain climate monitoring as its second major task. Jointly with national meteorological services in Europe, EUMETSAT has founded Satellite Application Facilities (SAFs) that have started or are close to starting monitoring of, for example, stratospheric ozone, sea surface temperature, sea ice, top of the atmosphere radiation fluxes, cloud parameters, etc. by using the new European operational satellite series (Meteosat Second Generation, launched in 2002, and MetOp, to be launched in 2005). The plans include regular reanalysis, maybe every five years, because when more sophisticated algorithms become available a new full improved time series has to be re-established. It is important in this context that the time series are as clean as possible of interference with models, as such time series are the key test set for models.

4.2.4
Fourth Challenge: Multisatellite and Multisensor Evaluation

Many geophysical quantities can only be retrieved from satellite data if several sensors on a single satellite or several sensors from several satellites are combined. Determining evaporation from the ocean surface, for example, requires wind speed, sea surface temperature and humidity in the lowest troposphere as input, which can be supplied by a combination of the Special Sensor Microwave Imager SSM/I on board the Defense Meteorological Satellite Program (DMSP) satellite series and the Advanced Very High Resolution Radiometer (AVHRR) on board the NOAA satellite series.

If the precipitation rate information of the SSM/I is used to detect severe storms over the Atlantic, all available SSM/I sensors on several satellites have to be used to get full coverage at least twice per day (Klepp et al. 2003). Unfortunately, multisensor evaluations for climate research are still rare. Hence, the full exploitation of all available series for climate monitoring is still not possible. In Europe the combining of sensors is hampered by the lack of responsibility of ESA for exploitation of satellite data and the comparatively short time series of polar orbiting satellites, together with the absence of satellite exploitation budgets for European research. Now, after 13 years of data from very successful experimental earth observation satellites (ERS-1, ERS-2, Envisat) we urgently need such an initiative for multisensor and multisatellite exploitation.

4.3
Variability of Biospheric and Climate Parameters from Space

For many biospheric parameters like vegetation period or phytoplankton chlorophyll concentration, and many climate parameters like cloud water and precipitation over the ocean, no observed variability on a global scale existed before the satellite era. It is, however, difficult to establish first climatologies or frequency distribution of parameters from time series shorter than 30 years, defined as a suitable period in climatology. Now that time series are frequently surpassing 10 years and sometimes have

reached 25 years, such preliminary climatologies emerge. Nearly all of them rely on data from operational meteorological satellites, mostly from the USA, where the NOAA satellite series started in 1979. In the following such first climatologies will be presented for cloud parameters, (ocean) surface parameters, and vegetation period.

4.3.1
Cloudiness

The least understood climate change agent is cloudiness. Its reaction to the enhanced greenhouse effect can be assessed only if at least an assessment of average cloud amount, cloud type, cloud optical depth, mean cloud droplet radius, their seasonal means and variability, as well as the annual cycle has been achieved. All these parameters together with their interannual variability and many more cloud related parameters are now available for the time since 1983, thanks to the International Satellite Cloud Climatology Project (ISCCP) of the Global Energy and Water Cycle Experiment (GEWEX) within the World Climate Research Programme (WCRP) (see web: *wrcp* and Rossow and Duenas, 2004). Some of the key findings are:

- the average global cloud cover is 62%;
- El Niño Southern Oscillation (ENSO) shifts cloud patterns nearly on a global scale;
- trend analyses are still not possible, as on board calibrations are mostly lacking in the solar radiation range;
- the Pinatubo eruption in 1991 had a discernible impact on global cloudiness.

Recently two new satellite sensors, namely the Moderate Resolution Imaging Spectroradiometer (MODIS) on Aqua, launched by NASA in 2000, and the Medium Resolution Imaging Spectrometer (MERIS) on Envisat, launched by ESA in 2002, have brought a major step forward as they allow cloud top pressure (height) determination at 300 m horizontal resolution once per day or every second day on a global scale to an unprecedented accuracy during daytime, using reflected solar radiation within and close to the O_2 absorption band at 0.76 μm wavelength. A further step, maybe a quantum leap towards understanding cloud microphysics, will come with active sensors (lidar and radar) on board the Cloudsat and Calipso satellites of NASA and CNES (to be launched in 2005) or on board Earth-CARE developed by ESA and JAXA. These satellites will provide profiling capability for aerosol and cloud parameters.

4.3.2
Ocean Surface parameters

Ocean-air interaction is *the* main climate variability generator. Therefore, all fluxes at this interface are of high relevance for a basic understanding of the Earth system, for weather forecasting, and climate model evaluation. While the assimilation of near surface wind speed data, derived from microwave back-scattering by ocean waves, into weather forecasting is already routine, heat and radiation fluxes at the surface just emerge. As shown in figure 4-2, for the long-term average incoming solar irradiance at the surface, the tropical oceans are the "filling stations" for the Earth system, due to low reflectivity (albedo) *and* comparably low cloudiness. Compar-

ing figure 4-2 with figure 4-3 and table 4-2 for the long-wave net flux density over ocean areas, it becomes clear that a large portion of the solar radiation absorbed by the ocean must leave the ocean via the heat fluxes (latent plus sensible), as long-wave net flux density values are less than half of the solar input, especially in the inner tropics where often only less than a fourth of solar input is lost by heat radiation. The reason is a strong downward atmospheric heat radiation often well above 400 Wm^{-2} at an ocean surface emission of up to 470 Wm^{-2} in ocean areas reaching 30°C surface temperature.

Table 4-2 shows the first long-term averages of global radiation budget components. In addition, by using two different algorithms it gives a hint about the accuracy reached. The following conclusions can be drawn:
- clouds cool the earth's surface because long-wave cloud forcing is significantly lower than short wave forcing;
- the variability from one year to the next is small,

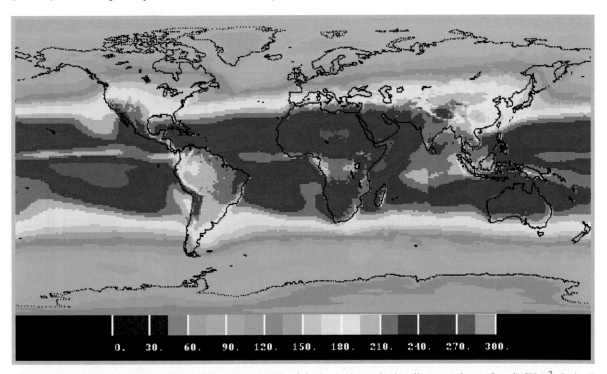

Figure 4-2: Long-term average (July 1983 to June 1995) of the incoming solar irradiance at the surface in Wm^{-2}, derived from geostationary and polar orbiting meteorological satellites by the Surface Radiation Budget (SRB) project of GEWEX within the WCRP; see web: *srb* for further information

Parameter	GEWEX Algorithm			Quality-Check Algorithm		
	Mean	Min.	Max.	Mean	Min.	Max.
SW Down	186.2	184.6	188.9	184.2	183.3	185.6
SW Net	164.6	162.0	168.1	160.9	160.0	162.1
LW Down	342.6	340.8	344.4	345.2	343.4	347.2
LW Net	-50.8	-51.9	-50.0	-47.2	-48.0	-46.0
Total Net	113.8	110.1	118.1	113.7	112.0	116.1
SW CRF	-56.9	-58.0	-54.8	-58.5	-59.4	-57.0
LW CRF	36.5	35.6	37.4	35.6	35.2	36.2
Total CRF	-20.4	-22.4	-17.4	-22.9	-24.2	-20.8

Table 4-2: GEWEX Surface Radiation Budget (SRB) project results: 12 year average (July 1983 – June 1995) of radiation budget components in Wm^{-2} (SW = shortwave, LW = long wave, CRF = cloud radiative forcing), Min = minimum yearly average of 12 years, Max = maximum yearly average; see also web: *gewex*

e.g., only up to 2 Wm^{-2} for net shortwave flux density at the surface for the quality check algorithms;

- more than 100 Wm^{-2} must leave the surface via latent and sensible heat flux;
- the downward solar energy flux of about 184 Wm^{-2} is by a factor of 7,000 larger than the energy throughput of mankind ($\sim$ 0.025 Wm^{-2} at present);
- the strong greenhouse effect of the atmosphere

leads to a back radiation average (342 Wm^{-2}) which is more than double the incoming net solar radiation flux density (160 Wm^{-2}).

Most of the solar input leaves the ocean via the latent heat flux, which can now be determined from space using operational meteorological satellites, here the Special Sensor Microwave/Imager (SSM/I) on board the satellites of the Defense Meteorological Satellite Program (DMSP) of the U.S. Navy and the Advanced Very High Resolution Radiometer

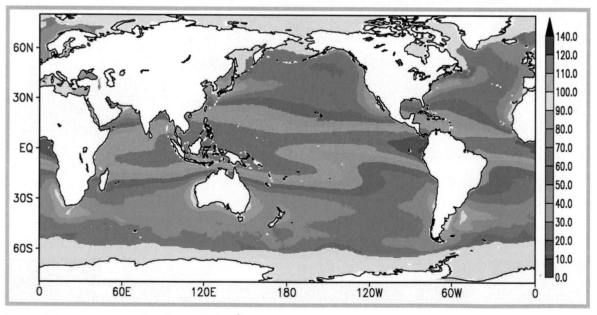

Figure 4-3: Long-wave net flux density in Wm^{-2} at the ocean surface derived from satellite data, following the algorithm by Schlüssel et al. (1995); this energy flux density is part of the Hamburg Ocean Atmosphere Parameters and fluxes from Satellite data (HOAPS) climatology (Jost et al., 2002); see web: *flux* for further information

(AVHRR) of the NOAA satellite series (for details of the algorithms see web: *hoaps*). If also precipitation can be estimated from space, as SSM/I and especially the Tropical Rainfall Measurement Mission (TRMM) of NASA and JAXA have demonstrated, the net freshwater flux density at the ocean surface can be estimated. The first such climatology is displayed in figure 4-4, taken from the Hamburg Ocean Atmosphere Parameters from Satellites (HOAPS) data set. Key features of the net freshwater flux density map in figure 4-4 are:

- more freshwater enters the ocean than leaves it in the Intertropical Convergence Zone (ITCZ), over the Kuroshio and the Gulf Stream extensions as well as in high latitudes;
- in the subtropics freshwater loss of the oceans reaches a maximum, in parts surmounting 5 mm/d, e.g., in the southern Pacific;
- the warm pool surrounding the "Maritime Continent" is the largest area where precipitation significantly surpasses evaporation.

4.3.3
Vegetation Period

Since December 1978 all NOAA satellites have carried the AVHRR, which allows derivation of the so-called Normalized Differences Vegetation Index (NDVI), the difference between two channels in the visible and near infrared divided by their sum. It is a measure of the greenness of a surface, because vegetated surfaces are brighter in the near infrared than in the visible part of the spectrum. Therefore, channel differences and thus NDVI increase with leaf area, and the vegetation period can be determined by applying thresholds. Droughts can be detected because they show up as reduced greenness. Although atmospheric turbidity is a disturbing factor for NDVI the daily observation capability leads to sufficiently cloud- and nearly turbidity-free situations for NDVI determination.

4.4
Emerging New Parameters

The increased understanding of the Earth system is also strongly driven by observations from experimental satellites that give first maps of hitherto inaccessible parameters, e.g., trace gas columns, optical depth of aerosols, and parameters related to severe weather. Here only a few such expectations will be presented.

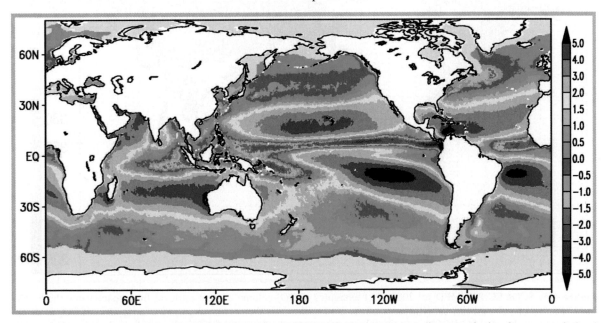

Figure 4-4: Net freshwater flux density in mm per day or litres per square meter per day over the ice-free ocean derived from satellite observations over the 1988 to 2002 period; see also Jost et al. (2002) and web: *flux*

4.4.1
Carbon Dioxide Column Content

The global carbon cycle is massively disturbed by mankind. About 3.2 Gt carbon are accumulated on average per year in the atmosphere. This has led to the United Nations Framework Convention on Climate Change (UNFCCC) and its Kyoto Protocol (enacted on 16 February 2005) that calls for global monitoring of carbon dioxide (CO_2) sources and sinks. This goal can be reached by the measurement of CO_2 column content and its use in general circulation models whose output is the net source (sink) flux distribution on a global scale.

Fortunately, we have the add-on payload SCIA-MACHY on Envisat (supplied by Germany and the Netherlands with a small contribution by Belgium). We can use reflected solar radiation in a CO_2 absorption band to derive CO_2 column contents. First attempts to get such overviews exist. One of these in figure 4-5 points to the feasibility and the problems

with clouds. However, this might help to get profile information later. Such analyses will boost our knowledge about the dynamics of the global carbon cycle.

4.4.2
Aerosol Optical Depth

The compounds with highest impact on the energy budget of the earth per mass unit are aerosol particles suspended as tiny droplets or solids in the air, mostly in the sub-micrometer radius range. Only about 0.1 gm^{-2} aerosol mass density in the atmospheric column attenuate incoming solar radiation by more than 10 Wm^{-2}. Their optical depth δ, related via $T = e^{-\delta}$ to transmission T is typically in the 0.05 to 1.0 range. The highest values are reached during desert dust storms, vegetation fires and in metropolitan areas. With the advent of the MODIS and MERIS sensors global monitoring with sufficient quality might have been started, since earlier at-

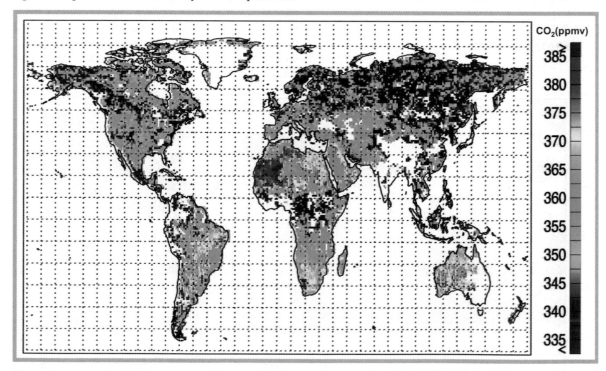

Figure 4-5: SCOs SCIA/WMFD Jul 2003; first examples of CO_2 column content retrieval from space using the SCIA-MACHY sensor onboard Envisat (Buchwitz et al., 2005). Shown is the column averaged mixing ratio of CO_2 as determined from the SCIAMACHY CO_2 and O_2 column measurements for cloud free pixels over land. For more details see web: *sciamachy*

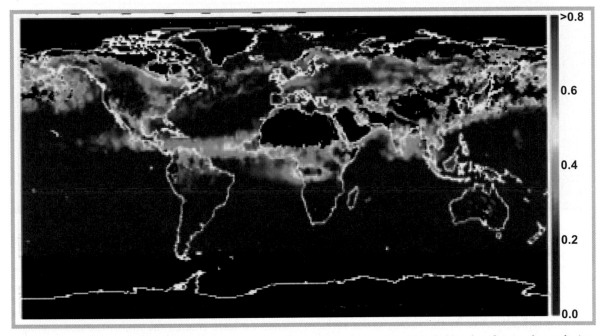

Figure 4-6: Aerosol optical depth observed from space with MODIS on Aqua for May 2003, taken from web: *modis* (see there for more details)

tempts using AVHRR visible and near infrared channels were restricted to ocean areas and were suffering from large error bars. Figure 4-6 convincingly shows the potential for global monitoring. We note basin-wide aerosol pollution over the North Atlantic and North Pacific, which might even have consequences for the average tracks of mid-latitude cyclones.

4.4.3
Other Trace Gas Column Contents

Whenever a trace gas shows medium to strong absorption in spectral windows of the atmosphere left by water vapour, CO_2 or ozone, it can be observed from space both in the solar and the terrestrial spectral region. This global mapping is now feasible at least for carbon monoxide (CO) and nitrogen dioxide (NO_2). As both are poisonous and the latter is especially strongly tied to surface and air traffic, it is of high interest to monitor them and to derive environmental protection options from such maps through modelling. The ESA image presented in figure 4-7 had a huge impact on the public in 2004. Industrialized nations dominate emissions of NO_2;

major air traffic routes become visible. In other words: we humans dominate the atmospheric NO_2 budget. Denoxification of power plants in central Europe as well as catalytic converters in personal cars are still insufficient environmental protection measures.

4.4.4
Early Detection of High Impact Cyclones over Sea

One of the most serious weather impacts on European societies remains winter-time cyclones causing wind damage, storm surges at the coasts, large-scale flooding, disruption of shipping, etc. Their forecast is hampered by the scarcity of surface observations and radiosonde profiles over the Atlantic and the inability to assimilate clouds and precipitation patterns into the starting fields of most weather forecasting models. Using all DMSP satellites, Klepp et al. (2003) could show that the detection of all precipitation fields and their strength is possible twice per day over the entire Atlantic, allowing warning of such rapidly moving and destructive cyclones, de-

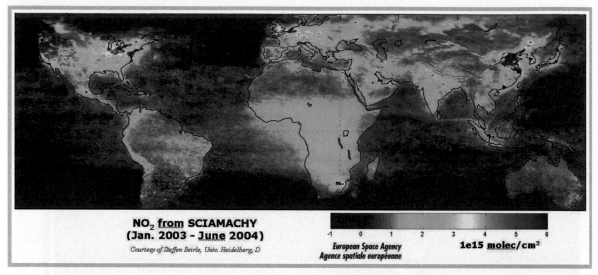

Figure 4-7: Nitrogen dioxide (NO_2) distribution derived from SCIAMACHY on Envisat; courtesy of ESA; for more details also visit web *no2_sciamachy*

veloping typically close to the strongest sea surface temperature gradients on our globe off the coast of North America near the Grand Banks. As demonstrated in figure 4-8, the severe storm later called Lothar could be detected very early over the Atlantic, characterized by nearly circular intense precipitation fields and moving across the Atlantic in just slightly more than one and a half days. One of the reasons for missing or underestimating such cyclones in weather forecasts is the elimination of ship observations by routine quality controls of starting fields in forecast models. The pressure tendencies are judged to be too large to be taken as real, and precipitation observations are not part of the starting fields.

4.5
First Studies of Environmental and Climate Change Using Satellite Data

Section 4.4 was devoted to average values and variability estimates of parameters whose distribution and variability was hitherto unknown and which became available with the satellite era. Here, after about 25 years of systematic observations, which is rather short for meaningful trend analyses of climate parameters, first such attempts will be presented.

Since we have entered the era of the anthropocene, trends may have accelerated and thus become visible rather early, justifying such trend analyses, if the detection is also followed by an attribution. The observed rapid average global warming during the recent decades will impact strongly on the cryosphere. Therefore, many trend analyses are related to snow, glaciers, sea ice, and sea level rise. The other fields with a possibility to detect anthropogenic trends are: stratospheric ozone depletion, air pollution decrease in OECD countries, and air pollution increase in industrializing countries like China and India.

4.5.1
The Cryosphere

Most cold areas are insufficiently monitored. Therefore, satellites brought a quantum leap in the observation of cryospheric parameters. For most parts of the cryosphere we already have first satellite based trend analyses.

Snow cover

Large proportions of the Northern Hemisphere are covered by snow in boreal winter. What impact did global warming have on this cover? The overall area with snow cover shrank by several percent during

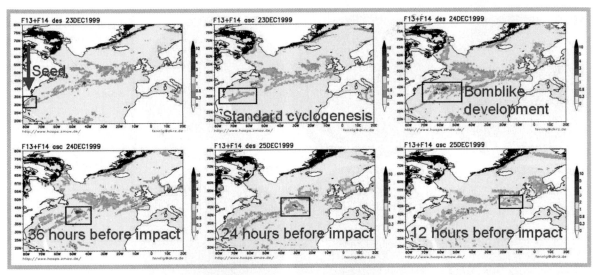

Figure 4-8: Precipitation fields of the severe storm Lothar derived from SSM/I on all available DMSP satellites. Please note the very rapid movement across the Atlantic; from Klepp et al. (2005)

recent decades, especially over North America, as derived from satellite data. But there are also areas, e.g., parts of northwestern Russia and Siberia, where the increase in precipitation has led to longer snow cover despite a mean warming also there. The number of days with snow cover has been reduced dramatically in western and central Europe during recent decades, as derived from in situ observations.

Glacier retreat

For only very few glaciers do detailed mass balance studies exist and for only a few percent of all glaciers, at most, are length changes registered. Therefore, the satellites with high spatial resolution radiometers, like the Landsat Thematic Mapper (TM) series, are an ideal means for glacier survey. As shown by Paul et al. (2004), Swiss glaciers have lost 22% of their area and 18% of their volume in just

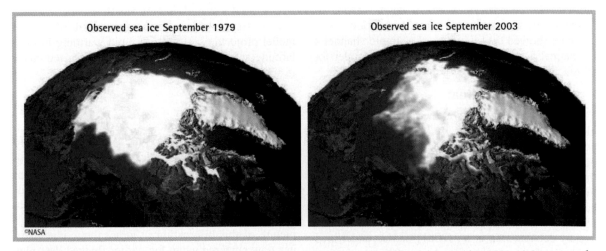

Figure 4-9: Observed northern hemisphere sea ice cover in September 1979 (left) and September 2003 (right); see web: *acia*

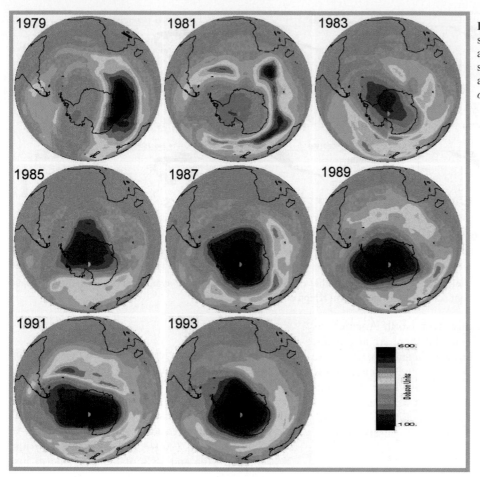

Figure 4-10: Stratospheric ozone depletion as observed from space since the late 1970s. For actual values see web: *ozone*

the last two decades; they call it down-wasting of glaciers (see also web: *glaciers*). Validation by in situ data showed the reliability of a simple channel 4 to channel 5 ratio for Landsat Thematic Mapper for all glaciers except those with massive debris cover.

Arctic sea ice retreat

The most sensitive part of the cryosphere with a global impact is sea ice. It largely insulates the ocean from exchange with the atmosphere; it is part of the positive ice albedo/temperature feedback of the water cycle; it is home of well adapted species. Since 1978, starting with the Nimbus 7 satellite, we have nearly global sea ice cover observations for all weather conditions. As shown already some years ago by Johannessen et al. (2001) and as underlined and extended now by the Arctic Climate Impact Assessment report (ACIA 2004) we are confronted

with a rapid loss both in area and thickness of Arctic sea ice (figure 4-9), confirming the result of climate model projections of a more rapid warming in high latitude areas, especially those losing sea ice cover in summer.

4.5.2
Ozone depletion

When Farman et al. (1985) published their discovery of the so-called ozone hole over Antarctica during austral spring, the satellite sensor TOMS (Total Ozone Monitoring System) had also observed it but because of unexplained strong variations the data remained unevaluated over the Antarctic. A full picture of this bifurcation in atmospheric chemistry has now emerged using the earlier flagged data plus data from many more satellites with ozone monitoring capability. As figure 4-10 demonstrates, stratospheric ozone depletion that is due to chlorine and bromine containing compounds, formed at sunrise after polar winter on the droplets or crystals of polar stratospheric clouds from about 12 to 20 km height, seems to have reached its maximum now; several years after the chemicals (CFCs and others) causing the chlorine and bromine containing compounds attacking the ozone molecule have reached their maximum concentration in the atmosphere. Satellite monitoring has clearly helped several times to enforce the Montreal Protocol of the Vienna Convention to Protect the (Stratospheric) Ozone Layer. For a near real-time overview of ozone column contents see also web: *ozone*.

4.5.3
Air Pollution Effects

Many trace gases in the atmosphere are part of global biogeochemical cycles that transform gases which result from the metabolism of plants, animals, microbes and industrialized societies into nutrients for plants. A typical example is the transformation of sulphur containing gases (often poisonous pollutants for animals and humans) like dimethylsulfide $(CH_3)_2S$, hydrogen sulphide (H_2S) and sulphur dioxide (SO_2) into sulfates and their condensation in the sunlit moist atmosphere into sulphuric acid particles (H_2SO_4) that exist as tiny solution droplets at high relative humidities. Because these particles form very efficient cloud condensation nuclei, the microphysical properties of clouds also depend on air pollution levels.

Using the so-called Pathfinder data set composed of intercalibrated NOAA-AVHRR spectral radiances since 1981 with daily global coverage, changes in cloud parameters can be detected if they exceed remaining calibration errors. We started such investigations in the area where air pollution effects have changed most strongly, in central Europe, because of the collapse of the Eastern Block and because of environmental protection measures in Germany and neighboring countries. The result (Krueger and Grassl 2002) was no surprise (see figure 4-11) as it confirmed earlier radiative transfer modeling results: drastically reduced emissions of SO_2 have led to lower cloud reflectance, by several percent, at about

1 µm wavelength. Fewer aerosol particles have caused fewer cloud droplets that backscatter radiation less strongly. When turning to an area with a dramatic increase in air pollution in recent years, namely China, we encountered a surprise (Krueger and Grassl 2004): also there cloud albedo and local planetary albedo shrank despite the strong increase in hazy conditions. The most plausible explanation is dominance of absorbing aerosol influence over the increased cloud droplet number effect. The reduction of local planetary albedo reaches values of about 6% over the Red Basin, a massive change in the radiation climate from the late 1980s to the late 1990s.

4.5.4
Northward Shift of Vegetation

With a prolonged vegetation period, as observed from Normalized Differences Vegetation Index time series in high northern latitudes since 1979, also a northward shift of vegetation zones might become visible. However, this long-term process needs an absence of local anthropogenic disturbance. This may only be the case at parts of the border between taiga and tundra.

4.6
The Transition to Earth Watch Missions for Successful Explorer Missions

As the European Space Agency has pointed out in its special report "The Living Planet" in 1997 (see ESA-SP 1227) successful experimental earth observation satellites, either Explorer Core or Explorer Opportunity missions, need to be continued in so-called Watch missions. These should allow monitoring on a global scale as a prerequisite for more intelligent decision making in our approach to sustainable development. While this transition took place already decades ago for meteorological satellites that observe at least the thermodynamic state of the atmosphere and of the surface, but also atmospheric composition in growing proportions, it did not make satisfactory progress in all other disciplines, besides oceanography, which is now taking the first steps. We are without any accepted observing strategy for operational systems in hydrology, agriculture, for-

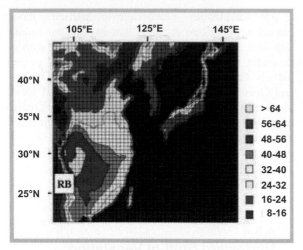

Figure 4-11: Cloud albedo decrease (in permille) from the late 1980s to the late 1990s over eastern China observed by AVHRR onboard the NOAA satellite series (RB = Red Basin); after Krueger and Grassl (2004)

estry, or regional development. Whatever steps have already been taken in these disciplines, they are often synergistic joint uses of meteorological operational satellites not built for this purpose.

However, recently we observe a real multiplication of activities through, firstly, the Global Monitoring for Environment and Security (GMES) initiative of both ESA and the European Commission (EC) and secondly, the initiative of the USA to form an intergovernmental Group on Earth Observation (GEO), which is now proposing a 10 year plan to create a Global Earth Observation System of Systems (GEOSS). In pilot projects GMES is trying to establish an end-to-end system for many potential services, e.g., Coast Watch, in Europe, which exploit already existing observing systems, with an emphasis on earth observation satellites.

4.6.1
Earth Observing Systems Development

Operational observing systems always had a research network as a forerunner. This will remain so. However, research observation networks are only transformed into operational ones if they hold promise of new applications with a benefit to cost ratio far above unity. Concerning such new applications

we see a new mode emerging: implementation of environmental conventions of the United Nations that are legally binding for the parties to the convention. We will soon see a push towards the monitoring of sources and sinks of many trace gases, as they are regulated in the Montreal and Kyoto Protocols. The main pillar of such monitoring is availability of satellite sensor data sets.

But also the detection of some useful predictability in the Earth system stimulates monitoring via both satellite and in situ observation. I will give examples for both applications below.

Exploiting Predictability

When the Tropical Ocean/Global Atmosphere (TOGA) project led in the early 1990s to seasonal climate anomaly predictions in areas affected by the El Niño Southern Oscillation phenomenon, the ingredients of this success, foremost coupled ocean - atmosphere models and the TOGA-TAO array of moored buoys, but also satellite altimeter data for ocean topography, had to be improved or continued. The TOGA-TAO (Tropical Atmosphere Ocean) array is now operational; radar altimeters for ocean topography seem to become operational and coupled models already exploit the extra-tropical Lagrangian drifters of the so-called Argo System (a research network), measuring profiles of temperature and in parts salinity in the upper ocean worldwide. There is hope that seasonal climate anomaly predictions will be useful also outside areas affected by the El Niño Southern Oscillation. If soil moisture estimates become available from satellites, the additional predictability of weather and climate anomalies on intraseasonal and seasonal time scales will be exploited, and if successful, the monitoring of soil moisture will probably become routine for even more reliable climate anomaly predictions.

Monitoring Atmospheric Trace Gas Composition

Knowledge of, for example, the three-dimensional ozone and carbon dioxide distribution would lead to several new applications. The ozone information is useful for "chemical" weather prediction, climate anomaly predictions and long-term ozone monitoring; the CO_2 information is urgently needed to derive the source and sink distribution of carbon on a global scale. This information is a must for the proper monitoring of the Kyoto Protocol (which became international law on 16 February 2005) and its envisaged enforcement, which has to be negotiated starting in 2005 for the second commitment period after 2012. Besides these two trace gas distributions, many others are also of high relevance for forecasting of chemical weather, evaluation of least-cost environmental protection measures, and climate-chemistry interaction studies. Therefore, at least carbon monoxide (CO) and nitrogen dioxide (NO_2) column content monitoring should become routine soon.

4.6.2
Operational Environmental Satellites

Urgent need for a certain type of information stimulates the supply of this information, as can be seen in meteorology. With the extension of weather forecasting timescales to about a week, ocean mixed layer as well as soil moisture information became necessary and are already being assimilated into models at a few numerical weather prediction centers. The possibility to predict climate anomalies with useful skill has led to new routine monitoring of upper ocean characteristics in the tropical Pacific and requests to have it on a global scale, as now delivered by the research network Argo. Also ocean topography is needed in this context and attempts to bring satellite altimeters with TOPEX-Poseidon accuracy onto operational meteorological satellites are under way.

As global weather forecasts have reached remarkable skill on time scales up to about five days, chemical weather, e.g., photosmog episodes, could also be forecast. Forecast models including atmospheric chemistry are therefore emerging at a few research centers. They need at least trace gas column contents as part of their starting fields. These are in principle available now from experimental satellites like Envisat, but they have to become operationally available before forecasting centers can get their research attempts funded for routine implementation. The funding would also depend on the infrastructure within services that are main users and that communicate with end users. Most services outside meteorology do not have the necessary infrastructure

and networks to make near optimum use of such new information. Therefore, the GMES initiative's value for Europe cannot be overestimated.

As Europe already has a leading operational organization in forecasting, the European Centre for Medium-range Weather Forecasts (ECMWF), and an operational agency for the exploitation of meteorological satellites, EUMETSAT, it is wise to build on their expertise. Even more so, as the basis for many new applications is the improvement of the meteorological starting field and the coupled models used for forecasting. In other words, we need the development of existing operational meteorological satellites into environmental satellites, and of weather forecasting centers into environmental forecasting and prediction centers. This implies at least the networking of services; however, their unification under at least a national umbrella may be even better.

4.6.3
European Contribution and Leadership in Earth Observation

With ERS-1, ERS-2, Envisat and Meteosat Second Generation the European contribution to earth observation has finally reached a technical level equivalent to the contribution of the United States of America. With the GMES initiative ESA and EU have in parts stimulated GEO and its strategy, called GEOSS. With the climate policy of the EU, e.g., Europe-wide emissions trading as a forerunner to global emissions trading under the Kyoto Protocol, the European Union has shown leadership in environmental policy making. Can the EU keep this leading position? Yes, but only if it also shows leadership in earth observation from experimental satellites via Earth Watch missions to operational environmental satellites. What does this mean?

- Firstly, implementation of Earth Explorer Core and Earth Explorer Opportunity missions on schedule, and rapid strengthening of the Earth Observation Envelope Programme of ESA;
- secondly, rapid decision for Earth Watch missions building on GMES networks, thereby using the EU's funding basis and ESA's expertise in satellite technology development;
- thirdly, establishment of an EU network of services in hydrology, agriculture, forestry and re-

gional planning with the aim to optimally use earth observation space data, and cooperation of these networks with the meteorological and oceanographic services;

- fourthly, continuous deployment of a polar and geostationary component of operational environmental satellites.

4.7
Conclusion

Homo sapiens, as the dominating species, has to use its progress in technology for sustainable development. The prerequisite is a basic understanding of the functioning of the Earth system. This understanding needs global continuous data, i.e., it needs satellite remote sensing or earth observation from space. It also needs Earth system models, whose development stage is just a measure of the sophistication of validation data sets. Thus satellite remote sensing is at the heart of progress in understanding and therefore acts not only as an early warning system for mankind but also offers problem solutions through better understanding.

4.7.1
Proper Mix of Explorative and Operational Missions

Progress in better understanding needs not only continuous earth observation but also new observation techniques. Therefore, the joint further development of explorative and operational missions is a must, as exemplified by the global meteorological community. To achieve it is, however, also a question of adapted infrastructure. At present we are certainly not close to an optimum. While a budget crisis always hits earth observation harder than, for example, subsidies for old-fashioned industries, just the opposite would be a future oriented policy making. Only two fully functioning earth observation systems exist for one discipline, namely meteorology: NASA + NOAA in the USA and ESA + EUMETSAT in Europe. We urgently need strategies to, firstly, improve the mentioned systems, secondly, to enlarge their disciplinary realm (environmental satellites), thirdly, to look at budget costs from a macro-economic perspective and not from a micro-

economic one. Why do nearly all countries shrink their meteorological service when the benefit to cost ratio is well above five, sometimes even above 10? Hopefully, the GEOSS initiative will cause a trend reversal.

In all disciplines and fields, e.g., volcanology, seismology, agriculture, and forestry, we need a proper mix between explorative and operational space missions, together with an adequate ground segment. This is often disregarded. It is not only important but also costly, if the proper harvest from space missions is to be secured. It seems easier, albeit still difficult, to launch experimental satellites than operational ones. A European strategy for a balanced approach seems to be emerging, but we still do not have a commitment by the EU to support operational environmental satellites and to deliver data freely to developing nations.

4.7.2
Global Data Sets as Basis for Environmental Policies

As mentioned above, the skill of climate or Earth system models depends on data used either for evaluation and subsequent improvement or the accuracy of starting fields for the models' forecasts or scenario calculations. Such global data sets will increasingly come from two new sources: earth observation satellites and proxies (from earth history). The forecasts or scenarios are the basis for environmental policy making. Two examples should clarify what is meant. Firstly, the implementation of environmental conventions of the United Nations has to be monitored in order to judge the compliance with international law of the parties to the conventions. Secondly, the impact of a variety of emission reduction measures has to be known in order to choose the one with highest benefit to cost ratio. In the first case, the shining example will become the correct implementation of the Kyoto Protocol to the Framework Convention on Climate Change (UNFCCC). Its monitoring needs the full information from observations (global atmosphere watch stations, carbon dioxide flux towers, satellites) assimilated into state-of-the-art circulation models that give the source and sink distribution for carbon. The results, judged by international science groups and then intergovernmental bodies, will undergird decisions about whether sanctions have to be applied. In the second case the impact of, for example, car emission standards on cloud properties, using circulation models validated with global earth observation data sets, will help to decide on best practice. Both examples are still at the forefront of science but illustrate what might become implementation reality soon, and where the fundament is earth observation from space.

4.7.3
Scenarios of Earth System Development

If humankind makes progress in understanding the Earth system, the management of that system, at least in a small range of its development, could be started, going well beyond first attempts like the Montreal and Kyoto Protocols. The prerequisite for such management steps is, however, the use of validated Earth system models based on global data sets predominantly from satellite remote sensing. The model output, namely scenarios for different human behaviour, will still not be able to show the corridor of probable human development, as the feedbacks of the socio-economic systems cannot be handled yet in these models. But these scenarios can point to appropriate measures to be taken to avoid dangerous climate change rates or continued or even accelerated biodiversity loss. In order to integrate socioeconomic feedbacks into the evolving Earth system models, social sciences have to take advantage of earth observation from space as well. I would like to call for research programmes to derive social science data sets from high resolution satellite sensor series, like Landsat TM or SARs on ERS-1/2 and Envisat.

4.8
Final Remark

As a European I hope that European integration will set the stage for a major contribution to GEOSS, whose key element is an end-to-end space component that ranges from innovative, more technology driven sensors via Earth Explorer Core and Opportunity missions to Earth Watch missions, also using the heritage from ERS-1/2 and Envisat, to operational satellites for continuous global environmental monitoring. This goal should also shape our infrastructure at the ground.

References

Arctic Climate Impact Assessment (2004) Impacts of a Warming Arctic. Cambridge, pp 1-144

Buchwitz, M, de Beek R, Noel S, Burrows JP, Bovensmann H, Bremer H, Bergamaschi P, Körner S, Heimann M (2005) Carbon monoxide, methane and carbon dioxide columns retrieved from Sciamachy by WFM-DOAS: Year 2003 initial data set. Atmos. Chem. Phys. (in press)

Johannessen OM, Skalina EV, Miles MW (2002) Satellite evidence for an Arctic sea ice cover in transformation. Science 286, pp. 1937-1939

Klepp CP, Bakan S, Grassl H (2005) Missing North Atlantic cyclonic precipitation in the ECMWF model detected through HOAPS II". Met. Z. (in press)

Klepp CP, Bakan S, Grassl H (2003) Improvements of satellite derived cyclonic rainfall over the North Atlantic. Climate 16, pp. 657-669

Krueger O, Grassl H (2002) The indirect aerosol effect over Europe. Geophysical Res. Letters, 29, Art. No. 1925

Krueger O, Grassl H (2004) Albedo reduction by absorbing aerosols over China. Geophysical Res. Letters 31, L2108-L2112.

Rossow W, Duenas BN (2004) The International Satellite Cloud Climatology Project (ISCCP). Web site – An online resource for research. Bull. Am. Met. Soc., 85, pp. 167-172

Wulfmeyer V, Bauer H, Di Girolamo P, Serio C (2005) Comparison of active and passive remote sensing from space: an analysis based on the simulated performance of IASI and space borne differential absorption lidar. Remote Sens. Environ. 2005, in press

Web References
gewex: http://srb-swlw.larc.nasa.gov/
glaciers:
  http://www.agu.org/pubs/crossref/2004.../2004GL020816.shtml)
hoaps: http://www.hoaps.org
wrcp: http://isccp.giss.nasa.gov/

srb: http://srb-swlw.larc.nasa.gov/
flux: http://www.hoaps.zmaw.de
sciamachi: http://www.iup.physik.uni-bremen.de/
sciamachy/ NIR_NADIR_WFM_DOAS/

no2_sciamachy: http://www.iup.physik.uni-
bremen.de:8083/doas/no2_icartt.htm

modis: http://modis-atmos.gsfc.nasa.gov
acia: http://amap.no/acia
ozone: http://www.temis.nl/protocols/03global.html

5 Weather Observations from Space

by Tillmann Mohr and Johannes Schmetz

Weather systems are large scale phenomena, therefore space is an ideal place to observe them. From the high altitudes above the Earth, satellites can provide images and other measurements that provide relevant information on current weather as well as data important for weather forecasting.

5.1
The Need to Observe the Changing Weather

The weather and its changes affect most people in their daily life and there is hardly any other specific topic that attracts the continuous attention of people in a comparable way. Most media feature a daily presentation of the actual weather and its expected change in the coming days. The reason for this interest is obvious: we live with the weather in the earth's atmosphere and more often than not the weather undergoes changes within short periods of time affecting us in one way or another. Weather ultimately determines what crops we grow, how we dress, what outdoor activity we pursue, whether and how we can travel, and last but not least how we feel.

Going back to first principles the weather is a result of the geographically uneven radiative energy input that the earth receives from the sun. High latitudes, closer to the poles, receive little or no solar radiative energy, whereas the lower latitudes such as the tropics and sub-tropics receive a lot of energy from the sun. The atmosphere, as a complex fluid system, tries to redistribute the energy from the areas of gain in the tropics and subtropics to the higher latitudes. This is accomplished by chaotic motions of the atmosphere which result in a net energy flux from lower to higher latitudes. The chaos and complexity of atmospheric motion exerts itself as physical weather phenomena ranging from very small scales,

such as the formation of clouds with droplets of radii smaller than 10 μm to low pressure systems with scales up to 2,000 km, which in tropical regions may appear as hurricanes. In between, in terms of spatial scale, there are many other processes and phenomena such as devastating tornadoes or thunderstorms. Obviously global satellite observations are also well-suited for climate observations. Those aspects of satellite observation are covered in chapter 4, "Climate and Environment".

5.2
The Global Meteorological Satellite Observing System

Until 1960 weather observations around the world were entirely provided by ground-based observational systems. This changed with the launch of the first U.S. meteorological satellite TIROS (Television and Infrared Observation Satellite) from Cape Canaveral, Florida, on 1 April 1960. TIROS was an experimental satellite in a polar orbit and provided for the first time regularly pictures of Earth's cloud and weather systems over large areas. Europe commenced to contribute to the global space borne observing system with the launch of Meteosat-1 on 23 November 1977. This satellite was the first in geostationary orbit to carry a water vapor channel in the 6.3 μm band. Since then the capabilities of operational meteorological satellites and of remote sensing satellites that can be used to observe parameters relevant to weather have grown considerably.

Today we look back on more than 40 years of meteorological satellites. They have proven to be the best way to observe the weather on a large scale, and recent analyses at the European Centre for Medium Weather Forecasts (ECMWF) even prove that the satellite information is the most important part of the global observing system for modern numerical weather prediction. Typically, operational meteorology utilizes two types of satellites to provide the required information. The two types of satellites differ considerably in terms of their orbit height:

Low Earth Orbiting Satellites fly at relatively low altitudes of around 800 kilometers above Earth, mostly in polar (sun-synchronous) orbits, and can provide information with high spatial resolution. The whole surface of the earth can be observed twice a day. More than one polar satellite with different equatorial crossing times is required in order to attain more frequent observations. Imaging frequency increases with latitude and corresponds at the poles exactly to the orbiting period (about 110 minutes).

Geostationary Satellites, flying in the equatorial plane at an altitude of about 36,000 kilometers above Earth, have the same revolution time as the earth itself and therefore always view the same area. They can perform frequent imaging, which, in animated mode, depicts the ever-changing atmospheric processes. The image repeat cycle is typically 30 minutes; however, Meteosat Second Generation, with Meteosat-8 as the first satellite of the series, scans the full Earth disk every 15 minutes. The disadvantage of the high altitude is that it limits spatial resolution and precludes the use of active instruments like radars.

Figure 5-1 shows the space based component of the Global Observing System of the World Meteoro-

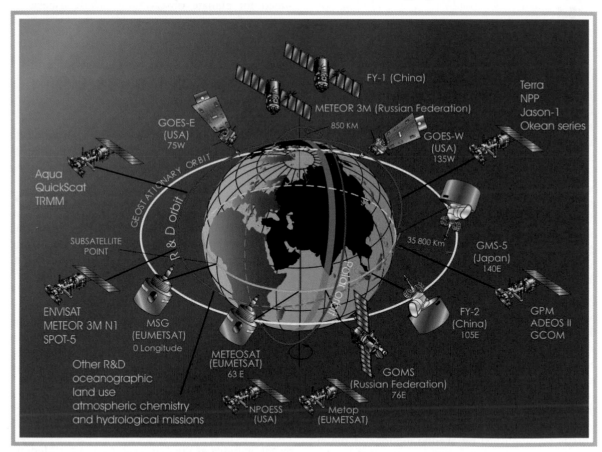

Figure 5-1: Space based part of the WMO's Global Observing System as of 2003, showing the typical orbits of operational meteorological satellites, namely the geostationary orbit and the solar-synchronous polar orbit. In addition, the low-inclination orbits of research satellites such as the Tropical Rainfall Monitoring Mission and other research satellites are indicated (WMO, 2003)

logical Organization's World Weather Watch program, consisting of polar and geostationary meteorological satellites that are operated by different countries and organizations. Figure 5-1 also includes examples of existing and planned research satellites contributing to the Global Observing System. International efforts are made such that the global satellite observing system is coordinated, ensuring that observations are complementary and meet the needs for both weather forecasting and climate observations.

A fundamental question that needs to be answered before we discuss geophysical parameters observed from satellites reads: *What do satellites observe?* When one refers to meteorological measurements one commonly has a specific geophysical quantity in mind such as temperature, humidity or the wind field. While those quantities and many others can be derived from satellite measurements, it is important to understand that satellites never measure such geophysical quantities directly. Satellite measurements utilize the principles of remote sensing where

the basic measured quantities are radiation fields, typically spectral radiances, or the radiant intensity into one particular direction. Figure 5-2 shows a typical radiance spectrum at the top of the atmosphere in the thermal infrared spectrum and the spectral channels of Meteosat-8. When the radiances carry information on the geophysical properties of the atmosphere, it is possible to infer, with adequate retrieval models, the geophysical parameters. This usually necessitates radiative forward calculations, or the simulation of the observed satellite radiances. In simple words, the simulated radiances help to interpret and utilize the measured radiances in order to infer geophysical parameters. Another way is to establish the relationship between observed radiance field and geophysical parameter in a statistical fashion: the geophysical parameters are inferred through empirical schemes that have been established on the basis of co-located and simultaneous observations of the radiance field and in-situ measurements of the relevant geophysical quantity.

5.2.1
Retrieval of Geophysical Parameters from Satellites

The potential for satellites to observe the Earth's atmosphere and surface is remarkable, with measurements that address many observational requirements through a variety of geophysical parameters. As mentioned above, the first one in 1960 was simple cloud imagery. Nowadays information on clouds encompasses many detailed quantitative parameters describing cloud fields in more detail. With regard to the atmosphere satellites measure the temperature and humidity profiles from the surface to over 40 km altitude, as well as winds, precipitation, aerosols, and the concentration of ozone and other gases. Over the oceans satellites measure surface temperature and winds, sea levels and waves, ice cover and ocean color. Over land surfaces they observe surface temperature, soil moisture, vegetation and snow cover, floods, forest fires and other parameters. Many of the weather relevant parameters are discussed below in more detail.

Most of the satellite derived products provide better data coverage than any alternative. In particular over the oceans, which cover about 70% of the planet,

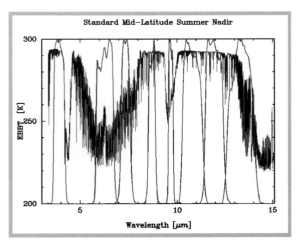

Figure 5-2: The blue curve shows the thermal terrestrial spectrum between 4 μm and 15 μm in terms of equivalent blackbody brightness temperatures (EBBT) at the top of the atmosphere simulated for a clear sky standard mid-latitude summer atmosphere and nadir view. The red curves depict the spectral response functions of the SEVIRI instrument on Meteosat-8. The observed bands are sensitive to water vapour, ozone, carbon dioxide respectively, or are little affected by gaseous absorption and hence measure the higher brightness temperatures of the surface

there are very few surface-based observations. As global data sets are essential for many applications, satellites offer a viable way to obtain the data. Concerning accuracy satellites are competitive to surface based and other in-situ measurements (such as from aircraft) for many parameters. One should, however recognize that satellite and in-situ instruments measure at different spatial and temporal scales. For instance a sensible comparison between a satellite observed cloud field and a surface observation usually requires some time integration of the surface observation and also needs to account for the different aspects of the observation, in our example the viewing angle and related distortions of clouds, and obscured cloud layers.

5.3
Observing the Atmosphere

Satellite observations are used to infer various geophysical parameters describing the atmospheric state. The following sections provide an overview of main applications such as cloud, temperature and humidity, precipitation, aerosol and trace gases. Emphasis is given to observations of the wind field.

5.3.1
Clouds

Information derived from cloud imagery such as cloud top height, cloud cover, cloud type, optical thickness, cloud phase (ice or water) and cloud microphysics is one of the key products of meteorological satellites. Cloud images observed in the visible and infrared parts of the spectrum are one element of the routine forecasts presented on television. The cloud images give a concise synoptic view of the weather systems and, animated as movie loops, they reveal the restless nature of the atmosphere on various spatial scales.

Cloud imagery is provided by both polar orbiting and geostationary meteorological satellites; figure 5-3 shows an example from Meteosat-8 (Schmetz et al. 2002), the first of a new generation of European geostationary meteorological satellites, with as many as twelve spectral channels taking images of the

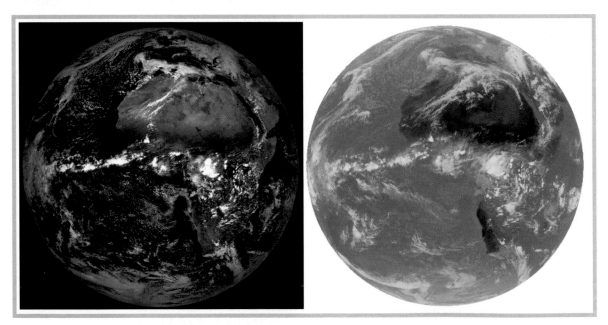

Figure 5-3: Two satellite images taken from Meteosat-8 on 29 March 2004 at 1200 UT. The left panel shows the earth observed with a channel at 0.6 μm in the solar spectrum. Clouds appear bright because of their high reflectance of solar radiation, and ocean surfaces dark. The right panel shows the same weather situation observed with the thermal infrared channel at 10.8 μm; the image has been inverted such that cold clouds appear bright and warmer surfaces appear dark

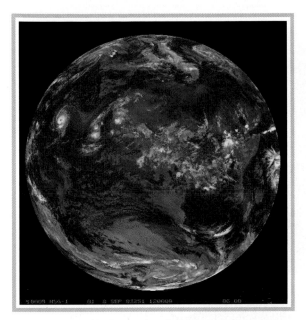

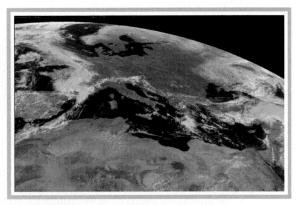

Figure 5-5: Part of a Meteosat-8 image taken on 2 April 2004. In this RGB image channel 1.6 μm has been assigned to the color red, 0.8 μm to green and 0.6 to blue. Ice and snow appear in blue because ice reflects significantly less at 1.6 μm in comparison to water clouds. Thus high level ice clouds, snow surfaces and the ice cover over the Baltic Sea are well depicted

Figure 5-4: In this figure six of the twelve channels from Meteosat-8 have been used to highlight high cloud and water vapor features. The image is based on the RGB technique with the following assignments: red to the channel difference 6.2 minus 7.3 μm; blue to 3.9 minus 10.8 μm; green to 1.6 minus 0.6 μm. High clouds appear orange and yellow, whereby yellow corresponds to smaller cloud particles. High level water vapour features appear as a purple veil

earth every 15 minutes. The left panel of figure 5-3 shows the 'full disk' observation at 0.6 μm, which is in the solar part of the spectrum, hence the satellite receives reflected solar light and therefore clouds and desert surfaces appear bright. The brightness of the cloud is also a measure of its optical thickness and with multispectral approaches the optical thickness can be estimated. The right panel shows the corresponding observation at 10.8 μm in the thermal infrared part of the electromagnetic spectrum. The image has been inverted; high radiative energy levels or high temperatures appear dark and low temperatures appear bright. Figure 5-4 depicts what can be achieved with advanced multispectral imaging where channel differences are cleverly combined to highlight certain aspects: channel differences are assigned to the colors red, green and blue in order to construct a so-called RGB image. In this manner

high cloud properties, including cloud particle size, and high-level water vapor features are highlighted.

Many other combinations of channels are possible in order to emphasize the feature or geophysical phenomenon of interest. Interesting and important is the application of observations around 3.7-3.9 μm, which are especially useful for the identification of forest fires. Figure 5-5 is an example for the utility of a channel at 1.6 μm where ice reflects little solar radiation in comparison to water clouds. The RGB image in figure 5-5 is based on three channels at 1.6 μm (assigned to red), 0.6 μm (green) and 0.8 μm (blue) in the solar spectrum. Ice clouds and surfaces with snow and ice covered sea (in the northern Baltic Sea) appear blue.

5.3.2
Temperature and Humidity

The basic thermodynamic weather quantities are temperature and humidity. Their three-dimensional structures are key parameters in numerical weather forecasting and other applications. Over land areas, from islands and from ships profiles of temperature and humidity are available with good accuracy from radiosonde data and other ground-based observations. However, over most of the world data coverage by conventional radiosonde measurements and

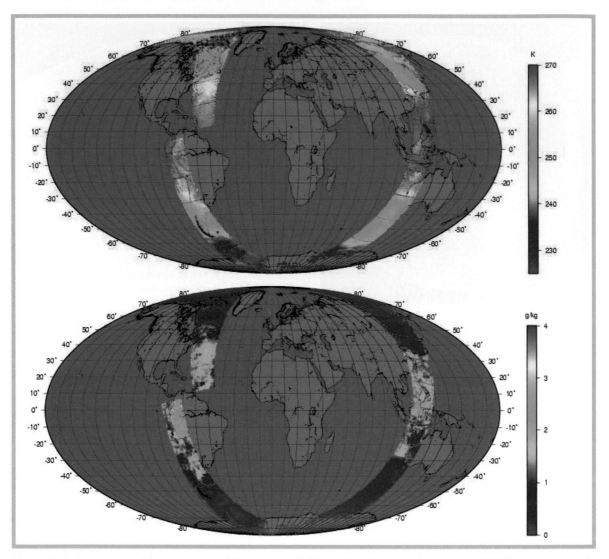

Figure 5-6: Temperature (above) and humidity (below) fields at 500 hPa derived from the ATOVS instrument suite aboard the polar orbiting NOAA-16 satellite. The fields were derived with the EUMETSAT prototype software for the EUMETSAT Polar System (EPS/Metop) (Klaes et al. 2004)

other ground-based remote sensing techniques is far from adequate. In these areas satellite data offer the only realistic possibility for filling the data void areas.

Satellite sounding instruments generally employ one of two different types of viewing geometry: scanning viewing or limb viewing. The scanning instruments look down at the earth at varying angles and measure radiances in a small solid angle. Spatial coverage is obtained by scanning (cross-track or conical). The satellite measures radiances in different channels (wavelengths). The different spectral radiances carry information on temperature and humidity at different altitudes. The simultaneous radiance observations at different wavelengths allow the retrieval of the vertical and horizontal structure of the temperature and humidity either through so-

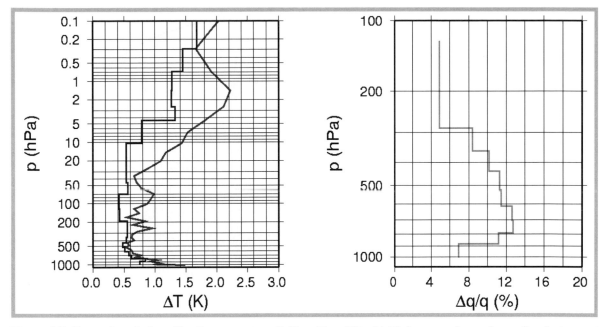

Figure 5-7: Error of vertical profiles for temperature (left) and humidity (right) from an advanced sounding instrument (IASI) on board EPS/Metop. The root-mean-square error has been simulated for a clear atmosphere including noise (Schlüssel and Goldberg 2002). The red curve gives errors for individual levels; the blue curve is the error for averages over layers. The goal of an accuracy better than 1 K per 1 km layer is achieved

called physical retrievals, explicitly considering physical principles, or statistical retrievals.

While satellites do have the advantage of providing truly global coverage they have the disadvantage of providing limited vertical resolution. This is inherent to the physics. Only with very high spectral resolution and a large number of channels (several thousand in the thermal infrared) can one significantly improve vertical resolution. This is achieved by modern technology such as NASA's AIRS instrument on the Aqua satellite (Aumann et al. 2003) or with IASI (Infrared Atmospheric Sounding Interferometer) (Schlüssel and Goldberg 2002) to be flown on EUMETSAT's operational polar orbiting satellites called Metop.

Sounding in the thermal infrared part of the spectrum (4-15 μm) has a disadvantage because clouds obscure the atmosphere below the cloud and prohibit retrieval of temperature and humidity there. The use of microwave radiances overcomes this problem because the microwave radiances are much less affected by clouds and allow, in principle, a retrieval of temperature and humidity in cloudy atmospheres.

An example of temperature and humidity fields measured from the polar orbiting satellite NOAA-16 during an orbit on 20 March 2002 is shown in figure 5-6.

Figure 5-7 demonstrates the good performance that can be obtained with an instrument like the Infrared Atmospheric Sounding Interferometer (IASI): shown is the vertical profile of the RMS error in temperature (left) and humidity (right). With its more than 8000 spectral channels, IASI allows for an unprecedented accuracy of vertical profiles, exceeding the required accuracy of 1K per 1 km layer of the troposphere and lower stratosphere.

It is interesting that many of today's Numerical Weather Prediction (NWP) models no longer use derived temperature and humidity profiles. Instead of converting the satellite radiances into temperature and humidity, the NWP model data are converted into simulated radiances, which can be done without ambiguity, though with some error. The simulated radiances are compared with the satellite observations which are also in the radiance domain. The

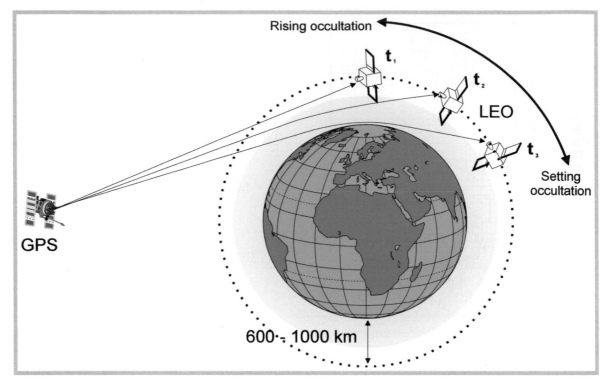

Figure 5-8: Geometry of a Radio Occultation measurement. The principle is to receive signals from GPS satellites orbiting at altitudes around 20,000 km by polar orbiting satellites at lower altitudes (800 km). A setting occultation starts at time t_1 and ends at time t_3, when the signal is cut off by the limb of the earth. In a rising occultation the signal is acquired starting at the time t_3

observed radiances are then assimilated into the NWP model (see also section 5.4 below on numerical weather prediction).

A promising way to obtain temperature or humidity profiles is through radio-occultation methods (Luntama and Wilson 2003). Radio Occultation measurement is based on an accurate measurement of the phase of a carrier wave transmitted by a GPS navigation satellite and received by a Radio Occultation receiver onboard a satellite in LEO (Low Earth Orbit). During an occultation measurement the signal propagation path moves through the atmosphere from a height of several hundreds of kilometers down to the surface due to the motion of the transmitting and receiving satellites (see figure 5-8). The movement of the ray path can be from the top of the atmosphere towards the surface (setting occultation) or from the surface towards the top of the atmosphere (rising occultation). The length of the signal

propagation path changes during the occultation due to the exponential vertical profile of the refractive index of the atmosphere.

This change can be seen in the measured carrier phase as a delay in comparison to the phase value calculated based on the accurately known geometrical distance. The change rate of the carrier delay provides data for the calculation of the vertical profile atmosphere index of refraction. The pressure, temperature, and humidity profiles can be retrieved from the refractive index profile using the equation of hydrostatic atmospheric equilibrium.

5.3.3
Atmospheric Winds

Atmospheric winds are probably the most important data for weather forecasting, notably for Numerical Weather Prediction (NWP) models. Why are the

Figure 5-9: Atmospheric Motion Vectors (AMVs) derived from Meteosat-8 images by tracking cloud features in the 10.8 µm channel on 23 February 2003. Colors indicate altitude, blue: low, green: mid level, yellow: high

wind data principally more important than temperature and humidity? This is because wind information contains, with the exception of tropical latitudes, information on the atmospheric mass field and therefore on temperature. Winds are especially important in the tropics, where this geostrophic wind relationship breaks down and winds cannot be inferred from temperature gradients. With radiosondes atmospheric winds are measured routinely by tracking the radiosonde from the ground. Other observations, like vertical sounding from the surface with profilers and in-situ measurements with specially equipped commercial aircraft, are also an essential ingredient of global wind observations. The disadvantage of those measurements is obvious; they are limited in terms of global coverage. Radiosondes are sparse and aircraft observe only along flight routes at a single level.

A well-known and extensively used wind product from satellite observations are the so-called Atmospheric Motion Vectors (AMVs) (Velden et al. 1997; Schmetz et al. 1993) They have been generated on an operational basis by all of the operators of geostationary meteorological satellites (see figure 5-9) for more than two decades. AMVs have clearly contributed towards improvements in NWP results with biggest impacts in the data-sparse areas of the tropics and the southern hemisphere. The method of deriving a wind vector is quite simple: cloud or water vapor features are tracked from one image to the next, thereby measuring a displacement vector. Assuming that the displacement vector represents a wind and assigning the vector to the correct height, one obtains a wind vector. The basic technique can be applied to clouds in both infrared and visible satellite imagery, as well as to other tracers, such as the patterns of atmospheric water vapor. It is anticipated that the AMVs and related products will continue to be a main source of satellite wind observations for many years to come. There are obvious limitations of AMVs. For instance, the wind blows through some stationary clouds generated by mountains, thus this type of cloud is not at all a passive tracer. Therefore correct target selection is important. Very difficult is also the height assignment of the displacement vectors, which often necessitates multispectral techniques (Nieman et al. 1993). Overall quality control is essential in order to produce a good AMV product (Holmlund 1998).

The geostationary satellites (orbiting over the equator) cannot be used to track features beyond latitudes of about 60°. Furthermore, winds can only be obtained at a few levels in the atmosphere, where suitable tracers exist. In short, AMVs from geostationary satellites do not meet the need for a full three-dimensional atmospheric wind field on a global basis, though they constitute an important data source.

More recently AMVs have been extended to high latitudes and polar regions using images from NASA polar orbiting research satellites (Key et al. 2003). Displacements of features, especially water vapor structures, can be determined at high latitudes because the separation of the images (about 110 minutes) is just short enough to allow for this. Recent impact studies of AMVs from MODIS instruments do show a clearly positive impact on NWP (Bormann and Thepaut 2004) and a continuous capability with operational missions would be desirable. Figure 5-10 provides an example of AMVs derived over

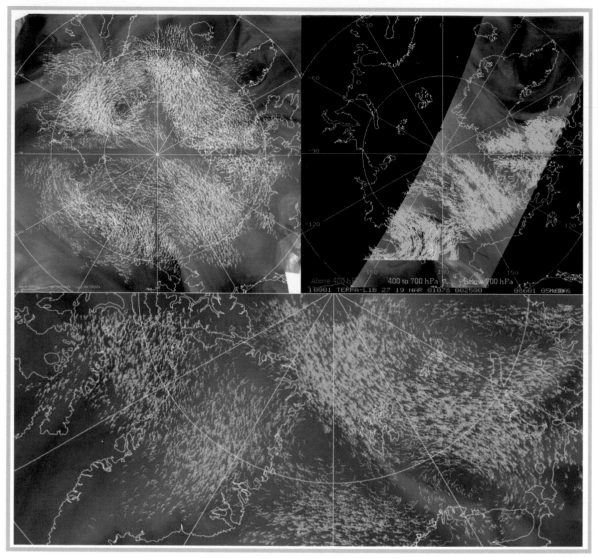

Figure 5-10: Atmospheric Motion Vectors over the arctic region from tracking of water vapour features detected with a water vapour channel of the MODIS instrument on polar orbiting satellites (Key et al. 2003). Top left: composite of several satellite overpasses; top right: single overpass; bottom: composite with different height categories (cyan 700-400 hPa, magenta above 400 hPa)

arctic regions from polar orbiting imaging observations.

5.3.4
Doppler Wind Lidar

A novel mission is wind determination with a spaceborne lidar (LIght Detection And Ranging). The European Atmospheric Dynamics Mission (ADM) sensor of the European Space Agency (ESA) will measure wind profiles using the Doppler Wind Lidar (DWL) principle. The lidar on the satellite emits a laser pulse towards the atmosphere and receives a signal frequency which is Doppler-shifted from the emitted laser light due to the spacecraft motion, earth rotation and wind velocity. The pulsed laser

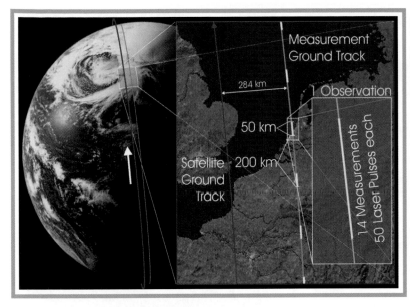

Figure 5-11: Winds from Doppler Wind Lidar: The lidar emits a laser pulse into the atmosphere and receives a signal frequency which is Doppler-shifted from the emitted laser light due to spacecraft motion, earth rotation and wind velocity. The lidar measures the wind projection along the laser line-of-sight which views at a viewing angle of 35° off the nadir or ground track. With an orbit height of 400 km this corresponds to an off-nadir distance of 284 km. Along the satellite track 14 measurements with 50 laser pulses each are averaged. This corresponds to a horizontal mean over 50 km (Ingmann 2004 personal communication and Ingmann et al. 2002). The mission is being developed by ESA

operates at 0.35 μm wavelength, utilizing both Rayleigh scattering from molecules and Mie scattering from thin clouds and aerosol particles. Measurement of the residual Doppler shift from successive levels in the atmosphere provides the vertical wind profiles. The lidar measures the wind component along the laser line-of-sight, which is 284 km off the nadir or ground track (see figure 5-11).

The orbit height of ADM is 400 km and the target date for launch is 2007. About 3,000 globally distributed wind profiles will be obtained per day, above thick clouds or down to the surface in clear air. The measurements are averaged over 50 km and the separation distance of measurements is typically 200 km along the satellite track. Improved knowledge of the global wind field will be crucial to many aspects of atmospheric research and notably weather prediction. In addition, a DWL will provide information on cloud top heights and on the vertical distribution of cloud and aerosol properties.

5.3.5
Ocean-Surface Winds

Scatterometers onboard satellites use radars to measure surface wind vectors over the ocean (see figure 5-12). The technique is based on the processing of microwave pulses backscattered by gravity-capillary waves on the ocean surface. Capillary waves are excited by the surface wind and hence the amount of back-scattered energy is related to wind speed. The backscatter is converted to wind speeds through a regression technique which considers the physical principles involved. Three simultaneous measurements are used to determine the wind direction. As the inversion process does not always provide a unique solution, ambiguity removal is needed, for instance through comparison with a first-guess background field obtained from a Numerical

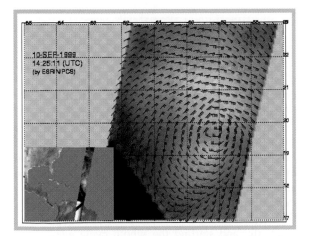

Figure 5-12: Wind vector fields at the ocean surface from the scatterometer aboard ESA ERS-1 satellite. The inset shows the instrument swath and its geographical location

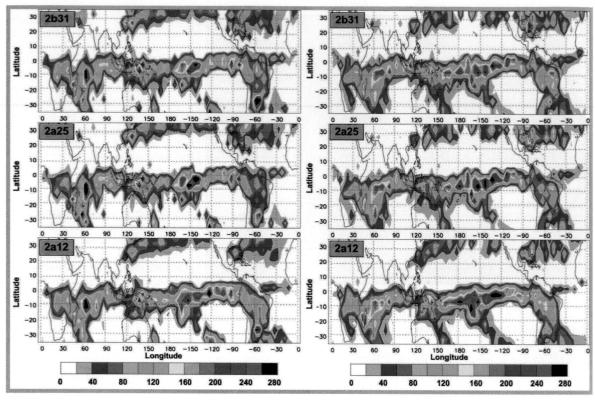

Figure 5-13: Results from the Precipitation Radar aboard the Tropical Rainfall Monitoring Mission (TRMM) from Yang and Smith (2004): Left panel shows distributions of daytime (0600 - 1800 MST) and right panels night time (1800 - 0600 MST) monthly rainfall accumulation over global tropics for February 1998 produced by: 1) upper panels: a combination of the Precipitation Radar (PR) with the TRMM Microwave Imager (TMI) in upper panels, 2) center panels: the PR-only rainfall, and 3) lower panels: the TMI-only precipitation. The colour code gives the precipitation in mm

Weather Prediction (NWP) model, or by assimilation of the possible multiple solutions into the NWP model (Stoffelen 1998).

It is also possible to estimate the wind speed over the ocean surface from passive microwave imagery. The emission of microwave radiation from the ocean surface depends on the surface roughness, which in turn depends on wind speed close to the surface. A drawback of methods using passive microwave radiances is that they are quite sensitive to the presence of rain clouds and have no information on wind direction, unless polarimetric information is available. The latter mission concept has already been realized with Windsat on the Coriolis satellite (Gaiser et al. 2004).

5.3.6
Precipitation

Clearly, observations of precipitation are important because of the immediate relevance to daily life. Precipitation abundance gives rise to floods while its absence causes droughts. It is also a vital parameter from the scientific standpoint, forming an essential link in the hydrological cycle in which water vapor evaporating from the ocean surface rises into the atmosphere, condenses into clouds, coalesces into rain or snow which falls into river catchment areas, flowing to the sea after contributing to the growth of plant life, and starting the whole cycle again from the oceans. When water vapor condenses into liquid drops it releases heat into the ambient air. Through this latent heating, global rainfall heats the atmosphere about three times more than the other major source of atmospheric heating, solar radiation, which is primarily absorbed at the earth's surface. Precipitation seems easily observed locally with a rain gauge, however this concept is deceptively simple because of the high temporal and spatial variability of precipitation.

Rain may fall continuously from a slow moving extra-tropical depression over a period of many days or more, while the same rainfall totals could accumulate locally from a small convective storm lasting only an hour or so. A conventional rain gauge might give a reliable indication of the former, but could entirely miss a convective storm.

Satellites offer the only prospect of measuring precipitation on a global scale, and provide useful results even though they are presently far from ideal for the purpose. With so much variability in precipitation there is a corresponding wide range of approaches to its measurement. The different approaches largely depend on the nature of the application. Great use is made of data from geostationary satellites, because they can observe frequently and thus should miss only few significant precipitation events. However geostationary satellites do not measure rainfall directly, they rather establish a somewhat loose relationship between cloud top temperature and precipitation (e.g. Levizzani et al. 1999). Passive microwave sensors on polar orbiting satellites measure signals that are "closer" to the precipitation and also have the capability of seeing through clouds. This means they can determine which parts of the cloud systems are precipitating; the passive microwave measurements are more closely correlated with precipitation than the thermal infrared measurements.

The Precipitation Radar is the first space borne instrument designed to provide three-dimensional maps of the intensity and distribution of the rain, the rain type, the storm depth and the height at which snow melts into rain. The Precipitation Radar is flown on the Tropical Rainfall Measuring Mission (TRMM), a joint mission of NASA and the Japan Aerospace Exploration Agency (JAXA), and has been designed to monitor and study tropical rainfall. The Precipitation Radar has a horizontal resolution

at the ground of about four kilometers and a swath width of 220 kilometers. It provides vertical profiles of the rain and snow from the surface up to a height of nearly 20 kilometers in the tropics. TRMM is in a so-called low-inclination orbit and provides accurate measurement of the spatial and temporal variation of tropical rainfall around the globe. Figure 5-13 shows results from TRMM (Yang and Smith 2004). It shows distributions of daytime (0600 - 1800 MST) (left panels) and nighttime (1800 - 0600 MST) (right panels) monthly rainfall accumulation over global tropics for February 1998 produced by: (1) a combination of the Precipitation Radar (PR) with the TRMM Microwave Imager (TMI) in upper panels, (2) the PR-only rainfall in center panels, and (3) the TMI-only precipitation in lower panels. During its mission, TRMM provides the first detailed and comprehensive data set on the four-dimensional distribution of rainfall and latent heating over vastly undersampled oceanic and tropical continental regimes.

5.3.7
Aerosols and Trace Gases

This section deals with satellite observations that are going to play an increasingly important role for future weather observations from space, namely aerosol and trace gases. Clearly both quantities are key climate variables in terms of their direct (aerosol and trace gases) and indirect (aerosol) radiative effects on the earth atmosphere (Haywood and Boucher, 2000). It is anticipated that these quantities also play an increasingly important role for weather observations and in NWP. The mapping and forecasting of the concentration of stratospheric ozone is already routinely pursued. It is vital because of the way in which ozone absorbs solar ultraviolet radiation, which is harmful to humans, animals and plants. Ozone products are already being assimilated in NWP models (e.g., at the European Centre for Medium Range Weather Forecasts). The need for integrating the satellite and ground-based measurements of ozone has been recognized (WMO/CEOS Report, see references) and future operational ozone monitoring missions will pave the way toward improved forecasts of air quality.

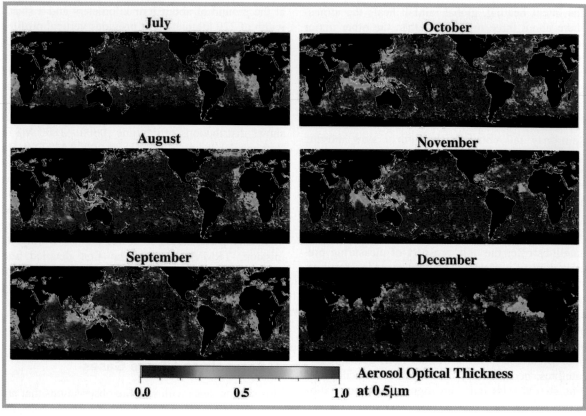

Figure 5-14: Aerosol optical thickness (at a reference wavelength of 0.5 μm) over oceans derived from two channels of the AVHRR (Advanced Very High Resolution Radiometer) instrument on NOAA satellites (Nakajima and Higurashi 1998)

Aerosol is an important quantity for air quality and also for NWP, the latter because of its profound impact on the solar surface radiation budget and the diabatic heating of the atmosphere. Satellite observations of aerosol have been established and especially over the oceans the methods are fairly reliable. Figure 5-14 shows an aerosol optical depth product over the world's oceans from Nakajima and Higurashi (1998). Aerosol injected into the atmosphere and stratosphere by volcanic eruptions also needs to be monitored carefully because it imposes a major threat to air traffic safety. Multispectral observations from satellites provide the means to observe volcanic ash clouds and to establish an adequate warning system (Prata 1989; Watkin et al. 2000).

5.4
Applications in Weather Forecasting

This section discusses the application of satellite data to major fields of operational meteorology, i.e. very short term forecasting (called nowcasting) and in numerical weather prediction. A separate section deals with the tracking of tropical storms.

5.4.1
Nowcasting Severe Weather

Nowcasting is weather forecasting over very short periods of time; such as just a few hours. Warnings of severe weather events are major applications of

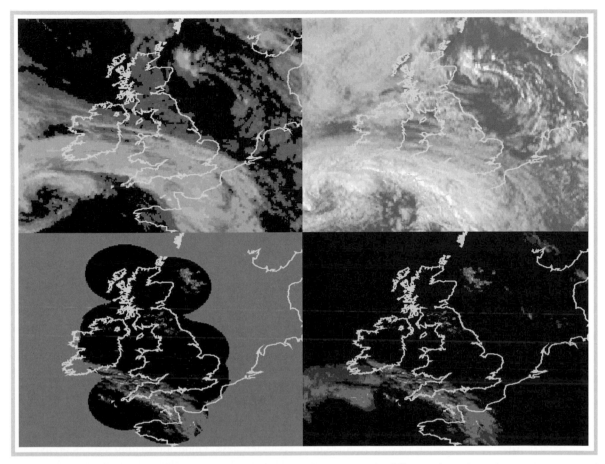

Figure 5-15: Nowcasting application of satellite data: satellite imagery in different channels, radar measurements of precipitation and the background (forecast) field of a numerical weather prediction model. The system is in use at the Met Office of the United Kingdom. Top: Left, infrared satellite imagery, right, Visible satellite imagery; down: left, Composite radar frame (corrected), right, Model background field

nowcasting. The challenge in nowcasting is to utilize the most recent observations, together with the output of regional numerical weather prediction models with high spatial resolution, to improve the short-term forecasts. Nowcasting can provide quantitative precipitation forecasts with more accuracy in space and time than the forecasts over longer ranges. Application areas are manifold and encompass aviation, prediction of weather conditions affecting road traffic, hail and thunderstorm warning for agriculture, flash flood forecast and information to power stations. Today's nowcasting techniques are computer assisted but do not yet rely on NWP models. Satellite imagery is essential for analyzing smaller scale

cloud and water vapor features and regional meteorological weather patterns. However, the satellite images contain far more information than can be absorbed or even detected by the human analyst in the fairly short time available to perform the weather analysis and issue the information or a potential warning. Therefore, image information is condensed into image products as shown in figure 5-15, which depicts satellite imagery in different channels, radar measurements of precipitation and the background (forecast) field of a numerical weather prediction model. That system is in use at the UK Met Office.

It is also interesting that for very short forecasting the extrapolation of cloud system displacement is

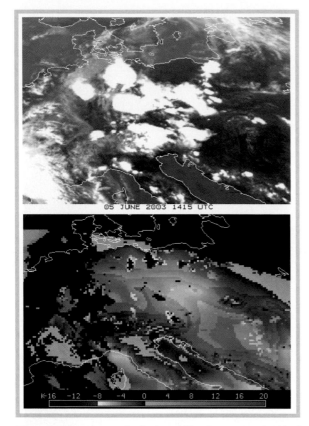

Figure 5-16: Atmospheric instability in terms of Lifted Index (bottom panel) as observed by Meteosat-8. Areas of high instability appear yellow and red and indicate the areas prone to convective activity. The upper panel is the satellite cloud image observed 5 hours later, showing that convective clouds have formed in areas depicted by the lifted index (König et al. 2001)

already a fairly reasonable forecast tool. For that purpose the atmospheric motion vectors (winds) derived from satellite images can be used. Another example of satellite monitoring important to now-casting is the observation of atmospheric instability. For instance, it is possible to derive a so-called lifted index representing the buoyancy which an hypothetical air parcel would experience if lifted mechanically from the surface to the 500 hPa level in the atmosphere. It is expressed as the difference in temperature between the ambient 500 hPa temperature and the temperature of the lifted parcel. A negative value (warmer than ambient) suggests positive buoyancy (and instability) while a positive differ-

ence suggests stability. The product is of particular value when used to forecast potential locations of severe convective storms (Menzel et al. 1998). Figure 5-16 shows an example of the lifted index derived from Meteosat-8 (König et al. 2001).

5.4.2
Numerical Weather Prediction

Most weather forecasts over more than a few hours are made by computer models. The computer models try to capture the salient physics that determine atmospheric processes and the weather. This is done by solving the equations of motion for the atmosphere, i.e. nonlinear partial-differential equations describing the atmospheric dynamics, thermodynamics, and conservation of mass, including the phase changes of water. Modern numerical weather prediction achieves this by finding approximate numerical solutions. However, many of the relevant physical processes like turbulence, radiation transport or cloud formation are not explicitly handled but rather parameterized because no computer would be powerful enough to allow for an explicit treatment of those physical processes. A forecast by a numerical weather prediction model is then the result of the computer calculations into the future. Typically the calculations are performed over a few days up to about ten days. Good forecasting skill is on average attained out to day seven although the skill significantly decreases with time.

Adequate observations of the atmosphere are key to a good forecast. This means, the start of the numerical model requires observations that describe the state of the atmosphere; the better the description of the atmospheric state at the beginning of the computer forecast calculation the better the quality of the forecast. A global meteorological observing system serves the task of providing the initial conditions for numerical weather prediction and satellites are nowadays the key element of the global observing system, complemented by radio soundings with balloons measuring wind, temperature and humidity, aircraft measurements, ground measurements, and remote sensing from the ground.

In the early days of the use of satellite data for Numerical Weather Prediction (NWP) profiles of temperature and humidity derived from the satellite radiances were ingested into the NWP models in a format that was as close as possible to that of the conventional radiosonde (RAOB) data. This approach had advantages; the models did not have to be adapted to use satellite data; the new data types could readily be checked and quality controlled against the old. However, it became clear that this approach also had many disadvantages. From the late 1990s the preferred way to use satellite data in numerical models has been through so-called variational analysis schemes using the observed radiance fields directly (Eyre 1997; Schlatter 2000). These use the Numerical Weather Prediction (NWP) model itself to extract the relevant information content. The starting point for any numerical forecast is a description of the current state of the atmosphere. Detailed radiometric models of the atmosphere and instrument are then used to generate "Forecast Observations" in terms of radiance fields, which is what satellites observe. The simulated radiances are then compared with the satellite observations. Essentially the NWP model is adjusted until it generates simulated radiances which match those observed by the satellite.

The satellites are now the most important element of the global observing system. This is for two reasons: a) satellites offer the only realistic opportunity for providing essential data for Numerical Weather Prediction on a truly global basis and b) after intensive research over more than two decades on how to use satellite data in the best manner, the utilization of satellite data has matured and satellite data have the highest impact on numerical weather forecasts. This has been achieved by assimilating radiance observations directly rather than making a detour via retrieved temperature and humidity profiles. In the southern hemisphere the impact of the satellite data is even more dramatic than in the northern hemisphere because of the lack of other data.

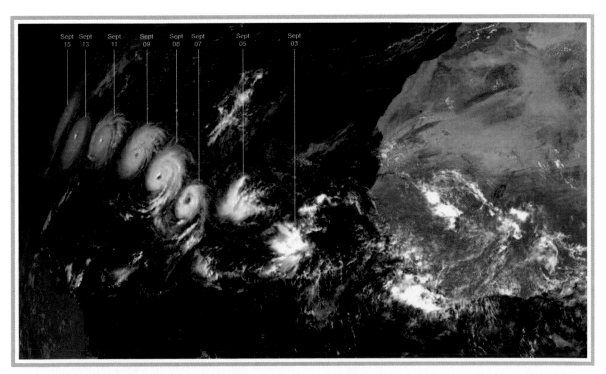

Figure 5-17: Hurricanes can be tracked well from geostationary satellites. This Meteosat-8 image shows a composite of the position of Hurricane Isabel over the North Atlantic from 3 September 2003, when it started developing from a tropical storm into a hurricane, until 15 September 2003

5.4.3
Tracking Tropical Storms

Typhoons, hurricanes, and other severe tropical storms known by various names around the world cause immense damage every year, often accompanied by many human deaths. These tropical storms and hurricanes form over tropical oceans; a prerequisite for their formation is a high sea-surface temperature above 26°C. The dramatic development of tropical storms is well depicted in sequences of satellite pictures, where the storms, once they have attained a mature stage, show a distinctive and easily recognized spiral cloud pattern. Thanks to satellite imagery the problem of detecting and monitoring tropical storms has been solved; it is unlikely that a tropical storm escapes detection anywhere in the world. Quantitative information on winds derived from satellite observations significantly helps to predict the track of devastating tropical storms (Velden et al. 1992). The factors which influence their formation and development are implemented in numerical models which, together with satellite derived products such as Atmospheric Motion Vectors, give fairly accurate forecasts of storm movement and intensity.

Figure 5-17 shows the movement of Hurricane Isabel in September 2003 over the North Atlantic ocean. The figure has been constructed from successive images in the visible spectrum observed from Meteosat-8.

5.5
Concluding Remarks

The utilization of meteorological satellite observations for short-term weather forecasting (nowcasting), numerical weather prediction and other applications has attained a mature level and is an indispensable ingredient of the global meteorological observing system. This space observation part of the Global Observing System of the World's Meteorological Organization's World Weather Watch program consists of complementary series of geostationary and polar satellites. The system is augmented by research satellites, which have the advantage of bringing new observing techniques and data, however, they have the disadvantage of not providing a sustained contribution to the Global Observing System. For numerical weather prediction the satellite observations are nowadays the single most important part of the global observing system. This has been achieved for two reasons: i) satellite observation capabilities have improved in terms of accuracy and spatial, temporal and spectral coverage, and ii) knowledge of how to use the satellite data for specific applications has made considerable progress. The latter is very important because the inference of a geophysical product from the satellite observed radiances is usually not trivial and the direct use of the observed quantities (i.e., the radiance fields) did necessitate a major breakthrough in the assimilation of satellite data.

Future benefit to global weather forecasting will come from instruments with better spectral resolution, which implies better vertical resolution. The Atmospheric Infrared Sounder (AIRS) instrument on the AQUA satellite of NASA (Aumann et al. 2003) covers three infrared wavebands 3.74-4.61 µm, 6.20-8.22 µm, and 8.8-15.4 µm with a nominal spectral resolution of $\lambda/\Delta\lambda = 1200$. The Infrared Atmospheric Sounding Interferometer (IASI) of the EUMETSAT Polar System (Metop) will be the first operational atmospheric sounding instrument with high spectral resolution (Schlüssel and Goldberg 2002). IASI has a spectral resolving power of about 2,000; this high spectral resolution is expected to yield an accuracy for temperature retrievals of 1 K per slabs only 1 km in vertical extent. Other exciting missions to come are the global observations of wind fields with a Doppler lidar (see above), and missions for measuring precipitation and atmospheric chemistry, which will be an important ingredient for the establishment of global operational air quality forecasts.

With regard to a general strategy toward the future space-based observing system, it is expected that the current cooperation amongst operational and research space agencies will continue and prosper. In the end a comprehensive and robust observing system will be established where contributions from different satellite operators truly complement each other in terms of their mission objectives. These mission objectives will be based on user needs and priorities, not only for weather observation and forecasting but also for operational oceanography, atmospheric chemistry and climate monitoring.

Acknowledgements

The authors thank Drs. E.A. Smith, T. Nakajima, P.M. Ingmann, R. Borde, J. Kerkmann, D. Klaes, M. König, J.P. Luntama, P. Schlüssel and S. Tjemkes for the provision of figures.

References

Aumann HH, Chahine MT, Gautier C, Goldberg M, Kalnay E, McMillin L, Revercomb H, Rosenkranz P, Smith WL, Staelin D, Strow L and Susskind J (2003) AIRS/AMSU/HSB on the Aqua mission: Design, science objectives, data products, and processing systems. IEEE Transactions on Geoscience and Remote Sensing, 41, pp. 253-264

Bormann N and Thépaut JN (2004) Impact of MODIS Polar Winds in ECMWF's 4DVAR Data Assimilation System. Monthly Weather Review, vol.132, pp. 929-940

Eyre J (1997) Variational assimilation of remotely-sensed observations of the atmosphere. Journal of the Meteorological Society of Japan, 75(1B) pp. 331-338

Gaiser PW, St. Germain KM, Twarog EM, Poe GE, Purdy W, Richardson D, Grossman W, Linwood Jones W, Spencer D, Golba G, Cleveland J, Choy L, Bevilacqua RM, and Chang PS (2004) The WindSat Spaceborne Polarimetric Microwave Radiometer: Sensor Description and Early Orbit Performance, IEEE Transactions on Geoscience and Remote Sensing 42 (11), pp. 2347-2361

Haywood J and O Boucher (2000) Estimates of the direct and indirect radiative forcing due to tropospheric aerosols; a review. Reviews Geophysics 38, pp. 513-543.

Holmlund K (1998) The utilization of statistical operties of satellite-derived atmospheric motion vectors to derive quality indicators. Weather Forecasting 13, pp. 1093-1104

Ingmann P, Endemann M and members of the ADM-Aeolus Mission Advisory Group (2002) Status of the Doppler wind lidar profiling mission ADM-Aeolus. Proceedings of the 6th International Winds Workshop, Madison, WI, USA, EUMETSAT Publication, EUM P 35, ISSN 1023-0416, pp. 231-238

Key JR, Santek D, Velden CS, Bormann N, Thépaut JN, Riishojgaard LP, Zhu Y, and Menzel WP (2003) Cloud-Drift and Water Vapor Winds in the Polar Regions from MODIS. IEEE Transactions on Geoscience and Remote Sensing 41, pp. 482-492

Klaes D, Ackermann J, Schraidt R, Patterson T, Schlüssel P, Phillips P, Arriaga A, and Grandell J (2004) The ATOVS and AVHRR Product Processing Facility for EPS. To appear in Advances in Space Research.

König M, Tjemkes S, and Kerkmann J (2001) Atmospheric instability parameters derived from MSG SEVIRI observations. Proceedings of the 11th Satellite Meteorology Conference, American Meteorol. Soc. pp. 336 – 338.

Levizzani V, Schmetz J, Lutz HJ, Kerkmann J, Alberoni PP, and Cervino M (2001) Precipitation estimation from geostationary orbit and prospects for Meteosat Second Generation (MSG). Meteorological Applications 8, pp. 23-41

Luntama JP and Wilson JJW (2003) GRAS Level 1 Processing and data products. Proceedings of the 2nd GRAS SAF User Workshop, Helsingør, Denmark. EUMETSAT report: EUM P 40

Menzel WP, Holt FC, Schmit TJ, Aune RM, Schreiner AJ, Wade GS, and Gray DG (1998) Application of GOES-8/9 soundings to weather forecasting and nowcasting. Bulletin American Meteorological Society 79, pp. 2059-2077

Nakajima T and Higurashi A (1998) A use of two-channel radiances for an aerosol characterization from space. Geophysical Research Letters 25, pp. 3815-3818

Nieman S, Schmetz J, and Menzel P (1993) A Comparison of Several Techniques to Assign Heights to Cloud Tracers. Journal of Applied Meteorology 32, pp. 1559-1568

Prata AJ (1989) Observations of volcanic ash clouds in the 10 – 12 μm window using AVHRR/2 data. International Journal of Remote Sensing 10, pp. 751-761

Schlatter TW (2000) Variational assimilation of meteorological observations in the lower atmosphere: a tutorial on how it works. Journal of Atmospheric and Solar-Terrestrial Physics 62, pp. 1057-1070

Schlüssel P and Goldberg M (2002) Retrieval of Atmospheric Temperature and Water Vapour from IASI Measurements in partly cloudy situations. Advances in Space Research 29 (11), pp. 1703-2706

Schmetz J, Holmlund K, Hoffman J, Strauss B, Mason B, Gaertner V, Koch A, and van de Berg L (1993) Operational cloud motion winds from METEOSAT infrared images, Journal of Applied Meteorology 32, pp. 1206-1225

Schmetz J, Pili P, Tjemkes S, Just D, Kerkmann J, Rota S, and Ratier A (2002) An Introduction to Meteosat Second Generation (MSG), Bulletin of the American Meteorological Society 83, pp. 977-992

Stoffelen A (1998) Scatterometry. Dissertation/Proefschrift Universiteit Utrecht

Velden CS, Hayden CM, Menzel WP, Franklin JL, and Lynch JS (1992) The Impact of Satellite-derived Winds on Numerical Hurricane Track Forecasting. Weather and Forecasting 7, pp. 107-118

Velden CS, Hayden CM, Nieman SJ, Menzel WP, Wanzong S, and Goerss JS (1997) Upper-tropospheric winds derived from geostationary satellite water vapor observations. Bulletin of the American Meteorological Society 78 (2), pp. 173-196

Watkin S, Ringer M, and Baran A (2000) Investigation into the use of SEVIRI imagery for the automatic detection of volcanic ash clouds. Proceedings of 'The 2000 EUMETSAT Meteorological Satellite Data Users' Conference', Bologna, Italy, EUM P-29, pp. 753-760

WMO/CEOS (2001) Report on a strategy for integrating satellite and ground-based observations of ozone. WMO TD No. 1046, pp. 128

WMO (2003) The role of satellites in WMO Programmes in the 2010s. WMO Space Programme SP-1, WMO/TD No. 1177

Yang S, and Smith EA (2004) Mechanisms for diurnal variability of global tropical rainfall observed from TRMM. Journal of Climate, submitted

6 Geodynamics

by Bert Vermeersen, Ron Noomen, Ernst Schrama, and Pieter Visser

The earth sciences underwent a revolution in the late fifties and early sixties of the 20th century. Whereas before that time continental drift was considered to be at best no more than an interesting alternative to existing theories—and was actually rejected by most geoscientists—it became the prevailing paradigm after that time. Symmetric patterns of magnetic anomalies around mid-ocean ridges, discovered in the fifties, heralded the new theory of plate tectonics by which such phenomena as mountains, volcanoes and earthquakes could be explained in a single coherent all-encompassing theory of a convecting Earth. Seismology consolidated our understanding of the earth's interior in the seventies and eighties in this framework of plate tectonics, notably by a technique that came to be known as seismic tomography. Subducting ocean slabs that were conjectured to be an essential part of plate tectonics showed up indeed in these "X-ray" images of the earth's mantle. Still, there were areas in the temporal-spatial domain for which no direct observations could be obtained. For example, seismological observations on the shortest time scales and geological observations on the other end of the time spectrum both hinted at ongoing plate motions on the order of a few centimeters per year. In the eighties, space geodesy opened these hitherto unexplored gaps in the temporal-spatial domain by directly detecting and measuring the cm/yr rates at which plates move by Very Long Baseline Interferometry (VLBI) and later by the Global Positioning System (GPS) of geostationary satellites. New measurement concepts for three-dimensional crustal deformation (e.g., Interferometric Synthetic Aperture Radar, InSAR), for ocean and surface ice topography (satellite altimetry), for gravity (e.g., satellite laser ranging and satellite gradiometry), for rotational variations and for magnetism have made satellite geodesy one of the cornerstones of modern geodynamic research.

6.1
The Changing Earth

Our Earth, as solid as it looks, is subject to continuous changes over timescales ranging from tidal periods to geological dimensions. In the following, a brief overview of the phenomena and processes will be presented.

6.1.1
Dynamics of the Crust and Lithosphere

The solid earth is continuously changing its shape and gravity field due to a broad range of geodynamic forcings. The central parts of Canada are rising due to postglacial rebound following the disappearance of the great ice domes that covered it during the last ice age; California is regularly shaken up by earthquakes, and while many fear the expected "big one" in this U.S. state, it was Southeast Asia that was struck during Christmas 2004 by a devastating earthquake and tsunami that killed about 300,000 people; mantle material rises up and solidifies at mid-ocean ridges; ocean plates dive into the mantle at subduction zones; volcanoes erupt; mountains rise; sea level varies due to solid-earth movements and continental ice-mass variations, and the rotation axis wobbles. These are just a few of the many examples whereby the solid earth is taking part in the figure and gravity field changes observed by space-geodetic techniques. Sometimes, temporal variations can be measured directly (e.g., crustal displacements due to earthquakes or postglacial rebound). Often, these temporal variations proceed in such slow motion that only the resulting spatial variations formed over geological time scales can be observed (e.g., low spherical harmonics of the geoid due to mantle convection). Why is the earth such a tectonically

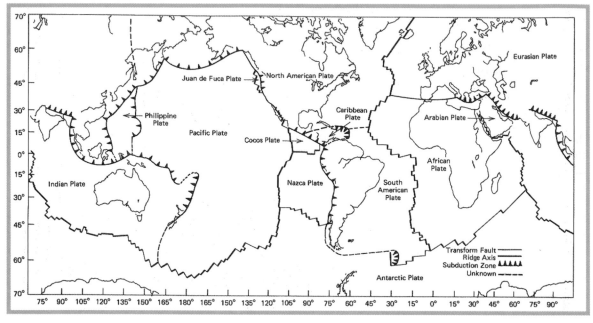

Figure 6-1: Tectonic plates of the earth. In total there are about 10-15 plates, depending on what one would still consider to be a separate plate. Their boundaries consist of both active (subduction zones and mid-ocean ridges) and passive margins. Oceanic (parts of) plates are generated at mid-ocean ridges and turn back into the mantle at subduction zones; continental (parts of) plates do not subduct

active planet and what are the causes for all these geodynamic processes?

The surface of the earth is split up into about 15 large plates, as depicted in figures 6-1 and 6-2.

In figure 6-2 the plate movements derived from geological observations are compared with the movements derived from GPS measurements. The geological data (derived from a particular model, the so-called NUVEL model) give the plate movements averaged over a time scale of about one million years. The GPS data give the plate movements as they are today. Figure 6-2 shows that most of the plate movements have hardly changed over the past

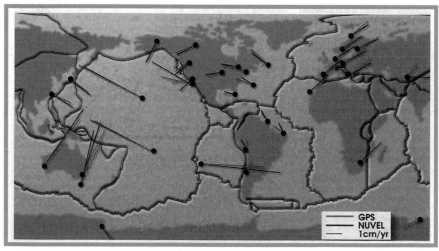

Figure 6-2: Comparison between GPS-observed (present-day) and geologically-modeled (NUVEL model, averaged over one million years) plate motions. Note the generally good fit between these two data sets; deviations are small but often do have significant geological causes on which much of current research is concentrated

million years, a truly remarkable fact, certainly as one realizes that the two types of data sets have been derived from completely different techniques (geological versus space-geodetic!). The boundary between two plates that move in the horizontal direction with respect to each other is called a transform fault (see figure 6-1). The movement is not smooth due to resistance, so stress builds up along the fault. Stress is, at least partially, released in catastrophic events which we experience as earthquakes.

Earthquakes mainly occur where two plates meet each other, as evidenced from figure 6-3. Not all plate boundaries are transform fault boundaries. Running along the middle of the Atlantic we see a lineage of earthquakes. Here, two plates diverge, the North American and the Eurasian in the north and the South American and the African in the south,. The lineage is called a mid-ocean ridge. Along the Pacific the earthquakes form a ring-shaped pattern. At the west coast of South America the Nasca and

South American plates converge. The Nasca plate dives under the South American plate; such a plate margin is called a subduction zone (see figure 6-1), and the Nasca plate is said to subduct.

The surface of the earth is composed of about 15 large plates which come in two types: oceanic and continental, with many consisting of a combination of the two types, such as the North American and Eurasian plates. Where an oceanic part adjoins a continental part inside a plate, the boundary between two parts of a plate is called a passive plate margin. The aforementioned boundary between the Nasca and South American plates is an active plate margin.

Where two plates separate (e.g., at the ocean ridge between the Pacific plate and the Nasca plate), material from the mantle wells up to replace the diverging material that is carried with the plates. This mantle material, molten as it reaches the surface, solidifies and becomes part of the oceanic plate. The oceanic plate thus continuously acquires material, which

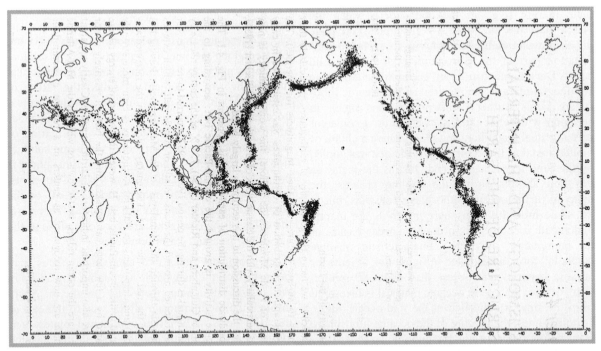

Figure 6-3: Position of 29,000 earthquakes over a six-year time interval. The earthquakes cluster at and around active plate boundaries, both at subduction zones (mainly around the Pacific) and at mid-ocean ridges. Note also the clustering along the line from the Mediterranean to Southeast Asia: ongoing seismic activity marks the position where the European-Asian continent collided with the southern continents (Africa, India) some 100 million years ago, closing the separating Tethys ocean

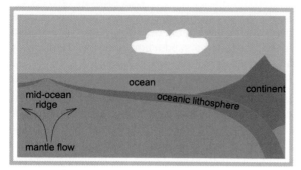

Figure 6-4: Creation and subduction of oceanic lithosphere. Mantle material adiabatically wells up at the mid-ocean ridge, solidifies, and is transported away towards the continent. Ongoing cooling results in a gradual densification of the oceanic lithosphere, thereby inducing the typical inverse-parabola bathymetry that is characteristic for the world's ocean floor. Subduction of the old ocean lithosphere at the continent is envisioned to be the principle driver of this convective motion

is transported away from the mid-ocean ridge with velocities of a few centimeters per year. At an active plate margin (e.g., at the boundary between the Nasca plate and the South American plate) the plate dives into the mantle, completing its surface manifestation part of the convection cycle of the solid earth. During the trip from ocean ridge to active continental margin, the material of the oceanic lithosphere cools further. This process has much in common with what can be observed on a conveyor belt at a steel manufacturing plant, where a chunk of molten metal at the beginning of the belt has cooled by the time it reaches the end. Just as the metal on this belt shrinks during the cooling process, the material of the oceanic lithosphere contracts, but at the same time the lithosphere grows as the mantle underneath it cools. The form of the ocean basins, or the bathymetry, thus has a characteristic "inverse parabola" form, as depicted in figure 6-4, at least for oceanic material that is less than about 80 million years of age (the age of oceanic material is measured from the time it solidified at the mid-ocean ridge). For older ages, the heating of the oceanic lithosphere from below compensates the cooling, so that effectively the oceanic lithosphere keeps its shape.

As the material of the oceanic lithosphere cools, it gets denser than the mantle material below. Therefore, the "old" oceanic material can sink into the

mantle near a continental margin. The initiation of such a sinking process is another question: it takes some special driver to start the whole process. One of these drivers could be sediment load on top of the ocean floor. Sediments are built up from the remains of marine animals (shells, bones) and all kinds of other material that sinks to the ocean floor, and from material that was eroded from the continents and transported by rivers to the ocean basins. These sediments also play another interesting role. While the oceanic lithosphere sinks into the mantle at active ocean margins with velocities of up to about 10 cm per year, it takes along with it part of the sediment layer on top. The sediments are thus transported to deeper regions inside the earth, squeezed between the subducting oceanic lithosphere and the continental crust and lithosphere. The melting temperature of this sediment layer is low. The temperature inside the earth rises as one goes deeper, which one can experience when descending in a mine shaft, so the sediments can form melts at certain depths. These melts percolate through the continental crust on top and give rise to volcanoes at the earth's surface. Therefore, most volcanoes (figure 6-5) are situated along the lines in figure 6-1 where subduction takes place, and it becomes understandable that the Pacific rim is sometimes called the "Ring of Fire." Note also that along mid-ocean ridges volcanoes are active (such as the well-known volcanoes of Iceland). Due to the processes of formation of oceanic lithosphere at a spreading ridge and subduction near continental margins, oceanic lithosphere is not older than about 200 million years. Continents, consisting of relatively light material, "float" on the convecting mantle and do not subduct. Therefore, the cores of the continents are very old, close to the age of Earth (about 4.5 billion years).

Over geological times two continental plates may meet. In fact, this is what is happening nowadays in the region stretching from the south of Spain through the Mediterranean all the way to the east of the Himalayas. When two continental plates converge, neither of them subducts, but in the clashing zone a mountain range will emerge. The Himalayas are the most prominent example of such orogeny (mountain building) today: the Indian plate ploughs into the Eurasian plate, as part of a closure of the Tethys on a grand scale. The Tethys was the ocean

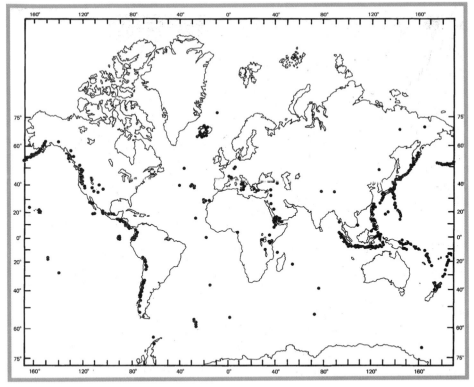

Figure 6-5: Position of active volcanoes. Note the strong correlation with the positions of earthquakes in figure 6-3. There are three main types of volcanism, two of which are related to the plate tectonic cycle depicted in figure 6-4: mid-ocean ridge volcanism and volcanism at subduction zones. The third type, to which the volcanic activities at Hawaii belong (hot spot volcanism), are related to plumes that arise from the deep mantle

that once separated the northern and southern part of the super-continent Pangea. Not only its height, but also its sharp-peaked topography indicates that the Himalayas is a young mountain range situated along an active plate boundary. Erosion will smooth the roughness of the topography with time, as has been the case with, for example, the Ardennes, which was formed hundreds of millions of years ago.

6.1.2
The Earth's Interior and its Dynamics

Based to a large part on seismic observations, the earth can be divided into a number of layers, as depicted in figure 6-6. The crust is the top layer of the solid earth. Its average thickness is about 35 km, but this varies: at a mid-ocean ridge it is virtually zero; beneath mountain ranges it is up to about 70 km. In general, continental crust is thicker than oceanic crust. The (continental) crust is often subdivided into an upper and lower part, whereby the lower crust is more ductile than the upper.

Below the crust is the lithosphere. The lithosphere is the strongest layer of the earth (least ductile), and extends to depths of near zero at mid-ocean ridges to about 200 km (or perhaps even more) below continents. The boundary between crust and lithosphere

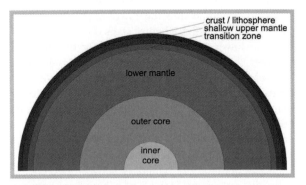

Figure 6-6: Radial profile of the earth (to scale, except for crust and lithosphere). The mantle mainly consists of silicates; the core mainly of iron. The inner core is solid, the outer fluid. The distinctions between the shallow upper mantle and the transition zone, and between the transition zone and the lower mantle, are mainly due to phase-change boundaries and partly also to chemical differences

is seismically detectable and is called the Mohoro-vičić discontinuity, more commonly known under the abbreviation "Moho."

Below the lithosphere is the upper mantle, which extends to about 670 km depth. The upper mantle can be subdivided into a shallow upper mantle (down to about 400 km) and a transition zone (400-670 km depth). This subdivision arises from a phase change which the material of the mantle undergoes as the pressure reaches a critical point. At about 400 km depth this is the so-called olivine-spinel phase change; at about 670 km depth the spinel-post-spinel phase change. These phase changes are seismically detectable. The top of the uppermost mantle (just below the lithosphere) is often more ductile than the mantle below it. This ductile layer is called the asthenosphere.

The 670 km discontinuity, forming the boundary between the upper and lower mantle, is a special and important one. It has been the subject of many studies. First of all, earthquakes do not seem to occur below it (actually, they sometimes do occur down to about 700 km depth, but this has to do with subducting slabs that warp the 670 km discontinuity downwards). Another point is that many of the subducting oceanic slabs do not seem to penetrate this 670 km discontinuity, so that at many places below continental marges there seems to be a "graveyard" of subducted material. However, recent seismic tomography results seem to indicate that some of the slabs do penetrate into the lower mantle at various places. The question whether slabs do or do not sink into the lower mantle is important for the question whether the earth convects in a single-layer mode (the whole mantle convects and thus the whole mantle mixes) or in a two-layer mode (upper and lower mantle convect separately, so there is no mixing between upper and lower mantle). A corollary is, of course, that the 670 km discontinuity might not be a pure phase change boundary, but partly also a chemical boundary (a boundary between two chemically distinct layers instead of a boundary between two structures of the same chemical composition). At the moment most geophysicists tend to the whole-mantle convection viewpoint, but the matter remains hotly debated.

Whereas the mantle is mainly composed of silicates, the core is mainly composed of iron and nickel. The region between mantle and core, the core-mantle boundary or CMB at a depth of about 2,900 km, is very interesting. It is generally known as the D" (pronounced "dee double prime") layer. In some way it is the "antipode" of the crust: not sharp, most likely very heterogeneous, and a collector of material that does not fit into the structure of the mantle and core. It could be the origin of so-called mantle plumes (see below). The core, which extends from about 2,900 km depth to the center of the earth at about 6,370 km, has a fluid outer core and a solid inner core (boundary between outer and inner core at about 5,150 km depth). The velocities of the "metallic fluid" in the outer core are on the order of 0.1 mm per second (or about one meter per three hours). The outer core is the source of the earth's magnetic field, which is continuously regenerated by dynamo action. The earth's magnetic field has a large dipolar component that has changed polarity many time (and quite irregularly!) over geological times. It is this switched polarity of the earth's magnetic field that has been at the base of the revolution in the earth sciences near the end of the fifties and the establishment of plate tectonics and continental drift in the fifties and sixties of the 20th century. We have seen that at mid-ocean ridges mantle material solidifies and becomes lithosphere. When this happens, the temperature falls below the Curie temperature and the magnetic field, in particular its direction, gets "frozen" into the lithospheric material. The lithosphere is transported away from its origin on the "conveyor belt." When the polarity of the earth's magnetic field switches after some time (this can be after 10,000 years, but also after several ten million years), the magnetic field direction gets frozen into the solidifying material at the mid-ocean ridge, and changes its direction. In this way, one gets a band of alternating magnetic polarities spread out over the sea floor from the spreading ridge towards the continental margin. The fact that this band is symmetrical with respect to the mid-ocean ridge is hard to explain by any other mechanism than sea floor spreading.

In general, seismic tomography shows positive density anomalies below subduction zones that extend quite deep into the mantle, some ending at about 700 km depth, others even deeper. Obviously, these positive density anomalies delineate subducting (old, cooled) oceanic lithosphere. Below mid-ocean ridges, however, seismic tomography generally shows negative density anomalies, but not below about 200 km depth. These negative density anomalies correlate with hotter mantle material, but obviously the source of the heating mechanism is quite shallow. Heating through adiabatic expansion by passive upwelling thus seems a better explanation for these anomalous low-density regions than does heating through deep-seated sources inside the earth.

Another indication that the source of the upwellings at mid-ocean ridges is shallow comes from hot spots. Hot spots are rather concentrated pointlike places on the earth's surface which show a surface heat flow higher than the average value. These are often associated with large basaltic outpourings, after which the hot spot leaves a track of volcanic events on the overriding continental or oceanic plate (the Hawaiian Emperor island chain is such a track). Although

still somewhat enigmatic, it is generally thought that these hot spots are associated with rising plumes of hot material that come from very deep inside the mantle, or even originate at the core-mantle boundary. Hot spot tracks cross plates relatively undisturbed, that is, no kink or discontinuation is visible where it crosses the boundaries. The nondisturbance of the tracks as they pass right through mid-ocean ridges over geological time scales is very difficult to explain if the source of the hot spot would be shallower than the source of the upwelling material of the ridges (the hot spots would be expected to drift with the conveyor belt system of the separating ocean plates in this case). The hot spots are, however, not completely fixed with respect to each other. They drift with respect to each other with typical velocities of about 1 cm per year, while the largest plates move relative to each other with typical velocities of about 10 cm per year. The relative stability of hot spots with respect to plate motions is an important issue in changes of the rotation axis of

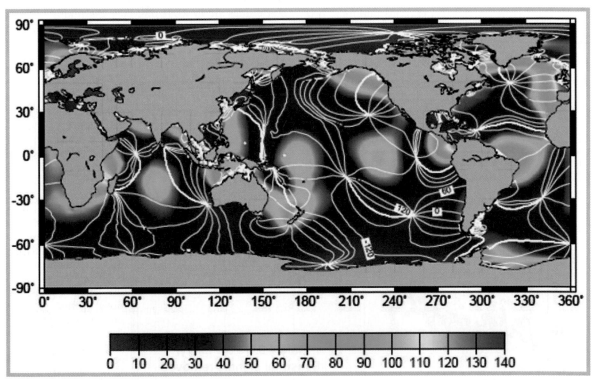

Figure 6-7: Amplitude and phase of the M2 ocean tide observed by TOPEX/Poseidon (Goddard Space Flight Center). It is clear that the half daily ocean tide due to the moon deviates considerably from a simple ellipsoidal pattern

the earth over geological time scales.

6.1.3
Coupling with Processes in the Ocean, Continental Hydrology and the Atmosphere

There are a number of geophysical processes that become relevant for geodynamics, we can mention the presence of ocean tides, sea level rise, variations in surface air pressure and wind, and the hydrologic cycle on land. All these processes describe variation in the distribution of masses loading on the earth's crust or are capable of transferring sufficient moment to Earth's angular motion or the position of the earth's rotational axis as will be explained in section

Tides

The physics of ocean tides has been well understood since Isaac Newton, who explained that the sun and moon are predominantly responsible for diurnal, semi-diurnal and some long periodic oceanic motions. The frequency of tides is of astronomical origin; tidal forcing is therefore well defined. This phenomenon can be measured with terrestrial gravimeters and space geodetic techniques. In addition

tidal corrections are taken into account when flight trajectories of satellites are computed.

Separate from the tidal forcing discussion is an elastic deformation of the earth and the response function of this elastic deformation as a result of astronomic tidal forcing. It was Love (1911) who formulated an elastic deformation theory which is based on the material properties of the earth. This discussion is distinct from ocean tide dynamics, which were precisely formulated by Laplace who explained that ocean tides are mostly linear in the deep oceans, meaning that they exhibit the same periodicity as the astronomical forcing except that the local amplitudes and phases differ.

The earth's oceans have been studied extensively with the help of various satellite altimetry missions that have collected height profiles with the help of microwave radar instruments. Figure 6-7 shows the amplitude and phase of the twice diurnal ocean tide caused by the moon, known as M_2. In this case the colors represent the amplitude of the ocean tide; the phase lines are in white and they essentially indicate the time at which the local tide obtains its highest level.

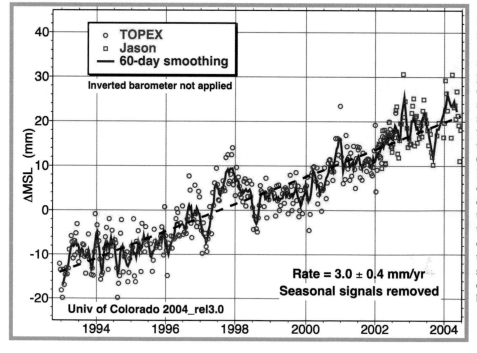

Figure 6-8: Sea level rise observed by satellite altimetry (University of Colorado). The plot is compiled from observations of two satellites: TOPEX/Poseidon and Jason. Note that the deviations from the linear trend of 3.0 ± 0.4 mm per year are large and that this deduced linear trend depends very much on the length of observational period (e.g., if one would only have the observations of Jason a markedly different trend would have been deduced)

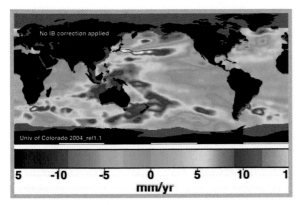

Figure 6-9: Sea level rise map as observed by satellite altimetry (University of Colorado). "No IB correction applied" indicates that this map has been made without an inverse barometer correction, i.e., the sea level changes were computed without taking into account the response of the sea surface to atmospheric pressure variations

Sea Level Rise

Sea level rise is a separate effect. On the one hand it is caused by present day melting of Greenland, Antarctica, and land glaciers. But in addition there is a contribution from the last glacial maximum, as dis-

cussed in section 6.4. The total observed global sea level rise estimates vary nowadays around 1.8 mm/yr according to Douglas (1991) based upon more than a century of sea level measurements recorded by 22 globally distributed tide gauges.

In parallel to the tide gauge sea level records there is an estimate of sea level rise by radar altimeter satellites, yielding a value of 2.8 +/- 0.4 mm per year according to Leuliette et al. (2004). Their results are shown in figures 6-8 and 6-9 in terms of a global and a geographic sea level change. Quite remarkable is the difference between both results. The TOPEX/Poseidon and JASON altimeter systems have shown significant geographic variations in sea level rise which are for instance caused by the presence of the 1997-1998 El Niño, see figure 6-10.

Not all of the observed sea level change is caused by ice sheet melting or relaxation of the earth's mantle as a result of ice sheet loading of the earth's crust during the last glacial maximum (see sections 6.2 and 6.3). The possibility of a thermal expansion of sea water is mentioned in, for example, Cazenave and Nerem (2004). A separate discussion is the pos-

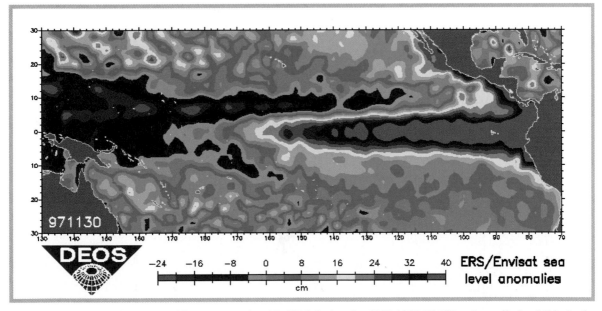

Figure 6-10: Sea level in the Pacific on November 30, 1997 during the 1997-1998 El Niño. Australia is visible in the lower left, the Americas on the right. Compare the amounts of the deviations (on the order of several dm) with the linear trend of 2.8 ± 0.4 mm per year that was deduced in figure 6-8: this multiyear regional variation is about 100 times larger, which illustrates the difficulty of deriving long-term global sea level change trends

sible acceleration of sea level rise for the next century as mentioned in Church et al. (2001). In this case the sea level is expected to change between 20 and 90 cm due to global warming, based on a number of model predictions.

6.2
Space Geodetic Observing Techniques

For a large variety of processes concerning Planet Earth, it is highly advantageous to gather observations from an elevated perspective: a satellite in space. Depending on the exact requirements on coverage and resolution, the altitude of such spacecraft may range from several hundreds of kilometers for the most demanding missions in terms of resolution, to intermediate satellite heights (1,000-1,500 km), to geostationary altitude (35,780 km). For a number of applications it is crucial that the location of the observing platform (i.e., the satellite) is known precisely. To this extent, the position of such spacecraft is typically observed from the surface of the earth or measured from other space platforms.

The following will give an overview of a number of such observation techniques: optical tracking, microwave tracking and intersatellite tracking. A schematic of the various options for satellite tracking is given in figure 6-11. In addition, the microwave observation of quasars and deformation monitoring by radar systems will be discussed; the latter is a further development of the application of space-based platforms to study Earth. The emphasis for all these techniques is on the measurement itself (concept, accuracy, etc.), rather than the analysis and/or scientific results. GPS and GLONASS (Global Navigation Satellite System) are treated in chapter 10.

6.2.1
Optical Tracking of Satellites

Optical tracking that is to say the use of signals in the visible part of the spectrum (wavelength 0.4-0.8 μm), is one of the most fundamental techniques used to observe satellites. Actually, it is based on methods dating many centuries if not millennia earlier than the start of our space age, with the optical observations of planets and stars by 16th and 17th century scientists, and even earlier by ancient cultures.

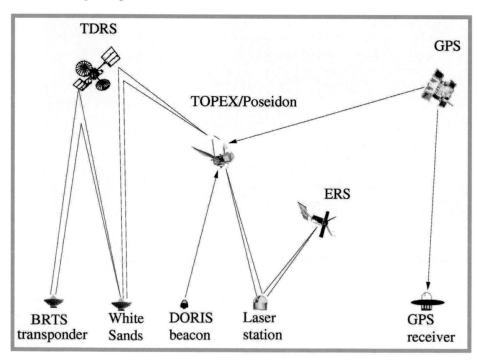

Figure 6-11: Overview of various satellite tracking systems. The signal paths for the various tracking techniques are indicated by the solid lines

Essentially, there are two options to express the (relative) position of the spacecraft with respect to the observer: direction (i.e., angles with respect to some reference) and distance. In the first decade after the launch of Sputnik 1 (October 1957), direction observations in particular were popular for orbit determination: an observer on Earth would visibly follow the target satellite, either directly or with supporting means like binoculars or a simple telescope. An observer could measure the exact moment that the target would cross a reference line in a particular stellar configuration and hence derive a directional observation. Using more sophisticated means, it is also possible to directly quantify the direction angles towards the target. Based on such simple and straightforward observations, the characteristics of satellite orbits (i.e., Kepler elements) could be estab-

lished, which led to remarkable scientific observations (e.g., King-Hele 1992). Baker-Nunn and Hewitt cameras, amongst others, were used for this purpose. This technique is not sufficiently accurate to meet present-day orbit requirements for geodetic satellites, however.

The Satellite Laser Tracking (SLR) technique developed significantly in the mid-1970s and early 1980s, and in particular the MERIT project ("Monitoring of Earth Rotation and Intercomparison of the Techniques of observation and analysis") were means to advance analysis and science applications. Further stimulation came from NASA's Crustal Dynamics Project, which aimed to observe crustal deformations and earth rotation, mainly using SLR and Very Long Baseline Interferometry (VLBI). In Europe, a stimulating initiative was the so-called WEGENER

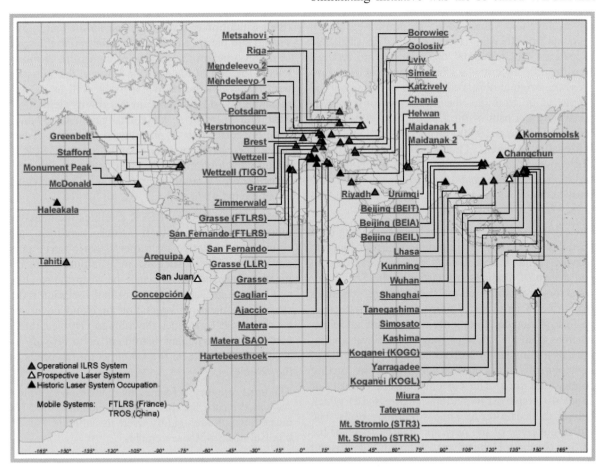

Figure 6-12: Overview of SLR stations (Web: *ILRS*)

project ("Working Group for the Establishment of Networks for Earthquake Research"), focusing on the assessment of crustal deformations in the central and eastern Mediterranean area (several groups from the USA also participated).

At the time of writing, the SLR technique has reached a mature state. Modern SLR stations acquire individual range observations with an absolute accuracy of a few mm. The travel-time observations are obtained with an accuracy of a fraction of a mm, whereas the corrections for station and satellite characteristics can be assessed with an accuracy of 2-5 mm (depending on the target satellite and station-specific aspects like signal strength, detector technique, editing, etc.). The Marini-Murray (1973) model is typically used to represent the effects of the troposphere; at zenith, this model is accurate to a single mm, whereas the uncertainty may increase at lower elevations (down to 15-20 mm). New developments are emerging (see below). The stochastic aspects of the individual observations are reduced significantly since it is a standard approach to compress the original observations into so-called normal points; as an example, the number of observations taken during a pass of the LAGEOS satellite can be reduced from near 10,000 to just 14, without loss of information content. Time-synchronization of the SLR observations is achieved (at ns level) by GPS techniques. These characteristics make SLR the most accurate technique in terms of absolute distance observation.

Since 1998, the various aspects of laser ranging have been coordinated by the International Laser Ranging Service (Web: *ILRS*). This covers the development and design of missions (in particular the SLR retroreflectors), organization of the tracking schedules (by virtue of its use of a transmit/receive telescope, an SLR station can track only one satellite at a time, whereas multiple satellites may be in view simultaneously), stimulation and coordination of technical developments, and coordination of analysis, quality control and such. The current network consists of about 40 active SLR stations (figure 6-12) which deliver their data on an almost real-time basis. The global coverage of the network appears to favor the earth's Northern Hemisphere,, although the situation for the Southern Hemisphere certainly has improved

over the past decades. Considering the spatial resolution of the network, the main contribution of the SLR technique is not so much on detailed phenomena, but more on global aspects of System Earth, in particular those that rely heavily on "absolute correctness." At this moment, about 25 satellites are operationally tracked by SLR (figure 6-13). Various types of missions and spacecraft can be distinguished.

The first category consists of geodetic satellites. Typically, these are passive (i.e., uncontrolled) cannonball-shaped satellites, with a more or less even cross-sectional area and a high mass (to minimize the effects of drag and surface forces like solar radiation pressure). Such satellites are typically used for the observation of the exact location of the geocenter (the center-of-mass of the earth with respect to the geometric center of the earth), global scaling (the assumption of a proper value for the length of a second and the speed of light translates into the scale of dimensions on the surface of the earth), to monitor the rotation of the earth (location of the spin axis with respect to the solid Earth), as well as to estimate the low-degree terms in the gravity field of the earth and linear or seasonal changes therein, which may be correlated to phenomena like post-glacial uplift, ice-cap changes and such). Clearly, SLR is the only tracking means for such satellites. The quality of the orbital solutions may be as good as about 1 cm in the radial direction.

The second category of SLR satellites includes the remote-sensing spacecraft. Such satellites are equipped with active instruments that observe phenomena on Earth, such as radar altimeter instruments to measure the distance between the satellite and the sea surface along the orbital footprint, radar systems to measure vegetation or cloud coverage, etc. Here, SLR provides the tools to very accurately calibrate the particular instrument or to provide an absolute constraint on the solutions for the orbits of the vehicles.

The third category is formed by navigation satellites. As was the case for the previous category, the SLR measurements are mainly used to calibrate the orbital solutions of these vehicles.

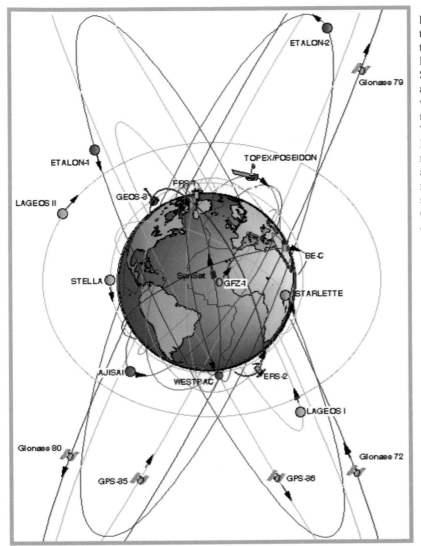

Figure 6-13: Overview of satellites that are operationally tracked by the SLR network (ILRS 2005). LAGEOS-1 and –2, Starlette, Stella, GFZ-1, Ajisai and Etalon-1 and –2 are geodetic satellites, for which SLR is the only tracking means. ERS-1 and –2, ENVISAT, TOPEX/Poseidon, Jason, GFO-1, ICESat, GRACE-A and B are remote-sensing satellites. GPS-35 and –36, and Glonass satellites are navigation satellites. Navigation satellites are being dealt with in Chapter 10. The other satellites are of the experimental type

Finally, the ILRS also tracks experimental satellites that carry exotic experiments and do not belong to any of the categories mentioned above.

Clear advantages of SLR are its simplicity and absolute quality. Drawbacks are its weather-dependency (SLR cannot operate in cloudy situations) and its costs (development and operations). To overcome the latter problem, the SLR community foresees more and more automation and stand-alone systems. Other interesting developments of recent years are simultaneous ranging at multiple wavelengths (to eliminate the troposphere delay in the observations,

and to develop new models to correct for these effects) and the use of high-frequency (kHz) systems to better deal with the satellite signatures.

6.2.2
Microwave Tracking of Satellites

An alternative option for tracking satellites is the use of radio signals. Such concepts use signals with frequencies between 0.4 and 8.5 GHz, for which the earth's atmosphere is in principle fully transparent (the radio window, comparable to the optical window in use by SLR signals). Microwave systems are not hindered by clouds or other meteorological circumstances. Also, the operational costs of microwave systems are considerably lower than those of optical systems. The use of radio signals also allows the combination of tracking signals with data and commands. A disadvantage is the complexity of the hardware and software required, both on the ground and for the space segment. This is a clear drawback with respect to the simplicity of the optical observations.

The measurement principle of microwave systems is the observation of a Doppler shift: a shift in frequency between the transmitted and the received signal. It is essential that the receiver has exact knowledge of the frequency with which the signal was broadcasted by the transmitter. To illustrate: a transmitter frequency of 2 GHz broadcast by a satellite moving at a relative velocity of 6 km/s would result in a Doppler shift of 40 kHz. For scientific investigations the Doppler shifts are not evaluated on a single epoch, but rather observed and integrated over a period of time, typically over an interval of 20 s. This results in the observation of the so-called Doppler count.

Like optical signals, radio signals are affected by the atmosphere. The most important contribution here arises from the ionosphere, the region with high concentrations of charged particles (electrons, ions). This part of the atmosphere causes two effects: a path delay (the signals are received later than they would in vacuum), and a phase advance (the apparent frequency of the signals becomes lower). As expected, the effect is proportional to the so-called Total Electron Content (TEC) the signals encounter on their way from transmitter to receiver. Fortunately, this effect is also frequency dependent, and by observing in two (slightly) different wavelengths it is possible to deduce and hence eliminate the ionospheric perturbation. To quantify these effects: the path delay is in the order of 10 m, whereas the effect on the Doppler observation is in the order of 1 cm/s.

A second source of perturbation is the troposphere. Unfortunately, this perturbation does not depend on frequency, so it cannot be eliminated by observing at different frequencies. Instead, a model is used to represent theses perturbations, such as the Davis model, which expresses the delay as a function of

local atmospheric parameters like pressure, temperature and humidity. Typical perturbation values are about 2 m for the "dry" component (the term which ignores the contribution from humidity), and about 10 cm for the "wet" component. For range-rate observations, the effect is quite small.

As mentioned before, the basic principle of microwave observations is the measurement of a Doppler shift which basically provides information on relative velocities along the line of sight. It is also possible to obtain range observations, but in that case a particular code has to be added to the signal. A number of specific microwave tracking techniques will be discussed below: X/S/Ku-band tracking, DORIS, GPS, GLONASS, and the future Galileo system.

6.2.3
X/S/Ku-Band

The majority of spacecraft are in need of some form of communication from the satellite to Earth (payload data, household data) and from Earth to the spacecraft (commands). This is typically achieved through a communication system which operates in S-band (2 GHz), X-band (8 GHz) or Ku-band (12 GHz). The majority of spacecraft have modest requirements in terms of orbit determination, so the communication system can also be used to provide the tracking information. In order to observe a Doppler shift and hence obtain useful information for orbit determination, one can either retransmit received range signals or assess the coherence between the transmitted and received signal (Larson and

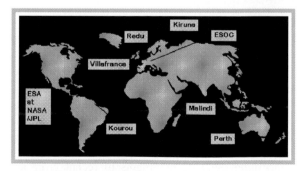

Figure 6-14: Overview of stations participating in the European Space Operations Center (ESOC) Station Network (ESTRACK)

Wertz 1992).

Various networks exist to perform the combined task of communication and tracking: ESA's ESTRACK; NASA's Ground Spaceflight Tracking and Data Network (GSTDN); NASA's Goddard Trajectory Determination System (GTDS); NASA/JPL's Deep Space Network (DSN, located in Goldstone, USA, Canberra, Australia and Madrid, Spain) which is fully dedicated to interplanetary missions; NASA's Polar Network (Poker Flat, Wallops, McMurdo, Svalbord); a network managed by CNES; as well as the private networks of industry (Space Systems Loral, Boeing). As an illustration, figure 6-14 shows the location of the stations in the ESTRACK network.

6.2.4
Doppler Orbitography and Radiopositioning

In the late 1980s, the Centre National d'Études Spatiales (CNES), the Groupe de Recherche en Géodésie Spatiale (GRGS) and the Institut Geographique National (Web: *IGN*) developed the Doppler Orbitography and Radiopositioning Integrated by Satellite (DORIS, Web: *IDS*) system. The DORIS tracking system consists of a receiver on board the host satellite, a permanent network of ground beacons and two master beacons (located in Toulouse, France and Kourou, French Guyana) which send commands and data required to operate the onboard system. In essence, DORIS is a one-way Doppler tracking system, where the receiver component is located on board the spacecraft. As usual for radio signals, the signal is distorted by the ionosphere. To overcome this effect, the ground beacons transmit two ultrastable frequencies: 2,036.25 MHz and 401.25 MHz. The latter frequency is also used for transmission of elementary data. DORIS works with integrated Doppler counts, measured over intervals of 10 s, and the observations (after correction for ionosphere and troposphere) have a noise value of about 0.2 mm/s.

The first version of DORIS was flown on SPOT-2 (table 6-1). Since then, it has proved to be a mainstay on many French, ESA and multinational missions. Particularly interesting has been the development of the DIODE navigator, where the satellite not only receives signals from the ground beacons (and is thus able to derive Doppler measurements), but is also able to compute its position in real time, an unprecedented and unique feature at the moment of writing. ENVISAT, the European satellite for remote sensing and climate studies, is equipped with another new feature of DORIS: it operates the second generation DORIS receivers, with half the mass of the older receivers, yet including the DIODE navigator and a dual channel capability which allows the simultaneous measurement of Doppler shifts with respect to two different ground beacons (so far, a DORIS receiver could handle just one beacon at a time).

The DORIS ground network consists of about 50 permanent beacons, which operate fully automatically and autonomously. They are evenly distributed

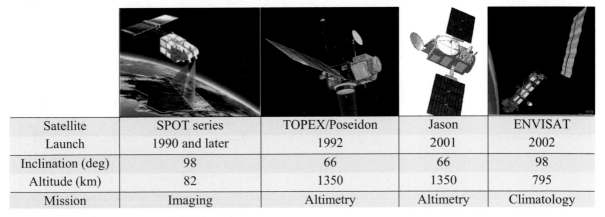

Satellite	SPOT series	TOPEX/Poseidon	Jason	ENVISAT
Launch	1990 and later	1992	2001	2002
Inclination (deg)	98	66	66	98
Altitude (km)	82	1350	1350	795
Mission	Imaging	Altimetry	Altimetry	Climatology

Table 6-1: Satellites that carry a DORIS receiver

over the entire globe, thereby providing a unique global facility for orbit determination.

6.2.5
Galileo

An interesting development taking place in Europe at this moment is the development of a third global navigation/positioning system, Galileo. Although the concept of Galileo shows many similarities with respect to that of GPS and GLONASS, it does have a number of characteristic differences with respect to its predecessors. First of all, it is purely designed for civilian users. Secondly, signal integrity has been considered a major aspect from initial design onwards.

6.2.6
Microwave Observation of Quasars

The technique of Very Long Baseline Interferometry (VLBI) is based on the observation of microwaves emitted by specific stellar sources. Because of its use of the same part of the spectrum, the VLBI technique shows some resemblance to other radio techniques, like GPS and DORIS. The main element of the VLBI technique is the use of a combination of two telescopes that observe the same stellar radio source. To get a reasonable angular resolution of the radio observations (e.g., at 8 GHz), a stand-alone system would need a diameter in the order of about 15 km. Combining the observations of two VLBI telescopes that are several hundred if not thousands of kilometers apart gives a clear advantage in terms

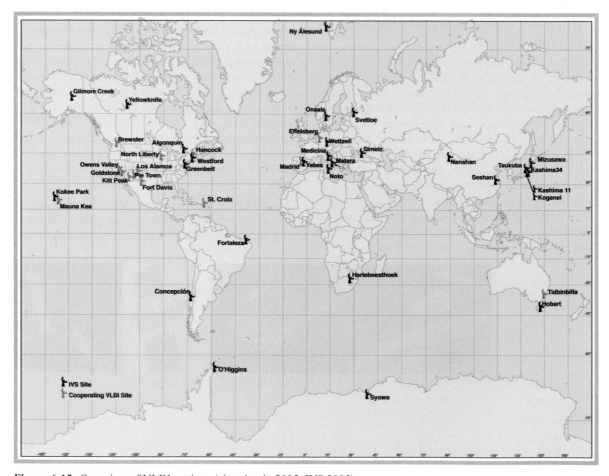

Figure 6-15: Overview of VLBI stations (situation in 2005, IVS 2005)

of effective angular resolution (in the order of 0.001 arcsec, Lambeck 1988).

The principle of VLBI can be explained as follows: two ground stations measure the difference in time of arrival of the wave front coming from the same radio source. A crucial element of this observation concept is the implementation of time in both receiving stations. First of all, each "local" time scale has to have a resolution of better than 100 ps (a typical arrival time difference is in the order of 0.02 s maximum), and secondly the clocks of the two stations need to be synchronized perfectly. The arrival time difference can be observed in two aspects: a phase delay or a group delay. As with the observation of GPS carrier phases, the phase delay observation is complicated by the absence of information on the integer number of cycles that are needed to cover the distance. However, this problem can be dealt with if a sufficient number of different stellar sources are observed.

Like GPS, the radio signals observed by the VLBI telescopes cannot be related to a reference point on Earth directly. First of all, one needs to model the phase center position exactly, at mm level. More important, the signals are hampered by ionospheric and tropospheric delays. Although the ionospheric delay may reach values of several tens of meters, it typically poses no problem, since the observations are accomplished using two different wavelengths (X-band (8 GHz) and S-band (2 GHz)). The delay that originates from the troposphere is much more problematic, however, since it does not depend on frequency and can only be eliminated by applying model representations. The nominal effect can be as large as 3 m, and models are typically restricted to 10-20 % accuracy (Lambeck 1988).

VLBI relies on perfect correlation of the stellar signals as they are received at two different ground stations. To this aim, correlators have been developed since the early days of VLBI (the late 1960s) to handle the large amounts of data efficiently; state of the art is the Mark 5A correlator (IVS 2005). In order to obtain accurate solutions for the three-dimensional interstation baseline, the stations need to observe quasars in different parts of the sky. It will be clear that such observations need to be organized well.

Like the other geodetic services, this community has organized itself in a global service: the International VLBI Service (Web: *IVS*). This organization has the

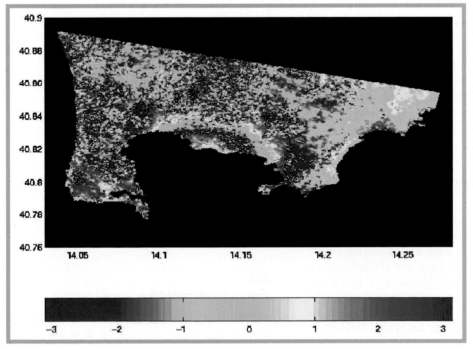

Figure 6-16: Deformation (in cm) near Mt. Vesuvius during the period 1995 to 1999. The city of Naples is situated in the lower left part of the colored area. The Vesuvius volcano, situated at the center of the image, is clearly undergoing vertical motion, as indicated by the concentric pattern

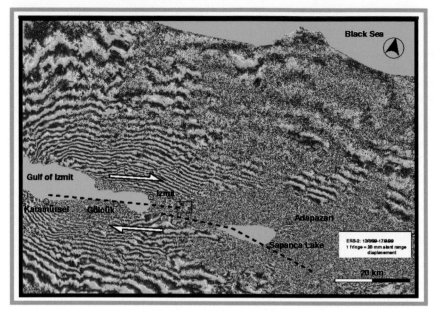

Figure 6-17: SAR interferogram of the August 1999 Izmit (Turkey) earthquake showing relative line-of-sight displacements. The part above the dashed line has shifted to the right with respect to the part below the dashed lines. Not everywhere does this InSAR image produce a clear pattern of fringes: especially places where there are forests or where there is agricultural activity do not produce coherent patterns related to the earthquake

same responsibilities and activities as the other services: coordination of measurement (campaigns), technique development, analysis coordination and control, outreach, etc.

Figure 6-15 shows the current network of VLBI stations, which encompasses about 28 stations (IVS 2005). The VLBI observations are used for a variety of fundamental space geodetic applications: crustal deformation studies, tidal effects and earth orientation investigations. The latter aspect is unique for VLBI, since it is the only technique capable of observing all aspects of the rotation of the earth with respect to an inertial celestial reference frame: polar motion, length-of-day variations, precession and nutation.

6.2.7
Deformation Monitoring by Radar Systems

Synthetic Aperture Radar interferometry (InSAR) is a technique by which radar images taken from satellites are combined to deduce high-precision (down to a few mm) images of crustal and ice deformation over relatively large (up to hundreds of km) areas. As such, it can be used on its own or in combination with other (point-)positioning techniques such as GPS. Two main applications for which InSAR has

successfully been applied are the observation of crustal displacements related to volcanic and seismic activity.

Figure 6-16 shows the deformation at Campi Flegrei, located near Mt. Vesuvius in Italy, over the period from 1995 to 1999 (Usai 2001). This area is subject to strong volcanic activity, as testified by the continuous alternation of phases of strong uplift with periods of strong subsidence (since 1985 the area has been undergoing subsidence). The deformation pattern depicted in figure 6-16 was derived from a set of interferograms derived from ERS satellite observations.

An example of InSAR applied to the study of seismic hazards is given in figure 6-17. It shows the relative deformation pattern of the August 17, 1999 Izmit earthquake obtained by comparing two radar images (of August 13 and of September 17 of that same year) that were observed by the ERS-2 satellite. Each full-color cycle ("fringe") represents a relative displacement of 2.8 cm in the line of sight of the satellite (Hanssen 2001).

6.3
Earth Orientation

In the 19th century it was already observed that the rotation of the earth is not regular. Both the rate of rotation and the position of the rotation axis were shown to be variable. Nowadays we know that these changes occur on all time scales: from shorter than a day to geological scales of hundreds of millions of years.

The changes in position of the rotation axis can be divided into two main categories: those in which the position of the axis changes with respect to the distant stars but not with respect to the earth's surface, and vice versa. For the latter category a hypothetical observer in space would see the earth shift underneath its rotation axis as a solid unit while the rotation axis itself remains fixed with respect to the stars, while an observer on Earth would see the rotation axis wandering over the earth's surface. Displacements of the axis of rotation with respect to the "fixed" stars (changes in which the whole planet is moving rigidly as a unit) are mainly due to external forces, notably the gravitational interactions between the earth and the sun, the moon and the other planets of the solar system. The astronomically well known precession and nutation are examples of this. The external forces exert a net torque on the equatorial bulge of the earth, as a consequence of which the rotation axis will spin. The most important periods are about 26,000 years (precession) and 18.6 years (nutation).

6.3.1
Motion of the Rotation Axis

Displacements of the axis of rotation with respect to a fixed position on the earth's surface are mainly due to mass displacements in the interior of the earth and in the hydrosphere and atmosphere. These mass displacements will generally induce changes in the moments and products of inertia. As the earth is a deformable body, the rotation axis will readjust itself to the new situation by shifting over the surface. The rotation axis does not change its position with respect to the stars, as during these mass displacements the angular momentum of the earth is conserved.

Apart from tidal interactions, there are a number of possible mechanisms responsible for the observed rotational variations, like growth and decay of ice sheets, changes in sea level, ocean currents, winds

and changes in the pressure distribution of the atmosphere, seasonal changes, earthquakes, tectonic plate movements, changes in convection of the mantle and core, and interactions between the core and mantle. Each of these mechanisms operates on specific time scales, and this is reflected in the frequency spectrum of the changes in rotation of the earth.

The observed changes in the position of the rotation axis with respect to the earth's surface, generally termed polar wander, consist of two kinds of movements: periodic and linear. The periodic motions consist mainly of two periods. The annual wobble is principally due to seasonally varying zonal winds. The cause of the Chandler wobble with a period of about 14 months is still largely unknown, though it is generally assumed that a combination of atmospheric winds and ocean currents is one of the principle contributors (Gross 2000). This periodic movement, which is essentially the free precession of the

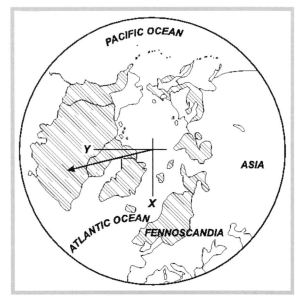

Figure 6-18: Present-day secular drift of the rotation axis (Sabadini and Vermeersen 2004) on the Northern Hemisphere. The arrow points in the direction of the observed present-day drift of the rotation axis. This direction is the same as is found from glacial isostatic adjustment models: present-day post-glacial rebound due to the melt of the great Pleistocene ice sheets (areas indicated by the shaded parts) induces mass displacements that shift the earth's rotation axis

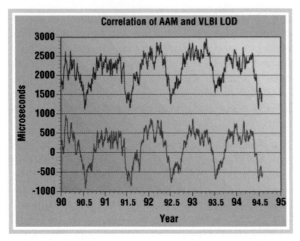

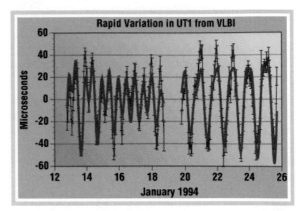

Figure 6-19: Length of day variations as recorded by VLBI (Ben Chao, Goddard Space Flight Center). The blue line in the upper part of the figure depicts the VLBI-observed length of day variation as a function of time; the red line in the lower part simulates values calculated from the effects of wind. Apparently, there is excellent correlation

Figure 6-20: Length of day variations observed by VLBI compared with a global ocean tide model (Goddard Space Flight Center). The blue line depicts the VLBI-observed values; the red line those derived from the global ocean tide model. The two seem to correlate quite well

earth, was theoretically predicted in the 18th century by Euler (therefore it is also called Eulerian free precession) but not observed until the end of the 19th century.

The contemporary secular drift has been determined by astrometric observation. Its direction on the Northern Hemisphere is illustrated in figure 6-18, which also shows the Laurentide and Fennoscandian ice sheets that covered the northern parts of North America and Europe, respectively, during the last Ice Age. Post-glacial rebound resulting from the decay of these ice sheets (see below) is thought to be the main cause of this secular drift, although mantle convection and plate tectonics could also make a contribution.

6.3.2
Observation by Space Techniques

Any satellite tracked from the earth's surface can be used to determine temporal variations in station coordinates, earth rotation pole positions, or rotation speed. High quality earth rotation results have been obtained from GPS and in particular satellite laser ranging (SLR) to both Lageos satellites and Star-

lette. In all these cases the center of mass of the planet and the modeling of satellite trajectories play an important role in the discussion. Conceptually a much different and in fact a more pristine technique to do the same job is Very Long Baseline Interferometry (VBLI) described above. This technique has the advantage of being independent of satellite orbit dynamics or center of mass knowledge, at best the center of figure of the earth is relevant in the discussion.

For earth rotation research the goal is to observe both polar motion values and variations in the length of days. Contained within their spectra is a wealth of geophysical information where the atmosphere, the oceans and the solid parts of the earth each transfer momentum to a dynamically rotating Earth.

6.3.3
Link with Solid Earth, Ocean and Atmosphere Dynamics

Two examples of a comparison between earth rotation variations observed by VLBI and the atmospheric and the oceanic tide signal are shown in figures 6-19 and 6-20.

In figure 6-19 the blue line represents the observed variation in the length of day (LOD) by VLBI, whereas the red line shows the model result matching this data. In this case the momentum transferred from the winds to the rotating earth are computed. In figure 6-20 the same is done for tidal currents as they are predicted by a global ocean tide model.

More recent is the result obtained by Richard Gross at JPL who found that a significant part of the 14 month Chandler wobble, which has been known for over a century, can be explained by ocean bottom pressure values obtained from a global ocean model (Gross 2000).

Long term drift phenomena in polar motion are often referred to as polar wander. Vermeersen and Sabadini (1999) explain that close connections exist between changes in the position of the earth's rotation axis with respect to the earth's surface (polar wander), variations in sea level, the rise and decline of ice ages, and global change on geological time scales.

6.4
Satellite Gravity and Geoprocesses

Gravity and the gravitational field of the earth constitute another class of observational signatures from which details about the structure and dynamics of the earth can be derived. Both static gravity and gravitational field as its time-varying parts can be used for such interpretations. The geoid is the gravi-

tational equipotential surface that represents a motionless mean sea level. It is a global height reference surface that also exists in continental areas. In this first decade of the 21st century three satellite gravity missions have or will be launched that already have improved our knowledge about the earth's gravity and gravitational field dramatically: CHAMP, GRACE and GOCE.

6.4.1
The Static Geoid and the Gravity Anomaly Field

A first glance at the geoid models derived from CHAMP and GRACE data in figure 6-21 shows that there is not much correlation with topography. At least, not for the dominating low spherical harmonics up to about degree 30 (the geoid depicted in figure 6-21 would hardly look any different if the spherical harmonic expansion would have been cut off at degree and order 30 instead of degree and order 150).

On the other hand, there is a strong correlation with tectonic and mantle convection features. For example, the geoid highs in the western Pacific and on the west coast of South America coincide with places where old, dense, ocean lithosphere subsides into the mantle. Other geoid highs are related to anomalously hot regions, for example near Iceland and southeast

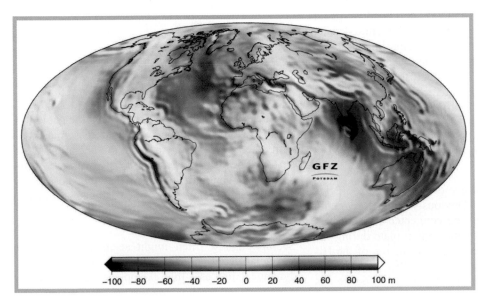

Figure 6-21: EIGEN-GRACE02S geoid model, complete to spherical harmonic degree and order 150. The acronym stands for the CHAMP-derived part (EIGEN) combined with the GRACE-derived part, while the 02S designates that this second release geoid map constructed from CHAMP and GRACE data has been derived from satellite observations only, implying that no terrestrial-based gravimeters have been included (GFZ)

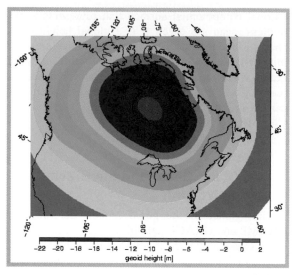

Figure 6-22: Numerically simulated geoid anomalies above Canada due to post-glacial rebound. The figure shows the modeled present-day deviation from isostasy (complete rebound). Apparently, at the center of the Hudson Bay there is still a depression of about 20-25 m with respect to the global mean level resulting from the enormous pressure the Pleistocene ice masses exerted on North America until melting started some 18,000 years ago. Still, the depression as observed by CHAMP and GRACE (figure 6-21) is about a factor 2 larger for this region

with a deviation from isostasy resulting from deglaciation of the Pleistocene Laurentide ice sheet. Some 20,000 years ago the great Pleistocene ice mass complexes above Canada, the Nordic countries in Europe, Russia and Antarctica started to melt. This process ended about 6,000 years ago, with the diminished Antarctic and Greenland ice sheets as "leftovers," leaving geoid lows above the formerly glaciated areas. Today, the earth has not completely rebounded to the isostatic condition, so the geoid low above Canada might be associated with the mass deficit in this region resulting from the termination of the last ice age. However, glacial isostatic adjustment (GIA) simulations produce geoid lows above Canada that are about a factor of two smaller than the geoid low of about 50 m at Hudson Bay that can be deduced from figure 6-22. The simulated geoid anomalies depicted in this figure are dependent on the ice and earth model parameters, but for realistic values of these parameters the geoid anomalies do not range deeper than about −25 to −30 m (Vermeersen et al. 2003).

The observed geoid low of about 50 m at Hudson Bay seems to be a robust feature, which is also visible with about the same value in the EGS96 model (Lemoine et al. 1998). In older geoid maps, e.g., in GEM of Lerch et al. (1979), this geoid low is even higher, about 85 m, so even further away from what GIA models predict. One can vary the parameters of such models to some extent, e.g., mantle viscosity which is the least constrained parameter, so that simulations and observations of the geoid deep agree, but then the simulations lead to large mismatches with respect to other observables like relative sea-level variations derived from raised beach lines around Hudson Bay.

Another look at figure 6-21 shows that the geoid low above Canada might be connected to some extent with other geoid lows, e.g., with the one at the western Atlantic near the eastern coast of the USA. Indeed, it is generally assumed that the most likely explanation for the discrepancy between simulated and observed geoid low is a contribution from mantle convection.

So, although it often might seem obvious that a particular geoid anomaly is related to a certain geodynamical process, this example of the geoid low

of South Africa. At these latter regions, the raised, dynamically sustained topography overcomes the negative geoid contributions of the lower (hotter) mantle densities in producing net positive geoid anomalies. Generally, effects of both mantle density anomalies and their induced dynamic topographies including those of internal mantle layers, must be taken into account to reproduce the geoid signals associated with mantle convection in general and subduction of oceanic lithosphere in particular. The question whether a mantle density anomaly produces a geoid high or a geoid low even depends on the rheology of the mantle, i.e., mantle viscosity (e.g. Vermeersen 2003).

The most conspicuous geoid low in figure 6-21 is centered below India. Obviously, this deep geoid low might in some way be connected with the fast northward movement of the Indian plate during the Cretaceous that resulted in the formation of the Himalayas during the Cenozoic. Another geoid low is centered near Canada. It is tempting to associate this

above Canada shows that these kind of associations should be made with caution. A similar problem of uncertainties concerning the causes of observed geoid features arises in studies on temporal secular variations of the low-degree geoid harmonics (see below).

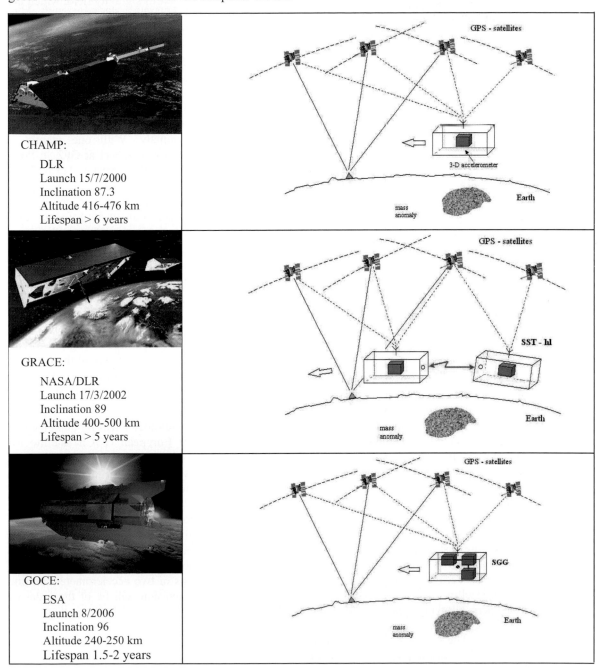

Table 6-2: Satellites observing the earth's gravity field and the respective measurement principle

6.4.2
Observation Scenarios – Present and Future

The motion of earth orbiting satellites is perturbed by the gravity field. When the satellite flies over a certain mass anomaly (table 6-2) the trajectory of this satellite will be affected. The trajectory can be observed very precisely and continuously by accurate tracking such as provided by GPS (section 6.2), referred to as high-low satellite-to-satellite tracking (SST-hl). The gravity field can then be reconstructed indirectly from GPS observations. However, GPS observes the total motion of the satellite affected not only by the gravity field, but also by nongravitational perturbations such as atmospheric drag and solar radiation pressure. By installing a three-dimensional accelerometer at the center of mass of the satellite, these nongravitational perturbations can be observed and separated from the gravitational perturbations. This principle was first realized by the German CHAMP satellite in 2000 (top of table 6-2), which is equipped with a very precise GPS receiver, enabling cm-level positioning, and an accelerometer, enabling the observation of nongravitational accelerations with a precision of 10^{-9} m/s^2 (CHAMP is also capable of observing the geomagnetic field: section 6.5).

More details of the gravity field can be observed by making use of so-called differential techniques. This led to the design of low-low satellite-to-satellite tracking (SST-ll), where the distance between two trailing satellites is observed with high precision, and to satellite gravity gradiometry (SGG), where the differences between the observed accelerations of the two satellites are observed. SST-ll was realized with the launch of the U.S./German GRACE mission in 2002, which consists of two CHAMP-like satellites, one flying behind the other in the same orbit at an approximate distance of 220 km (center of table 6-2). The initial orbit at an altitude of about 500 km is continuously decaying due to atmospheric drag. The low-low satellite link is established by a microwave instrument allowing observation of the inter-satellite position once per second with a precision of about 1 μm/s. The GPS receivers on board the two GRACE satellites are improved versions of the CHAMP receiver. Also, the GRACE acceler-

ometers have an improved precision of about 10^{-10} m/s^2.

The purpose of the observations made by GRACE is to chart variations of Earth's gravity field, which are for a large part of geographic origin due to the presence of density differences within the earth's core, mantle and crust. The measurement principle is that the relative distance between both spacecraft, which increases as the first passes over a density anomaly, decreases as the second satellite is approaching the anomaly while the first one is pulled back, to reduce to the original distance once both satellites have passed over the anomaly. With one month of GRACE measurements researchers at GFZ in Potsdam, Germany and at CSR in Austin, Texas, USA were able to demonstrate an accuracy of their gravity model better than all previous versions that incorporated tracking data to about 50 different satellites collected since the beginning of space flight.

The accuracy of the GRACE gravity models applies to long wavelengths of the field associated with length scales greater than 500 km. This should not be confused with high resolution gravity models that provide resolutions to better than 3 km which are collected from radar altimetry, marine gravity measurements collected by ships, and terrestrial gravity data collected on land. The accuracy of the GRACE gravity models reaches approximately 1 cm on the long length scales, where this number should be interpreted as geoid.

A satellite gravity gradiometer will be implemented on board the future European GOCE satellite together with a high-quality GPS receiver (bottom of table 6-2), which is scheduled for launch in August 2006. The gradiometer will consist of an orthonormal triad of pairs of accelerometers (six in total) allowing observation of the local gravity gradient, further enhancing the capability of observing higher details of the global gravity field. Gravity gradient observations are obtained by taking the difference between observations of two accelerometers on one axis. The relative precision will be of the order of 10^{-12} m/s^2.

6.4.3
Temporal Gravity Field Variations and their Link to Earth Processes

With the help of the GRACE geoid it is also possible to study the effect of geophysical processes that leave a gravity signature. An annual variation of continental water level is essentially such a change in mass distribution. Tapley et al. (2004) demonstrated this effect in the Amazon area. The results are represented in figure 6-23 as monthly geoid height changes. This effect can be translated into the equivalent water height responsible for this observed geoid change. The amplitude of the annual water level variation can be inferred from this observed

geoid change, and this turns out to be approximately 10 cm.

Twenty-five years of satellite laser ranging (SLR) to dedicated satellites as Lageos have taught us that many short-term effects can perturb the outcome of secular solutions (e.g., El Niño). Even nowadays only the degree two term has been well established. On the interpretative side, all is not clear either. It is generally thought that two geodynamical processes are responsible for these secular variations: Glacial Isostatic Adjustment (GIA) and present-day continental ice sheet melt or decay. To what extent tectonic and mantle convection processes might contribute to the secular changes is an open question.

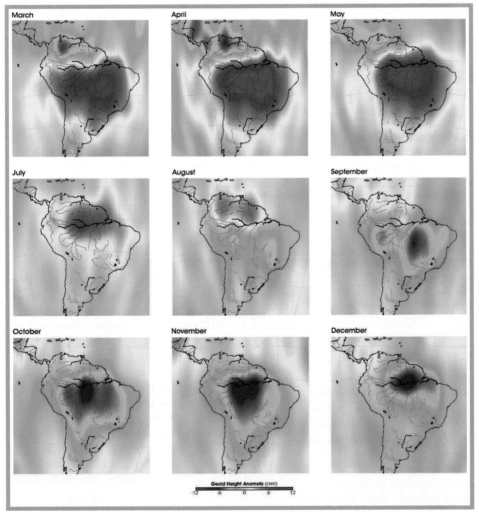

Figure 6-23: Geoid height changes observed by the GRACE system over the Amazon area in 2003 (Tapley et al. 2004). The panels for each of the nine months (January, February and June are lacking) were obtained for a smoothing radius of 400 km. GRACE data have an increased error for such a low smoothing radius, but were nevertheless chosen since the large signals for the Amazon area allow a high resolution. The harmonic degree two coefficients (ellipticity of the earth) were not included

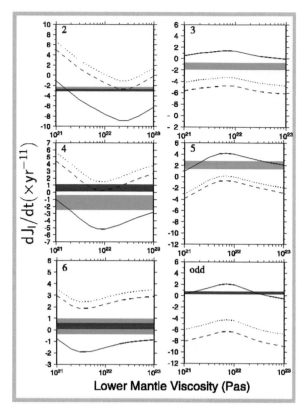

Figure 6-24: dJn/dt as a function of lower mantle viscosity (in Pas) for GIA and contemporary ice sheet changes in Antarctica and Greenland. The value of n is given in the upper right corner in each of the six panels, with "odd" denoting a combination of odd degrees (see Devoti et al. 2001). The green areas denote the observed ranges by Cheng et al. (1997); the red areas denote the observed ranges by Devoti et al. (2001). The solid line is the Pleistocene-only case; the dashed line denotes the effect of including a present-day Antarctic melt scenario and the dotted line the effect of including a present-day Antarctic and Greenland melt scenario

Given all these observational and interpretative uncertainties, it still makes sense to deduce information from the comparison of forward models of GIA and present-day ice mass variations with observational data of the secular changes in the low geopotential harmonics (Sabadini and Vermeersen 2004). In figure 6-24 the secular changes dJ_n/dt of degree n derived from the observations of Cheng et al. (1997) and Devoti et al. (2001) and forward models for GIA and present-day ice mass changes are depicted as a function of the lower mantle viscosity. Only the GIA

models are dependent on this parameter; the observations and the present-day ice mass variations are not dependent on mantle viscosity, but the elasticity of the earth has been taken into account for the effects of the present-day ice mass variations on the geoid.

An interesting pattern emerges from the panels of figure 6-24. In some cases the solid lines, delineating the effects of GIA derived from GIA relaxation models with the Pleistocene ICE-3G ice melt history model of Tushingham and Peltier (1991), cross one or both of the observations. The same is the case if a value of about 5×10^{14} kg per year contemporary ice melt of the Antarctic ice sheet is added (the dashed lines). However, the majority of the solutions cross the observations in all cases. Note also that although the J_4 observations of Cheng et al. (1997) and Devoti et al. (2001) are not in agreement (even having opposite signs); the solutions with and without present-day Antarctic ice sheet decay cross both observations. This remains true if present-day ice sheet decay of the Greenland ice sheet of about $1.5 \cdot 10^{14}$ kg per year is added.

Apparently, present-day ice sheet decay of Antarctica (and perhaps of Greenland) is necessary to reconcile the geodynamic simulations with the SLR-derived observations in figure 6-24, but with a value that is less than $5 \cdot 10^{14}$ kg per year. Formal inversions show that the same lower mantle viscosity solution of about 10^{22} Pa s is found for all panels in figure 6-24 if the Antarctic ice sheet decays presently by a value of $(2.50 \pm 0.80) \cdot 10^{14}$ kg per year.

A contribution from the Greenland ice sheet cannot be discerned from these data: the dotted lines in figure 6-24 represent the case in which a value of about 1.5×10^{14} per year contemporary ice melt of the Greenland ice sheet is added. The dotted line remains close to the dashed line in all panels. The reason is quite simple: the center of the Pleistocene Laurentide ice sheet complex above Canada, which plays a dominant role in the GIA models, was geographically quite close to Greenland. A change in mantle viscosity which causes a change in the dJ_n/dt contributions from GIA can therefore have the same effect on the total values of dJ_n/dt as a contemporary mass variation of the Greenland ice sheet. Low-degree values n < 6 are simply not discriminative enough to break this trade-off effect.

If it would turn out to be possible to discern secular variations of the J_n with CHAMP or GRACE for higher values of n, then such a discrimination between GIA and present-day ice mass variations of the Greenland ice sheet might be possible.

6.5
Earth Magnetic Field and the Geosphere

The magnetic field of the earth largely resembles that of a bar magnet with the South Pole currently directed to Canada and the North Pole to Antarctica, which is slightly off the earth's rotational axis. The magnetic field at the surface of the earth is largely produced by a self-sustaining dynamo operating in the fluid outer core, causing the so-called internal field (figure 6-25). In addition, an important part of the magnetic field at the earth's surface is caused by processes in the atmosphere, referred to as the external field. The magnetic field acts as a sort of cocoon and protects life from a continuous flow of charged particles, the solar wind, before it reaches the atmosphere. Moreover, it plays an important role in controlling many of the physical processes in the earth's environment that directly affect the increasing utilization of highly technological systems in space: the particularly low magnetic field intensity

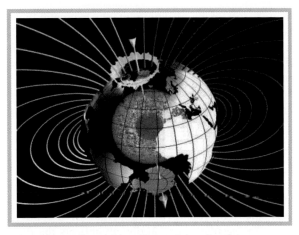

Figure 6-25: The earth's self-sustaining dynamo

in the so-called South Atlantic Anomaly area has caused a series of radiation-induced satellite failures in recent years (Heirtzler et al. 2002).

The geomagnetic field is a vector quantity and in order to describe it at any point a magnitude and direction are required. It varies in magnitude or in direction both spatially and in time. Over geological history, the magnetic field has frequently reversed its polarity allowing recording of Earth's paleomagnetic field and in conjunction plate tectonic motion over time scales of millions of years. It has decreased by nearly 8% over the last 150 years (Jackson et al. 2000), a figure comparable to that seen at times of magnetic reversals (Hulot et al. 2002). Al-

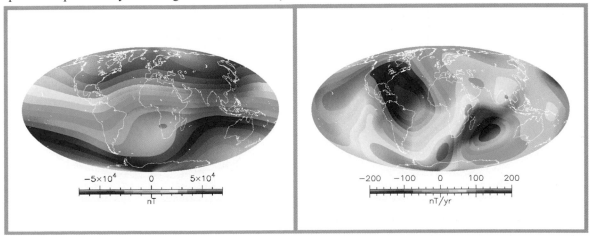

Figure 6-26: Radial magnetic field at the earth's surface of the core for 2000 (left) and its secular trend change (right) according to IGRF 2000. Units are in nanoTesla (left) and nanoTesla per year (right), respectively

though systematic mapping of the magnetic field began already some five hundred years ago with work by the early explorers of the Atlantic Ocean (Fowler 2004), continuous global monitoring can be achieved only by satellites.

6.5.1
The Internal Field

The biggest component of the magnetic field is caused by the earth's core, which is a highly conductive medium. The core field and, in particular, its temporal changes are among the very few means available for probing the properties of the liquid core. The secular variation directly reflects the fluid flows in the outer core. The magnitude of the core field is of the order of 50,000 nT and its secular change varies between –200 and 200 nT per year (figure 6-26). The magnetic field is very weak over the South Atlantic and is decaying even more. The secular trend of the IGRF 2000 model indicates a possible reversal within about 2000 years (Olsen et al. 2000).

The crust and uppermost part of the mantle (referred to as the lithosphere) also contribute to the internal field, with a magnitude of about 50 nT at the earth's surface (figure 6-27, left). The lithospheric magnetic anomaly field is produced by spatial variations in the magnetization carried by crustal and some mantle rocks. This magnetization is partly induced by the ambient field, and is therefore proportional to its strength, and to the susceptibility of the rock. It can

also be a remnant magnetization acquired during the formation of the rock. The lithospheric magnetic field provides information about the history of the core field and about geological activity.

In the oceans, the motion of the electrically conducting seawater generates electromagnetic fields with a magnitude of a few nT. Knowledge about this component of the geomagnetic field helps in better understanding ocean currents and complements observations collected by altimeter and gravity field satellites (sections 6.2 and 6.4).

6.5.2
The External Field

The geomagnetic field is an effective shield against charged particles impinging from outer space onto the earth. The external magnetic field is caused by magnetospheric and ionospheric electrical current systems. The magnetosphere (which is the region encompassing the earth's magnetic field) diverts most of the solar wind around the earth. The particles that do enter into the magnetosphere are guided by the magnetic field and form fundamental structures like the radiation belts and the ring currents (figure 6-28). The motion of charged particles causes electric currents, which can be traced by the magnetic fields. The ionosphere is the part of the atmosphere ionized by solar radiation. The interaction between the solar wind and the ionosphere induces energy into the earth's magnetic field. The magnetospheric and ionospheric magnetic field

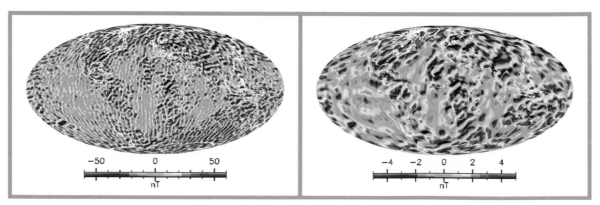

Figure 6-27: Radial magnetic field of the lithosphere from the CHAMP-based MF3 model at the surface of the earth (left) and at 400 km altitude (right). The MF3 model was provided by GFZ Potsdam, Germany. Units are in nanoTesla for both panels

strengths are typically of the order of 60 and 30 nT, respectively, and depend, among other factors, on the solar activity level, which exhibits an 11 year cycle.

6.5.3
Applications

Global surface maps of the geomagnetic field can help not only in the prospection stage of petroleum and mineral exploration, but also to delineate the composite structure of the crystalline basement and to investigate the chemical and thermal evolution of the lithosphere. For many purposes in the fields of ionospheric, magnetospheric and cosmic-ray physics and in studies of crustal fields, particularly in exploration geophysics, internationally produced and agreed global models of the core field and its secular variation, such as IGRF 2000 (Olsen et al. 2000), are widely used.

Geomagnetic information is used with increasing importance as a navigational tool in the drilling of deviated (i.e., nonvertical) wells in the oil and gas industries. It is common for 50 or more deviated wells to be drilled from a single rig and it is therefore necessary to be able to control the dip and azimuth of the drilling tool to within close tolerances

(0.1°). Gyroscopic devices can be used to supply this information, but they are sensitive to vibration, and drilling operations must therefore be suspended while measurements are made. This is expensive and the preferred method of navigating the drill string is, in most cases, to use the geomagnetic field as the source of directional reference.

Geomagnetic models are indispensable for studying Earth's rotation, time-variable gravity, and core-mantle coupling, and for modeling the heat flux under the ice caps (which is of considerable importance for climatology as it provides insight into the ice cap's long-term stability).

Knowledge of the magnetic field and its variations is important for applications such as magnetic compass corrections and navigation; orientation of satellites; guidance and detection systems; bio-magnetism and animal navigation. The magnetic field is the dominant controlling factor regarding the external environment of the earth, space weather. Radiation damage to spacecraft and radiation exposure to humans in space is a matter of increasing concern. In particular, over the South Atlantic Anomaly, the low magnetic field strength allows high-energetic particles from the radiation belts to penetrate deep into the upper atmosphere, creating intense radiation. Recent

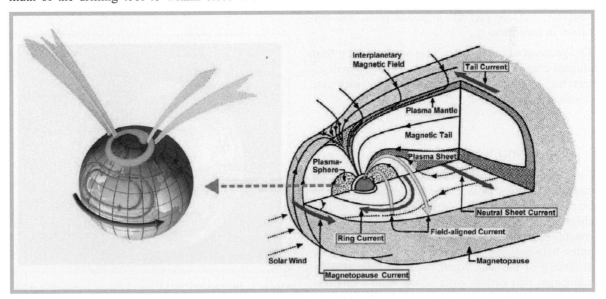

Figure 6-28: Schematic drawing of major current systems in the ionosphere (left) and magnetosphere (right) (ESA, 2004)

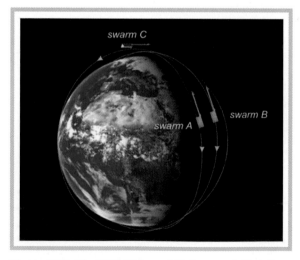

Satellite	MAGSAT	ØRSTED	CHAMP	SAC-C
Agency	NASA (US)	Denmark	DLR (Germany)	Argentina
Launch	30/10/1979	23/2/1999	15/7/2000	21/11/2000
Inclination	96.8	96.1	87.3	98.2
Altitude (km)	352-578	730-850	416-476	628-701
Lifespan/EOL	11/6/1980	8 years	> 6 years	> 4 years

Table 6-3: Satellites carrying a magnetometer

instrument failure statistics suggest that this anomaly has shifted to the Northwest. Geomagnetic models derived from satellite data have confirmed this (figure 6-26, right). Accurate and timely geomagnetic field models clearly play a pivotal role in space mission planning and operation. Better understanding of its geographical distribution and its time variations, due to internal dynamics as well as to the changes introduced by solar variability, may help in understanding and mitigating effects regarding damage to satellite systems, disruption of satellite communications, GPS errors, varying orbital drag on satellites, induced currents in power grids, and corrosion in pipelines.

The geomagnetic field may have a role in long-term climate changes since the secular variation will affect the geographical distribution of the incoming cosmic ray flux.

6.5.4
Monitoring from space

Already at the start of the space age, early after the launch of Sputnik in 1957, the first American spacecraft, Explorer 1, carried a Geiger counter and an altimeter to study parts of the earth's magnetosphere. Explorer 1 flew in an eccentric orbit at altitudes between 358 and 2,550 km. This led to the discovery of the Van Allen radiation belts and the realization that certain parts of the earth's environment, which are influenced strongly by its magnetic

field and interaction with the solar wind, are particularly hazardous to spacecraft.

Observing the earth's magnetic field globally can be done efficiently by low earth orbiting satellites. MAGSAT, launched in 1979, was the first satellite capable of mapping the earth's magnetic field with a spatial resolution of about 1,500 km (table 6-3). ØRSTED, launched in 1999, is the second satellite with a similar capability offering the first opportunity to study (secular) changes in the core field. With the launch of CHAMP and SAC-C in 2000, three satellites capable of observing the geomagnetic

Figure 6-29: The SWARM constellation consisting of one pair of satellites flying en echelon and one satellite flying at a higher altitude

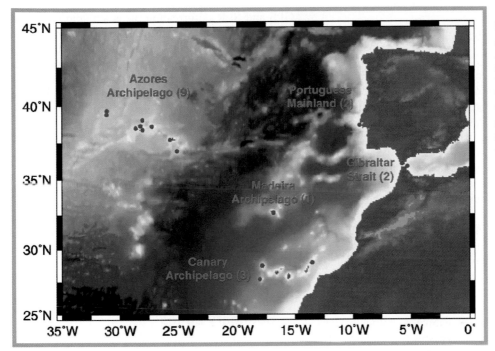

Figure 6-30: North American-Eurasian-African plate boundary zone and TANGO network. The numbers behind the geographical indications denote the number of GPS stations

field are flying simultaneously (status February 2005), allowing better spatial and temporal sampling of the complex geomagnetic field. However, the orbital geometry of these satellites was not optimized to this aim (the satellites are owned by different agencies and also serve other objectives, such as observing the earth's gravity field).

In 2004, the European Space Agency (ESA) selected the Swarm mission, which is a mission especially optimized for separating the spatial and temporal contributions to the many different sources of both the internal and external magnetic field. Swarm will consist of one pair of satellites flying an echelon at about 450 km altitude with an inter-satellite distance around 160 km for observing the high-resolution part of the geomagnetic field, and a satellite flying at about 550 km altitude improving the temporal sampling (all in near polar orbits, figure 6-29). The objectives of Swarm are to observe the core field and its secular changes down to length scales below 3,000 km, and the lithospheric field with a resolution of around 150 km. The low altitude of the Swarm satellites is required to reduce the damping effect of the high resolution internal geomagnetic field, which is already considerable at 400 km altitude (see figure

6-27, right), necessitating high precision instruments. Swarm will be equipped with both a scalar and a vector magnetometer enabling three-dimensional observations of the magnetic field with a precision of about 1 nT at 4 Hz (allowing us to observe for example also the small contribution coming from ocean currents). The Swarm instrument suite will be complemented with a high precision, dual frequency GPS receiver for precise positioning, an accelerometer for deriving atmospheric drag (and in conjunction atmospheric density) and an electrical field instrument (EFI) for observing ion and electron density (ESA, 2004). The unique Swarm constellation allows the separation to a large extent of the several sources that contribute to the internal and external field, enabling better modeling, monitoring and understanding of the magnetosphere and ionosphere as well. Swarm is foreseen to be launched in 2009 and will have a lifetime of at least four years.

Swarm will extend the observation periods of ØRSTED, CHAMP and SAC-C. With Swarm, it is anticipated that for the first time the geomagnetic field can be monitored globally with high precision for a period of at least one full 11-year solar cycle, covering a full spectrum of solar activity levels.

161

6.6
Plate Tectonics

Satellite tracking data, in particular SLR and GPS, in combination with precise satellite orbits have proven to be a powerful means to accurately measure the extremely small present-day motions of the earth's crust, which are the result of geodynamic processes (see section 6.1). The data are used to compute long time series of precise coordinates of the tracking systems, both from permanent and from campaign type stations. Careful analyses of these time series provide detailed information about the absolute positions and relative motion of the stations. Therefore, the main objectives are establishing accurate global and regional geodetic reference frames, improving the representation of the present-day global tectonic plate motion model, studying plate boundary processes and intra-plate deformations, identifying and localizing tectonic micro-blocks, assessing strain accumulation in seismically active regions, and analyzing co- and post-seismic motions. The results also have the potential to provide significant constraints for the development of geophysical models.

Contemporary investigations include both analyses of global GPS and SLR networks as well as studies in selected regions with a clear deformation signal.

GPS and SLR deformation maps are complemented by high-resolution, high-accuracy maps of surface deformations provided by InSAR (see section 6.2). Currently, an important focus is on developing models and algorithms for estimating of surface deformations from InSAR data and on assessing of the quality of the derived deformation maps by analyzing potential error sources and error propagation. In particular, the atmospheric signal in InSAR observations and the problem of temporal de-correlation deserve special attention. The latter is the main limitation of InSAR for monitoring slow deformation processes. This space-geodetic research area is strongly related to natural hazards (earthquakes, landslides, volcanoes, etc.) and has become of great social importance.

6.6.1
Deformation Observations in Plate Boundary Zones

As was already shown in figure 6-2, velocity vectors of the global plate motions as derived from present-day space-geodetic observations are in very good agreement with those obtained from geological observations. This is remarkable for more than one reason: it shows that space-geodetic techniques give highly reliable results and it shows that global plate motions have not changed much over the past few

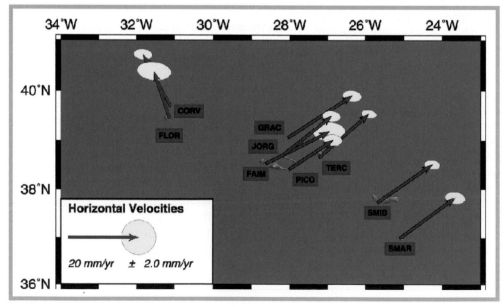

Figure 6-31: GPS velocity vectors of the TANGO sites for the Azores Archipelago

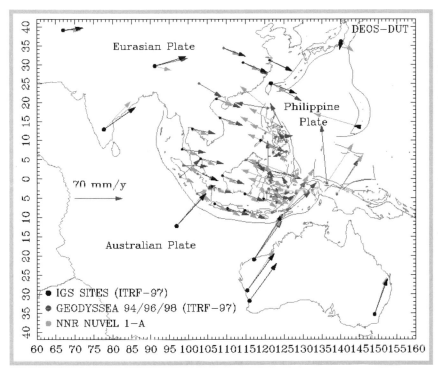

Figure 6-32: GEODYSSEA GPS velocity observations in Southeast Asia. The black and red arrows denote GPS-derived values; the green arrows are derived from the No-Net Rotation NUVEL model (see also figure 6-2 for a global view). The NUVEL model has been determined from geological observations and thus represents values on million-year time scales

million years. Still, we know that on regional scales plate motions can show huge deviations on short time scales from these long-term trends: for example, the great earthquakes of 1960 in Alaska and of 2004 in Southeast Asia have resulted in local plate motions at the rupture zones of more than 10 m in less than a minute. Between these cataclysmic events and the smooth one million year motions there is an often complex pattern of deviating motions at plate boundary zones. Two examples are given below.

6.6.2
The North American – Eurasian – African Plate Boundary Zone

The Azores-Gibraltar region is an area of complex tectonic boundaries. It includes the triple junction between the North American, Eurasian and African tectonic plates and the western segment of the Eurasian-African plate boundary. A detailed study of the geokinematics of this region was carried out under the scope of the TANGO (Trans-Atlantic Network for Geodynamics and Oceanography) project, using

precise observations of relative motions acquired by means of GPS (figure 6-30). Several island groups and the coasts of Europe and Africa provided relatively stable reference points for this study, which gave a better understanding of the neotectonics of this region.

An example of the observed velocity vectors is provided in figure 6-31 for the Azores Archipelago.

A comparison of this velocity field with the global tectonic model NUVEL (see figure 6-2) has shown significant discrepancies. For the Azores region, with the exception of the two western sites FLOR and CORV located on the North American plate, the sites' motions are not in agreement with the Eurasian and African motions predicted by the NUVEL model. The same is the case for the motions of the sites located in the Iberian Peninsula and in the Madeira Archipelago. For the entire network, the estimated velocities are higher than those predicted by this model, which has even led to an alternative tectonic global model.

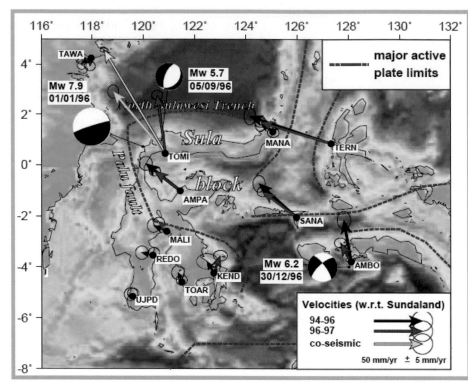

Figure 6-33: GPS velocity observations for Sulawesi hinting at two tectonic blocks separated by the dashed red curve that contains the Palu fault: the Sula block to the northeast and the Makassar block to the southwest. Such differential motions between blocks as determined by GPS observations are of great importance for interpreting regional tectonic motions and therefore for seismic risk and other natural hazard assessments

6.6.3
The Eurasian-Australian-Philippine Plate Boundary Zone

Before the great earthquake of December 26, 2004 stroke the region, accurate GPS measurements in the framework of the GEODYSSEA (Geodynamics of South and South-East Asia) project were made that gave more insight into the complicated plate motions at the boundaries between the Eurasian, Australian and Philippine plates. An example of observed GPS motions is given in figure 6-32.

The observations show that most velocities in the triple junction region are stable. The obtained steady-state velocities with respect to the Sundaland block are shown in figure 6-33.

The observations furthermore enabled the identification of two tectonic blocks in Sulawesi. These are the Sula block with a rotation pole east of Manado (MANA) with clear limits along the Palu-Koro fault in the west and the North Sulawesi arm to the north, and the Makassar block rotating around a pole west

of Makassar (UJPD). The observations of the Palu-Koro fault transect were combined with previous results to determine the velocity profile across the Palu-Koro fault with greater confidence. At present this rapid left lateral strike-slip fault (with a total velocity of about 4.5 cm/yr) is locked at a depth of about 12 km. These observations might be of importance in interpreting future seismic risks in this region.

6.7
Conclusions and Outlook

Satellite geodesy has very rapidly become an indispensable component of modern geodynamical research, and yet it seems that we are only beginning to understand its myriad applications and possibilities. Only twenty years ago who could have foreseen that with artificial satellites we would be able to detect tectonic movements on the order of one mm/yr over distances of hundreds of kilometers? Or that the earth's gravity field would be known globally with an accuracy of one cm over distances of a mere 100 km, even in the most remote and unreachable regions? In this chapter we have only given a few of the many possibilities and applications of recent and upcoming satellite missions and technologies that have made these amazingly accurate observations possible or are expected to make them possible within the next few years. Our understanding of the earth, its structure, its dynamics and the many interrelated processes that take place in its interior and on its surface has increased dramatically over the past few decades and this would never have been possible without the use of artificial satellites and their many applications.

References

Cazenave AA, Nerem RS (2004) Present-Day sea level change: Observations and causes, Rev. Geophys 42, RG3001, doi:10.1029/2003RG000139

Cheng MK, Shum CK, Tapley BD (1997) Determination of long-term changes in the Earth's gravity field from laser ranging observations. J. Geophys. Res. 102, pp. 22377-22390

Church JA, Gregory JM, Huybrechts P, Kuhn M, Lambeck K, Nhuan MT, Qin D, Woodworth, PL (2001) Changes in sea level, Intergovernmental Panel on Climate Change 3rd Assessment Report, Cambridge University Press, pp. 638-689

Devoti R, Luceri V, Sciaretta C, Bianco G, Di Donato G, Vermeersen LLA, Sabadini R (2001) The SLR secular gravity variations and their impact on the inference of mantle rheology and lithospheric thickness, Geophys. Res. Lett. 28, pp. 855-858

Douglas B (1991) Global sea level rise, J. Geophys. Res. 96(C4), pp. 6981-6992

ESA (2004) Swarm – The Earth's Magnetic Field and Environment Explorers, ESA SP-1279(6) (plus Technical and Programmatic Annex)

Fowler CMR (2004) The Solid Earth – An Introduction to Global Geophysics (2nd Edition), Cambridge University Press

Gross RS (2000) The excitation of the Chandler wobble, Geophys. Res. Lett. 27, pp. 2329-2332

Hanssen R (2001) Radar Interferometry – Data Interpretation and Error Analysis, Kluwer Academic Publishers, Dordrecht

Heirtzler J (2002) The future of the South Atlantic anomaly and implications for radiation damage in space, J. Atmospheric Solar-Terrestrial Phys. 64(16), pp. 1701-1708

Hulot G, Eymin C, Langlais B, Mandea M, and Olsen N (2002) Small-scale structure of the Geodynamo inferred from Ørsted and Magsat satellite data, Nature 416, pp. 620-623

Jackson A, Jonker A, Walker M (2000) Four Centuries of Geomagnetic Secular Variation from Historical Records, Phil. Trans. R. Soc. Lond. 358, pp. 957-990

King-Hele DG (1992) A Tapestry of Orbits. Cambridge University Press

Lambeck K (1988) Geophysical Geodesy; the Slow Deformations of the Earth. Oxford University Press

Larson WJ, and Wertz JR (1992) ed.: Space Mission Analysis and Design (2nd Edition), Microcosm Inc. and Kluwer Academic Publishers, Space Technology Series

Lemoine FG, Kenyon SC, Factor JK, Trimmer RG, Pavlis NK, Chinn DS, Cox CM, Klosko SM, Luthcke SB, Torrence MH, Wang YM, Williamson RG, Pavlis EC, Rapp RH, Olson TR (1998) The development of the joint NASA GSFC and the National Imagery and Mapping Agency (NIMA) Geopotential Model EGM96, NASA/TP Report 206861, GSFC Greenbelt Maryland

Lerch FS, Klosko SM, Laubscher RE, Wagner CA (1979) Gravity model improvement using GEOS 3 (GEM 9 and GEM 10). J. Geophys. Res. 84, pp. 3897-3916

Leuliette EW, Nerem RS, Mitchum GT (2004) Calibration of TOPEX/Poseidon and Jason Altimeter Data to Construct a Continuous Record of Mean Sea Level Change, Marine Geodesy 27(1), pp. 79-94

Love AEH (1911) Some Problems of Geodynamics, Cambridge University Press (reprinted as Dover edition in 1967)

Marini JW, Murray Jr. CW (1973) Correction of laser range tracking data for atmospheric refraction at elevations above 10 degrees, NASA-TM-X-70555

Olsen N, Sabaka TJ, Clausen LT (2000) Determination of the IGRF 2000 model, Earth, Planets Science 52, pp. 1175-1182

Sabadini R, Vermeersen LLA (2004) Global Dynamics of the Earth: Applications of Normal Mode Relaxation Theory to Solid-Earth Geophysics, Modern Approaches in Geophysics Series 20, Kluwer Academic Publishers, Dordrecht

Tapley BD, Bettadpur S, Ries JC, Thompson P, Watkins MM (2004) GRACE measurements of mass variability in the Earth system, Science 305, pp. 503-505

Tushingham AM, Peltier WR (1991) ICE-3G: A new global model of late Pleistocene deglaciation based upon geophysical predications of post-glacial relative sea level change, J. Geophys. Res. 96, pp. 4497-4523

Usai S (2001) Bypassing the decorrelation problem in SAR interferometry. In: Vermeersen LLA, Klees RAP, Ambrosius BAC eds.: "Earth-Oriented Space Research, DEOS Final Report 1996-2000", pp. 92-94

Vermeersen LLA (2003) The potential of GOCE in constraining the structure of the crust and lithosphere from post-glacial rebound. Space Sci. Rev. 108(1-2), pp. 105-113

Vermeersen LLA, Sabadini R (1999) Polar wander, sea-level variations and Ice Age cycles, Surv. Geophys. 20, pp. 415-440

Vermeersen LLA, Schott B, Sabadini R (2003) Geophysical impact of field variations. In: Reigber CH, Lühr P, Schwintzer P eds.: "First CHAMP Mission Results for Gravity, Magnetic and Atmospheric Studies", Springer Verlag, Heidelberg, pp. 165-173

Web References

IDS: http://ids.cls.fr/welcome.html, status March 15, 2005

IGS: http://igscb.jpl.nasa.gov, status March 15, 2005

ILRS: http://ilrs.gsfc.nasa.gov, status March 15, 2005

IVS: http://ivscc.gsfc.nasa.gov, status March 15, 2005

Looking up: Stars and Planets

Overleaf image: Dust clouds in the Orion nebula (NASA)

7 Astronomy and Astrophysics

by Ralf-Jürgen Dettmar

The human fascination for space is to a significant extent based on the fundamental questions relating to our very existence in the universe. The starry night sky thus always was an important source of creative inspiration not only for scientists and engineers but also for artists and philosophers, poets and magicians.

The advent of space exploration in the second half of the 20th century has led to a completely new view of the sky resulting in radically new concepts for our understanding of the universe. It was mainly the new possibility to observe electromagnetic radiation from astrophysical sources in spectral regimes not accessible at the base of the earth's atmosphere which changed our view. In the following, the physical background is briefly reviewed. Some constraints given by the space environment will be discussed in

Figure 7-1: The bubble of interstellar material surrounding the variable supergiant star V898 Mon imaged with the Advanced Camera for Surveys (ACS) on board the Hubble Space Telescope (HST). Three integrations using different filter path bands are combined for this true color image. The object is a wind blown bubble of gas around a star in a late phase of stellar evolution. This bubble is illuminated by a light echo from an outburst of the central variable star (NASA/ESA/STScI)

section 7.2 before a few examples of the most challenging questions of today's astronomy are compiled. Then typical examples of space astronomy missions are given and a perspective for the future is developed[1].

7.1
Astronomy from Space

The case for Space Astronomy was made immediately after the availability of launchers right after World War II. Among the very first astronomical observations was the attempt to record the ultraviolet (UV) spectrum of the sun with a spectrograph launched by a group of scientists at Naval Research Laboratory (NRL), headed by Richard Tousey. They were using a V-2 rocket in October 1946 with first results reported immediately thereafter by Baum et al. (1946). The early history of space astronomy is subject of the book by DeVorkin (1992). Already at that time the value of a "general purpose" telescope outside the earth's atmosphere was first noticed (Spitzer 1990), an idea followed up more than two decades later with several astronomy satellites. Since the 1970s astronomy satellite projects have been part of the programs of all major space agencies. The selection of missions is made in close collaboration

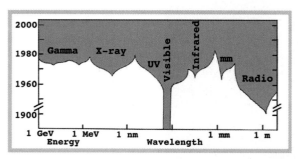

Figure 7-2: The development of wavelength coverage during the past century. Only radiation in the small optical window of approximately one octave around the center of the visible wavelength regime at 550 nm is accessible with traditional optical astronomy. The exploration of the universe at all other frequencies started in the second half of the 20th century - most of them are accessible from space only (AIRUB/Haberzettl, adopted from Léna et al. 1998)

with the scientific communities. Typically, a few large-scale projects (> €500 million) are accompanied over a period of 10-15 years by a couple of medium-size and small missions (of only a few million euro). While the larger missions are frequently operated in an observatory-style mode with open access for the community to a broad spectrum of science instruments, the smaller missions are more focused scientifically and more limited in their scope.

Today the enormous success of the Hubble Space Telescope (HST)—which actually was started in the framework of the *Great Observatories* already in the 1970s and originally called LST (Large Space Telescope)—in popularizing its beautiful high resolution images to a broader public has to be acknowledged. This very convincingly demonstrates the outstanding data quality that can be achieved with space observatories. One of the deepest images of the universe, i.e., showing the faintest objects, obtained by HST was already presented in chapter 1 (figure 1-6). Another of the beautiful HST images is reproduced in figure 7-1. It shows the light echo caused by the luminosity outbreak of a supergiant star being reflected by circumstellar material, probably blown out in an earlier active phase. An inspiring collection of HST images with accompanying interpretations of the scientific results can be found in several popular illustrated books (e.g., Fischer and Dürbeck 2003) or on the Web pages of the Space Telescope Science Institute in Baltimore, USA (Web: *STSCI*) and the Space Telescope European Coordination Facility of the European Space Agency (ESA) at the European Southern Observatory (ESO) in Garching, Germany (Web: *STECF*).

The scientific reasoning for space astronomy arises from several lines of argument, most of them related to properties of the earth's atmosphere preventing or at least severely degrading ground based observations. Among these properties are the effects of absorption and scattering, but also turbulence and both line and thermal emission. A few other arguments for space astronomy arise from other physical principles. Some specific observational techniques profit from using satellite observations simply because of the large physical distance between detectors. For completeness it also deserves mentioning

---

[1] a more detailed, but a bit outdated, account is given by Davies (1997)

that future probes reaching the interstellar medium outside the solar heliosphere are under consideration. These astrophysical in situ experiments are, however, beyond the scope of this contribution.

7.1.1
Absorption

One of the main effects of the earth's atmosphere on astronomical observations is caused by the wavelength dependent transparency or absorption.

The arriving attenuated intensity I of the incident radiation I0 is related to the integrated opacity κ (cm^2g^{-1}) along the line of sight s via the *optical depth* τ,

$$\tau = \int \rho(s)\, \kappa\,(s)\, ds$$

with $\rho(s)$ describing the mass density distribution of the air:

$$I = I_0 \cdot exp(-\tau).$$

The frequently used atmospheric transparency T is defined as the ratio of I/I_0 (see, e.g., figures 1-13 and 7-3). Since very different physical processes from molecular line absorption to nuclear reactions contribute to κ and thus to the attenuation (absorption and scattering) at different wavelengths, the dependence of the atmospheric transparency on altitude

varies strongly with wavelength. In the far infrared (FIR), e.g., the main contribution for absorption arises from water vapor in the lower layers of the atmosphere. It is, however, the wavelength regime for which the flux distribution of galaxies peaks, as shown in figure 7-3. It was already mentioned in chapter 1.3.1 (figure 1-13) of this volume that the most bothering absorbing layers at these wavelengths can be overcome and the transparency significantly increased by using balloon experiments or airplanes in the stratosphere at altitudes higher than 15 km. Balloon experiments thus have frequently been a valuable test bed in the development of space experiments. Especially for (far-)infrared (FIR/IR) observations the Kuiper Airborne Observatory (KAO)—with a 1 m telescope on board a C-141 Starlifter airplane—was very successfully operated until 1995 over a period of 20 years. This is followed from 2005 onward by the Stratospheric Observatory for Far Infrared Astronomy (SOFIA), a Boeing 747 airplane with a 2.7 m telescope run jointly by NASA and Germany's DLR (Titz and Röser 1999, Web: *SOFIA*).

7.1.2
Emission

Even the layman notices on a moonless clear night that the sky background observed in the visual range by the eye is not really dark but rather grayish. A substantial part of this background originates within the atmosphere.

Fluorescent emission is due to the recombination of atoms which are photo dissociated by the solar irradiation during daytime at altitudes typically upwards of 100 km. This leads to a continuous blend of emission lines, the so-called airglow. The emission intensity in the brightest lines reaches, e.g., 2000 R^2 in the geocoronal Lyman-α line of hydrogen in the UV (at $\lambda = 121$ nm), typically a few 100 R in the most prominent optical night sky lines of oxygen (such as the OI line at $\lambda = 630$ nm), and even several 10^5 R for the prominent molecular OH bands in the near infrared. This airglow or night sky brightness limits the sensitivity of ground based observations since

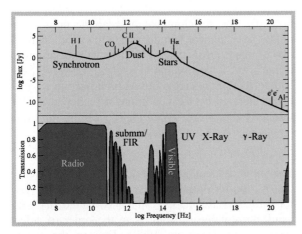

Figure 7-3: A typical spectrum (log flux) of a galaxy in comparison to the transmission of the atmosphere. It is noteworthy that The absolute peak in emission in the FIR is not observable from the ground. The continuum is labeled with the generating radiation processes and a few important emission lines are indicated

[2] the emission intensity is measured in Rayleigh, $1R = 10^6$ photons $cm^{-2}s^{-2}sr^{-1}$

the statistical effects of the photons (shot noise) contribute to the noise level of any measurement, in particular as it undergoes significant intensity fluctuations on time scales of a minute and spatial scales of an arcmin. Although the background level outside the atmosphere can still be significant with an important contribution coming from light scattered by zodiacal dust in the solar system, observations from space can gain from the much better estimates of the stable background.

With a typical mean temperature of T = 250 K the atmosphere is also radiating as a blackbody described by a Planck spectrum. This description is valid as a crude approximation only, but important conclusions and consequences can be drawn from it. For the spectral region between $\lambda \sim 5$ µm (thermal infrared) and the mm regime the thermal background is orders of magnitudes higher than the flux of any astronomical source. Therefore the choice of the observational site is critical for observations at these wavelengths and only differential measurements are possible. In this wavelength regime space missions are again favorable for the most sensitive experiments.

7.1.3
Turbulence

As discussed above, the properties of the atmosphere are far from ideal even in the classical atmospheric window at optical wavelengths where transparency and background emission certainly have impeding effects. Another significant limiting factor, in particular for optical and near infrared (NIR) observations, is the turbulent motion of the atmosphere. Temperature gradients deviating from the adiabatic condition cause the mixing of air of different temperatures. This causes temperature to fluctuate locally, which in turn causes the refractive index to fluctuate. These statistical fluctuations of the refractive index along the line of sight induce small shifts in phase and small changes in amplitude of the incident wave front, resulting in small changes of the position and intensity of an object. Although more subtle in detail (see Woolf 1982), this phenomenon in general is called seeing and is well known to the layman from the flickering of starlight.

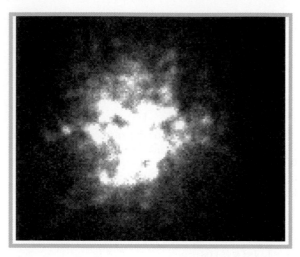

Figure 7-4: Short integration (1/30 sec) of a stellar image in the focal plane of a telescope showing the typical speckle structure due to individual turbulent cells. The angular size of individual speckle spots corresponds to the diffraction limited resolution of the telescope of ~0.1arcsec (MPIA)

In the framework of Kolmogorov turbulence the atmosphere can be characterized with a single variable, the Fried parameter r_0 describing a typical cell size, which is strongly wavelength dependent
$$r_0 \sim \lambda^{6/5},$$
a typical value being $r_0 \approx 10$ cm for $\lambda = 500$ nm. In principle, the development of adaptive optic techniques over the last two decades allows for a correction of phase errors caused by the small variations in refractivity. It is possible to analyze the wave front and correct phase errors in real time over an angular extent defined by the conditions of coherence. This so-called *isoplanatic patch* Θ_0 is given by the Fried radius at a typical altitude h for the turbulence:
$$\Theta_0 = 0.31 \ r_0/h.$$

The value of h can again be derived from turbulence theory, and for optical wavelengths the field of view for which phase errors can be corrected, e.g., by using deformable mirrors, is in the order of a few arcsec[3] only (Beckers 1998). Due to the strong wavelength dependence described above, the situation is much better in the near infrared. Yet, even

---

[3] 1 arcsec = 4.8×10^{-6} rad, equivalent to the angle under which a 1€ coin is seen at 5 km distance

here only few objects have sufficiently bright point sources nearby that can be used for phase referencing. In principle, this limitation can be overcome by the simultaneous use of many such overlapping corrected sight lines. This technique is known as multiconjugate adaptive optics and, if necessary, artificial laser guide stars can be used. However, until this expensive and technically very challenging concept is fully realized, diffraction-limited images of large fields can only be obtained from space, especially in the optical regime and at short wavelengths.

To demonstrate the advantage of (wide field) imaging from space let us recall that the (angular) resolution α of a telescope is given by diffraction theory as $\alpha[rad] = 1.7 \, \lambda/D$. A useful rule of thumb now gives us a reference: a telescope of ~20 cm diameter used at optical wavelength ($\lambda = 550$ nm) has a resolution of ~1 arcsec. Since this is the typical atmospheric seeing at most observatories the angular resolution achieved by any ground based telescope with larger diameter is limited by seeing rather then by diffraction. The HST with a primary mirror of 2.5 m produces diffraction-limited images with a resolution just under 0.1 arcsec. If we now compare HST images, like the one reproduced in figure 7-1, to ground based images we have to consider that the resolution elements are two dimensional. Therefore the improvement depends on the square of the angular resolutions, resulting in a gain of more than 25 if even the very best ground based images are considered.

For completeness it should be mentioned here that radio-astronomical observations suffer from a similar effect since ionospheric fluctuations lead to phase errors at radio frequencies.

7.1.4
Extended Baseline

To increase the resolution one can use combined beams of two or more telescopes in an interferometer; the resolution then is given by the maximum distance between two telescopes. The use of interferometric methods in astronomy is best known from applications in radio astronomy, in particular in Very Long Baseline Interferometry (VLBI). The detection of electromagnetic waves by coherent receivers at the long-wavelength end of the spectrum allows the phase information of the wave pattern to be recorded. Correlation of the signal can then be achieved in a subsequent off-line process which is completely independent of the observation. This enables a combining of telescopes in one hemisphere of the earth in order to increase the angular resolution α according to the formula for the diffraction limit given in the previous section. D is given by the largest distance between stations, defining the diameter of the "synthesized" telescope. With transcontinental distances of thousands of km, typical angular resolutions of milliarcseconds (mas) can be achieved with cm and dm receivers. At a given wavelength the resolution in this situation is obviously limited by the largest distance between telescopes which can still simultaneously observe the same astronomical source.

One possibility to increase this so-called baseline of the interferometer is to use an antenna in orbit. Early experiments of this kind used the TDRS communication satellites (see chapter 9.4), before the dedicated Muse-B or Halca satellite was launched in 1997 and operated for several years by the Japanese space agency ISAS in the framework of the VLBI Space Observatory Programme (Web: *VSOP*).

7.2
The Space Environment

In the effort to overcome the effects of the atmosphere by using an astronomical satellite one has to consider a couple of new effects caused by the space environment, some of which were already discussed in chapter 1.3.3. Details of the space environment depend on the specific orbit of the satellite and, since all the different classes of orbits as described in chapter 2 are used by astronomical satellites, the situation is different from case to case. To give one example, the influence of cosmic rays and charged particles on the detectors on board HST is demonstrated in figure 7-5. The low earth orbit (LEO) of HST is determined by the fact that it was launched using the space shuttle.

At an altitude of ~600 km the satellite is exposed to energetic elementary particles to the effect that some very sensitive instruments even have to be shut

Figure 7-5: Several integrations of the galaxy NGC891 obtained in an R band filter with the Wide Field / Planetary Camera 2 (WFPC-2) on board HST are added up without (left) and with (right) cosmic ray rejection algorithms, demonstrating the effect of the charged particles on CCD detectors (AIRUB/Rossa)

down during the regular passages through the so-called South Atlantic Anomaly, a region of increased cosmic ray density caused by the complex shape of the earth's magnetosphere (Stassinopoulos 1970). This particle radiation not only contributes to the background and thus to the signal-to-noise ratio of an astronomical observation, it also deteriorates detectors (e.g., Meidinger et al. 2000) and solar panels, limiting the lifetime of individual instruments and the infrastructure of satellites. In (the unique) case of the HST with its relatively low orbit the servicing missions (see figure 2-16) were used to regularly exchange the instruments and the solar panels.

The earth's magnetosphere actually introduces yet another effect since a satellite moving through the magnetic field can experience a relatively strong magnetic torque. This has to be taken into consideration for the design of the satellite and can, in turn, be used for attitude control. Using the information provided by the on-board magnetometers it was possible to significantly extend the lifetimes of several astronomical satellites after the original control by gyros had failed for mechanical reasons.

Finally the effect of the solar irradiance (see 1.3.3) should be discussed here with regard to the operation of astronomical satellites. If one wants to make use of the diffraction-limited imaging quality, e.g., at optical wavelengths, the guiding accuracy has to be fractions of the aspired resolution. For HST the requirement is on the order of milliarcseconds. On

this level, however, the thermal stress of the spacecraft during the day-night terminator passage is considerable, leading to a significant *jitter* that has to be compensated.

7.3
Astronomical Themes to Come

Progress in observational astrophysics is—like in other empirical sciences—very closely linked to progress in measurement techniques. More specific, it is increased sensitivity and resolution with regard to the incoming (mainly) electromagnetic radiation that allows for the detection of new phenomena and for the better measurements which change our perception[4].

Over the past decades two developments have substantially changed both the quantity and quality of astronomical observations. Continuous development in detector technology on the one hand and new telescope technologies on the other hand have contributed in this respect to ground-based as well as to space astronomy. To give just a few examples, with Charge Coupled Devices (CCDs) of more than 95% detective quantum efficiency the almost perfect detector is now available for radiation energies of ~1 eV (the optical and NIR). Also X-ray observations benefit from specially trimmed CCDs by significantly increased resolution and sensitivity. For telescopes the development towards a new generation is mainly based on the availability of new mirror technologies. Mirrors using active optics for ground based observations have led to the construction of so-called 10 m-class telescopes, with the ESO Very Large Telescope (VLT) (Web: *ESO*) and the Keck telescopes on Hawaii (Web: *Keck*) being the most prominent examples. In a similar development X-ray astronomy has benefited from using the Wolter design (see 7.4.6) of many nested mirrors at the Bragg-condition to substantially improve the imaging properties of satellites, as very successfully demonstrated by ROSAT (Bradt et al. 1992).

At the turn of the century all major national and international science and space agencies summarized the resulting achievements in astronomy and astro-

---

[4] For astronomy this has been studied and analyzed in detail by Harwit (1984).

physics over the last decades in order to define the challenges for the future in these fields. Based on some particularly outstanding aspects of recent astrophysical research these studies unanimously identified a few issues of general importance which are briefly described in the following.

7.3.1
Cosmology and Structure Formation

One of the most dramatic changes in our understanding of the universe is related to its beginnings. Here the physics of the big bang is connected with the cosmological model and the question how all smaller scale structures such as galaxies and galaxy clusters emerged over the age of the universe from the homogeneous distribution of matter in the beginning. In recent years the combined analysis of galaxy redshift surveys, the distances to high redshift supernovae of type Ia, gravitational lensing studies, the statistics of Lyman-α absorption lines in quasar spectra, and the anisotropies in the cosmic microwave background resulted in a cosmological model with cold dark matter and the *cosmological constant* Λ determining the shape and fate of the universe. Space astronomy has substantially contributed to several of the above mentioned observations; the use of faint supernovae as distance indicators, for instance, depends on follow-up observations with HST.

The Cosmic Microwave Backgroud (CMB) anisotropy measurements, first detected with the Cosmic Background Explorer (COBE) in 1992 (Smoot et al.), are of particular value in this respect since they allow determination of density fluctuations present in the universe at the time when it became neutral and thus transparent with the recombination of hydrogen. These early density fluctuations cause angular fluctuations in the temperature of the CMB at a level of $\Delta T/T \sim 2 \times 10^{-5}$. This angular fluctuation spectrum contains a wealth of cosmological information and thus describes very accurately the initial conditions for any further calculation of the structure formation process. The better angular resolution and the higher sensitivity of the Wilkinson Microwave Background Anisotropy Probe (WMAP) resulted in the significantly improved all-sky map of the CMB presented in figure 11-15. The detailed analyses of

the angular fluctuation spectrum as shown in figure 7-6 allowed various cosmological parameters to be measured with unprecedented accuracy. This is now called the beginning of *precision cosmology* (Spergel et al. 2003).

The surprising implication of the result now is that the universe is observed to be very close to flat, with the density parameter Ω close to unity, while at most 5% of the matter can be attributed to the known baryonic material. Dark matter and dark energy contribute ~70% and ~25% of the critical density, respectively. With the unknown nature of dark matter and the currently even more mysterious dark energy, frequently associated in a phenomenological sense with the famous cosmological constant Λ originally introduced by Albert Einstein, the unknown dark components present one of today's greatest challenges to fundamental physics (see also

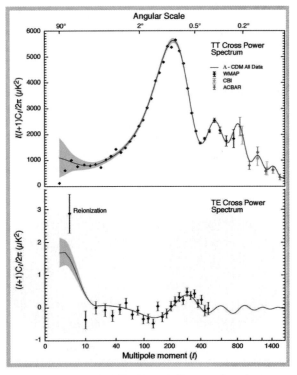

Figure 7-6: The distribution of the cosmic microwave background emission on different angular scales measured by WMAP and other experiments. The peaks in the multipole development constrain various cosmological quantities. The most important results are summarized in table 7-1 (NASA/WMAP Team)

175

Precision Cosmology Parameters after WMAP		
Ω_{total}	1.02 +/- 0.02	total density
Ω_Λ	0.73 +/- 0.04	dark energy density
Ω_{Matter}	0.27 +/- 0.04	matter density
Ω_{Baryon}	0.044 +/- 0.004	baryon density
H_0	71 +/- 4 km/s/Mpc	Hubble constant
t_0	13.7 +/- 0.2 Gyr	age of the Universe

Table 7-1: WMAP results

chapter 11).

If we look at the numerical calculations of structure formation we see yet another interesting aspect demonstrating the importance of space astronomy. These calculations follow the path of baryons within the dark matter dominated gravitational potentials over cosmic history and predict the fate of this only known component of the universe. Since much of the potential energy of the baryons will be converted into kinetic energy and then dissipated into heat, most of the baryons are expected to reside in hot gaseous phases known as hot intergalactic medium or warm and hot intergalactic medium at temperatures of 10^7-10^8K. Figure 7-7 shows the current

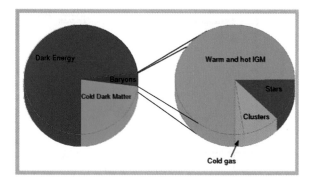

Figure 7-7: The main constituents of the universe contributing to the total mass density $\Omega = 1$. It is noteworthy that of the total predicted baryonic content only a small fraction can be observed directly without space astronomy (AIRUB/Bomans, adopted from NASA/WMAP team, Fukugita et al. 1998, Cen and Ostriker 1999)

notion of the relative importance of the different constituents in the universe. The 5% of baryonic matter are split up into different components with the warm and hot intergalactic medium making up for ~75%. This largest reservoir of baryons is, however, currently hardly accessible to observations and future X-ray missions will be designed to map this hot gas.

Other key questions in this area are obviously related to a better understanding of the fundamental and dominant ingredients of the universe. The amount and nature of dark matter as well as its changing distribution over cosmic time is just one of these questions. Using the only known form of matter, the baryons, as test particles to trace structure formation of the dark matter in the early universe, we rely on the light generated by fusion in stars. Stellar light generated in the early universe is, however, significantly redshifted due to the cosmic expansion. This is one of the reasons the next generation space telescope, called the "James Webb Space Telescope" (JWST), is designed to preferentially observe in the infrared. This exemplifies how the next generation of space astronomy missions is designed in response to today's most urgent astrophysical problems.

7.3.2
Black Holes and Compact Objects

The enormous power output of quasars (or quasi-stellar objects) exceeds that of entire galaxies. It is generated in volumes much smaller than the solar system and is frequently associated with the ejection of matter (jets) at nearly the speed of light. Together with spectroscopic evidence and studies of a variety of other types of so-called active galactic nuclei the observations have led to a *unified picture* of active galactic nuclei in which material falling into the gravitational potential of a black hole is heated by friction and swirls around the central black hole in an accretion disk.

In recent years, kinematical observations with high angular resolution using radio VLBI methods as well as HST spectroscopy in optical or X-ray spectroscopy allowed determination of the masses of the central black holes in galaxies. With masses in the range from several 10^6 times the mass of the sun

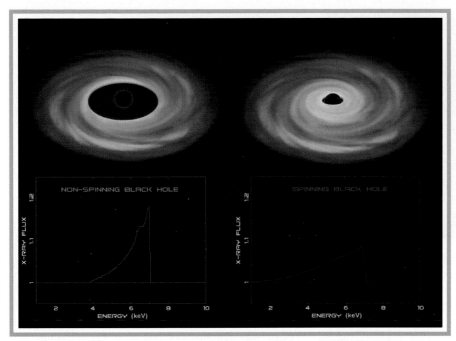

Figure 7-8: X-ray spectra taken with XMM/Newton and AXAF/Chandra are sensitive enough and have sufficient energy resolution to show from the analysis of emission line shapes the kinematical difference of gas in orbit around nonrotating and rotating black holes (NASA/CXC/SAO/M.Weiss/J.Miller)

($M_\odot$) up to almost $10^9\,M_\odot$ they belong to the class of supermassive black holes. Yet another puzzle in the structure formation history of the universe comes from the fact that the masses of these central black holes seem to be correlated with the larger scale mass distribution in the host galaxy as measured by the central light distribution (the bulge). The question which of the two—black hole or bulge—was formed first is now one of the crucial questions in galaxy evolution.

Black holes are, of course, also of special interest due to their direct connection to general relativity. Here the interest is not just limited to supermassive objects; stellar mass black holes as remnants of supernova explosions at the end of stellar evolution also allow us to study aspects of general relativity. Some recent observations of NASA's Rossi X-Ray Timing Explorer (RXTE) have discovered oscillations in the X-ray emission of a stellar black hole candidate with a frequency that could be characteristic for the last stable Keplerian orbit around a central black hole. This frequency can be used to estimate also the angular momentum of the black hole and thus, whether a Schwarzschild or Kerr geometry is applicable. Some first spectroscopic evidence for the difference in gas kinematics for two stellar black

hole candidates is schematically shown in figure 7-8. In this regard it is also exciting that flux fluctuations on short timescales of the central black hole in the Milky Way have been seen in the optical, infrared, and X-ray regimes now also constraining the physical conditions of the accretion disk.

Further progress in studying matter under strong gravity can be expected from the next generation of X-ray satellites. With the proposed increase in collecting area by a factor of 100 these telescopes will allow high signal-to-noise observations with high time resolution that could discriminate the kinematics of gas around rotating and nonrotating black holes for many more cases.

Yet another cosmic mystery, the *γ-ray bursts* were discovered already several decades ago serendipitously by space experiments as flashes with a duration of only seconds. First detections by military satellites designed to monitor clandestine nuclear tests were followed subsequently by thousands of detections by the Compton Gamma Ray Observatory (CGRO), one of NASA's Great Observatories in operation from 1991 to 2000. For many years the crude angular resolution of the instruments available for γ-ray observations did not allow accurate determination of the positions of the flashes fast enough

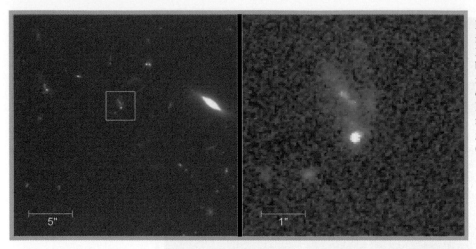

Figure 7-9: Optical identification of Gamma Ray Burst GRB990123 by STIS on board HST in imaging mode. This observation is a clue in linking GRBs to supernova events in distant galaxies (Fruchter/STScI/NASA)

to correlate or associate them with any other astronomical source. Only in recent years could dedicated space experiments link the phenomenon to explosions in galaxies at high redshift, which makes γ-ray bursts the most energetic explosions known.

This success story of observational astrophysics used in a first step a network of satellites to triangulate the position of the burst. A larger area of the sky at the rough position of the burst was then followed up very rapidly with the specially designed Italian-Dutch X-ray satellite (BeppoSAX). This satellite was able to monitor an *afterglow* of the flashes in soft X-rays and thus could establish positions sufficiently accurate to allow comparisons with known objects on arcsec scales. Meanwhile the follow up observations at other frequencies such as the optical and radio wavelengths are also sufficiently speeded up to establish afterglows. A very important step was taken by using HST in identifying the burst position with a possible supernova in a distant galaxy (figure 7-9). The exact timing of the flash at different wavelengths now constrains the models and favors the formation of a black hole in a special kind of core collapse supernova, called *hypernova*, under many other possibilities. Common to most of the models is the fact that they require highly relativistic effects to explain the energetics and geometry of the phenomenon. The energy required to explain the observed luminosities suggest a directed—so-called *beamed*—explosion at work. To further constrain possible models of γ-ray bursts more time resolved observations of afterglows are required.

The Swift satellite launched in 2004 is specifically designed for this kind of observations (Web: *SWIFT*). It not only carries a γ-ray burst alert telescope but also specific instruments to measure the afterglow *swiftly* within ~1 min after detection at other wavelengths. For this an X-ray telescope and an ultraviolet and optical telescope are on board.

7.3.3
Formation of Stars and Planetary Systems

A field of astrophysical research which has made enormous and even spectacular progress in recent years is the observation of the formation of stars and planetary systems out of the interstellar medium. The most important development in this area is the discovery of extrasolar planets (Marcey and Butler 1996). This is certainly exciting to the general public and at the same time also of great scientific interest since now we have within grasp answers to long time questions about the formation and evolution of planetary systems, the existence of solar systems similar to ours, and even about the presence of other habitable planets in the universe.

Meanwhile more than 100 extrasolar planets are known. By far most of them were detected by measurements of the radial velocity variations that they introduce on the central star, a method that also biases detection towards more massive objects. Therefore most of the planets detected so far are Jupiter-like giant planets. The ultimate ambition for future missions is to directly observe extrasolar

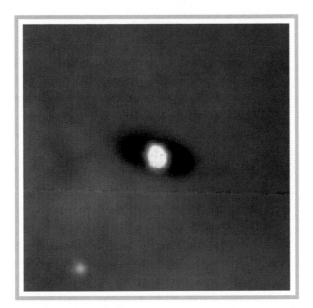

Figure 7-10: A protoplanetary disk or proplyd surrounding a young star in the Orion nebula as observed with WFPC2 on board HST. The size of the gas and dust disk is about twice the size of the solar system (STScI)

planets and to overcome the hitherto existing selection bias to make observations of earth-like planets possible. These very ambitious goals require new observational techniques. While some of these techniques may also be possible with ground-based telescopes in the future, the most promising approaches depend on space experiments. Specific photometric as well as interferometric missions are being considered for this problem, and also the planned astrometry mission (see 7.5.2) will contribute to the search for planetary systems.

The existence of planetary systems is closely related to the process of star formation. Circumstellar disks of gas and dust are now observed using molecular emission lines in the (sub)mm regime from the ground and also in a few cases in the optical and NIR with the high-resolution capabilities of HST (figure 7-10). The problem with observing sites of star formation is ultimately linked to the physical process. Due to the fact that the contracting and cooling interstellar medium becomes opaque, stars are born deep inside cold, dust-enshrouded cores of molecular clouds. It is possible to overcome the absorption by observing at infrared or even longer wavelengths, where dust and gas have much smaller

optical depth. Observations in the infrared are crucial also to probe the physical conditions of the interstellar matter on larger scales since the most important emission lines controlling the heating and cooling balance of the medium are in this regime (see figure 7-3). However, at these wavelengths Earth's atmosphere is opaque, as discussed in 7.1.1, making such observations almost impossible from the ground.

The understanding of star formation has therefore benefited from infrared satellite missions such as the Infra-Red Astronomical Satellite (IRAS) or the Infrared Space Observatory (ISO), which were operated by NASA and ESA, respectively. Now a new generation of infrared satellites with increased sensitivity and angular resolution is operational or will be operational soon. The SIRTF (Space Infra-Red Telescope Facility) mission, the latest of NASA's Great Observatory missions, now called the Spitzer Observatory, has been in orbit since 2003 and the ESA satellite Herschel, originally called FIRST (Far Infra-Red Space Telescope) will be launched soon.

7.4
Current Space Astronomy Missions

As mentioned above in section 7.1 the whole electromagnetic wavelength spectrum has been covered by astronomical satellites since the mid-eighties. This does not, however, mean that simultaneous coverage at all wavelengths is guaranteed at all times. This is understandable from the very different lifetimes of satellites with very diverse technologies. Infrared satellites used to have very short lifetimes of a few years since they required cryogenic cooling by liquid helium, which is used up after a given period. In contrast, the International Ultraviolet Explorer (IUE) was operated continuously for 18 years and HST has already been in operation much longer than the originally anticipated 10 years mission.

In this section some of the currently operating space astronomy missions are described together with missions under construction. For each of them some technical background and a few astronomical applications are given. It is appropriate to start with HST, not only because it is the most famous astronomical satellite; it also specifically exemplifies the idea of an observatory mission.

7.4.1
The Hubble Space Telescope and the Optical Wavelength Range

HST can be considered the prototype of an astronomical space observatory. It was constructed for servicing by astronauts, with a modular design that allows easy exchange of aging, malfunctioning, or outdated parts. This concerns infrastructure like the solar panels as well as scientific instruments. The value of this concept can be easily demonstrated if the timescales of the project are brought to attention.

HST has been under construction since 1977; the 2.4 m mirror finished in 1981, and the satellite was ready in 1984. The Challenger accident caused the launch to slip into 1990 with the result that much of the design was 15 years old and outdated when the satellite reached orbit. This concerned many different parts including some of the scientific instruments as well as the on-board computer.

Figure 7-11 shows some general design features of HST, identifying in particular the four axial and the radial science instruments. Three radial positions are

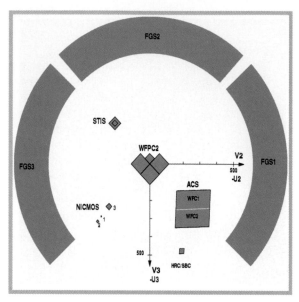

Figure 7-12: The focal plane of HST is used by several science instruments. Given here is the status after servicing mission 3B with two cameras in the optical regime (WFPC2 and ACS), one infrared camera with spectroscopic capabilities (NICMOS), a high-resolution spectrograph and camera (STIS), and three fine guiding units (STScI)

used for the fine guidance sensors, the fourth is occupied by the wide field planetary camera (WFPC). Meanwhile three more servicing missions have overhauled all other science instruments as well as parts of the satellite subsystems, such as control electronics, magnetometers, gyroscopes, solar arrays, data recorders, reaction wheels, etc.

When the orbital verification after launch revealed that the mirror suffered from spherical aberration[5], the advantages of the modular construction and the availability of servicing and maintenance during "space walks" or Extra Vehicle Activities (EVAs) of space shuttle crew members became very obvious. This concept allowed for a servicing mission late in 1993 that replaced one of the axial science instruments by a complex optical bench called COSTAR (Corrective Optics Space Telescope Axial Replacement). This system allowed corrective optics to be brought into the beam of the remaining old axial

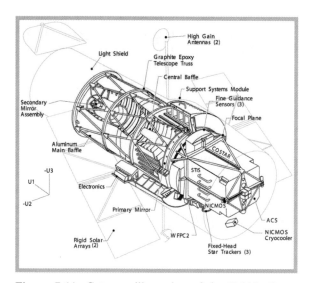

Figure 7-11: Cutaway illustration of the Hubble Space Telescope showing the scientific instruments in the rear. (STScI/AURA/NASA)

---

[5] a discussion of the circumstances is found in Chaisson (1994)

instruments, thus correcting for the aberration of the mirror.

Figure 7-12 gives the status of the focal plane usage after servicing mission 3B in March 2002. With this servicing mission all first light instruments were replaced. Since all new instruments are designed to correct for the spherical aberration, COSTAR could be removed to allow for the originally planned set of four axial instruments. In a very typical way for an observatory mission with an imaging telescope the focal plane is split up into sections for access by the various instruments. The Space Telescope Imaging Spectrograph (STIS) is designed specifically for UV spectroscopy. The Advanced Camera for Surveys (ACS) is the main new imaging device in the optical range; the Near Infrared Camera and Multi Object Spectrograph (NICMOS) was designed for use in the NIR.

The very versatile instrumentation has contributed a lot to our astrophysical understanding today. To mention just a few of the many important results:
- the measurement of masses for supermassive black holes in the centers of galaxies by high resolution spectroscopy,
- the identification of the host galaxies of γ-ray bursts by rapid follow-up observations (target-of-opportunity),
- quantifying galaxy evolution by observations of changing properties of galaxies over a significant fraction of the age of the universe, e.g., in the deep fields,
- the exact photometry of light curves for distant supernova leading to the concept of dark energy,
- the detailed modeling of the dark matter distribution through gravitational lensing in clusters of galaxies,
- the detailed structure of outflows from young stellar objects.

An example of a result from HST is given here to demonstrate the science case for two smaller space missions in the optical regime. Figure 7-13 shows the light curve of the star HD 209458 with photometric accuracy reaching a fraction of ‰ as obtainable only from space. The light curve shows the occultation of the stellar disk during the transit of a much smaller planet. The satellites COROT (COnvection ROtation and planetary Transits) and Kepler

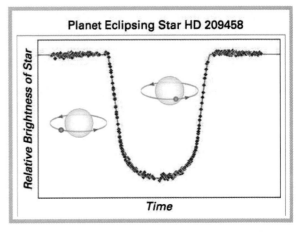

Figure 7-13: The light of the star HD 209458 is dimmed by 0.015% when a planet orbiting this star is occulting the small part of the stellar disk. More than 1000 observations with HST are used to synthesis this light curve (ESA/NASA/T.M. Brown)

are two small missions specialized in high precision stellar photometry and thus able to detect more planets through transits. COROT is a small French-led international satellite using a 30 cm telescope designed for several applications in stellar photometry. The mission of the Kepler satellite planned by NASA is dedicated to the search for transits, in particular since this is currently the only feasible technique to find earth-size planets in orbit around nearby stars (Web: *Kepler*). Equipped with 42 CCD detectors covering more than 100 square degrees this wide-field telescope with a 95 cm Schmidt corrector is large enough to monitor the brightness of 100,000 target stars.

7.4.2
UV Missions: FUSE, Galex, and More

For the wavelengths range covered by HST this satellite is indeed unique with regard to sensitivity and angular resolution due to the relatively large mirror. Some smaller dedicated missions, however, complement the HST capabilities in the ultraviolet part of the spectrum.

The Far Ultraviolet Spectroscopy Explorer (FUSE, Web: *FUSE*) is a dedicated spectroscopy mission in the λ = 90-120 nm range consisting of four independent spectrographs with ~37 cm primary mirrors each. One of the many important results is related to the halo of hot gas in galaxies. This thin hot gas of 10^5 K is currently only detectable by absorption lines of highly ionized atomic species such as fivefold ionized oxygen (O VI). Figure 7-14 shows one such absorption line spectrum with *local gas* in the disk of the galaxy close to the velocity of the sun, clearly separated from gas at negative velocities, i.e., gas that is moving towards our position in the galactic plane at more than 200 km/sec. The analysis of such UV spectra led to the conclusion that the Milky Way as well as other galaxies are surrounded by a halo of hot gas containing more than 10^9 M$_\odot$.

The Galaxy Evolution Explorer (GALEX) is yet another small NASA mission for UV astronomy (Web: *GALEX*). Here the goal is to image for the first time a large sample of galaxies in the UV. The emission of galaxies in this spectral range is characterized by the UV photons of hot stars which are mainly residing in star formation regions. The UV emission is thus a measure for the rate of current star formation in a galaxy. The importance of the GALEX observations rests in the need to establish a local reference of the UV radiation from galaxies for comparison with the highly redshifted observations, as observed, e.g., in the Hubble deep field observations.

The spectral energy distribution of redshifted galaxies observed in the optical range with HST is equivalent to rest wavelengths in the UV for nearby objects (the local universe) and it can therefore be used to deduce the evolution of the UV radiation field with redshift or age of the universe. For the last 10^{10} years a substantial decrease of the star formation rate in galaxies has been observed. To phrase this important finding differently, most stars in the universe were formed already 10 billion years ago.

UV astronomy can also serve as an excellent example how dedicated small missions contribute to very specific scientific questions. As an example, the relatively small mission STSAT-1 (Space and Technology Satellite 1) by the Korean Advanced Institute of Science & Technology carries a UV spectrograph built at the Space Science Laboratory of the University of California in Berkeley to map UV lines created by the complex processes in the interstellar medium of our Milky Way galaxy. This SPEAR

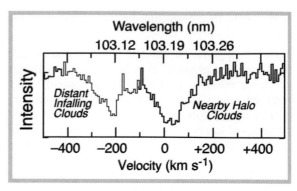

Figure 7-14: An ultraviolet absorption line spectrum of the λ = 103.19 nm line shows gas (blue) approaching the galaxy at more than 200 km/sec (and therefore blueshifted at shorter wavelengths) (JHU/NASA)

(Spectroscopy of Plasma Evolution from Astrophysical Radiation) spectrograph specifically measures the line of threefold ionized carbon (C IV) at λ = 102 nm from the 10^5 K component of very extended hot gas which results from shocks caused by supernova explosions in the Milky Way.

7.4.3
Covering the Infrared: the Spitzer Observatory

For the first time since the Infrared Space Observatory (ISO) finished operation in 1998, the infrared sky is again observable under space conditions with the Spitzer Observatory. Launched in 2003 this satellite is designed as a facility instrument within NASA's Great Observatories series (see 7.3.3). It enables measurements from near-infrared (NIR) wavelengths at λ = 3 μm to the far-infrared (FIR) regime at λ = 180 μm.

The technological progress over the past decade in detectors in this wavelength range is enormous and the high resolution, high sensitivity images delivered are spectacular (Web: *Spitzer*). Besides the new detectors the success of the mission is also due to other innovative conceptual ideas. While previous (F)IR satellites were operated in an earth orbit, thus subject to the radiation field of the earth, the choice of an earth-trailing heliocentric orbit cools the shielded satellite passively to ~40 K while the satellite is slowly drifting away from the earth at a rate of

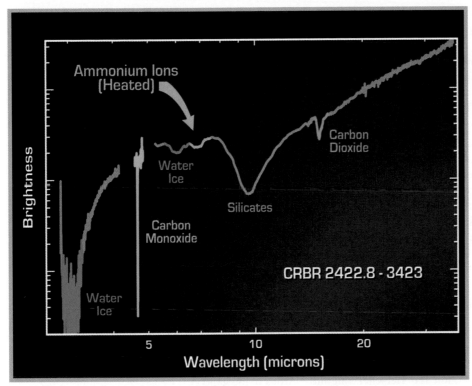

Figure 7-15: Three settings of the Infrared Spectrograph IRS on board the Spitzer Observatory of the protoplanetary disk CRBR 2422.8-3423 reveal features of ices of water, ammonium, and carbon dioxide (NASA/ JPL-Caltech/K. Pontoppidan-Leiden)

~0.1 AU/year[6]. This choice of orbit with passive cooling also allows for a warm launch which not only substantially reduces the amount of liquid helium needed for cooling, it also results in a much longer mission lifetime. With only 360 liters of liquid helium the expected lifetime is about five years at an operational temperature of only 5.5 K. For comparison, the first U.S.-British-Dutch Infrared Astronomical Satellite (IRAS) in the mid-eighties used 520 liters for a 10 month mission and the earlier mentioned ISO mission was operating for almost 2.5 years in the mid-nineties using 2,140 liters.

The science cases for the Spitzer Observatory are closely related to the problem of star formation, on galactic as well as on extragalactic scales. As with other observatory or facility missions the focal plane of the 0.85 m telescope is shared by different science instruments, here the four channel Infrared Array Camera (IRAC), the Multiband Imaging Photometer for Spitzer (MIPS), and the Infra-Red Spectrograph (IRS). One of the early spectroscopic results covering the near and middle infrared gives insight into the very complex chemistry of a protoplanetary disk with evidence for ices of water, ammonium, and carbon dioxide in the inner planet forming region near the central young star (figure 7-15).

7.4.4
The Far Infrared and the Sub-mm: Herschel

The main limitation of the Spitzer Observatory is the relatively small primary mirror of only 0.85 m. This considerably limits the angular resolution in particular at FIR wavelengths. The ESA Herschel Space Observatory will overcome this limitation by a 3.5 m primary mirror. The aperture of the Herschel satellite is the largest of any space telescope for several years to come. This satellite also extends the wavelength coverage of space observations into the submillimeter range. Since the measuring principles and techniques are similar to the Spitzer Observatory, the mission design also shares similarities with Spitzer. The radiation shielding of both satellites

---

[6] 1 Astronomical Unit (AU) is the average distance between Sun and Earth of 1 AU $= 148 \times 10^6$ km

makes them look similar (figure 7-16) and also for Herschel an orbit far from the earth was chosen. In this case the satellite will be brought close to the second Lagrange point L_2 of the Earth-Sun system (see figure 1-14 in chapter 1.3.1) at a distance of 1.5 million km from the earth. The detectors used require, however, much better cooling with superfluid helium.

Three instruments will cover the wavelength range between $\lambda = 60$ µm and $\lambda = 670$ µm:
- "Heterodyne Instrument for the Far Infrared," a very high resolution spectrometer
- "Photodetector Array Camera and Spectrometer," an imaging photometer with medium resolution spectroscopic capabilities, and
- "Spectral and Photometric Imaging Receiver,", an imaging photometer combined with a Fourier spectrometer.

The spectral distribution of a typical galaxy (see figure 7-3) shows the importance of the FIR and sub-mm wavelength range. The radiation is typically characteristic for the cool interstellar medium in star forming regions. With its unprecedented high sensitivity and angular resolution the Herschel satellite will certainly enable most significant steps in understanding star formation, both locally as in molecular clouds of the Milky Way as well as in galaxy evolution by observing the most distant objects (Pilbratt et

al. 2001, Web: *Herschel*).

7.4.5
Looking at the Microwave Background with Planck

Following the great success of the WMAP experiment the medium size ESA Planck mission (Web: *Planck*) will provide a more detailed look at the recombination era of the universe by an extended wavelength coverage and by measuring the polarization of the background radiation. The 1.5 m telescope will be equipped with radio receiver arrays using high electron mobility transistors for the low frequency and bolometer arrays for the high frequency range. These detectors are cooled to 20 K and 0.1 K, respectively. The large frequency coverage will not only allow the cosmic microwave background radiation to be separated from galactic and extragalactic foreground components in order to look for anisotropies, it will also provide a database for a broad range of astrophysical applications, from radio-loud active galactic nuclei to interstellar magnetic fields and the propagation of cosmic ray electrons in the galaxy.

While WMAP has already led to a set of precise

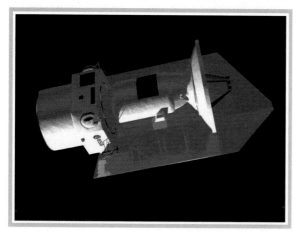

Figure 7-16: Artist's view of the Herschel Space Observatory. The satellite has a total length of 9 m and a 4x4 m overall cross section. A characteristic design feature for IR satellites is the radiation shielding seen here in the back (ESA)

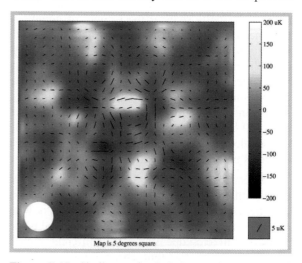

Figure 7-17: Similar to the 5x5 degree field observed here from Antartica the Planck satellite will measure the CMB polarization over the whole sky. Black lines indicate orientation and strength of the polarization, the CMB temperature fluctuations are shown in color. (DASI collaboration)

cosmological parameters, the results are not yet unique solutions. An inherent degeneracy of the solutions can be solved for some parameters by measuring polarization anisotropies with Planck. This capability of the Planck satellite could also help with another important implication of the standard cosmological picture. To reconcile all observational cosmological facts, such as the flatness of the universe and the observed expansion within one theoretical standard model of the big bang, a phase of rapid expansion or inflation of the universe is required. It is directly related to unified theories in particle physics and describes the universe $\sim 10^{-33}$ sec after the big bang at typical energies of 10^{15} GeV. Quantum fluctuations in this inflationary phase are expected to be the seed of any structure formed later on and thus to be seen in cosmic background fluctuations. One of the predicted consequences of inflation is the polarization of the microwave background observable with Planck (figure 7-17). Anisotropy measurements thus demonstrate the close relation of observational cosmology with central questions in fundamental physics (Puget 2005, see also chapter 11).

At even longer wavelengths the radio window of the atmosphere opens up, and for most applications ground based observatories cover this domain. Only at the lowest frequencies below 30 MHz radio waves cannot pass through the earth's ionosphere. This remains the only unexplored frequency domain for space astronomy. Although science cases exist for research in this frequency range currently no specific space missions are foreseen.

7.4.6
The X-ray Sky Seen with Newton and Chandra

The X-ray window to the universe was opened up accidentally in the early 1960s with the discovery of the first X-ray point source in the constellation Scorpius, called Sco-X1, later identified with a compact object and possible black hole candidate (Giacconi 2003). The many discoveries made in this wavelengths range over the last forty years were considered such an important contribution to physics that the Nobel prize 2002 was awarded to Riccardo

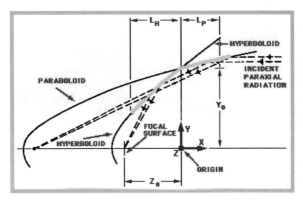

Figure 7-18: Design of an imaging X-ray mirror (Wolter telescope) combining the Bragg reflection at a hyperboloid and a paraboloid to focus (ESA)

Giacconi for his early discovery and the following life-long contributions.

According to the before mentioned general principle, progress in X-ray astronomy has also been very closely linked to instrumental developments in the main components, telescopes and detectors. Detectors for X-ray satellites have developed from traditional nuclear physics devices like proportional counters to solid-state detectors which are now also being used in nuclear and elementary particle physics. State of the art technology in this regard is the European Photon Imaging Camera (EPIC) on board ESA's X-ray Multi-Mirror (XMM) satellite[7]. The design of X-ray telescopes also changed dramati-

Figure 7-19: A total of 58 Wolter mirrors, the largest of .70 cm diameter, are nested in a coaxial and cofocal configuration to build one of three similar X-ray mirrors on board the ESA XMM satellite (ESA)

Figure 7-20: The design of the Wolter X-ray mirror leads to large focal lengths which in turn require the typical very long structure of X-ray satellites. Here NASA's Chandra (AXAF) satellite is shown (NASA/HEARC)

cally with the possibility to built nested Wolter mirrors (figures 7-19 and 7-20) with sufficient accuracy. This technique now enables the construction of X-ray imaging optics with an angular resolution of only a few arcseconds. The XMM satellite has three similar Wolter telescopes on board which image the sky on two different type of CCD detectors within the EPIC camera. The CCDs are either Metal-Oxide Semi-conductor (MOS) chips or specially developed pn-CCDs. This combination of telescope and detectors is sensitive in the energy range from 0.2 to 12 keV.

The relatively large field of view of 30 arcmin observable with XMM is complemented by the Advanced X-ray Astrophysical Facility (AXAF), one of the Great Observatories, now called Chandra X-ray Observatory (Web: *Chandra*). Here a few Wolter mirrors of larger diameters (up to 1.2 m) provide better angular resolution for a smaller field of view in comparison to XMM. Two detector types are used in the focal plane of the Chandra satellite. The CCDs of the Advanced CCD Imaging Spectrometer (ACIS) are complemented by *Multi Channel Plates* in the High resolution Camera (HRC) providing the possibility for high time resolution observations.

In comparison to earlier X-ray satellites, both XMM and AXAF have a much increased sensitivity. The resulting improvement in photon statistics therefore allows in particular analysis of spectral information. Thus it is now possible to observe X-ray emission lines which provide strong constraints on the physical conditions in the observed astrophysical plasmas. To give just one example related to the above mentioned (see 7.3.1) baryon budget in the universe: a

significant fraction—up to 40%—of the baryons in clusters of galaxies was found to reside in the hot *intra cluster medium*. With the much increased spectral sensitivity and resolution it is now possible to study the temperature dependence and content of heavy elements[8]. These elements are found to be surprisingly abundant in the intra-cluster medium, which raises the question how and when they were produced and dispersed into the medium.

7.4.7
The High Energy End: γ-Rays

Similar to the terminology in atomic and nuclear physics we talk about γ-rays at even higher energies. It should be noted, however, that γ-rays span more than five decades in energy, from ~100 keV to ~30 GeV. For comparison, the classical optical window covers half a decade in frequency only. In this energy range the radiation results from completely different physical processes of non thermal nature, such as inverse Compton scattering, synchrotron radiation of relativistic electrons in the extremely strong magnetic fields of some compact objects, nuclear reactions, or even electron-positron annihilation radiation (Schönfelder 2001).

At these energies, detectors and telescopes are of completely different design. They rather are highly specialized particle physics experiments, and techniques vary over the large energy range covered, with scintillation counters and anti-coincidence logics for detection and coded masks for directional information as typical examples.

By the nature of this high energy radiation, research on γ-rays is directed to compact objects, either of stellar (neutron stars, black holes) or extragalactic origin. This includes the γ-ray bursts discussed earlier as one of the most challenging problems. For research on γ-rays the highly specialized Swift satellite was already mentioned. The International Gamma-Ray Astrophysics Laboratory (INTEGRAL) is an observatory style facility in orbit since 2002 and designed to study lower energy γ-rays. Following up on results of the previous Compton Gamma-

---

[8] in astrophysics the terms *heavy element* or *metals* refer to all elements resulting from nuclear fusion in stars, i.e., elements with A > 9

Ray Observatory, INTEGRAL has measured γ-lines of elements recently synthesized in massive young stars (figure 7-21). INTEGRAL observations have also confirmed an exciting finding of the Oriented Scintillation Spectrometer Experiment (OSSE) on board the Compton-Observatory (Web: *Compton*): the central region of the Milky Way is filled with radiation from electron-pair annihilation at 511 keV. Ideas of its origin are highly speculative and include nuclear reactions in galactic γ-ray bursts as well as the possibility that this radiation is connected to the nature of dark matter particles (Web: *Integral*).

The INTEGRAL mission is complemented by the Gamma-ray Large Area Space Telescope (GLAST) which in particular extends the capabilities into the high energy regime with its large area telescope. It will surpass the very successful Energetic Gamma Ray Experiment Telescope (EGRET) of the Compton observatory by more than a factor of ten in sensitivity (Web: *GLAST*).

7.5
Future Missions

The relatively long development times for space astronomy missions require planning ahead for more than a decade. In the following we will therefore briefly summarize goals and concepts for some of the most important astrophysics satellites to be launched or under construction until about 2015.

7.5.1
The James Webb Space Telescope

With the success of HST, planning a next generation space telescope began in 1996 with a report entitled HST and Beyond edited by Dressler (1996) which outlined goals for the first decade of 21st century astronomy. The discovery of galaxies at very high redshifts in the Hubble deep fields, i.e., at large distances or large cosmic look-back times, initiated a shift of interest into the IR. The particular challenge is to look at galaxies as probes of structure formation in the cosmic period between the recombination epoch of cosmic microwave background generation and the epoch of the most redshifted objects observable with HST and ground based large telescopes. This era of cosmic history is meanwhile known as

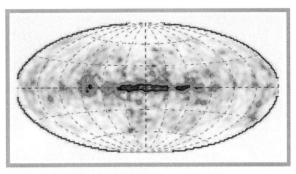

Figure 7-21: The distribution of γ-ray emission in the 1.809 MeV line of ^{26}Al measured with the Imaging Compton Telescope (COMPTEL) experiment of the Compton Observatory. This emission is due to the radioactive decay of ^{26}Al with a lifetime of only 106 years. In comparison to the typical timescales for galactic evolution it must be freshly produced by massive young stars (ESA)

the dark ages. Shifting the focus to the IR will also considerably improve observing conditions for studies of star formation, the second of the outstanding problems in today's astrophysics. And there are, of course, many more applications (Benvenuti 1998) for a sensitive IR telescope, including searches for the number density of the faintest stars, the so-called brown dwarfs. Meanwhile named after the former NASA Administrator James Webb (JWST), the observatory will be operated with a 15% share of ESA following the example of the joint effort for HST. This kind of international collaboration has a long tradition in space astronomy missions and with the likely increase of costs for future missions this may become even more common.

Currently the design of a 6 m class telescope (Stiavelli 2004) with good imaging quality from $\lambda = 0.6$ to 28 μm is under further consideration (figure 7-22). Similar to other IR satellites the preferred orbit for JWST is a Lissajou figure around the second Lagrange point L_2. Accordingly, the telescope will also be shielded against direct sunlight. With the size of the satellite given by the diameter of the primary mirror, the mission requires new techniques. The mirrors, some spacecraft appendages like the high gain antenna or the solar panels, and the sunshield have to be deployable. The sunshield is designed to reduce the radiation from 300 kW on the illuminated side to only 23 mW on the back side. This cold envi-

Figure 7-22: The chosen design for the James Webb Space Telescope by Northrop Grummann and Ball Aerospace (NASA/ESA)

ronment is in particular required for the intended mid-infrared instrumentation.

Like other facility missions, JWST will carry a number of different experiments. Besides the fine guiding sensors, a near infrared camera and spectrometer, as well as the mid-infrared instrument are under development. The JWST instrumentation will have novel features, many of which will certainly also find use in ground based instruments. This includes a micro-electromechanical system for use as a dynamical aperture shutter mask that will provide simultaneous spectroscopy for hundreds of different objects in a single field of view (Web: *JWST*).

In preparation for JWST a small satellite is supposed to provide a new all-sky map in the IR from $\lambda = 3.5$ μ to 23 μm. This Wide-field Infrared Survey Explorer (WISE, Web: *WISE*) will be operated for only six months in a LEO orbit to catalog objects in the infrared 1,000 times more sensitive than the IRAS mission.

7.5.2
Astrometry: Precise Positions with GAIA

As one of the most fundamental astronomical techniques, *astrometry* is the measurement of celestial objects to determine their positions and movements.

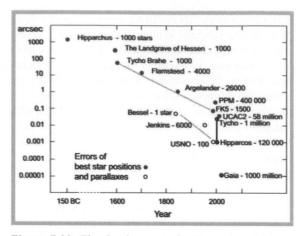

Figure 7-23: The development of astrometric catalogues over time, from Hipparcos 150 BC to the Hipparcos satellite 2000, and in comparison the expected output from GAIA (ESA)

The variation of the position over time contains information on the tangential velocity component. Measuring, in the course of a year, the parallax caused by the earth's orbit around the sun allows also the distance to be deduced. Some examples should help to define a reference: the stars in the direct neighborhood of the sun with distances of typically a few parsec[9] show a parallactic ellipse with semimajor axis in the order of fractions of an arcsec. At typical tangential velocities of nearby stars of 25 km/sec the displacement or *proper motion* measures a few 0.01 arcsec per year. In general, these two effects can only be measured differentially with the position of one star being defined relative to a local reference frame and, of course, the two effects being superimposed. The limiting astrometric accuracy is defined by the turbulence of the atmosphere (Lindgren 1980) and much higher precision can be obtained by well designed space missions. One advantage for space missions is due to the fact that a much larger fraction of the sky is accessible at any time. This allows for a global astrometric solution taking into account all available differential measurements of all measured stars. The proposed ESA Corner Stone mission GAIA (Global Astrometric Interferometer for Astrophysics) builds on the

---

[9] 1 pc (parsec) = 3.08×10^{16}m

great success of the concept tested by the former ESA Hipparchos astrometry mission.

The GAIA satellite (Web: *GAIA*, Battrick 2000) will be continuously scanning with two simultaneous fields of view which are separated by a well defined angle of 106°. This allows measurement of accurate one dimensional coordinates along great circles. The global analysis of all measurements will result in distances and proper motions of stars and, as a by-product, an unprecedented amount of information will become available on double and multiple stellar systems, photometry, variability, planetary systems, etc. A dramatic step in astrometric accuracy with mean parallax errors of only 5 to 10 μarcsec for stars brighter than magnitude 15 is expected from GAIA, as can be seen in figure 7-23. Not only will the measurement precision increase by two orders of magnitude, the number of objects with high precision astrometric positions will increase even more dramatically. Basically, the content of the Milky Way Galaxy will be mapped out by this mission.

To illustrate the targeted accuracy it is noteworthy to mention that relativistic effects have to be taken into account since basically all positions will be affected by the gravitational effects of solar system objects. This also open the possibility to use GAIA to constrain predictions from general relativity. Another aspect for fundamental physics questions comes from the possibility to constrain time variations of the gravitational constant G.

7.5.3
Interferometry in Space

To increase the astrometric accuracy even further one has to operate an optical interferometer outside the atmosphere. While many astrophysical applications will profit from an even higher precision than provided by GAIA, the motivation clearly comes from the idea to search for earth-like planets by astrometric techniques. One may recall that the objects in bound gravitational systems orbit around the center of mass or *barycenter*. This makes the star wobble around this common center of gravity, which may be inside the central star. Jupiter's influence on the sun results in positional changes on the order of 500 μarcsec if seen from a nearby star in 10 pc distance. To measure the influence of a much smaller

earth-like planet the NASA Space Interferometry Mission (SIM) is designed to monitor the positions of ~250 nearby stars with a precision of ~1 μarcsec. With less stringent requirements one will still be able to detect and study planetary systems containing *giant planets* (Web: *SIM*). Since up to now most detections have been made by radial velocity variations, a different technique is required to avoid strong selection effects in order to get a representative census of planets. This is important since the size and orbit distribution is considered a clue to the formation and evolution of planetary systems.

To achieve these goals two telescopes of 0.3 m diameter will be mounted on the spacecraft at a distance of 10 m. Many technological challenges in particular with regard to the extreme requirements on metrology have to be faced in this development. Similar to the Spitzer Observatory, SIM will be brought into an earth-trailing orbit. The space environment allows the optical fringes to be maintained over time scales of several minutes, and thus it will be possible to use SIM for interferometric imaging very similar to the techniques known from radio interferometers.

SIM will also have the possibility to make another significant step by measuring the proper motions of ~25 nearby galaxies. This would complement our knowledge of the six-dimensional position and velocity vector for these galaxies. A dynamical analysis based on these data will then provide strong constraints on the total masses of the galaxy ensemble and thus on the dark matter content.

7.5.4
A Closer Look at Earth-Like Planets: Darwin and TPF

Space interferometry is a key to the perhaps most challenging goal in going beyond the detection of earth-like planets with missions such as SIM or GAIA, namely analysis of their atmospheres in a search for tracers of life. ESA as well as NASA are considering missions to accomplish this, called Darwin and Terrestrial Planet Finder (TPF), respectively (Friedlund and Henning 2004). The requirement for a space mission comes from the preferred wavelength range, the mid-infrared, to minimize the contrast between planet and star and to obtain spec-

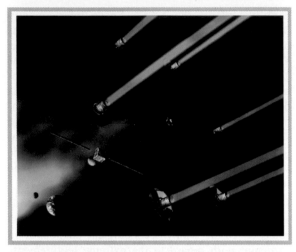

Figure 7-24: An artists view of the free flying Darwin satellites (ESA)

troscopic information of prominent spectral features considered to be good indicators for life influencing the composition of the atmosphere, like O_3, H_2O, and CO_2. A look at the earth-like planets in the solar system gives a guideline. Infrared spectra of Venus, Earth, and Mars show CO_2 absorption at 17 μm, indicating the presence of an atmosphere. However, only the earth's spectra show water bands at 5-8 μm and ozone bands at 9.8 μm.

Both missions have similarities in design, with Darwin currently being planned to consist of six 1.5 m class telescopes on free flying satellites linked interferometrically over distances of ~50 m. The resulting typical largest baselines of 100 to 200 m then would provide an angular resolution of ~0.01 arcsec. This is sufficient to resolve the earth's orbit from a distance of 10 pc. The interferometric combination of telescopes will make use of a special design, the nulling interferometer. Here the beam combination is used to bring the central star to destructive interference, thus nulling out its intensity to enable the detection of faint, close-by planets.

This most challenging mission also has a new requirement on the orbit. To reach the best possible signal-to-noise ratio at the wavelengths under consideration one has to avoid the thermal emission of dust in the plane of the solar system. Ideally, a mission of this design therefore would be operated in

the outer solar system at a distance of several AU from the sun (Web: *Darwin*; Web: *TPF*).

7.5.5
The Next Generation of X-Ray Telescopes

To make significant progress with regard to the energetic phenomena related to black holes and compact objects, but also in relation to the hot gas expected to reside in the filamentary large scale structure of the universe, the sensitivity of the next generation X-ray telescopes has to be increased considerably. One of the main factors in this problem is the collecting area of the telescope. As already mentioned in chapter 1.3, ESA is considering the X-Ray Evolving Universe Spectrometer (XEUS) mission. The major new step taken in the XEUS design is a mirror with a collecting area of 30 m^2. This mirror size will result in a focal length of ~50 m for the telescope (as discussed in 7.4.6), making a single spacecraft impractical. Therefore, two satellites flying in formation are considered. The mirror spacecraft will be followed by the detector spacecraft, with their relative position established by laser metrology, L_2 being again a preferred orbit for the formation flight. Even with this "two spacecraft" solution the mirror spacecraft has to be put together in space and therefore first mission scenarios considered using the International Space Station as a base for robotic assembly. Meanwhile new techniques such as high precision pore optics for the construction of X-ray mirrors are under consideration and the mission concept may slightly change before its realization in 2015. To reach the science goals, the required sensitivity has to surpass existing X-ray satellites by a factor of ~100 and for the first time in the X-ray domain one will have a spectral resolution of a few eV only, equivalent to $\Delta E/E \sim 1000$ (Web: *XEUS*). Development of more sensitive detectors is also a continuing effort that will contribute to the higher sensitivity of future missions. Superconducting transition edge sensor calorimeters and cadmium zinc telluride (CdZnTe) based high energy detectors are just some examples. In preparing for this big step in sensitivity additional surveys are required to identify sources suitable for spectroscopic follow-up observations. In this context sev-

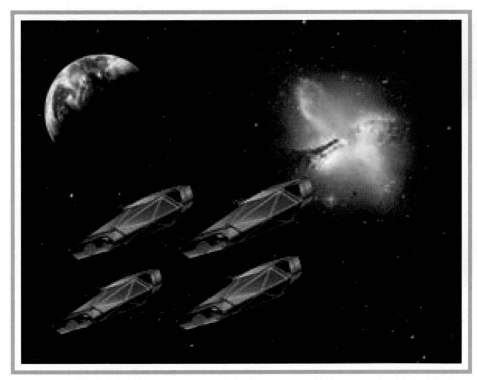

Figure 7-25: Artist's view of the NASA concept for Constellation-X (NASA/HEASARC)

eral smaller missions are under consideration, to name some examples: ROSITA (ESA/DLR), DUO (NASA/DLR), or Lobster (ESA/UK).

Science goals and mission requirements similar to XEUS found a slightly different solution in the NASA concept of Constellation-X. Here the increase in total collecting area is realized by combining several smaller telescopes in a fleet of telescopes to be operated also at L_2 (Web: *Constellation-X*). The smaller mirror diameters allow for a feasible spacecraft length of ~10 m. Figure 7-25 gives an impression of one of the designs under study.

7.6
Perspectives and Conclusions

Space astronomy has significantly changed our understanding of the universe and thus the perspective of mankind with regard to the question of its origin. This makes today's astrophysics one of the most popular fields in science. The quest for a more general understanding of matter and energy in the universe also brings astronomy in close relation to research in particle and fundamental physics. To give

just one example following the discussion in 7.4.5: observing the signature of the inflationary phase of the big bang in cosmic background anisotropies can constrain models describing the unification of all forces of nature. A very generalized approach to problems in modern physics including observational astrophysics is described by Turner et al. (2003). Here the close link to other important questions in fundamental physics is demonstrated, such as the challenge to observe gravitational waves directly. The frequency range of gravitational waves of astrophysical relevance cannot be completely covered on the ground. The space interferometer LISA, discussed in more detail in chapter 11, is thus designed to extend the frequency coverage of ground based laser interferometers. The LISA mission design underlines once more that the concepts for several of the next generation space astronomy mission require formation flying at L_2, still considered a technological challenge for space utilization.

Several other important aspects of modern astrophysical techniques have not been addressed in this contribution since it has concentrated on the analysis of electromagnetic radiation. As an example of in-

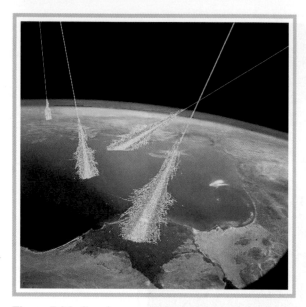

Figure 7-26: Cosmic ray particles generate air showers of secondary particles after nuclear reactions in the upper atmosphere. These secondary particles propagate faster than the speed of light in air and produce Cherenkov radiation. This radiation can be observed to trace the particles. From space EUSO could monitor the night side of the atmosphere

situ astrophysical experiments a possible future mission passing through the heliosphere into the interstellar medium was mentioned in section 7.2. There is yet another class of carriers of information from outer space, elementary particles. One particular class are cosmic rays (Schlickeiser 2002), which partly originate within the solar system and partly in the interstellar medium, most likely in shocks generated by energetic events such as supernova explosions. Their composition and energy distribution is to be explained by generation and acceleration processes together with the path through interstellar space where they interact with magnetic fields and particles in collisions and scattering processes. Here one of the exciting and challenging questions is related to particles with the highest observed energies, the ultra-high energy cosmic rays. They are found with energies exceeding 10^{20} eV. These most energetic particles are seen so far only in very rare events and no correlation with potential sources is currently known. The physical processes to acceler-

ate these particles to the highest known energies remain a mystery.

Since these particles are so rare, extremely large areas are required to detect them. To monitor high energy cosmic rays, air shower experiments are very useful. The particles interacting with the atmosphere cause particle reactions which lead to a cascade of secondary particles. The secondary particles can either be detected by particle detectors on the ground or simultaneously by the *Cherenkov* radiation that they emit passing through the atmosphere. The ultimate detector would use the largest detector volume conceivable, the atmosphere of Earth's night side (figure 7-26), to observe the light flashes of Cherenkov radiation. One possible approach would be an earth-observing satellite, the Extreme-Universe Space Observatory (EUSO). This joint effort by ESA and NASA has an interesting aspect if seen in the context of this book: it describes looking up by looking down.

The scientist's dream may already reach much further than what seems now possible for the "next" generation of experiments for space astronomy. In this kind of dream even the far infrared proto-galaxy imager or a new telescope that could image the sky in γ-rays (Gamma Ray Imager GRI) is taking shape.

The study of structure formation in the universe - from big bang to galaxies and from planets to life - under the influence of gravitation, dark matter, and dark energy has become the central question for astronomical research and will remain important for the next decades. To make progress requires using all the available information across the wavelength spectrum, possible only with space astronomy.

References

Battrick B, Edwards L, Sword R (2000) GAIA: composition, formation and evolution of the galaxy. ESA Publ. Division

Baum WA, Johnson F S, Oberly JJ, Rockwood CC, Strain CV, and Rousey R (1946) Solar Ultraviolet Spectrum to 88 Kilometers. Phys. Rev. 70, pp. 781

Beckers JM (1993) Adaptive Optics for Astronomy: Principles, Performance, and Applications Ann. Rev. Astron. Astrophys. 31, p. 13

Benvenuti P et al. (1998) eds: The Next Generation Space Telescope, Proc. 34th Liege International Astrophysics Colloquium, ESA SP-429

Bradt HV, Ohashi T, and Pounds KA (1992) X-ray astronomy missions, Ann. Rev. Astron. Astrophys. 30, p. 391

Cen R, OstrikerJP (1999) Where are the baryons? Astrophys. J. 517, p. 31

Chaisson E (1994) Space Trials: The Hubble Space Telescope Story. Harpercollins, New York

Davies JK (1997) Astronomy from space: the design and operation of orbiting observatories. Wilcy, New York

DeVorkin DH (1944) Science with a vengeance: How the military created the US space sciences after World War II. Springer, New York

Dressler A (1996) HST and Beyond. AURA, Washington, D.C.

Fischer D, and Dürbeck H (2003) Hubble revisited, Springer, New York

Fukugita M, Hogan CJ, and Peebles PJE (1998) The Cosmic Baryon Budget. Astrophys. J., pp. 503 - 518

Fridlund M, and Henning T (2003) eds: Towards Other Earths: DARWIN/TPF and the Search for Extrasolar Terrestrial Planets. ESA SP-539

Giacconi R (2003) Nobel Lecture: The dawn of X-ray astronomy. Rev. Mod. Physics 75, p. 993

Harwitt M (1984) Cosmic discovery. The search, scope, and heritage of astronomy. MIT Press, Cambridge

Léna P, Lebrun, F, and Mingnard F (1998), Observational Astrophysics, Springer, Berlin

Lindegren L (1980) Atmospheric Limitations of Narrow-field Optical Astrometry. Astron. Astrophys. 89, p. 41

Marcy GW and Butler RP (1998) Detection of Extrasolar Giant Planets. Ann. Rev. Astron. Astrophys. 36, p. 57

Meidinger N, Schmalhofer B, and Strüder L (2000) Nucl. Inst. Measurem. A439, 2-3, p. 319

Pilbratt GL, Cernicharo J, Heras AM, Prusti T, and Harris R (2001) eds: The promise of the Herschel space observatory. ESA SP-460

Pudget JL (2005) CMB polarization and early universe physics. In: Zensus et al. eds: "Exploring the Cosmic Frontier", Springer, Heidelberg

Stassinopoulos EG (1970) World Maps of Constant B, L, and Flux Contours W. NASA SP-3054

Schlickeier R (2002) Cosmic Ray Astrophysics, Springer, Heidelberg

Schönfelder V (2001) ed: The Universe in Gamma-rays. Springer, Heidelberg

Smoot GF, Bennett CL, Kogut A et al. (1992) Structure in the COBE differential microwave radiometer first-year map. Astrophys. J. 396, p. 1

Spergel DN, Verde L, Peiris HV et al. (2003) First-Year Wilkinson Microwave Anisotropy Probe (WMAP) Observations: Determination of Cosmological Parameters. Astrophys. J. Suppl

Spitzer L (1990) The First Known Report Concerning the Astronomical Importance of an Extraterrestrial Telescope. Astr. Quarterly 7, p. 129

Stiavelli M, et al. (2004) JWST Primer Vers. 1.0. STScI, Baltimore

Titz R, and Roeser H-P (1989) eds: SOFIA Astronomy and Technology in the 21st Century, Wissenschaft und Technik, Berlin

Turner MS et al. (2003) Connecting Quarks with the Cosmos, The National Academies Press, Washington, D. C.

Woolf NJ (1982) High resolution imaging from the ground. Ann. Rev. Astron. Astrophys. 20, p. 367

Web References

Chandra: http://chandra.havard.edu

Compton: http://cossc.gsfc.nasa.gov/

Constellation-X: http://constellation.gsfc.nasa.gov

Darwin: http://ast.star.rl.as.uk/darwin

ESO: http://www.eso.org/paranal/
http://www.eso.org/projects/vlt/

EUSO: http://aquila.lbl.gov/EUSO/

FUSE: http://fuse.pha.jhu.edu

GALEX: http://www.galex.caltech.edu

GAIA: http://sci.esa.int/science-e/www/area/index.cfm?fareaid=26
http://astro.estec.esa.nl/SA-general/Projects/GAIA/gaia.html

GLAST: http://glast.gsfc.nasa.gov/

Herschel: http://www.rssd.esa.int/SAgeneral/Projects/Herschel

Integral: http://astro.estec.esa.nl/SA-general/Projects/Integral/integral.html

JWST: http://ngst.gsfc.nasa.gov/

http://www.stsci.edu/jwst
http://sci.esa.int/science-e/www/area/index.
cfm?fareaid=29
Keck: http://www2.keck.hawaii.edu
Kepler: http://www.kepler.arc.nasa.gov/
LISA: http://lisa.nasa.gov, http://lisa.jpl.nasa.gov/
NASA: http://spacescience.nasa.gov
NASA-Universe: http://universe.nasa.gov
Newton: http://sci.esa.int/science-e/www/area/index.
cfm?fareaid=23
NSSDC: http://nssdc.gsfc.nasa.gov/astro
Planck: http://sci.esa.int/science-e/www/area/index.
cfm?fareaid=17
RXTE: http://heasarc.gsfc.nasa.gov/docs/xte/xte_1st.
html

SIM: http://planetquest.jpl.nasa.gov/SIM/sim_index.
html
SOFIA: http://www.sofia.arc.nasa.gov
Spitzer: http://www.spitzer.caltech.edu
STECF: http://www.stecf.org
STScI: http://www.stsci.edu
Swift: http://swift.gsfc.nasa.gov
TPF: http://planetquest.jpl.nasa.gov/TPF/tpf_index.
html
VSOP: http:// www.vsop.isas.ac.jp
WISE: http://www.astro.ucla.edu/~wright/WISE/
WMAP: http://map.gsfc.nasa.gov
XEUS: http://sci.esa.int/science-e/www/area/index.
cfm?fareaid=25

8 The Solar System
by Tilman Spohn and Ralf Jaumann

Humanity has been fascinated for eons by the wandering stars in the sky, the planets. With the advent of the space age, the planets have been transferred from bright spots to worlds in their own right that can be explored. In particular the terrestrial planets are interesting because comparison with our own planet allows a better understanding of our home. Venus, for example offers an example of a runaway greenhouse that has resulted in what we would call a hellish place. With temperatures around 450°C and a corrosive atmosphere that is also optically nontransparent, Venus poses enormous difficulties to exploration. Mars is a much friendlier planet to explore, but a planet where greenhouse effects and atmospheric loss processes have turned a world with rivers running on its surface and moderate climate into a cold dusty desert. But aside from considerations of the usefulness of space exploration in terms of un-

derstanding Earth, the interested mind can visit astounding and puzzling places. There is the dynamic atmosphere of Jupiter with a giant thunderstorm that has been raging for centuries. There is a volcanic moon, Io, that surpasses Earth and any other terrestrial planet in volcanic activity. This activity is powered by tides that twist the planet such that its interior partially melts. There is another moon, Titan, that hides its surface underneath a layer of photochemical smog in a thick nitrogen atmosphere, and there are moons of similar size that lack any comparable atmosphere. Magnetic field data suggest that icy moons orbiting the giant planets may have oceans underneath thick ice covers. These oceans can, at least in principle, harbor or have harbored life. There is Saturn with its majestic rings and there are Uranus and Neptune with complicated magnetic fields and with more satellites that are astounding.

Figure 8-1: The nine planets Mercury, Venus, Earth, Mars, Jupiter, Saturn, Uranus, Neptune, and Pluto are shown in this compilation of NASA images with their correct relative sizes and ordered according to their distance from the sun. Mercury is barely visible at left close to the arc of the surface of the sun and Pluto is barely visible at the outer right. The dark spot on Jupiter is the shadow of Io, one of its major satellites (C.J. Hamilton)

Moreover, there are asteroids with moons and comets that may still harbor the clues to how the solar system and life on Earth formed.

8.1
Our Planetary System

The solar system contains a myriad of bodies rang-

ing in size from the sun to miniscule dust particles. The Encyclopedias of Planetary Sciences (Shirley and Fairbridge 1997) and the Solar System (Weissmann et al. 1999) are invaluable and almost inexhaustible sources of information about the solar system. The nine planets Mercury, Venus, Earth, Mars, Jupiter, Saturn, Uranus, Neptune, and Pluto are shown in figure 8-1. The sun, a middle-aged

	Terrestrial, earth-like Planets				Giant Planets				
	Mercury	Venus	Earth	Mars	Jupiter	Saturn	Uranus	Neptune	Pluto
Radius [km]	2438	6052	6371	3390	71492	60268	24973	24764	1152
Mass [10^{24} kg]	0.3302	4.869	5.974	0.6419	1899	568.46	86.63	102.4	0.0131
Density [10^3 kg/m^3]	5.430	5.243	5.515	3.934	1.326	0. 6873	1.318	1.638	2.050
Uncompressed Density [kg/m^3]	5.3	4.0	4.05	3.75	0.1	0.1	0.3	0.3	2.0
Rotational Period [d[a]]	58.65	243.0[b]	0.9973	1.026	0.4135	0.4440	0.7183[b]	0.6713	6.387[b]
Inclination [°]	0.5	177.4	23.45	25.19	3.12	26.73	97.86	29.56	122.5
Orbital distance [AU[c]]	0.3871	0.7233	1.000	1.524	5.203	9.572	19.19	30.07	39.54
Orbital Period [a[d]]	0.2410	0.6156	1.001	1.882	11.87	29.39	84.16	165.0	248.8
Magnetic Moment [10^{-4}T · Radius^3]	$3 \cdot 10^{-3}$	$< 3\ 10^{-4}$	0.61	$< 6\ 10^{-4}$	4.3	0.21	0.23	0.133	?
Effective Surface Temperature [K]	445	325	277	225	123	90	63	50	44
Specific heat flow or luminosity [pW/kg]	?	?	7	?	176	152	4	67	?
Known Satellites	0	0	1	2	16	33	17	8	1

	Moon	Io	Europa	Ganymede	Callisto	Titan	Triton
Primary	Earth		Jupiter			Saturn	Neptune
Radius [km]	1737	1821	1560	2634	2400	2575	1353
Mass [10^{20} kg]	734.9	891.8	479.1	1482	1077	1346	214.7
Density [10^3 kg/m^3]	3.344	3.53	3.02	1.94	1.85	1.881	2.054
Specific heat flow [pW/kg]	8	890	300 (?)	?	?	?	?
Orbital Period	27.32	1.769	3.551	7.155	16.69	15.95	5.877[b]

Table 8-1: Properties of the Planets (above), and Properties of Major Satellites (below) (Lodders and Fegley 1998, Spohn 2001. Remarks: [a]A day [d] is equivalent to 24 hours. [b]The motion (rotation, revolution) is retrograde. [c]AU is one astronomical unit or 149.6 million km. [d]A year [a] is equivalent to 365 days [d]

main sequence star, contains 98.8% of the mass of the solar system but only 0.5% of its angular momentum. The next smaller body, Jupiter, still 300 times more massive than Earth (compare table 8-1), contains more than 60% of the mass of the rest. Jupiter is the biggest of the *giant planets*, a group of gaseous planets that constitute a major subgroup of the solar system. Among the giant planets are, in addition to Jupiter, Saturn, Uranus, and Neptune. The latter two are sometimes called the *subgiants* because they are notably smaller than Saturn and Jupiter. Earth, the biggest member of the other major subgroup of family members, the *terrestrial planets*, is the only planet on which we know to date that life had a chance to develop. Among the members of this group are Mercury, the innermost planet, Venus, Earth's twin with respect to size and mass, and Mars. The latter planet has the best chance of having developed some primitive forms of life, which makes it the prime target of present day space missions. The terrestrial planets together have about 0.005% of the mass of the solar system. Table 8-1 collects some data of general interest on the planets and some major moons.

Comparative planetology is the science of studying the planets by comparing and finding general properties and common lines of evolution as well as

Figure 8-2: The four Galilean satellites of Jupiter Io, Europa, Ganymede, and Callisto, shown from top left to bottom right (NASA/JPL)

features that are specific. It uses the methods of the natural sciences and is strongly interdisciplinary. Comparative planetology is not restricted to the planets *sensu strictu*, but also considers the major moons of the planets such as Earth's moon, the major satellites of Jupiter (Io, Europa, Ganymede, and Callisto) (figure 8-2), the major Saturnian satellite Titan and, finally Triton, Neptune's major satellite. Other bodies of interest include the yet largely unexplored members of the asteroid belt, the asteroids. The asteroid belt is found between Mars and Jupiter and its members are speculated to be the parent bodies of most meteorites (stones from space found on Earth's surface). These contain rich information about the formation of the solar system, its evolution, and even the evolution of matter before the formation of the solar system.

8.2
Planetary Missions

Comparative planetology and planetology as such are made possible at their present levels only through the exploration of the solar system with space missions (see, e.g., Friedman and Kraemer 1999). All planets and major satellites except Pluto have been visited at least by a flyby of a robot spacecraft. Others, such as Venus, the moon, Mars, and Jupiter have been explored by orbiters and/or landers. The moon has even been visited by astronauts. It can be said without any doubt that space exploration has turned dim discs in the sky observable with telescopes into worlds in their own right waiting for further, perhaps eventually human, exploration. Space exploration began with missions to the moon in the 1960s, characterized by the race between the U.S. and the then Soviet Union to the moon. This race culminated in the landing of men on the moon. For planetary sciences, the significance of the Apollo program lies—among other reasons—with the first sample return, the first seismic exploration of a planetary interior, the first heat flow measurement, and the first geologic field trip on another planet. Although Apollo failed to solve the puzzle of the origin of the moon and the solar system, it has provided a wealth of data and has made the moon the best known planetary body other than Earth. The planet most frequently visited by space craft aside

Figure 8-4: Maat Mons on Venus. An eight km high volcano at 194.5°E and 0.9°N. Lava flows extend for hundreds of kilometers across the fractured plains shown in the foreground. Radar images are monochromatic (black and white), of course. The simulated hues are based on color images recorded by the Soviet Venera 13 and 14 spacecraft (NASA/JPL)

Figure 8-3: The Magellan spacecraft in orbit around Venus (Artist's conception). On October 11, 1994 the Magellan spacecraft finally descended into the atmosphere and was presumably destroyed. During its mission, Magellan mapped the surface of Venus with a Synthetic Aperture Radar (SAR) with better than 1 km resolution, and the gravity field (NASA/JPL)

from the moon is Mars (compare table 8-1). Highlights of Martian exploration include the Viking missions, the first landing on Mars, the Pathfinder mission, the first vehicle introducing mobility, Mars Express, the first European planetary mission *sensu strictu*, and the Mars Exploration Rovers, MER for short. Venus has been the target of very successful Soviet missions that even included eight landers (Venera 7-14). These landers survived in the highly corrosive atmosphere of Venus (see below) only for a few hours, but transmitted color photos from the surface that shape our image of the Venusian surface. A further highlight of Venus exploration was the Magellan mission, which provided the first systematic mapping of the surface with radar (figures 8.3 and 8-4).

Since the atmosphere of Venus is optically thick, cameras are of little use there for surface explora-

tion. Mercury, the innermost planet, has been visited only during two flybys of the American Mariner 10 mission targeted for Mars. Mercury is the target of the Messenger launched in 2004 and the ambitious European BepiColombo mission scheduled for launch in 2012.

The outer solar system was first explored by the Pioneers that flew by Jupiter (Pioneer 10 and 11) and Saturn (Pioneer 11) and subsequently by the very successful Voyager I and II missions that flew by all four major planets of the outer solar system. In addition to exploring the planets these spacecraft discovered rings at Jupiter and Uranus, explored the satellite systems, and found Io to be the most volcanically active body in the solar system. The Voyagers for the first time allowed a detailed view of the rich phenomenology of icy satellite surfaces. After the Pioneer and Voyager flybys, orbiters were the next logical step in outer solar system exploration. The first mission in this class was Galileo, which orbited Jupiter and plunged into Jupiter's crushing atmosphere in September 2003. The spacecraft was deliberately destroyed to protect the Jovian system, in particular Europa, from being polluted. Galileo discovered the magnetic field of Ganymede and evidence for oceans underneath the ice crusts of the

Galilean moons Europa, Ganymede, and Callisto. On its route to Jupiter Galileo discovered for the first time a moon (Dactyl) that is orbiting an asteroid (Ida) and observed comet Shoemaker-Levy crashing into Jupiter. An instrumented descent probe released from the Galileo orbiter entered the atmosphere in December 1995 and provided the first in-situ measurements of the state and the chemistry of a giant planet atmosphere shroud. At the time of writing Cassini is orbiting Saturn. Cassini is expected to top the success of Galileo through its more advanced payload capability. The Huygens probe supplied by the European Space Agency ESA descended through the atmosphere of Saturn's biggest moon, Titan, and successfully landed on the surface on January 14, 2005 to transmit the first images (color) from Titan's surface. Titan's atmosphere is optically thick, like that of Venus. Huygens landed on a solid surface strewn with ice boulder. On its descent it imaged rivers in which hydrocarbons are flowing or have flown in the past, and a possible shoreline. Cassini itself will explore Saturn, the rings, and its 33 moons in a four year orbiting journey through the system. Among the most spectacular planned missions for the outer solar system is a landing mission for Europa. The latter mission is driven by the perception that an ocean on Europa may be covered by a thin ice shell that can be drilled and by the possibility that life developed in this ocean.

Asteroids and comets are believed to be remnants of accretion and to consist of largely primitive matter. Although this perception is questionable to some extent—comet nuclei that have repeatedly flown by the sun may have experienced thermal metamorphism; some asteroids show signs of endogenic activity and may be differentiated—asteroids and comets are minor solar system bodies of great interest. The Dawn mission will explore the two largest asteroids, Vesta and Ceres. The first is believed to be a differentiated body while the latter is believed to be undifferentiated and mostly primitive. Comet nucleus Churyumow-Gerassimenko is the target of the ESA Rosetta mission (launched in 2004). After a ten year cruise the Rosetta orbiter will begin to orbit the nucleus while the lander Philae will dock onto the nucleus and explore it in situ. Both the orbiter and Philae will collect data as the comet approaches the sun.

Planetary exploration has taken and is still following a stepped approach by which flybys are followed by orbiter missions. These are followed by lander missions including networks of landing stations and rovers. Sample return to Earth is then provisioned, followed by human exploration. The latter step is often met with skepticism because of its prohibitive cost and, except for the moon, is still facing huge technical and medical problems.

The payloads of orbiters and flyby spacecraft are specifically designed for their science goals. This is even more true today after some major space mission failures resulted in smaller, more dedicated missions. It is still possible, however, to characterize generic payload elements that are typically found on these spacecraft. Almost every mission has at least one camera, often two for wide-angle and for high resolution. Early missions had panchromatic cameras but recent missions have color capabilities, although not necessarily at the same resolution. Stereo is a useful feature that can be found on missions like Mars Express. Stereo will allow three-dimensional imaging and digital terrain models. Digital terrain modeling is helped by laser altimetry, as the most successful Mars Observer Laser Altimeter MOLA on Mars Global Surveyor has so convincingly demonstrated. MOLA has made laser altimeters almost a must on a modern exploration spacecraft. As a geodesy tool providing the topography of a planet the two are complemented by radio science instrumentation to measure the gravity field. In its simplest form, radio science instrumentation uses the communication devices of the spacecraft and a suitably accurate clock to measure Doppler shifts of radio signals traveling to and from the craft. More advanced systems use their own radio channels and accelerometers to measure the changes in spacecraft motion. If the surface is hidden underneath an optically thick atmosphere, radar can be used to map the surface, as has been successfully done at Venus by the Magellan mission. Radar is further used to map the surface roughness and to penetrate into the first kilometers of the near surface layers, for instance to look for ground ice, as is done by Mars Express. Other microwave sounders have been proposed that use different wavelengths and have differing depths of penetration.

While cameras, laser altimeters and radio science instrumentation are indispensable instruments for planetary geology and geodesy, spectrometers are the tools of cosmochemistry and mineralogy. These instruments typically analyze particles (photons, phonons, neutrals, ions) emitted from surfaces and atmospheres by the incident solar radiation. Visible, near and thermal infrared (0.5–15 μm) spectrometers are used to characterize the mineralogy of surface rock (silicates, carbonates, sulfates, etc.) and to search for water ice. Mapping infrared spectrometers such as Themis on Mars Odyssey and Omega on Mars Express allow a mapping of minerals on the surface. The Planetary Fourier Spectrometer on Mars Express works in a similar but broader spectral range (1.2–45 μm) but has a much higher spectral resolution at the expense of spatial resolution. Ultraviolet spectrometers measure neutral metals such as Al, Na and S, ozone, and OH radicals. Gamma-ray spectrometers can measure radio nuclides (K, Th, and U), major rock forming elements (O, Mg, Al, Si, Ca, Ti, and Fe), hydrogen (from water), and carbon, depending on the energy resolution and the elemental concentration in the surface. X-ray spectrometers are suitable to detect the major elements Mg, Al, Si, Ca, and Fe.

The particles and electromagnetic fields surrounding planets and other solar system objects are measured with magnetometers and plasma detectors. Dust analyzers, finally, collect dust particles and analyze the direction and strength of dust flux in space. Landing missions carry cameras and have been equipped with spectrometers and other sensors to measure composition and mineralogy. Mobility is an important issue since exploration of just one landing site may lead to biased results. Recent examples are the two MER rovers that carry a panoramic stereo camera, an alpha-particle X-ray spectrometer for elemental composition, a Mössbauer spectrometer for the identification of iron-bearing minerals, a miniaturized thermal infrared spectrometer for mineralogy, and a rock-abrasion tool to allow the measurement of "fresh" rock. Network landers have not been flown yet but are generally agreed to be important. These landers will operate the same payload simultaneously at several locations. Applications are seismology to explore the interior of planets, meteorology to explore planetary atmospheres, electro-

magnetic induction studies, and planetary geodesy. But single landers still have important applications such as the ill-fated Beagle Lander on ESA's Mars Express that was geared to search for traces of extinct and extant life on Mars and that failed upon landing in late 2003. Although landing missions still carry a significantly higher risk as compared to orbiter missions, it is generally agreed that they will be indispensable tools of future space exploration.

8.3
Planet and Satellite Orbits and Rotation States

There are some interesting commonalities between the orbits and rotation states of the planets and their major satellites. The orbits lie mostly in a plane that is defined by Earth's orbital plane, termed the ecliptic, and which is close to the equatorial plane of the sun (figure 8-5).

The normals to the actual orbital planes have inclinations relative to the ecliptic normal that are small, only a few degrees. Pluto is the only major exception with an orbital inclination of ~17°. This, together with its orbital eccentricity, small size, and density has led some to believe that Pluto may actually be the biggest known member of the Kuiper belt. The latter is a group of icy bodies mostly between the orbits of Neptune and Uranus and is the source of the short period comets. The rotation axes of most planets and major satellites are within a few tens of degrees to the vertical to their orbital planes. Notable exceptions are Uranus, whose rotation axis lies almost in its orbital plane, Venus, whose retrograde rotation can be expressed as an inclination of almost 180°, and Pluto, whose inclination is between those of the former two. The reasons for these anomalies are unknown, but are sometimes speculated to be attributable to impacts of planet-size bodies on the young planets during the late stage of accretion.

The rotation and revolution of most planets and moons are prograde, that is counterclockwise if viewed from above, from the celestial north pole. Exceptions are Venus, Uranus, and Pluto, whose retrograde rotations can also be described as inclinations of more than 90° of their rotation axes to their

Figure 8-5: View of the solar system (Artist's conception). Except for Pluto, the orbits of the planets lie close to the ecliptic plane. The planet sizes and distances in the image are not to scale (NASA)

orbital plane normals. Another exception is Triton, whose retrograde orbital motion about Neptune is unusual. Triton, like Pluto, is speculated to have originated in the Kuiper belt and to have been captured into its retrograde orbit by Neptune. There are many more satellites with retrograde orbits. However, these are typically much smaller than Triton, objects in a size range of only a few hundred kilometers. The known rotation periods of the major satellites and of most satellites in general are equal to their orbital periods. This is termed a 1:1 spin–orbit coupling and is believed to be the result of tidal evolution. Tidal evolution occurs because planets raise tides on their satellites, just as the moon raises tides on Earth. The gravitational force of the planet then pulls on the tidal bulge of the satellite. This torque will reduce the rotation rate of the planet—should it be greater than the average orbital angular speed of the satellite—and the interaction will end with the observed 1:1 coupling. The satellite's orbit expands during this process. As a result, the satellites present their primaries with mostly the same face at all times.

The planet Mercury is particularly interesting with respect to spin–orbit coupling since it is the only planet that is in such a resonance state with the sun. However, its coupling is not 1:1, as one may expect, but 3:2 (compare table 8-1). The reason for this somewhat odd ratio is believed to lie with the unusually large eccentricity of Mercury's orbit that has a value of 0.2. (Values of less than 0.1 are more typical.) Only Pluto's orbital eccentricity of 0.25 is larger than Mercury's. The large eccentricity causes significant differences to arise between the constant rotational angular velocity (of the sun) and the orbital angular velocity that varies along the orbit. The two angular velocities could only be exactly equal at all times if the orbit were perfectly circular with zero eccentricity. The orbital velocity on an eccentric orbit increases towards the *pericenter* (the point of the orbit closest to the primary) and decreases towards the *apocenter* (the point farthest from the primary). Mercury's 3:2 resonance now causes the orbital angular velocity to be equal to the rotational angular velocity at perihelion, the pericenter of an orbit around the sun. This minimizes the tidal torque on Mercury and stabilizes the resonance.

The orbital distances of the planets from the sun roughly follow a law with the distance of one planet to the sun being roughly twice the distance to its

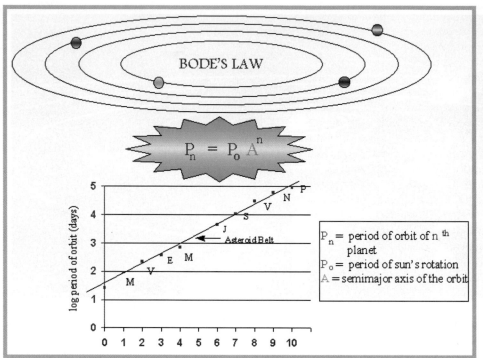

BODE'S LAW

$$P_n = P_o A^n$$

P_n = period of orbit of nth planet
P_o = period of sun's rotation
A = semimajor axis of the orbit

Figure 8-6: Illustration of the Titius-Bode law. This is a rough rule to predict the spacing of the planets in the solar system. It was first pointed out by Johann Titius in 1766 and was formulated as a mathematical expression by J.E. Bode in 1778. (web: *astrosun*)

inner neighbor (see figure 8-6). This rule is called the *Titius-Bode law* and works with Jupiter and Mars only if the asteroid belt is counted as a planet.

The origin of the law is little understood and it is a condition to be used as a test for accretion models. Regular relations between orbital distances and periods (the latter are coupled through Kepler's third law of orbital motion) are not unusual in the solar system, however. The most prominent example is the Laplace resonance between the innermost three major satellites of Jupiter: Io, Europa, and Ganymede. A comparison of the orbital periods in table 8-1 shows that these are in ratios of 1:2:4. The origin of the Laplace resonance and its stability is attributed to tidal interactions between Jupiter and its three resonant moons. The orbits expand in the resonance as rotational energy is transferred through tidal interaction from Jupiter to Io and passed on in part to Europa and Ganymede (see, e.g., Spohn 1997 for a review). The age of the resonance and how it formed are unknown. From the amplitude of libration it has been concluded that the resonance is relatively young. Thermal modeling, however, suggests an age of at least two billion years. The tidal interac-

tion and, in particular, the dissipation of tidal energy are believed to be the cause of volcanic activity on Io and of a subsurface ocean on Europa. This ocean is completely covered by icebergs and ice shields that may slowly move relative to each other, and the ocean may even harbor, or may have harbored, primitive forms of life.

8.4
Composition and Interior Structure of Planets

The chemical composition of the solar system is mainly dominated by the composition of the sun, which has 98.8% of the mass of all bodies in the system. Although the compositions of the planets are remarkably different from that of the sun they are related to the sun's composition and reflect varying degrees of depletion in volatile elements (e.g., Hubbard 1984). The sun (figure 8-7) is mainly composed of hydrogen and helium but contains all the other elements found in the solar system in abundances that are characteristic of the so-called *solar composition*.

These abundances are obtained from solar spectroscopy and from the analysis of primitive meteorites, the *CI chondrites*. Hydrogen and helium are important for Jupiter and Saturn, which are massive enough to keep these elements in molecular form against their tendency to escape into space. The potential of a planet to keep a specific element is measured by its escape velocity, which is proportional to the square root of the ratio between the planet's mass and its radius and must be compared with the thermal speed of an element, which is inversely proportional to its molecular weight. While Jupiter is believed to be of a composition similar to the sun's, a conclusion drawn from the deuterium to hydrogen ratio measured in its top atmosphere by spacecraft, Saturn appears to be depleted in both hydrogen and helium. The latter conclusion is based on spectroscopy data from the Saturnian atmosphere in which elements heavier than helium are substantially enriched relative to the solar composition. In addition to hydrogen and helium, Jupiter and Saturn contain substantial amounts of water (H_2O), ammonium (NH_3), and methane (CH_4), compounds collectively known as the planetary ices because of their occurrence on the surfaces of the major planets' icy satellites. In the deep interiors of the giant planets, these compounds are to be found in their supercritical forms, in which there is no difference between the gaseous and liquid states. In addition to hydrogen, helium, and planetary ices Jupiter and Saturn have deep sitting cores of rock and iron. These components, hydrogen and helium, ices, and rock/iron are the main elements of the recipe of the members of the solar system. Cosmochemists have learned how to infer the abundances of elements even in the deep interiors of planets from element correlations that are typical of solar system objects.

Models of the interior structure of planets can be constructed from a sufficiently detailed knowledge of their *figures* and their *gravity fields*. The construction of these models is a primary task of planetary physics. Rotation deforms the planets from spheres into prolate spheroids. The flattening of both the figure and the gravity fields is dependent upon the variation of density—and thus composition—with depth. Unfortunately, these data do not yield unique models. The most accurate method of exploring interior structure is *seismology* (Lognonne and

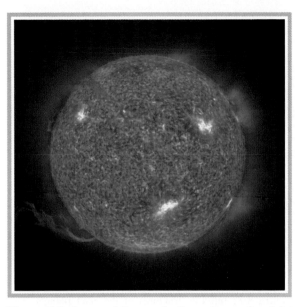

Figure 8-7: Image of the sun in the extreme ultraviolet (304 Å) taken by the Solar & Heliospheric Observatory (SOHO) on June 14, 1999. The sun is composed of hydrogen, helium, oxygen, carbon, neon, nitrogen, magnesium, silicon, iron, and traces of all other elements (NASA/ESA)

Mosser 1993), the study of sound waves that travel through the interiors of planets. This method has been extensively used on Earth to provide models of the variation of density with depth and also laterally by a technique that is know as *seismic tomography*. Seismometers, the instruments to record the waves, were also placed on the moon during the Apollo program and the interpretations of the results have provided important insights, though the results are not yet completely satisfactory. The method is not entirely restricted to the terrestrial planets, but can also be applied, at least in principle, to the giant planets, although seismometers cannot be placed on their surfaces. Here, cameras are used to study global modes of oscillation.

The models (e.g., Guillot 1999 and Hubbard et al. 2002 for reviews) suggest that the outer layers of both giant planets are mostly hydrogen and helium. Hydrogen is molecular at moderate pressures but becomes metallic at pressures higher than 170 GPa (figure 8-8). This pressure is equivalent to depths of about 14,000 km on Jupiter and 27,000 km on Saturn. The phase transformation occurs at greater depth on Saturn because the pressure increases at a

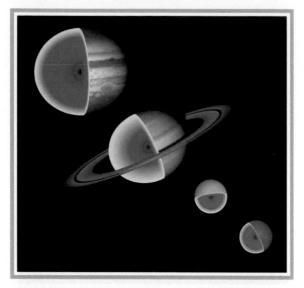

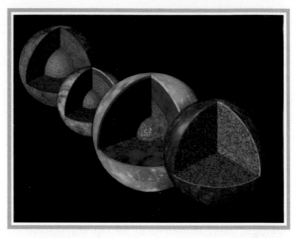

Figure 8-9: Interior structures of the Galilean satellites of Jupiter: Io, Europa, Ganymede, and Callisto (background to foreground) (NASA/JPL)

Figure 8-8: Interior structures of the giant planets. The yellow color indicates the molecular H, He shroud. Orange indicates the metallic H, He layers, blue is the range where the ice components are found and the dark colors indicate the rock/iron cores (T. Guillot)

smaller rate in this planet due to its smaller mass. The deeper interiors of both contain the denser ices in a shell above the densest iron/rock core. The cores plus the ice shells are believed to have masses equivalent to about 15 to 20 Earth masses and radii of a few Earth radii.

Uranus and Neptune are further depleted in hydrogen and helium (as compared with Saturn and Jupiter) and consist mainly of ice and the rock/iron component in addition to some hydrogen and helium. Their structure, as interpretations of the gravity field suggest (Hubbard and Marley 1989), is not as clearly layered as those of Jupiter and Saturn. Rather, there appears to be a gradual increase of density with depth accompanied by a gradual increase of the abundance of ice at the expense of hydrogen and helium followed by a gradual increase of the abundance of rock/iron at the expense of ice.

It is worth noting that most satellites of the outer solar system and Pluto also contain substantial amounts of ice, as their densities around 2,000 kgm^{-3} suggest (compare table 8-1). Examples to the contrary with substantially greater densities are Io and Europa, the two innermost Galilean satellites of

Jupiter. It is possible that these two also started their evolution with a substantial abundance of ice. Io may have lost the ice completely and Europa to a large extent as a consequence of heating by impacts or as a consequence of tidal heating. Since Io and Europa are closer to Jupiter than are Ganymede and Callisto, both the energy of impactors and the tidal dissipation rate will have been larger and may have caused the loss of water. However, it is also plausible that the temperature in the Jovian nebular from which the satellites formed was too hot at Io's and Europa's orbital distance to allow the accumulation of substantial ice shells. In any case, the four Galilean satellites are the only satellites of the outer solar system yet for which the present data gathered by the Galileo mission allow reasonable modeling (e.g., Sohl et al. 2002 and Schubert et al. 2004 for recent reviews) of their interior structure (figure 8-9). Accordingly, Io has an iron-rich core of about half the satellite's radius and 20% of the satellite's mass with a rock shell above. Europa has a (water) ice shell about 150 km thick overlying a silicate rock shell and an iron-rich core. The nonuniqueness of the data allows a wide range of sizes of these reservoirs, but the most likely radius values are 500–600 km for the core and 800–1,000 km for the rock mantle. It is widely believed that a large part of the ice shell may actually be liquid, allowing for an internal ocean. This is suggested by the results of the Galileo mission magnetic field measurements that indicate a field induced by Europa's motion in the magnetic field of Jupiter. This interpretation of the data requires a near surface electrical conductor, and water with some abundance of salts is a good candidate for this conductor. Moreover, model calculations show that tidal dissipation in the ice may produce enough heat to keep an ocean underneath an ice lid a few tens of kilometers thick. It is also possible, depending on little known values of *rheology* parameters and thermal conductivities, that radioactive decay and tidal heating in the rock mantle may keep the ocean fluid.

Ganymede has an ice shell around 800 km thick, a 900 km thick rock mantle shell, and a metallic core with a radius about the same. The case for an iron-rich core is particularly appealing for Ganymede because the magnetometers on Galileo have proven a self-generated field for this satellite (we will discuss magnetic field generation in section 8.6 further below). The magnetic field data also suggest an induced component, although the evidence for the induced field is much weaker. If the latter interpretation is correct, then Ganymede may also have an ocean of perhaps 100 km thickness at a depth where conditions are around the triple point for ice, water, and water vapor. This will be at a depth of around 150 km for this satellite. Since tidal heating is negligible at present, the heat that keeps the ocean molten must either be derived from radioactive decay in the rock mantle or from heat stored in the deep interior during earlier periods of strong tidal heating (e.g., Malhotra 1991) or even from accretion. The latter possibilities are highly speculative, however. In any case, the existence of the self-generated field is evidence enough for a molten core or, at least, a molten core shell.

Callisto is unusual among the Galilean satellites because the gravity data suggest that its interior is not completely differentiated. It is likely that there is an ice shell a few 100 km thick overlying a mixed interior of ice and rock also containing iron. It appears as if Callisto has traveled an evolutionary path different from that of Ganymede. The layering may have formed as a consequence of the slow unmixing of the ice and rock/iron components (Nagel et al. 2004). The model requires that Callisto has never been heated above the melting temperature of the ice. It is possible if not likely that the iron is no longer present in its metallic form but is oxidized to form magnetite and fayalite. The oxidation of the iron, if it occurred, would have precluded the formation of a metallic iron core. As for Europa and Ganymede, the magnetic field data suggest an induced field and an ocean for Callisto as well. As a matter of fact, the evidence is more clear-cut for Callisto than it is for Ganymede. Since Callisto is not in the Laplace resonance, tidal heating can be discarded as a heat source. The remaining candidates are radiogenic heating and accretional heating.

The terrestrial planets, finally, are almost completely devoid of hydrogen and helium, have ice in only modest concentrations, and mostly consist of the rock/iron component. The rock/iron component in these planets, just as in Io, Europa, and Ganymede, is differentiated into a mostly iron core, a silicate rock mantle, and a crust consisting of the low-melting point components of rock (compare figure 8-10). The crust, a layer of some tens to, at most, a few 100 kilometers thickness, forms by partial melting of and melt separation from the mantle. Crust growth is an ongoing process on Earth (together with crust recycling), but was most likely almost completed early on Mars and Mercury.

Seismology has provided us with a detailed image of the interior structure of Earth. It is expected that future missions will provide us with similar data for the other terrestrial planets. The most likely candidate here is Mars, for which mission scenarios with a seismological network have been repeatedly studied. The seismological data for Earth show that there are phase transformations mostly at moderate pressures and depths and chemical discontinuities. The existence of these layers, and their thicknesses and depths will vary between the planets. For instance, it is likely that Mercury's mantle will not have major phase-change layers simply because the pressure in this small planet does not reach the levels it reaches within Earth but it may be chemically layered, as is the case with the moon. Although the latter is even smaller than Mercury, the limited seismic data available suggest a layered structure. Much must have depended on the very early evolution and on the sizes of the planets. For instance, there is evidence that the planets were once covered by what is called *magma oceans*, rock molten by the dissipation of the energies of infalling *planetesimals* during the late stages of accretion. The moon possibly became chemically layered as the magma ocean fractionally crystallized. Earth may have escaped that layering because vigorous convection mixed the products of fractional crystallization. The vigor of convection depends on the size of the body or layer undergoing convection.

8.5
Surfaces and Atmospheres

The surfaces of most planets and major size satel-

lites have been observed optically, in the visible wavelength range of the electromagnetic spectrum, by cameras on board spacecraft. The gaseous giant planets and subgiants show the top layers of their gaseous envelopes in which clouds may form. Since these bodies lack solid surfaces, their surface radius has been defined to be the radius at which the atmospheric pressure is 10^5 Pa. Mercury and the major satellites except Titan show their solid surfaces. These bodies are lacking substantial atmospheres. The ability of a planet or satellite to keep an atmosphere depends on the planet's gravity. Equating the kinetic energy of a molecule of a species to the potential energy defines the escape velocity, which is

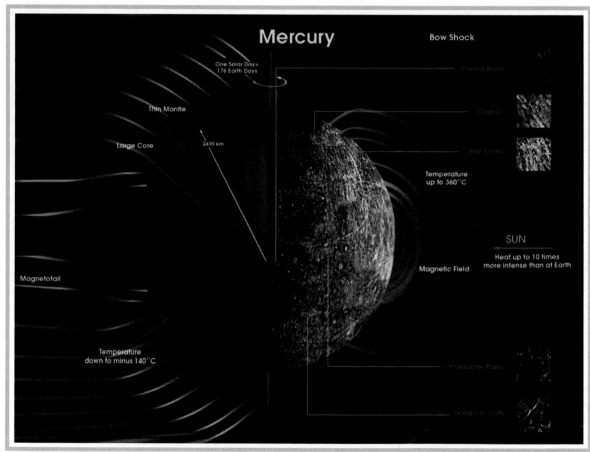

Figure 8-10: Characteristics of Mercury's surface, magnetic field and interior structure. The interior is likely to be differentiated into a solid rock shell, the crust and the mantle, and an iron-rich core. In comparison with other terrestrial planets, the rock shell is relatively thin and the core is comparatively large. The core itself is likely to be stratified with a liquid outer shell and a solid inner core. Growth of the inner core likely powers the dynamo that generates Mercury's magnetic field (ESA)

independent of the mass of the particle. Atoms, ions or molecules move at their thermal speed—which obeys a Maxwell-Boltzmann distribution and is proportional to the square root of the ratio of the temperature to the particle mass—but can be accelerated by other processes much beyond their thermal speeds. Among these processes are impacts, photoionization and pickup by the solar wind, hydrodynamic effects and sputtering (e.g., Chamberlain and Hunten 1987). The presence of a magnetic field will help to keep an atmosphere by blocking the solar wind from eroding the atmosphere. Escape due to the thermal speed (at the long end of the Maxwell-Boltzmann distribution) exceeding the escape velocity is called Jeans escape and is most relevant for light species such as hydrogen and helium. Jeans escape can be used for a systematic discussion in terms of planetary mass, radius and temperature. It explains why Mercury with its comparatively small mass and high surface temperature is prone to losing an atmosphere and why massive Jupiter can bind hydrogen and helium, but it cannot explain why Titan has a substantial atmosphere while Ganymede, similar in mass, radius and temperature, has none. The difference may lie with Titan being rich in ammonia from which the mostly nitrogen atmosphere may have formed by photodissociation (Coustenis and Taylor 1999). The compositions of the atmospheres of the terrestrial planets suggests that any solar-type primordial atmosphere has been replaced by a secondary atmosphere that is the result of outgassing and perhaps cometary impacts. It is also possible that the difference between Titan and Ganymede lies with their early differentiation and outgassing histories.

The surface temperatures of the planets are mostly determined by the solar radiation and therefore decrease with increasing distance from the sun. The actual surface temperature will depend on the reflectivity or *albedo* in the visible wavelength of the electromagnetic spectrum, on the emissivity in the infrared, and on the thermal and optical properties of the atmosphere. A useful quantity, the *effective temperature*, can be calculated assuming thermal equilibrium of the surface with the solar radiation neglecting the effects of the atmosphere. The effective temperature decreases from 445 K for Mercury to 44 K for Pluto. In the case of an atmosphere, the energy balance is more complicated. Visible light passes through the atmosphere and reaches the surface where it is in part reflected and in part absorbed and reradiated. The infrared radiation emanating from the surface is in part absorbed by CO_2 and/or H_2O in the atmosphere and raises the atmosphere's temperature. This effect, the so-called greenhouse effect, is particularly important for Venus due to the large amount of CO_2 in its atmosphere. Here, the surface temperature is about 740 K, about twice the value of the effective temperature. It is possible that Mars had a much denser atmosphere in its early evolution that allowed for higher temperatures through the greenhouse effect and liquid water. This would have helped to develop life on this planet. The effective surface temperatures of the planets are listed in table 8-1.

The solid surface of Earth is partly covered by clouds in the atmosphere and partly visible, depending on the extent and the pattern of the cloud cover. Venus' surface, however, is entirely covered by an optically dense atmosphere, and the same is true for the Saturnian satellite Titan. Venus' atmosphere is mostly CO_2 and the atmospheric pressure is 9.2 MPa. The pressure at the bottom of Titan's atmosphere is 150 kPa. The orange color haze that hides the surface from view is most likely smog produced by the photodissociation of methane. Venus' surface has been explored by radar observation. Titan is also explored by radar through the Cassini mission, which, by the time of this writing, is orbiting the planet and has repeatedly flown past Titan at a distance of about 1,000 km. On January 14, 2005, the European Huygens probe on board Cassini successfully landed on Titan and returned the first images from its surface (figures 8-11 and 8-12). The images reveal a complex surface morphology and suggest rivers, perhaps oceans. The surface upon landing was softish and muddy, as has been reported by J. Zarnecki, the principal investigator of the surface science package. The available data suggest a complex meteorology with hydrocarbon precipitation. Scientific experiments on board Huygens include a gas chromatograph and mass spectrometer for chemical analysis, an atmosphere structure instrument, a wind sensor, an imager and spectral radiometer, and an aerosol particle collector and analyzer.

Figure 8-11: The Huygens landing site as viewed from a height of about 8 km. The image is a composite from images returned by the Huygens probe during its successful descent to land on Titan. It shows the boundary between the lighter-colored uplifted terrain, marked with what appear to be drainage channels, and darker lower areas. The resolution is about 20 meters per pixel (ESA/NASA/JPL/University of Arizona)

Figure 8-12: First color image from Titan's surface. This image was processed from images returned on 14 January 2005 by ESA's Huygens probe during its successful descent to land on Titan. The rock-like objects just below the middle of the image are about 10 centimeters across. The surface consists of a mixture of water and hydrocarbon ice. There is also evidence of erosion at the base of these objects, indicating possible fluvial activity (ESA/NASA/JPL/University of Arizona)

The surface of Mars is visible most of the time. The main component of the Martian atmosphere is CO_2 but the atmospheric pressure is much smaller than Venus', of the order of 600 Pa. However, the surface may be hidden during times of global dust storms, which occur repeatedly on time scales of a few years. Ancient river beds, outflow channels and erosion features suggest that the Martian atmosphere was more massive in the past and the climate was wetter and warmer. The pressure on Mars is close to the triple point pressure of water (611 Pa, figure 8-13). Thus there is little room for liquid water except for low lying regions like the bottom of impact basins such as Argyre and Hellas. However, meltwater may exist for some time metastably on Mars and longer as its surface freezes over and it is covered by ice (see articles in Tokano 2005 and references therein). The early disappearance of the magnetic field of Mars may have been partly responsible for the escape of the early Martian atmosphere. The atmosphere of Earth differs from that of its neighbors Venus and Mars. The atmospheric pressure of 0.1 MPa is in between, and the main components are N_2, O_2, and H_2O.

The outer surface of Uranus is bland, greenish in color, and mostly featureless. The greenish color is attributed to methane and high altitude photochemical smog. At the other extreme is Jupiter, whose surface features a large number of bands or stripes largely parallel to the equator interspersed with dots and vortices. Particularly remarkable is the giant red spot, a vortex that covers about 1/10 of the planetary disk (compare figure 8-1). These features point to a highly dynamic atmosphere. The dynamics are dominated by rotation, as witnessed by the band structure, but the vortices and dots show that these become unstable at various scale lengths. The

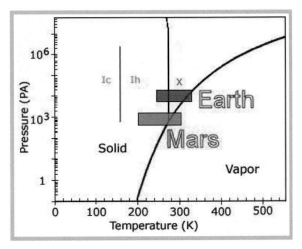

Figure 8-13: Phase diagram of water and typical atmospheric pressure and temperature ranges on the Martian and Earth surfaces. The diagram illustrates why water can be stable as ice and in the liquid phase on the Earth, but not easily as a liquid on Mars

Figure 8-14: The topography of Mars as measured by the Mars Observer Laser Altimeter (MOLA) (NASA/JPL)

brownish to reddish color of some of the features is attributed to ammonium-hydrogen-sulfate, NH_4SH, while the whitish colors are attributed to ammonium. Saturn's atmosphere is much calmer than Jupiter's, as the surface patterns suggest. These patterns are stripes similar to those in the Jovian atmosphere but the dots and vortices are missing, although wind speeds are extremely high. Neptune finally is similar to Uranus in that its surface is bluish in color, which is attributed to methane. It resembles Saturn in its stripe pattern. But in addition, there are a few vortices and a large spot that resemble features on the Jovian surface.

The solid surfaces of the terrestrial planets have some common features. Most prominent among these are impact craters that occur on a very wide ranging scale. Craters are believed to be remnants of the early evolution of the planets and satellites when the young bodies were bombarded by planetesimals during the late stage of the accretion process (see section 8.7 below). This phase is often termed the phase of heavy bombardment. The large range in size of the craters shows that at that time the size range of bodies in the solar system was likely similar to that of today. The major difference is that the number of bodies was much larger. The larger bodies swept up most of the smaller ones and the impact

rate declined with time to its present value, which is small but not zero. The distribution and density of craters on planetary surfaces is an important indication of the age of the surface or parts thereof. A surface is the older the higher the density of impact craters. For instance, Mars shows a dichotomy in surface age (and topography) between its northern lowland and southern highland hemispheres (compare figure 8-14). The age difference on average is about one billion years. Some of the most cratered and therefore oldest surfaces in the solar system are found on Mercury, Earth's moon, and the Jovian satellite Callisto. There are surfaces that are saturated with craters, implying that for every new crater formed by an impact one existing crater will be destroyed. Geologists and planetary scientists use the term *exogenic dynamics* to characterize the processes modifying the surface by impacts. Many impact craters on terrestrial planets have been partly or completely destroyed by erosion. Some have been buried by extensive resurfacing in the early history of the planet. On icy satellites long term relaxation of the ice may cause craters to flatten out. Some craters on Mars (figure 8-15) exhibit internal structures indicating the erosion of crater sediments. Cracks as exposed in figure 8-15 may be due to lava cooling, sediment drying or permafrost processes.

Planetary surfaces are also modified by atmosphere–surface interactions (erosion) and by processes related to *endogenic dynamics*; that is by processes that originate from within the planet. The most prominent of the latter processes is *volcanism*. Volcanism is the consequence of (partial) melting of the planetary interior and the rising of the buoyant melt to the surface. It is not restricted to rock and silicate volcanism as we know it from Earth; there may also be ice volcanism on the surfaces of the icy satellites. Volcanism has shaped at least parts of the surfaces of the terrestrial planets and silicate volcanism is the one important element of crust growth (compare section 8.4 above). Prominent volcanic features are the giant Tharsis dome on Mars, the island arcs on Earth, and the mares on the moon. The surface of Venus is dotted with volcanic cones. As we will discuss further below, endogenic activity scales with the mass of a planet. This partly explains why the activity on smaller planets usually dies off earlier than on the bigger planets.

A special case is the Jovian satellite Io. This satellite is of the mass of the moon, yet it is the most volcanically active planetary body in the solar system. Volcanic features cover its surface, but impact craters have not been detected. This shows that the surface is very young and is permanently renewed by the volcanic material (figure 8-16). The surface is to a large extent covered by allotropes of sulfur and by sulfur dioxide. These deposits cause the yellow to whitish color of the surface. The dark spots are likely volcanic vents and sulfur lakes. Sulfur deposits are often found associated with volcanic structures on Earth. The reason for the unusual activity on Io lies with an unusual heat source: Io is flexed by tides raised by its massive primary Jupiter. The deformation energy that is dissipated as heat is sufficient to make it more volcanically active than the largest terrestrial planet, Earth. A further measure of the enormous energy that is dissipated in Io's interior is the surface heat flow that is at 2-3 W m^{-2} , 20 to 30 times larger than the surface heat flow from Earth. Io's heat flow is rivaled only by Jupiter, which radiates about 5.4 W m^{-2}.

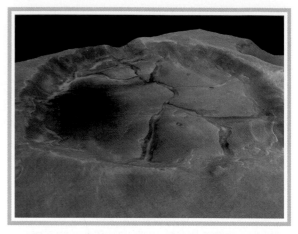

Figure 8-15: Crater (diameter 27.5 km) with eroded filling north of Valles Marineris (0.6° S, 309°E). The Mars Express HRSC image has a resolution of 12.5 m/pixel and was taken during orbit 61 (ESA/DLR/FU, G. Neukum)

Earth features a style of endogenic activity, *plate tectonics*, that according to our present knowledge is unique to this planet (Cox and Hart 1986). Plate tectonics involves the continuous production of basaltic crust along volcanically active linear ridges at the bottom of the oceans. Prominent examples are the Mid-Atlantic ridge and the Eastern Pacific rise. It also involves subduction of this crust underneath island arcs and continental margins. Prominent examples are the islands of Japan for the former and the western continental margin of South America for the latter. These subduction zones are the loci of most of the seismic activity of the planet. Both processes cause Earth's surface to be divided into seven major plates that drift across the surface. Plate tectonics is driven by extremely slowly revolving convection currents in its deep interior. Although convection is not unique to Earth but is expected to occur in the silicate rock shells of the other terrestrial planets as well, this feature is usually hidden underneath a thick stagnant lid (see section 8.5). The reason why convection on Earth extends to the surface and includes the crust at the bottom of the oceans is not entirely known. It is speculated that this is caused by the presence of water and there may even be links between plate tectonics and life. In any case, plate tectonics causes the surface of the ocean basins to be completely renewed on a time scale of a few hundred million years. Plate tectonics

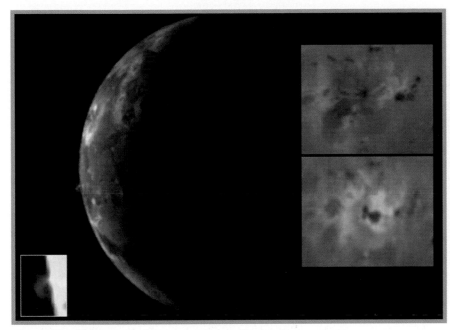

Figure 8-16: Volcanic plume on Io imaged by the Galileo spacecraft. The blue-colored plume extends to about 100 km above the surface. The blue color is consistent with the presence of sulfur dioxide gas and sulfur dioxide snow condensing as the volcanic gas in the plume expands and cools. The images on the right show a comparison of changes seen near the volcano Ra Patera since Voyager 1979 (top) and Galileo 2001 (bottom). An area of 40,000 km^2 was newly covered with volcanic material between the two observations (NASA/JPL)

together with erosion even incorporates recycling of the more stable continents. Material denudated by erosion from the continents is transported to the ocean basins where it is incorporated into the plate tectonics cycle. That loss of continental material is balanced by the production of continental rock through volcanic activity. It has been speculated that Mars went through a phase of early plate tectonics very early in its history. However, while this is a possibility it remains speculation. Although plate tectonics seem to be unique to Earth, there are other processes of crust recycling. These involve sub-crustal erosion, as is assumed for Io, and the foundering of the surface, as has been speculated to have happened on Venus at the sites of *Coronae*.

The best studied example of a single-plate planet is Mars. Two instruments on board spacecraft have recently allowed major progress in detailed studies of the tectonics of this planet. The first is the laser altimeter MOLA on Mars Global Surveyor and the second is the HRSC camera on Mars Express. While MOLA stopped operating as an altimeter in 2001 due to the finite lifetime of its laser pump diodes, HRSC is in the midst of its campaign at the time of writing. The major progress associated with these instruments lies with the establishment of an accurate geodetic network of surface features. The scientific exploration of Mars has been furthered by the successes of the American Mars Exploration Rovers, MER.

The topography of Mars exhibits a clear dichotomy which divides the surface into a southern highland hemisphere as heavily cratered as the lunar highlands and rising several thousands of meters above the zero level and a northern lowlands hemisphere that lies well below the datum (figure 8-14). The origin of the dichotomy is ascribed variously either to long-wavelength mantle convection or to post-accretional core formation sweeping up most of the crustal material into one large protocontinental mass. Another hypothesis interprets the dichotomy as a result of a giant impact. Whether an endogenic or exogenic process caused the dichotomy remains an unanswered question of Martian geology. The highlands cover about 60% of the planet, including almost all the southern hemisphere. The surface has survived the heavy bombardment prior to 3.8 billion years ago with only minor modifications and thus records the early history of the planet.

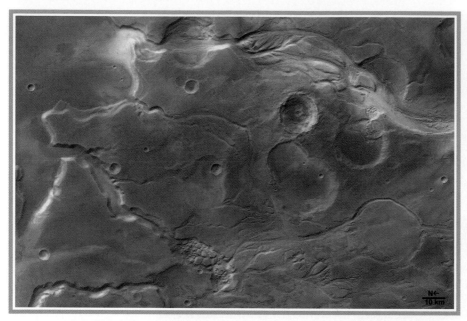

Figure 8-17: Mangala Valles extending roughly south-north near the top of the image and Minio Valles running from southwest to northeast near the bottom of the image. Streamlined islands indicate the fluvial character of the valleys and the chaotic structure in the lower left part shows where the valleys originate. The Mars Express HRSC image has a resolution of 28 m/pixel and was taken during orbit 299; (center of image 5°S, 209°E) (ESA/DLR/FU, G. Neukum)

Networks of small valleys are common in the highlands. The valleys are a few kilometers up to 200 km in length and are several hundreds of meters wide. Some of the shorter valley networks show a dendritic pattern. The morphological development of these valley networks mainly supports the hypothesis of a warm and wet early Mars. The valleys may have been eroded by surface runoff involving precipitation, but it is more likely that groundwater processes supported by mass wasting have played a major role in their formation. The plains of the northern lowlands are diverse in origin. Most common around the volcanic centers of Tharsis and Elysium are lava plains with clearly developed flow

Figure 8-18: Hecates Tholus (31.7N, 150°E). The asymmetric caldera shows different levels that are due to eruption events which emptied the underlying magma reservoir. The volcano is 5,300 m high and the caldera diameter is 10 km with a maximum depth of 600 m. On the flanks of the volcano several elongated depressions and riles indicate collapsed subsurface lava tubes. The Mars Express HRSC image has a resolution of 12.5 m/pixel (ESA/DLR/FU, G. Neukum)

fronts. However, the vast majority of the northern plains lacks direct volcanic characteristics, but are textured and fractured in a way that has been attributed to the interaction with ground ice and sedimentation as a result of large flood events, and finally modified by wind. The northern plains are younger by about 1Gy than the southern highlands and range in age from Hesperian to Amazonian. (The Martian history is divided into three stratigraphic epochs, the Noachian extending from 4.5 Gy before present to about 3.7 Gy, the Hesperian to roughly 3.1 Gy, and the Amazonian from the end of the Hesperian to present. The southern highlands are mostly Noachian.) Channel systems cut through the northern highlands from Valles Marineris (figure 8-17) to the Chryse Planitia and along the Mangala Valles, or emanate on the flanks of the Elysium volcanoes and terminate in the lowlands.

Most of these channels are developed as outflow channel and originate from collapsed depressions called chaotic terrain. The channels are interpreted to be formed by catastrophic flooding involving water or water and ice. It is assumed that several episodes of subsequent flooding were involved in the channel formation process.

Tharsis and Elysium are the two prominent volcanic provinces on Mars. The Tharsis bulge is of enormous size, covering about one third of the planet and rising 10 km above the datum. The three large volcanoes Arsia, Pavonis, and Ascraeus Mons, are oriented along the southwest-northeast tending top of the bulge with their summits 27 km above the

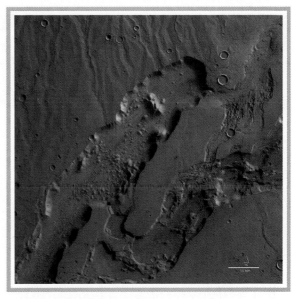

Figure 8-19: Source region of the outflow channels Dao and Niger Valles (32°S, 93°E) at the lower flanks of the volcano Hadriaca Patera. The valley system is up to 30 km wide and 2,400 m deep. Terraces and mesas show remnants of flooding events. The Mars Express HRSC image has a resolution of 40 m/pixel and was taken during orbit 528 (ESA/DLR/FU, G. Neukum)

datum. Olympus Mons, the largest volcano in the solar system, is located on the northwest flank of Tharsis. Each volcano has a large and complex summit caldera, and lava flows and channels are visible all over Tharsis. Smaller volcanoes on Tharsis show the same characteristics as the bigger ones. The flanks of some volcanoes exhibit lava flows

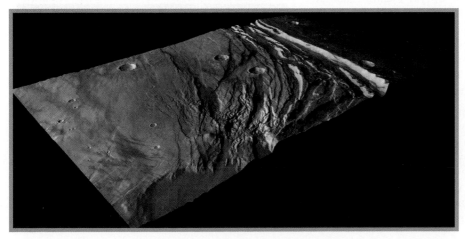

Figure 8-20: Acheron Fossae (35°-40° N, 220°-230° E) are located about 1,000 km north of Olympus Mons. These faults are part of a radial net of extension features originating from the center of Tharsis. The Mars Express HRSC image has a resolution of 40 m/pixel and was taken during orbit 528 (ESA/DLR/FU, G. Neukum)

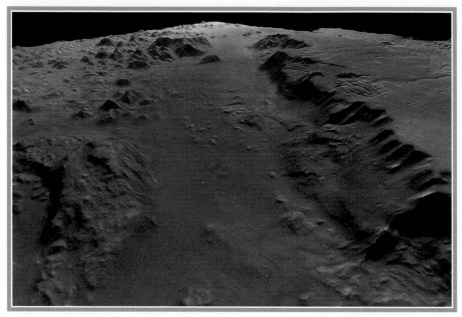

Figure 8-21: Eos Chasma (11°S, 322° E) . Eos Chasma is located at the eastern end of Valles Marineris where the canyons enter the chaotic terrain source region of the large outflow channels. The 5,000 m high highland plain steeply drops down to the canyon floor. Mass wasting and extensive erosion dominates the canyon walls and floor. The Mars Express HRSC image has a resolution of 40 m/pixel and was taken during orbit 528 (ESA/DLR/FU, G. Neukum)

with only a few superimposed impact craters, indicating a very young age, whereas extended flows between the large volcanoes are obviously older. This, the huge dimension of the Tharsis complex, and the low effusion rates of Martian volcanoes demonstrate that volcanism was an ongoing process throughout most of the planet's history. The Elysium volcanoes (figure 8-18) are smaller but similar to those on Tharsis.

Huge channels start at the western flank of the volcanoes and extend to the northwest into Elysium Planitia. These channels resemble flow-like features like the ones extending to Chryse Planitia. Similar channels are located adjacent to Hadriaca Patera on the rim of the Hellas basin. Because of the association of these channels with volcanic regions their formation is interpreted as volcano-ice interactions resulting in massive and catastrophic release of water by melting of ground ice (figure 8-19).

As Martian volcanism evolved, tectonics began to play a major role in the development of the crust. The evidence for surface deformation is faults, indicating extension, and wrinkle ridges, indicating compression. Most of the tectonic pattern on Mars is associated with the Tharsis bulge, which is surrounded by a vast system of grabens and fractures (figure 8-20), whereas wrinkle ridges are most common east of Tharsis. This tectonic pattern is obviously induced by the stress in the lithosphere caused by the presence of Tharsis.

Outside of the Tharsis region, tectonic structures are rare on the planet and occur mostly where the crust was differentially loaded, such as around large impact basins or in the Elysium volcano complex. The most spectacular tectonic process on Mars was the formation of the Valles Marineris which radially extend from Tharsis to the east where they merge with the chaotic terrain of the outflow channels. This

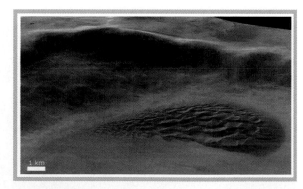

Figure 8-22: Dunes in a crater of 45 km diameter (43,3°S, 303.2° E). The dune field covers about 90 km2. The Mars Express HRSC image has a resolution of 16.2 m/pixel and was taken during orbit 427 (ESA/DLR/FU, G. Neukum)

system of canyons is about 4,000 km long, tens to a few hundreds of kilometers wide, and about 3-7 km deep. The canyons are incised into plateau rock (figure 8-21). The plateau-material is probably in part solidified lava that extruded along previously formed faults. During canyon formation faulting played a major role, however other processes such as mass wasting and possibly leakage of groundwater aquifers cut by faults have further modified the canyons. Fluvial erosion and deposition evidenced by thick sequences of layered materials caused the channel fillings. Finally, wind contributed to the sculpturing of the channel floors. Throughout Martian history its surface has undergone volcanic, tectonic, fluvial periglacial and also aeolian processes (figure 8-22) which are still modifying the Martian surface morphology.

8.6
Energy Balance and Evolution

Although the evolution of the planets seems to have followed some common general lines, there are significant differences among individual planets. The early evolution of all planets seems to have been dominated by impacts from the debris left over from planet formation. Moreover, it is believed that the planets differentiated early within a few tens of million years. Isotopic evidence is available for Earth and for Mars to suggest early differentiation by iron-core formation for these two planets. Of course, the evidence for Mars rests with the very well-founded assumption that the so-called *SNC meteorites* are in fact rocks from Mars. It is further believed that the planets started hot, the terrestrial planets heated to near melting temperatures of their rock components and to temperatures well above the iron melting temperatures by the energy deposited in the interior during accretion and by heat dissipated upon differentiation. The planets then cooled from this initial hot state and the cooling drove their evolution. It is possible, for instance, to explain the present rate of infrared radiation from Jupiter simply by cooling. This is not possible for Saturn, for which an additional heat source is required. The continuing gravitational settling of helium may provide enough energy to explain the observed present *luminosity* of this planet. The energy balances of most other planets and satellites seem to invoke both cooling and heat generated by the decay of radioactive elements, in particular of uranium, thorium, and potassium (the isotope ^{40}K) which generates 4–5 pW per kg of rock. A further possibility is tidal heating, but this mechanism seems to be relevant at present only for Io and, perhaps, Europa.

It should be stressed that the energy balances of the planets are not very well known. A crucial quantity, the surface heat flow or the intrinsic luminosity, is known for the giant planets, which radiate enough energy in comparison with the solar insolation incident on their surface so that their luminosity can be measured from orbit. This is also true for Io, whose enormous luminosity is remarkable. On the terrestrial planets and the other satellites, as far as we know them today, the surface heat flow must be measured in situ. This measurement requires a borehole deep enough to avoid the influence of the daily and seasonal temperature variations of the atmosphere. This difficult measurement has been done at many locations on Earth and at two sites on the moon. In table 8-1 we list the specific luminosities or surface heat flows for the planets and satellites for which the heat flows and luminosities have been measured. Since heat production depends on mass we have divided the heat flows by the masses. A comparison of the entries in table 8-1 shows that Earth and the moon are within a factor of two of the specific radioactive heat production rate of rock (about 4.5 $pW kg^{-1}$), suggesting that about half of the heat flow can be attributed to cooling and half to heat production. This conclusion is known to hold for Earth even on the basis of more specific data and more careful energy balances. An educated prediction of the heat flows for Mars, the other terrestrial planets, and the major satellites could be based on the above observation. The specific luminosity of Uranus also is close to the radiogenic heat production rate per unit mass of rock given above. It is likely that Uranus has a substantial rock core in which that heat is generated.

Neptune, Io, and Europa stand out in comparison. In the case of Neptune, the extra heat flow has been attributed to cooling and whole-planetary contraction. For Io, the extraordinarily large heat flow is with little doubt due to tidal heating, although ohmic dissipation of energy carried by electric currents between Io and Jupiter is sometimes quoted. The value for Europa in table 8-1 is highly uncertain and has been derived from indirect observation of tectonic surface features. If the estimate is correct then tidal heating would be the best explanation for its value.

The dynamic or geologic, chemical, and magnetic evolutions of the planets are mainly governed by their thermal evolutions. Planets can be regarded as heat engines that convert heat into mechanical work and magnetic field energy. Mechanical work will be performed by the building of volcanic structures and mountain belts, for example, or during the movement of plates over Earth's surface. The agent for these processes is convective heat transport, a mode of heat transport that involves movement of hot material to cold surfaces and cold material to hot surfaces (see, e.g., Schubert et al. 2001). Convection is driven by sufficiently great temperature differences across a layer or planetary shell and occurs whenever the convective heat transfer rate exceeds the heat transfer rate by heat conduction. Convection is possible even in the solid rock mantles of the planets because of the long time scales that are involved. On geological time scales of tens to hundreds of millions of years, rock behaves as a very viscous fluid. The convection flows at speeds of centimeters per year still transfer more heat than does conduction. This is in part due to the enormous masses involved in the flow and in part due to the very low thermal conductivity of rock. There is even a thermostat principle at work in the silicate and ice shells since the viscosities of these solids are strongly temperature dependent: If the flow is too slow to transport the heat, the temperature will increase until the flow is strong enough to transfer the heat. If the temperature is so high that the convective heat transfer rate largely outbalances the heat production rate or the heat flow into the layer from below, then the temperature will decrease, viscosity will increase and convective vigor will decrease. Chemical differentiation is mostly a consequence of melting and the density difference between melt and solid that will again lead to movement. A particularly good example for the chemical differentiation of the terrestrial planets that occurs in addition to core formation is the growth of the crust through volcanic activity. However, crust growth likely involves much longer time scales than does core formation.

The material in the crust is derived from partial melting of the mantle. Partial melting is possible when a material of complex chemistry lacks a simple melting temperature. Rather, in materials such as rock (and ice of complex chemistry) there will be a temperature termed the *solidus temperature* at which the material begins to melt. The melt will consist of that component of the entire assemblage that has the lowest individual melting point. The remaining solid will be depleted in that component. As the temperature rises, other components begin to melt until, finally, at the *liquidus temperature*, the whole assemblage will be molten. In the rock mantles of planets the component to most easily melt is basalt. Once a basalt liquid forms it will tend to rise to the surface by virtue of its density, which is lower than the density of the remaining rock. In a water–ammonium ice the low melting component will be a water–ammonium mixture of a particular composition. In most cases the melt is likely to be produced within the top few 100 km of a planet by a process termed pressure-release melting. Pressure-release melting occurs when the melting point gradient is steeper than the average temperature gradient. Relatively hot uprising convection currents will emanate from depths where the temperature is below the melting temperature. Upon rising to the surface adiabatically (that is with little heat exchange) the flow will hit a depth where the solidus temperature is exceeded and partial melting will ensue. Because melt is usually much more compressible then solid, there will be a pressure and a depth below which melt will no longer be buoyant and rise to the surface, but may actually sink towards the deeper interior. Melting in the interior of planets requires a heat source. This source is mainly the heat that is stored in the planets during accretion and the heat that is generated by the decay of radiogenic elements. Both are finite reservoirs that will not be replenished. That is why the endogenic activity decreases with time.

Since melting occurs when hot material from the deep interior rises in convection currents, the surface distribution of volcanism is speculated to give an indication of the planform of the convection under-

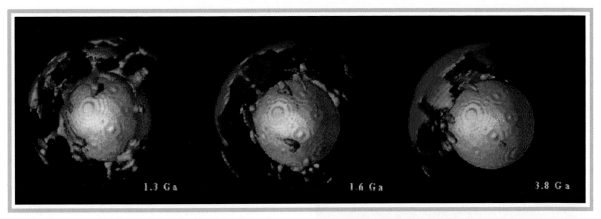

1.3 Ga   1.6 Ga   3.8 Ga

Figure 8-23: Development of a superplume. This rendering shows superhot material in the model Martian mantle that rises towards the surface. In this model by Breuer et al. (1998) the upwellings concentrate as a function of time into one superplume because of the interation of the flow with mantle phase transitions. It is possible that superplumes can also result in models with depth dependent material properties

neath. The timing of the volcanic activity can be used as a guide for assessing the time evolution of the convection. For instance, the differing surfaces of Mars and Venus may have recorded differing planforms of mantle convection and flow history. The dominance of Tharsis on Mars (compare figure 8-14) suggests that there is or once was a giant superplume, a very large upwelling underneath this volcanic dome (compare figure 8.23). The Mars Global Surveyor mission data suggest that Tharsis formed early, and the HRSC data from Mars Express suggest volcanic activity only a few million to ten million years ago. This long term stability of Tharsis as the major center of volcanic activity on Mars is puzzling, all the more so since model calculations suggest that the superplume's lifetime should not be much more than one billion years. Although Tharsis apparently formed early, the photogeologic evidence also suggests that volcanic activity on Mars started globally and retreated to Tharsis over time and may then have decreased in vigor. For a discussion of Martian thermal and volcanic history see, e.g., Spohn et al. (2001). On Venus (e.g., Bougher et al. 1997), there is indication that the volcanic activity has been global even recently (on geological time scales recently involves the past 10 to 100 million years). There is further evidence for a global volcanic resurfacing event a few hundred million years ago. This suggests that the thermal history of a planet may have involved episodes of more and of less activity that most likely were linked to episodes of greater and less vigor of convection. Volcanism may also have influences on a hydrosphere and atmosphere by degassing the deep interior of volatile elements. At the time of strong volcanism, an increase of atmospheric gases or water on the planet's surface is expected.

There is a fundamental difference between the evolution of Earth and most other terrestrial planets and major satellites that is related to the occurrence of plate tectonics on Earth (e.g., Cox and Hart, 1986). Since the viscosity, or more generally, the rheology of rock is strongly temperature dependent and since the surface temperatures of these planets are much lower than the melting temperatures, it follows that there must be outer layers that are comparatively stiff. These layers are termed the lithospheres. Usually the lithospheres are connected lids that are pierced here and there by volcanic vents. Convection currents flow underneath the lithosphere and deliver heat to the base of the lid through which it is then transferred to the surface by heat conduction. Since these lids are stagnant, this form of tectonics is often termed stagnant-lid tectonics (figure 8-24). On Earth, however, the lithosphere is broken into seven major plates that move relative to each other driven by the convective flow underneath. Volcanic activity occurs along some plate margins and results in the growth of plates. Other plate margins are destructive and are loci where plates are forced under their own

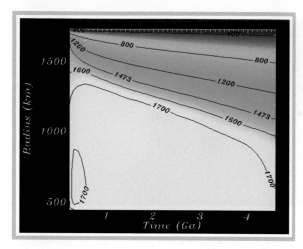

Figure 8-24: Growth of a stagnant lid in the lunar mantle. The growth of a cold, thermally conductive lid on top of the lunar mantle is shown as resulting from convection model calculations by Konrad and Spohn (1997)

weight to subduct into the mantle. There is no convincing evidence that plate tectonics occurs on any other planet or satellite, although it has been speculated that it may have occurred on early Mars and on the Jovian satellites Europa and Ganymede. There is evidence in the images of the surface units of lateral movement on these two satellites. The Jovian satellite Io does not seem to be undergoing plate tectonics. However, its present resurfacing rate, which is larger than the resurfacing rate of Earth, calls for some recycling of Io's crust volcanic material with the underlying mantle. This recycling probably occurs through delamination of the base of the crust. Delamination of the base of the crust may have been or may be operative on other planets, such as Venus.

The interior evolution of the gaseous planets is even less well constrained. It must be assumed that the dynamics of their atmospheres is related to the vigor of convection underneath, but the solar insolation also matters. Thus models of the evolution are mostly constrained by their luminosities, but as we have seen above these leave enough puzzles to be solved, as witnessed by the differences between the luminosities of Neptune and Uranus.

8.7
Magnetic Fields and Field Generation

Some of the planets, in particular, Mercury, Earth, and the giant planets, have largely dipolar magnetic fields that are supposed to be produced by dynamo action in their interiors. Of the major satellites, Ganymede is known to produce a magnetic field. The Galileo and Cassini data indicate that Io and Titan have no self-generated fields.

The icy Jovian satellites Europa and Callisto are surrounded by magnetic fields that vary along with their movement through the magnetosphere of Jupiter (e.g., Kivelson et al. 2004). These fields are, therefore, interpreted to be induced in electrically conducting layers in the satellites interiors as the satellites orbit Jupiter and are subject to a time varying magnetic flux. The geometry and the strength of these fields suggest that the conducting layers are at depths of some tens (Europa) to a few hundred kilometers. The most likely candidates are salty oceans. Thermal considerations suggest that these oceans are feasible (e.g., Spohn and Schubert 2003). Ganymede may have an induced field on top of its magnetic field, although the Galileo magnetometer data are not as conclusive in this case. An alternative interpretation associates the signal with higher order terms in the dynamo field.

Self-sustained magnetic fields are generally thought to be enigmatic to planets, part of their interior evolution during which thermal (and potential energy) is converted into mechanical work and magnetic field energy. The Mars Global Surveyor data have recently confirmed this hypothesis by showing that this planet has a remnantly magnetized crust (figure 8-25). This crust must have been magnetized during an epoch when Mars was generating its own magnetic field. (For a recent discussion of Martian magnetism see Connerney et al. 2004). Interestingly, it is mostly the oldest crust units that are magnetized. It has long been speculated that Mars, Venus, and Earth's moon once produced magnetic fields by dynamo action in their cores (e.g., Stevenson 1983). The observation of magnetized rock on the surface of Mars now confirms this general notion.

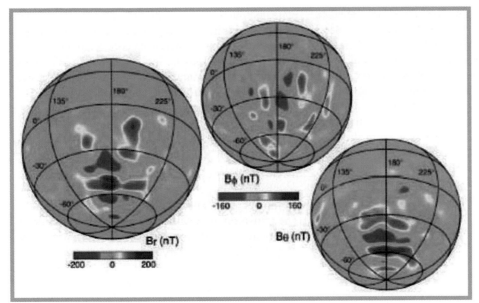

Figure 8-25: Orthographic projections of the three components of the magnetic field in spherical coordinates (r: radius, Θ: latitude, and Φ: longitude) onto the surface of Mars at a nominal mapping altitude of 400 km (From Connerney et al. 2004)

The dynamo mechanism (e.g., Busse 1989) invoked to explain magnetic field generation is similar among the planets although the source regions differ between the terrestrial planets and major satellites, the giant planets, and the subgiants, Uranus and Neptune. Required is an electrically conducting and fluid region that undergoes turbulent flow driven by thermal or compositional buoyancy. If a magnetic field exists, for instance the external magnetic field of the sun or a primary planet, a field will be induced in the source region. The flow will distort the field lines and thereby generate magnetic field energy by induction. If the flow satisfies certain conditions, the interference between the generated field and the pre-existing field can be constructive, powering and maintaining a magnetic field against dissipative losses. Dynamo theory, a complicated mathematical theory still in its infancy at the end of the twentieth century, elucidates the importance of rotation for the twisting of the field lines and the induction. Indeed, balancing the major dissipation mechanism, Ohmic dissipation, or Joule heating, against rotational inertial forces gives a scaling law for planetary magnetic fields that is supported by the observations.

For the terrestrial planets and the major satellites the candidate source regions are the fluid metallic cores or outer core shells. Because of their low viscosities

(around 1 Pa) these regions are often unstable with respect to thermal convection. The stability is believed to depend on the rate of heat removal from the core by the mantle convection flow. If the latter is too low (below about 10 mW m^{-2}), the heat flow in the mantle can be balanced by heat conduction in the core and the core will be stably stratified. Dynamo action is then not possible. If the heat flow into the mantle is larger than the critical value, convection in the core must be invoked to balance the heat removal rate. The situation becomes more favorable for convection in the core if the core begins to freeze and grow a solid inner core. First, latent heat liberated upon inner core growth helps to power the dynamo. Second, the core may contain sulfur and/or oxygen as light alloying elements. The light alloying elements expelled from the solid inner core can drive (chemical) convection and a dynamo very effectively because this dynamo will not be subject to a Carnot efficiency factor that by the second law of thermodynamics limits the efficiency at which a heat engine can do work. It is believed that the freezing of the inner core drives Earth's core dynamo.

Dynamo action in a terrestrial planet thus depends on the efficiency of mantle convection, on the composition of the core and the core material phase diagram. Plate tectonics, a mode of mantle convection associated with present-day Earth, is very effective at cooling the deep interior of a planet. Because the cold plates sink through the mantle, the deep mantle and core will cool effectively. It is therefore conceivable that Earth's core has been cooling to temperatures below the liquidus of its core alloy. Venus, a planet of similar size, appears to have been lacking that efficient cooling mechanism and appears to be cooling by convection underneath a stagnant lid. Consequently, the heat flow from the core is subcritical and during the evolution of the planet the core has not been cooling enough to reach liquidus temperatures. As a consequence of this and the general decline of mantle convective vigor, the core became stably stratified and a possible early dynamo ceased to operate. A similar scenario is likely for Mars. Early plate tectonics may have helped in generating the early field that has long become extinct but is recorded in the remnantly magnetized regions of the crust. There is little hope to find remnantly magnetized crust on Venus should landing missions be able to overcome the forbidding operating conditions on this planet. The surface temperature of around 450°C is above the temperatures at which candidate minerals become remnantly magnetized. The moon may be telling basically the same story, although here the evidence for early magnetic field generation may have partly been lost. There is doubt though that the small lunar core could have supported a dynamo. Alternative explanations for the recorded remnantly magnetized crust units invoke plasma clouds generated by major impacts. Some of the strongest magnetic anomalies are suspiciously located at the antipodes of major impact basins. Mercury differs from both the moon and Mars because of its extraordinarily large core and thin rocky mantle. The thin mantle may be quite effective, as model calculations suggest (Schubert et al. 1988), at removing core heat and driving a dynamo that produces a magnetic field, albeit a weak one. The weakness of the field has been discussed to be difficult to explain with conventional dynamo theory. An alternative dynamo mechanism invokes the thermoelectric effect. Ganymede is puzzling because it still produces a magnetic field, although a comparison with Mars and the other terrestrial planets would have suggested that its dynamo should also have ceased early. There is no evidence for Ganymede to have an effective cooling mechanism such as Earth's

plate tectonics and its core is only about 900 km (roughly a third of the satellite's size) in radius. Perhaps Ganymede formed its core relatively slowly and late, and we may be looking at a core in a phase of evolution that Mars and the moon passed through during the first billion years of their histories.

Dynamo action in the gaseous giant planets (e.g., Stevenson 1982, 1983; Russell 1987) most likely happens in the regions where hydrogen becomes metallic due to the extraordinary large pressure. The magnetic fields of these planets are most likely powered by the energy that has been stored in the planets during accretion and that is removed by convection. The transition to metallic hydrogen occurs at a depth of roughly 15,000 km in Jupiter and almost twice as deep in Saturn. The larger depth of dynamo action may partly explain why Saturn's field is much more ideal in terms of a dipole than is Jupiter's. In the source regions, the fields are likely to be very complex and not at all similar to a dipolar planetary field. The field in the source region can be thought of as a superposition of many multipolar fields of various amplitudes. Distance from the source region affects the higher order multipolar fields much more strongly then the lower order fields. If n is the order ($n = 2$ for a dipole), then the strength of the field components decreases with radial distance r from the source regions as r^{-n}. Thus, a planet with a comparatively deep source region should feature a more ideal dipolar field as compared with one where the source region lies less deep. This does not explain entirely the characteristics of the Saturnian magnetic field for which screening by helium rain has been additionally invoked.

Pressure in the interiors of the subgiants Uranus and Neptune is not sufficient to cause a transition to metallic hydrogen. Rather, it is believed that there are ionic oceans at relatively shallow depths in these planets. In these ionic oceans the magnetic fields can be generated. The fields of the subgiants have been measured only by a single flyby each and are therefore not well known. Nevertheless, it appears that these fields are very complex in their topologies. This fits in nicely with the idea of the fields being generated close to the surface at relatively shallow depths.

8.8
Origin of the Solar System

The origin of the solar system is still not completely understood. The basic elements of the theory are similar to the Kant-Laplace hypothesis of formation from a gaseous nebula but the details are debated (see e.g., Wuchterl et al. (2000) and Bodenheimer and Lin (2002) for recent reviews including the evidence from extrasolar planets). It is widely accepted that the nebula collapsed to form a central mass concentration—the proto sun—surrounded by a spreading thin gaseous disk. The most widely held view postulates that temperature in the inner part of the nebula soon became low enough—about 1500 K—to allow the condensation of silicate and iron grains. In the outer solar system, beyond about 5 AU, temperature became so low as to allow for the condensation of ice phases. Within the first few million years, these solid grains agglomerated to form bigger grains that, in turn, agglomerated to form even bigger ones. The cascading scenario of ever bigger and fewer planetesimals led to the formation of solid bodies the size and mass of the terrestrial planets in the inner solar system (see, e.g., Wetherill 1980 for a review of numerical simulations of this process) and 15 to 20 Earth masses in the outer solar system. In the outer solar system these solid proto-planets became the cores of the future giant planets. Growth of the solid proto-planets slowed to almost a standstill as the solid matter became exhausted, at least in the feeding zone of the proto-planets which is determined by the competition between a proto-planet's gravity and the gravitational pull of the sun. At about one Earth-mass a proto-planet begins to accrete gas onto its surface. This happened at a time when in the outer solar system the solid cores were still accreting solid planetesimals. The heat dissipated during the influx of planetesimals was radiated from the surface of the forming gaseous envelope. After the influx of solid matter ceased, the proto-giant began to cool and contract, thereby increasing its potential to accrete more gas. The growing mass led to even more accretion and runaway growth was set in place. Runaway accretion came to a halt when a gap formed in the nebula around the planet. This gap can form as a consequence of the competing gravities of the proto-planet and the sun and as a consequence of the finite viscosity of the gas.

This model offers an explanation for the grand features of the solar system: The inner planets are solid and of refractory composition because the temperature in the inner nebula favored the condensation of refractory phases. The inner planets are small because the feeding zone was smaller with less mass of solid particles. The bigger of the inner planets—Earth and Venus—have atmospheres with masses as expected, although the present atmospheres are not likely to be the primordial ones which were lost and replaced by degassing of the interiors. The present atmospheres are therefore gas that was originally deposited as solids in the growing proto-planet. In the outer solar system we find planets with cores of 15–20 Earth masses and massive gaseous envelopes. In addition, we find a wealth of satellite systems witnessing accretion in orbit around the growing proto-giants. There are a number of features that can be explained by fine-tuning the theory, for instance, the differences between Jupiter and Saturn and Uranus and Neptune, but this will not change the grand picture. A further consideration is the formation of the asteroid belt and the formation of relatively small Mars, which must be attributed to near-by massive Jupiter.

Giant impacts, a mechanism of great importance for some bodies, are the near final events of the accretion scenario. Giant impacts are thought to have caused the formation of the moon (e.g., Cameron 1997) and may be responsible for the high density of Mercury. A giant impact is a collision of almost grown proto-planets. For instance, it is thought that the moon formed after a Mars sized proto-planet hit the proto-Earth. The outer layers of the Earth vaporized during the impact and the moon formed from the condensed vapor cloud. Among the arguments for the giant impact hypothesis for the formation of the moon is the geochemical closeness of the bulk moon to Earth's mantle. Mercury may have suffered a similar giant impact that removed a substantial part of the original planetary mantle. This may explain why present Mercury has a comparatively big core and a thin mantle. A moon did not form around Mercury because of the closeness to the sun. Giant impacts may also affect the rotation and it has been proposed that Venus' retrograde rotation may thus be explained. Similar explanations have been brought forward for Uranus.

8.9
Concluding Remarks

The chemical compositions, interior structures, surfaces, evolutions, and magnetic fields of the planets and the major moons have been compared in this article. The planets are largely of solar composition but differ in their depletion of volatile elements. The degree of depletion increases with decreasing mass and with decreasing distance from the sun with Jupiter being closest in composition to the sun and with the terrestrial planets being most depleted in volatile elements. The planets are mostly internally differentiated with the heavy elements tending to be found near the center and the most volatile elements near the surface. The gaseous planets present the observer with the top layers of their envelopes, while most of the smaller planets present their solid surfaces. Relicts of very early impacts characterize these surfaces to varying degrees with volcanic and tectonic activities and with erosion. The volcanic activity and the magnetic fields of those planets that have self-generated fields are due to convection in their interiors driven by cooling and by heat generated by radioactive decay or by the dissipation of tidal energy. The source regions of the magnetic fields are metallic or, in the cases of Uranus and Neptune, ionic fluids. Life has had a chance to develop on Earth but possibly also on Mars and Europa. Finding evidence for life outside Earth is a challenge for the future. The bio-geo system of Earth may be a self-regulating system that sets this planet apart from its sister planets. Its position in the solar system, together with a moderate greenhouse effect in its atmosphere, has provided the planet with habitable temperatures and with an atmospheric pressure allowing for liquid water. Life plays an important role here by removing CO_2 from the atmosphere and moderating the greenhouse effect. Water, on the other hand, may be instrumental for plate tectonics since its effect on the rheology may pave the way for surface plates to subduct to the deep interior, thereby cooling the core. The cooling of the core is speculated to be instrumental for magnetic field generation. The magnetic field is a feature that Earth's brother and sister planets Mars and Venus lack. The magnetic field protects the environment, and thereby life, from radiation. It will be important to see whether or not life developed in the less friendly environments of Mars or even Europa or early Venus.

Exploration has started to look beyond the solar system for other planetary system. More than one hundred extra-solar planets have been discovered to date that are in the mass range of about 1/10 to 10 Jupiter masses, or, assuming the same density, roughly one half to two times the size of the latter. Of course, the interest is directed towards Earth-like planets in other solar systems. These planets are beyond the reach of present observational tools, and space agencies are planning to launch space based telescopes that have more favorable seeing conditions, such as the European Corot satellite to be launched in 2006 and the NASA terrestrial planet finder. But the question is, is the Earth typical? May other solar systems have planets with thick ice covers such as the icy moons of Jupiter, but bigger? Is it conceivable that there are planets of about Earth's size with primordial, Jupiter-like atmospheres? From a theoretical point of view, these worlds are possible (a statement which may simply reflect our ignorance). In the icy planets, radiogenic heat may support oceans and it is conceivable that these oceans harbor life. Could life exist in gaseous envelopes?

References

Bodenheimer P, Lin DNC (2002) Implications of extrasolar planets for understanding planet formation. Annu. Rev. Earth Planet. Sci., 30, pp. 113-148

Bougher SW, Hunten DM, Philips RJ (1997) eds: Venus II. Arizona University Press, Tucson

Breuer D, Yuen DA, Spohn T, Zhang S (1998) Three dimensional models of Martian mantle convection with phase transitions. Geophys. Res. Lett., 25, pp. 229-232

Busse F (1989) Geomagnetic field, main: theory. In: James DE ed: The Encyclopedia of Solid Earth Geophysics, van Nostrand, New York, pp. 511–517

Cameron AGW (1997) The origin of the Moon and the single impact hypotheses V. *Icarus*, 126, pp. 126-137

Chamberlain JW, Hunten DM (1987) Theory of Planetary Atmospheres, second edition, Academic Press, New York

Connerney JEP, Acuna MH, Ness NF, Spohn T, Schubert G (2004) Mars crustal magnetism. Space Sci. Rev. 111, pp. 1–32

Coustenis A, Taylor F (1999) Titan: The Earth-like Moon. World Scientific Publishing, London

Cox A, Hart RB (1986) Plate Tectonics: How it works. Blackwell Scientific, Palo Alto

Friedman L, Kraemer R (1999) Planetary exploration missions. In: Weissmann PR, McFadden L, and Johnson TV eds: Encyclopedia of the Solar System, Academic Press, New York, pp. 923-940

Guillot T (1999) Interiors of giant planets inside and outside of the solar system. *Science* 286, pp. 73-77

Hubbard WB (1984) Planetary Interiors. Van Nostrand, New York

Hubbard WB, Burrows A, Lunine JI (2002) The Theory of Giant Planets. Ann. l Rev. Astron. Astrophys. 40, pp. 103 – 136

Hubbard WB, Marley MS (1989) Optimized Jupiter, Saturn, and Uranus interior models. Icarus 78, pp. 102-118

Kivelson MG, Bagenal F, Kurth WS, Neubauer FM, Paranicas C, Sauer J (2004) Magnetospheric Interactions with Satellites. In: Bagenal F, Dowling TE and McKinnon WB eds: Jupiter, Cambridge University Press, Cambridge, pp. 513-536

Konrad W, Spohn T (1997) Thermal history of the moon: Implications for an early core dynamo and post-accretional magmatism. Adv. Space Res. 19, pp. 1511-1521

Lodders K, Fegley B Jr. (1998) The Planetary Scientist's Companion, New York: Oxford University Press

Lognonné P., Mosser B (1993) Planetary Seismology, Surveys in Geophysic. 14, pp. 239-302

Malhotra R (1991) Tidal origin of the Laplace resonance and the resurfacing of Ganymede. Icarus 94, pp. 399-412

Nagel K, Breuer D, Spohn T (2004) A Model for the Interior Structure, Evolution and Differentiation of Callisto. Icarus 169, pp. 402-412

Russel CT (1987) Planetary magnetism. In: Jacobs JA ed: Geomagnetism, Vol. 2, Academic Press, London, pp. 457-523

Schubert G, Anderson JD, Spohn T, McKinnon WB (2004) Interior composition, structure and dynamics of the Galilean satellites. In: Bagenal F, Dowling TE and McKinnon WB eds: Jupiter, Cambridge University Press, Cambridge, pp. 35-57

Schubert G, Olsen P, Turcotte DL (2001) Mantle convection in the Earth and Planets, Cambridge University Press, Cambridge

Schubert G., Ross MN, Stevenson DJ, Spohn T (1988) Mercury's thermal history and the generation of its magnetic field. In: Vilas F, Chapman CR, and Matthews MS eds: Mercury. Univ. of Arizona Press, Tucson, pp. 429-460

Shirley JH, Fairbridge RW (1997) eds: Encyclopedia of Planetary Sciences, London: Chapman and Hall

Sohl F, Spohn T, Breuer D, Nagel K (2002) Implications from Galileo observations on the interior structure and chemistry of the Galilean satellites. Icarus 157, pp. 104–119

Spohn T (1997) Tides of Io. In: Wilhelm H, Zürn W, and Wenzel HG eds: Tidal Phenomena, Lecture Notes in Earth Sciences Vol. 66, Springer Verlag, Heidelberg, pp. 345-377

Spohn T, Acuna MH, Breuer D, Golombek M, Greeley R, Halliday A, Hauber E, Jaumann R, Sohl F (2001) Geophysical constraints on the evolution of Mars. Space Sci. Rev. 96, pp. 231–262

Spohn T, Schubert G (2003) Oceans in the icy satellites of Jupiter? Icarus 161, pp 456-467

Stevenson DJ (1983) Planetary magnetic fields. Reports on Progress in Physics 46, pp. 555-620

Tokano T (2005) ed: Water on Mars and Life. Springer, Heidelberg/New York

Stevenson DJ (1982) Interiors of the giant plantes. Annu. Rev. Earth Planet Sci. 10, pp. 257-295

Weissman PR, McFadden L, Johnson TV (1999) eds: Encyclopedia of the Solar System, Academic Press, New York

Wetherill, GW (1980) Formation of the terrestrial planets. Ann. Rev. Astron. Astrophys. 18, pp. 77-113

Wuchterl G, Guillot T, Lissauer JJ (2000) Giant planet formation. In: Mannings V, Boss AP, and Russell SS eds: Protostars and Planets IV, Arizona University Press, Tucson, pp. 1081-1109

Web reference
astrosun: http://astrosun2.astro.cornell.edu

Between Space and Earth

Overleaf image: Space Shuttle Columbia on its way into space (NASA)

9 Communications

by Edward Ashford

This chapter explores the field of communications satellites in some depth, starting with an introduction and a brief history of communications. It then goes on to explore, in a nontechnical fashion, the general principles underlying radio communications, together with explanations of the terminology needed to understand the subsequent discussion on communications via satellite. Some pros and cons of satellite versus terrestrial means of communication are given, followed by a summary of the various types and services being provided by communications satellites today. Finally, some predictions are made for the future evolution of satellite communications.

9.1
Background and Objectives

Communications satellites earn learn large amounts of money for their operators. They are, arguably, the only subset of the satellite field today about which that statement can be made. Because of their financial viability, communication satellites are therefore the only type of satellite that has been able to attract sizeable amounts of private investment. Scientific satellites, meteorological and earth observations satellites, and manned space ventures all generally require some form of government or intra-government support for their development and operations. Even launch vehicle manufacturers have difficulty, at least when launches are few and far between, making enough money to commercially finance their continued development.

9.1.1
The Need for Means of Communication

Before delving into the intricate field of satellite communications, we should first define just what we mean by the term "communications."

Almost all creatures communicate in some fashion with others. This may be to broadcast their availability for mating, to warn of impending danger, or perhaps to notify others of a food source. The means of this communication can, however, vary considerably. Single-celled animals communicate typically on a chemical level, through exchanges of chemical "messages." With animals of higher order, chemical messages may also play a part (since the sense of smell is based on chemical processes), but other senses generally play a more important role. The use of sounds to convey warnings is commonplace, and of course human speech is based upon the sense of hearing. Touch can play a large role in transmitting sexual messages, indicating the willingness or not to mate, and vision is used extensively, ranging from, for example, the "dance" of honeybees to indicate the location of nectar-bearing flowers to the use by humans of pictures or printed text.

All of these methods involve the exchange of some form of information between two individual animals through an intervening distance. The distance may be very small, where touch is the medium of transmission, larger perhaps in the case of sound and hearing, and even larger with visual communications means. Thus we can, in a nonrigorous definition, consider "communication" to be just that, the exchange of information through a distance.

This definition begs the issue, however, since it immediately brings up the question, "what do we mean by "information"? Again, nonrigorously for the moment (a more rigorous definition will be given later), we can define information to be some form of knowledge or data known or available to the sending individual that is not known (or cannot be known, at least with absolute certainty) by the individual receiving a communication.

This chapter, of course, is about satellites providing communications between humans, so we need to consider another medium for communication, electromagnetic waves. The original information may be in another form of course, be it sound, picture, or text, but in the case of satellite communications, it must first be transformed into a modulated electromagnetic wave (a radio frequency wave, or possibly, into a modulated light wave) before it can be transmitted via a satellite.

9.1.2
Radio Frequency Communications - from Maxwell to Marconi and Beyond

In 1873, the Scottish physicist, James Clerk Maxwell, published "A Treatise on Electricity and Magnetism," which documented some of his theoretical research over the previous decade, in which he presented four equations that, collectively and in a very concise and elegant manner, described the behavior of electric and magnetic fields, and the interrelationships between them. His equations stated that whenever there is a time-varying electric field, it has associated with it a magnetic field, and vice versa. More importantly, he pointed out that the equations demonstrated the existence of "Electromagnetic Waves" of coupled undulating electric and magnetic fields, traveling through space at the speed of light.

A German physicist, Heinrich Hertz, demonstrated the first experimental confirmation of Maxwell's waves in his laboratory in experiments made between 1886 and 1888, when he showed that an oscillating electric current in a length of wire, generated by a discharge through a spark gap connected to a resonant circuit consisting of a simple inductor and a capacitor, could produce waves that left the wire,

traveled through space, and could be detected in another part of his laboratory.

Figure 9-1 shows a schematic diagram of Hertz' original experiment. When the switch on the left is closed, current flows through the primary of the induction coil, generating a high voltage in its secondary windings. The voltage generated is high enough to cause an arc to jump between the two electrodes of the spark gap. The capacitor and the inductance of the coil form a relatively high frequency resonant circuit that causes the current to flow back and forth across the spark gap. The interrupter in the primary circuit of the induction coil alternately makes and breaks the circuit (at a relatively low frequency compared to that in the secondary), ensuring that there is continuous generation of a high voltage in the secondary.

The rapidly oscillating flow of current in the secondary circuit produced electromagnetic waves which, just as Maxwell had predicted, radiated away into space. That this was occurring was demonstrated by a loop of wire placed some meters away which also had a (very narrow) spark gap. Sparks could be seen jumping across this gap when the experiment was operating, demonstrating that energy was being transmitted through space, and that the coil of wire was intercepting enough of this to induce sparks across the "receiver" spark gap.

The credit for putting the Hertz experiments to practical use goes to an Italian inventor, Guglielmo Marconi, who, in 1895 succeeded in generating "Hertzian Waves" outside the laboratory and detecting them as much as a mile or more away. This accomplishment was the starting point for the development of radio as a means of communication, and it, and subsequent experiments and demonstrations by Marconi, led to his being awarded the Nobel Prize in Physics in 1909.

9.1.3
The Road to Satellite Communications - a Brief History

The development of terrestrial radio, television, and radar grew directly out of the foregoing predecessor activities. Thoughts of using satellites to extend these means of communications outside the earth's

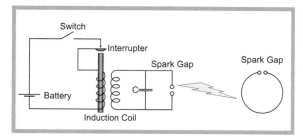

Figure 9-1: Experimental confirmation of Maxwell's equations

atmosphere only came under serious discussion in 1945 when a young British journalist and science fiction writer, Arthur C. Clarke, published the seminal article on satellite communications, "Extra-Terrestrial Relays - Can Rocket Stations Give Worldwide Radio Coverage?" in the October issue of the British magazine, "Wireless World." In that article, Clarke put forward the concept that satellites orbiting over the equator at radii of some 42,000 km (now called, interchangeably, either a "Clarke Orbit" or a "Geostationary Earth Orbit – GEO") would have periods identical to that of the earth, and would therefore appear, to an observer on the ground, to hover in one position in the sky. They could thus be used as relay stations to receive radio signals from anywhere on the earth from which they were visible and transmit those signals down to any other location on the ground visible to them. In this way, as few as three such satellites, separated around the equator by 120 degrees of arc, could provide complete coverage of the globe (except for the far northern and southern polar regions, from which the satellite would not be visible).

Clarke's prophetic views were, however, only that, a prophesy, because at that time, no launcher existed that could reach the necessary altitude, much less bring any sizeable payload there. With the advent, however, of the launch by the USSR in October 1957 of Sputnik I, the world's first artificial satellite, that situation began to change quickly. Sputnik had the effect of kick-starting the so-called space race, in which the USA and the USSR competed to develop bigger and more capable launch vehicles. Each side also began to develop the technology to build larger satellites, to keep pace with the development of launchers.

As part of the space race, the U.S. instrumented the upper stage of one of its Atlas B military missiles with an active repeater payload, and successfully launched it into orbit in December 1958 (Ashford 1986). While the batteries on the upper stage failed in less than two weeks, a limited number of communications experiments were able to be carried out before the failure, and the "satellite," named SCORE (for Signal Communications Orbit Relay Experiment) demonstrated for the first time that satellites could be useful for long-distance communications.

SCORE, and several other satellites that followed, were all low-earth orbiting satellites. The first successful 24-hour geosynchronous orbit satellite was the U.S. Syncom-II, launched in July 1963 (a predecessor, Syncom-I, launched in February the same year failed upon orbit injection). Syncom-II was geosynchronous, but not geostationary, since its orbit had an inclination with respect to the equator of some 33 degrees. As a consequence, to an observer on the ground it appeared to trace out a figure-8 path in the sky every 24 hours. It wasn't until August of 1964 that a sister satellite, Syncom-III (see figure 9-2) (Martin 2000), was injected into a truly geostationary orbit, and finally demonstrated both the feasibility and the usefulness of the Clarke orbit.

Also in August 1964, representatives of eleven founder member countries signed an accord forming the organization "International Telecommunications Satellite Consortium – Intelsat." In April 1965, that organization launched its first satellite, Intelsat-1 (also known as "Early Bird") into GEO, which ushered in the first truly commercial communications satellite service.

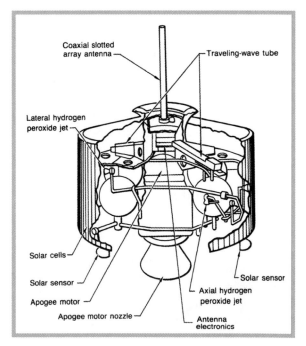

Figure 9-2 Syncom satellite details (The Aerospace Corporation)

The first communication satellites were developed to provide long distance trunking services. That is, they were intended for point-to-point transfer of relatively high bandwidth data streams, consisting either of many individual telephone connections multiplexed together into a single stream, or, alternatively, of television channels (originally only one per satellite, which was all the first satellites could carry). The points which were connected by these early satellites were generally on either side of the Atlantic or Pacific oceans since, in the early 1960s, the wire-based technology of the transoceanic cables which existed was of limited capacity, and only a few such cables were then in operation. Expansion in the number of transoceanic cables, the development of glass fiber technology that began to replace wire based cables, and their subsequent encroachment onto land areas soon began to change the competitive marketplace. As a result, satellites have evolved and developed in many different directions since then. This will be further discussed later in this chapter, after we first explore some of the basics of radio communications so that we can then understand just how such satellites operate.

9.2
Principles of Radio Communications

In the slightly more than one century since radio transmissions were first demonstrated, the technology in the field has advanced tremendously. The basic principles determining just how a radio system works, however, have changed very little, and are still in use in modern communication systems, whether terrestrial or satellite based. The principles involved are the following: in a transmitter, a *signal* carrying some form of *information* is used to *modulate* an RF current in some fashion. The modulated current may then be coded, is next intensified by a *power amplifier* and then radiated away from the transmitter. The radiated wave spreads out, decreasing in intensity as it travels away from the transmitter, until a (usually very small) fraction of it is received and amplified some distance away in a receiver. The received modulated signal is then *demodulated*, amplified and, if all has gone well, the resulting signal at the output of the receiver contains the same information that was fed into the transmitter.

Let's examine all these terms to see what they really mean in the context of a communications system. We will first consider a simple terrestrial RF communications link, such as that illustrated in figure 9-3, having a single transmitting station and a single receiving station, separated by a distance that might be in the order of some tens of kilometers or so.

Signal

By *signal* we mean a varying electrical (or possibly, optical) parameter, such as a voltage or current, the variations of which represent information being carried by the signal. The output from a telephone microphone, for example, is a varying voltage that is an electrical analog of the sounds spoken into the mouthpiece. Not all signals are, however, analog.

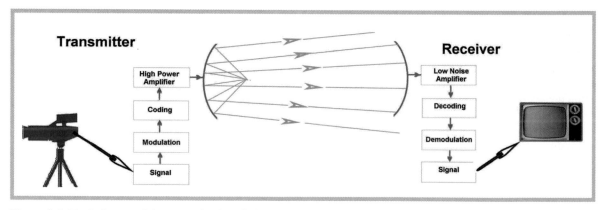

Figure 9-3: Typical terrestrial communications link - TV broadcasting direct to users

The long and short impulses of electrical current (representing Morse code dots and dashes) carried on the telegraph lines that were so prevalent at the beginning of the electronic communications revolution are examples of digital communications signals. Although analog signals were commonly used in the early days of satellite communications, most satellites now being developed, and a high proportion of those in operation, now communicate exclusively by digital means, and it is this that we will concentrate on throughout the rest of this chapter.

Information

Information, as the term is used in communications theory, is a rather difficult concept to understand without resorting to a number of mathematical and probabilistic equations. However, as we mentioned earlier, it can be defined intuitively as anything known at the transmitting end of a communications link which is not known (or at least not fully known) at the receiving end of the link until after it has been transmitted and received. Thus, for example, sending someone a message "It is nine o'clock in the morning where you are" could be conveying a certain amount of "information" to the recipient if they weren't in possession of a watch or clock, but would have little or no real information content if they already knew the time. Likewise, if a telephone caller you know very well and whose voice you easily recognize states his name at the beginning of a phone call to you, the amount of information conveyed in so doing is minimal, while if someone totally unknown to you does so at the beginning of his call, the information content is significant (Richharia 1999).

The amount of information in a message varies in an inverse fashion with the probability of its occurrence. Engineers use a binary logarithmic function to define the precise amount of the information in a message, and write (Richharia 1999):

$$I = \log_2 (1/p_k)$$

(measured in "bits"), where p_k is the probability of occurrence (which can have any numerical value between zero and one) of the k-th message. Thus, if the probability of a particular message being sent is very low (p_k near zero), its information content is very large. Conversely if the probability is very high (p_k near one), then I is close to zero, and very little information is conveyed.

An important aspect of the concept of information is that one can determine numerically just how much information (in bits/second) can be sent through any communications channel (defined as its *Capacity,* "C") having a signal strength "S," in the presence of interfering *Noise,* "N") by using what is known as the Shannon-Hartley theorem:

$$C = B \log_2 (1 + S/N)$$

where "B" is the bandwidth (in Hertz) of the channel. While this provides only an upper limit to what can be done, it does allow communications engineers to make tradeoffs between bandwidth and signal power when designing a satellite communications system.

A second importance of considering the information content in a message is that it can be shown that almost all actual messages have a large amount of inherent redundancy in them. That means in practice that they contain far more bits of message than actual bits of information. This leads to the conclusion that it should be possible to *compress* a message by removing redundant bits, and therefore send it in a narrower bandwidth, or alternatively, to reduce the transmitter power when sending the message, and still have it received successfully. This is, in fact, a most powerful technique, and has been a major factor in making satellite television broadcasting both feasible and economical.

Modulation

The term *modulation* refers to a process whereby an information-containing signal is used to vary a physically measurable parameter of a radio frequency (or an optical) beam. The beam itself is referred to as the *carrier,* and the physically measurable parameter can typically be one of either the amplitude of the carrier, its frequency, or its phase. Early radio broadcasts primarily used amplitude modulation (AM – see figure 9-4), as this was the easiest to produce with the technology then available. Later, when more sophisticated circuitry became possible, frequency modulation (FM) became increasingly popular. In this instance, the changing amplitude of the input signal is used to vary the frequency of the carrier rather than its amplitude.

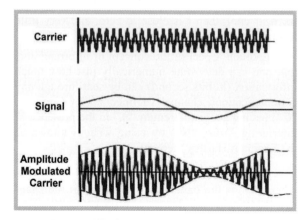

Figure 9-4: Amplitude modulation

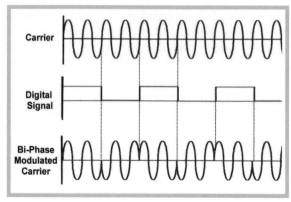

Figure 9-5: Bi-phase phase shift key modulation

FM signals became popular because it was found that they are less susceptible to interference by extraneous electrical *noise* than are AM signals.

With the advent of satellite communications systems, a simplified form of FM modulation was initially used, called Frequency Shift Keying (FSK). In FSK the output of the modulator shifts between two discrete frequencies, depending upon whether the input signal to the modulator is a digital "zero" or a "one." FSK soon began to be supplanted by a type of *phase modulation* known as Phase Shift Keying (PSK). This form of modulation works by shifting the phase of the transmitted carrier in response to a digital input signal. In a receiver, the phase of the received carrier can be compared to that of an internally generated RF frequency with constant phase, and the difference extracted as representing the original signal at the transmitter.

The amount by which the phase of a carrier is shifted by the input signal can vary, depending upon the specifics of the modulation used. In the simplest version, known as BPSK (the "B" standing for "bi") and shown in figure 9-5, a phase shift of either zero degrees or 180 degrees is used, depending upon, for example, whether the signal sent at that instant of time is a binary "zero" or a binary "one." Another form of PSK, used more commonly in satellites today, is called QPSK, for "quadrature" phase shift keying, in which sets of two binary *bits* are combined at the sending end into *symbols*, and a signal shifted in phase by zero degrees, 90 degrees, 180

degrees, or 270 degrees is sent, depending upon, for example, whether the symbols are 00, 01, 10, or 11.

A form of modulation that is sometimes used in terrestrial communications links, known as QAM (for Quadrature-Amplitude Modulation), combines the two concepts of Amplitude Modulation and Phase Shift Keying. In simple QAM, both the phase and the amplitude of a carrier are modulated by a signal. Suppose that four possible phase angles, separated by 90 degrees, are combined with two different possible values of power level (amplitude). Then eight different possible states can be sent with each change in the modulation envelope, and we would have what is known as 8QAM. Another version of the same technique, where eight possible phase shifts (in 45 degree increments) are combined with two possible amplitudes, is termed 16QAM.

Since higher orders of modulation are possible (8PSK, 16PSK, 8QAM, 16QAM, ...), where 3, 4, 5, or even more bits of information can be sent per symbol, why aren't these used rather than the QPSK that is commonly used with satellite transmissions? The answer is simple, and is an example of the old adage, "you can't get something for nothing." The smaller the increment of amplitude and/or phase shift that is used, the more susceptible a modulated beam becomes to being contaminated by extraneous noise, unless its overall power level is increased significantly. Satellite transmissions are, unfortunately, susceptible to interference and degradation from a variety of sources, and providing extra power on board a satellite is a very expensive undertaking, so the higher orders of modulation for satellite signals are generally avoided, unless the situation is particularly interference free.

There is another type of modulation that makes use of the tradeoff possibility mentioned above in the discussion of the Shannon-Hartley theorem. The equation defining that theorem shows that, if you make the bandwidth large enough, you can reduce the level of the signal and still transmit the same quantity of information per unit time across a communications link. This is one of the underlying principles behind what is known as *Spread Spectrum* modulation. By "spreading" a signal over a wide frequency band, the level of the signal per unit bandwidth can be reduced to very low levels, until it is comparable to or even less than the level of the interfering noise.

One method of expanding the bandwidth of a signal to take advantage of this possibility is to transmit the signal in a narrow band only slightly larger than needed, but to vary the actual frequency of the band used as a function of time over a much wider range of frequencies. If the specific frequency bands used, the order in which they are used, and the rate at which the frequencies are changed are known only to the transmitter and the intended recipient, then it would be extremely difficult for another receiver, without knowledge of these parameters, to intercept and demodulate the signal. This was actually the original intent of this concept, which was patented in 1941 during World War II by an Austrian born Hollywood actress, Hedy Lamarr (under her married name of H. Markey) and a colleague, as a means, for example, of safeguarding the military communications used to control the guidance of a torpedo against interception and/or jamming (U.S. Patent 1942).

An alternate method of spreading a signal over a wider bandwidth is where a digital code, extending over the entire available bandwidth (known as a *spreading code*), is used along with the narrower band signal to digitally modulate a carrier. In this case, if the spreading code is known at the receiving end, it can be used to extract the very low level received signal, even in the presence of other signals occupying the same overall bandwidth but with different spreading codes. This ability of many signals to occupy the same spectrum with little if any interference is known as Code Division Multiple Access (CDMA).

Coding

The susceptibility of transmissions to interference leads directly to another important concept in communications systems, that of *coding*. If a transmitted digital signal becomes contaminated by sufficient noise, what is received may be so corrupted so that transmitted binary "ones" become interpreted in a receiver as "zeros," and vice versa. In a properly designed communications system, it is generally possible to make the probability of such an error occurring low, but there always remains a small chance that it will happen. How then is the receiver to know whether it has received a correct signal or an incorrect one? One solution is to add *redundancy* to the signal.

For example, consider a lengthy digital message that consists of a very long string of ones and zeros ("bits"). It is common to divide such a string into segments of eight successive bits, with such a segment referred to as a "*byte*." Suppose then, (although as we will soon point out, this is not very practical), we send each byte not once, but twice in succession. Since the probability of a bit error occurring in any byte is small, then the probability of two successive bytes both being corrupted in exactly the same way should be sufficiently small as to be negligible. Thus, if the receiver compares two successively received bytes, bit by bit, and finds them to be the same, it can conclude that no error has occurred, and then go on to a similar comparison of the next two received bytes.

If, however, such a bit by bit comparison does detect an error, and depending upon the communications *protocol* being employed (more about this later), the receiver end of the link can send a message back to the transmitting end to ask it to retransmit the byte in question. That retransmitted byte can then be compared in the receiver to the two previously received bytes, which usually allows the receiver to determine which of the two had been in error and continue to process the received message using the correct byte.

Such a process is an over-simplified example of what is known as *"Error Detection Coding."* Such a coding process would not be very efficient, however, since it would require every byte to be sent twice, effectively halving the amount of information that could be sent across the link per unit of time. A much simpler and more efficient method of coding would be to insert what are known as *parity bits* into a message.

With this approach, again perhaps over-simplifying, the succession of ones and zeros in the message can be divided into segments each containing seven bits. To each of these segments, the encoding device then adds an additional "check bit," which can be either a zero or a one, determined as follows. The encoder counts the number of ones in the seven bits. If this is an even number, the encoder adds a zero at the end. If the number of ones is odd, however, the encoder will add a one.

Thus, every successive eight-bit byte to be transmitted will now have an even number of ones in it. All the receiver needs to do is to count the number of ones in each byte it receives (called a *parity check*). If this is an even number (called *even parity*), the receiver can conclude that the byte is correct; if it is odd, the receiver can send a message to the transmitter asking it to retransmit that (encoded) byte again. Thus, by only adding a single bit to a seven-bit segment of a message (i.e., 1/7 or about 14% redundancy), one can accomplish what would have needed eight additional bits for every eight bits (8/8 or 100% redundancy) in the previously described coding method. Parity coding is thus a much more efficient method.

Of course, there is always the possibility that a byte will experience two bit errors in transmission. In this case, the receiver will not detect it with a simple parity check. While this is uncommon, it can occur, so even more rigorous means of coding have been developed that can detect multiple errors if they in fact occur.

There are even codes that have been developed that not only detect when a bit error has occurred, but can even determine what bit is in error. Such a code, called a Forward Error Correction (FEC) code, does not require the receiver to ask for any retransmission of erroneous bytes. They can be corrected in the receiver based solely on what has been received. A

		Bits								
		Information Bits							Row Parity Bit	
W o r d s		1	0	0	1	1	1	0	0	
		0	0	0	1	1	1	0	1	
		1	0	0	0	0	1	1	1	
		1	0	1	1	1	1	0	0	
		1	1	0	0	0	0	0	0	
		1	0	0	1	0	0	1	1	
		0	0	1	1	1	1	0	0	
Column Parity Bit		1	1	0	0	0	1	0	1	

Figure 9-6: One-bit error detection block code

simplified example of such a code is illustrated in figure 9-6, whose explanation is as follows.

Suppose we select seven successive words in a message. To each word, we append a parity bit as explained above, so that we obtain a set of seven words, each now of eight bits, and each of which is of even parity. Then let us line these words up in columns, and add an additional eight bit word to the set whose elements are such as to make each of the columns have even parity. We then transmit this eight bit by eight bit *Block* as a sequence of 64 bits. When these bits are received, the decoder arranges them back into an eight by eight array, and checks the parity of each word and each column.

Suppose for example, that during transmission a single bit error has occurred in the fourth bit of the fourth word. Then the parity bits for both that row and that column will be found to be in error, and their intersection will identify the specific bit in error. It can then be automatically corrected in the decoder, without having to ask the transmitter to resend any data.

It is easy to see that the ability of such a code to correct a single error would hold true even if one of the parity bits were the bit in error. Also, as was the case for the simple error detection codes, more rigorous FEC codes have been developed that can detect and correct multiple errors. If, however, the probability of multiple errors becomes too large, the "overhead" of the code itself may have to be so large that it would limit the rate at which the actual information content of a message can be sent to unacceptable levels. In such a case, it would be necessary to increase the signal power of the transmission, so the received signal would have fewer errors. The increase in the corresponding signal to noise ratio would, as dictated by the Shannon-Hartley theorem, also increase the maximum information rate.

With the proper combination of modulation type and coding, satellite channels can be made highly reliable, with their links providing error-free signals through all but the most severe propagation environments.

Transmission

The next step in the process is to transmit the modulated signal. As predicted by Maxwell, and as demonstrated by Hertz and Marconi, an alternating electrical current can be made to propagate through space as a radiated electromagnetic wave. At low frequencies, the amount of energy in such a wave is small, but at the frequencies employed in satellite communications, the strength of the radiated wave can be considerable. All that is needed is to amplify the power of the modulated signal sufficiently and then to direct it to a suitable antenna, where radiation occurs naturally. The power amplifier in most common use in satellite payloads is a device called a Traveling Wave Tube Amplifier (TWTA), which can amplify signals over a wide bandwidth with relatively high efficiency. Alternatively, it is also not uncommon, particularly at the lower satellite communications frequencies, to use Solid-State Power Amplifiers (SSPAs) which, although they are not as efficient as a TWTA, can be made smaller and lighter, and with at least potentially a higher intrinsic reliability.

The output of whichever type of power amplifier is used is fed to the antenna, from which the actual radiation takes place. The antenna is most commonly a parabolic shaped dish, similar to that used on the ground for direct reception of satellite television. Its purpose principally is to focus the radiated beam in the direction of the intended receiver(s), so sending it in directions where it is not needed doesn't waste satellite power. Unfortunately, antennas cannot produce an absolutely parallel ray output, so all radiated beams diverge, resulting in a decrease in *flux density* (radiated power per unit area) as the beam gets further and further away from the source antenna. A receiving antenna some distance away,

itself having an aperture area of only a few square meters, therefore receives only a very low signal level.

Reception

When such a signal is received, the inverse process that was used at the transmitting end begins. The received modulated wave must first be amplified greatly by what is known as a Low Noise Amplifier (LNA) and the information extracted from the modulated wave by the process of demodulation. If, as is usually the case, the signal was coded at the transmitter, then the process of decoding must be undertaken to determine if any errors have occurred, and these must be corrected if they are detected. The method by which this is done, as was said earlier, depends upon what type of coding was used, and to a great extent, on what communications *protocol* is used.

A communications protocol is basically a set of rules that govern how data is to be formatted and sent over a communications link. For example, the well-known (but often not so well understood) Internet Protocol, IP, defines a set of rules to be used to break messages into smaller sized packets, and to add addressing information to each packet, so that Internet routers will know to where individual packets should be sent. This is commonly bundled together with another protocol, the Transmission Control Protocol, TCP, which has rules governing how the packets should be sent, and what should happen if a receiver doesn't send an acknowledgement signal indicating that it has received a packet.

Compression

The last concept we will cover before delving into the specifics of satellite communications is one mentioned earlier, that of data compression. This refers to the process whereby the redundancy inherent in most messages is reduced or eliminated before they are sent, thereby reducing the bandwidth required for transmission.

Consider, for example, a "message" consisting of a "3 megapixel" digital camera's color photograph of a distant airplane outlined against a clear blue sky. Such a camera typically has a resolution of 2,048 by 1,536 pixels = 3,145,728 actual pixels, and each pixel usually consists of 24 (or more) bits of information (8 bits each for the sensed level of each of three primary colors) (web: *digital cameras*). Each of the three colors in a particular pixel then is assigned one of the 256 values between 0 and 255 ($2^8 = 256$), and the complete pixel has any of $2^{24} = 16,777,216$ values assigned to depict its overall color and intensity. The digital representation of that photo would then appear to have more than 50 million megabits of "information" associated with it. Or does it?

Earlier in this chapter we defined the information content of a message (or, in fact, any part of a message) as being the logarithm of the reciprocal of its probability of occurrence. In the example given here, where most of the picture is composed of the clear blue sky, the probability of a blue pixel occurring is very high, so the amount of information a blue pixel conveys must be very low. It should therefore be possible to reduce considerably the amount of data needed to convey such a picture.

For example, rather than sending each successive 24 bit pixel in a string of many such "blue" pixels, one could instead send the data corresponding to the color and intensity of only the first pixel, followed by the number of times it occurs in a string until a pixel of another color or intensity is encountered. This way, by removing redundant data, the amount of data that must actually be sent can be reduced by a significant factor, thereby greatly reducing the bandwidth required to send it.

Removal of redundant data from a message before sending it is an example of what is termed "lossless" compression. That is, within the limits imposed at least by characterizing each pixel in the picture to have only a discrete number of digital values assigned to it, the compressed message sent would contain all the real information that existed in the picture before it was compressed. In the contrived example given above, the reduction in data might be one or more orders of magnitude, but in practice, compression by a factor of two or slightly higher is more typical.

There are, however, further steps that can be taken to reduce the amount of data that must be transmitted to send a picture over a communication link. Instead of analyzing each line of a picture to look for runs of the same color and intensity on a line-by-line basis, one can instead characterize each area of a picture by a mathematical formula based on the way your eyes actually perceive color. For example, the Joint Photographic Experts Group (JPEG) has devised a compression algorithm that looks at a pixel and its neighbors in all directions and determines certain factors in a formula that best represents all those pixels (web: *JPEG*). Then, instead of sending the data associated with each pixel, one can send only these derived factors. When a picture compressed and transmitted this way is received, the JPEG process plugs the factors back into the formula to generate new pixels that "best represent" the original picture. Since less data is sent than is actually inherent in the picture, such a compression type is known as "lossy" compression.

While the algorithms used for actual digital data compression of pictures are much more complicated than what is described above, one can get a feel from the example how one can "compress" the amount of data contained in a static picture. The same type of consideration also holds true for other types of messages. While the concepts involved are not quite the same, one can also compress both sound and written text into narrower band channels.

The percentage by which the data in a picture can be compressed by the JPEG process is variable, but providing that one does not try to compress too far, the received picture will look "almost" like the original. The amount of data "lost" in the process is not detectable by the naked eye until higher levels of compression are achieved.

Suppose we wish to send a video instead of a static picture. Of course, each frame of the video could be compressed by a process like that of the JPEG, and each sent in turn, but there are other tricks that can be used to compress video data much further than for a static picture.

Typically, a video frame consists of data defining its background (the blue sky in the example given previously), which remains relatively static, and foreground data (the moving airplane in the previous example). Suppose we compare two subsequent frames in a video on a pixel-by-pixel basis. If the background hasn't changed, there is no need to send any background data for the second frame at all. All that may need to be sent is information depicting the movement of the airplane itself (information depicting its size and shape may remain sufficiently constant to treat it the same way as background information).

As was the case for static images, video data compression can be done with either lossless or lossy approaches. By taking good advantage of such considerations, so called Moving Picture Experts Group (MPEG) lossy compression methods can (normally) reduce the amount of data that must be sent, and hence the bandwidth of the communications channel needed to send it, by several orders of magnitude with little discernible degradation in video quality.

9.2.1
The Pros and Cons of Satellite Communications

Why are satellites used for communications at all? Why not rely upon terrestrial transmitters? The reasons for using communications satellites can be categorized as technical, economic, and historical.

Technical: As will be explained in more detail later, the radio-frequency spectrum available for communications is a limited resource. With few exceptions (which will also be explained later), only one user at a time in any particular geographical area can make use of a particular segment of the spectrum without causing and/or experiencing significant interference. Lower frequencies also tend to curve around the surface of the earth by effectively "bouncing off" the charged particles making up the ionosphere, so that the interference problems at these frequencies can cover a considerable area, even at relatively low transmitter powers. As the lower frequency bands became crowded, new users were forced to make use of higher and higher frequencies to avoid such interference. At higher RF frequencies, however, the ability to send radio signals "over the horizon" becomes limited, since the higher frequencies tend to penetrate the ionosphere without significant reflection. At the higher frequencies, only line of sight point-to-point communications become practical. While this can limit the area over which these signals will cause interference, it also limits the range over which signals can be sent without using an array of "repeaters" to receive and retransmit signals over the horizon. Since satellites lie outside of the

atmosphere, and can see wide areas of the earth from orbit, the use of only a few satellites for such tasks can avoid the need for having many such repeaters scattered all over the earth.

Economic: The above explanation leads directly into the second category, economics. Sending RF signals at those higher frequencies over long distances, particularly if one wants to broadcast over wide areas, may require a very large number of signal repeater stations. Individually, these may not be too expensive, but collectively, their costs soon become excessive. A study conducted by the European Space Agency in the early 1970s showed for example that, for telephony signals (requiring a large number of point-to-point links) for calls over ranges greater than 800 kilometers, it was less expensive at that time to put up and use a GEO satellite than to install the needed network of terrestrial repeaters. One repeater on a GEO satellite can effectively do the work of thousands of repeaters on the ground.

Historical: As mentioned earlier, when communications satellites were first being developed in the 1960s, there were only a few intercontinental undersea cables in existence, and those that did exist had only a limited bandwidth capacity. The first trans-Atlantic telephone cable, called TAT-1, had the capacity to carry only 36 simultaneous telephone calls when it was put into operation in 1956 (web: *telephone*). When Intelsat-1 went into operation in 1965, it could carry 240 simultaneous calls. Satellites therefore became a preferred means of providing intercontinental connectivity where there were no cables, and, where there were existing cables, as a backup means of restoring communications in the event of a cable failure.

The capability of satellites developed rapidly. When Intelsat-3 went into operation in 1968, it had the capacity for 1,200 simultaneous voice circuits plus two television channels. Intelsat-4 raised the ante still further in 1971 with 4,000 voice circuits, and Intelsat-5, in 1980, yet again with 12,000 circuits (Ashford 1986).

It seemed at first that undersea cable capacity would never catch up, but this was not to be the case. In 1988, the first trans-Atlantic optical fiber cable was put into operation between the USA and France, having the capability of carrying 40,000 simultane-

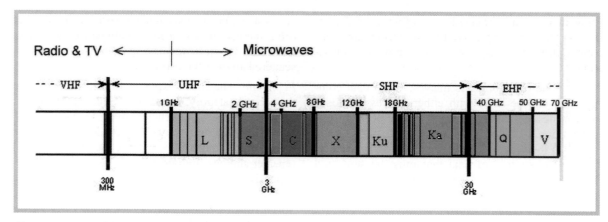

Figure 9-7: Part of the electromagnetic spectrum in the radio and microwave region indication the communication bands

ous telephone calls (web: *capacity*). Subsequent fiber cables and technology developments that brought about increases in fiber capacity soon caused such cables to greatly surpass their satellite competition for such point-to-point applications. The bandwidth available when using optical means of communication is so much greater than that available in the RF spectrum for satellites that satellites find it difficult to compete head-on with the fiber cable industry for such applications. As we will see below, however, there are many other areas in which satellites can and do excel, and will continue to do so for many years.

The foregoing discussions have highlighted some of the advantages of using satellites compared to terrestrial means of communications. In summary these are the wide area coverage that can be provided from even a single satellite, and the relatively low cost involved in developing and orbiting a satellite system in comparison to that involved with setting up an equivalent coverage terrestrial network. To this must be added the relatively short time to provide 100% coverage by satellite compared to the more gradual evolutionary growth of terrestrial coverage. Satellites can also provide mobile connectivity to ships and planes over ocean regions, which would not be feasible by purely terrestrial means. As discussed hereafter, however, there are also limitations on the extent to which satellites can provide communications capacity and on the extent to which the provision of that capacity can be assured.

9.2.2
The Drawbacks of Satellites - Limited Frequency Spectrum

Radio, television, and microwave frequencies occupy only a relatively small portion of the electromagnetic spectrum. This small portion, moreover, is a finite resource that satellites must share with a wide range of terrestrial users. Long before communication satellites were even seriously envisaged, there were already a large number of users in the lower frequency bands that often created interference and congestion. To establish order in this area, at a Plenipotentiary Radiotelegraph Conference in 1927, the International Telegraph Union (ITU), which itself had been formed in 1865 to establish standards for international telegraphy, set up the International Radio Consultative Committee (CCIR) to, inter alia, establish standards and rules for the use of the RF spectrum. It also allocated specific discrete frequency bands to the different types of radio services that were then being provided (these were fixed, maritime, aeronautical, and broadcasting services) (web: *ITU*). These allocations were made separately in each of what are known as the three "ITU Regions," which are defined roughly as: Region 1 – Africa and Europe(including all of the previous USSR territory), Region 2 – North and South America (including Greenland), and Region 3 – Asia and Oceania (including Australia). The allocations in each of these regions are embodied in a set of documents called the "Radio Regulations," which have been updated and expanded through the years and now consist of several complicated and rather massive documents. The frequencies that have been allocated by the ITU for use by communications satellites fall within segments of the RF spectrum referred to commonly as "bands" (see figure 9-7). The bands principally used by commercial satellites, which are L-band (in the range of 1-2 GHz, used for MSS), S-band (2-4 GHz, used for mobile and satellite telemetry services), C-band (4-6 GHz, used for FSS), Ku-band (12-18 GHz, used for FSS and BSS), and Ka-band (18-40 GHZ, for FSS). Within these band designations, however, only a number of relatively narrow "sub-bands" are allocated to the satellite services mentioned. The size and exact frequencies of the sub-bands varies somewhat, depending upon which of the three ITU regions is considered, but in general, the higher the frequency band used, the more bandwidth that has been allocated to satellite use. Thus, at L-band, the mobile satellite services have an allocation only 25-35 MHz wide, but at Ka-band, there is a full gigahertz of bandwidth available.

The ITU (now known as the International Telecommunications Union) is now an arm of the United Nations. Of the 191 UN member countries, 189 have joined and enjoy full signatory membership in the ITU. In addition, there are more than 640 Sector Members, consisting of telecommunication operators and organizations, manufacturers, research and development organizations, etc. The ITU is tasked, inter alia, with holding periodic "World Radio Conferences - WRC" (previously known as World Administrative Radio Conferences – WARC), at which the Radio Regulations can be revised and/or expanded. The ITU also acts as a coordinating body to which entities proposing to make use of specific frequency bands must apply (through the governmental Member Organization in their country) to establish a "Coordination Process" with other users of the same frequency bands and/or positions in the Clarke orbit who might be adversely affected by the proposed new usage.

Propagation Problems

In order to understand the following and subsequent sections of this chapter, a small technical digression is needed for readers who are not yet familiar with the jargon of communications. When one looks at the amount of power that is received, for example, by a small residential satellite television receive dish, and compares it to the power that is radiated from the satellite, one finds that there are factors of many, many orders of magnitude between them. Communications engineers conventionally measure such differences in decibels, which are defined by an exponential equation, $P(dB) = 10 \cdot \log_{10}(P/P_0)$, where P_0 is a reference level, usually taken as one watt. If a power level of 100 watts is radiated from an antenna, a communications engineer would say that its radiated power level is $10 \cdot \log_{10} 100/1 = 20$ dBW. In comparing two signals, if one signal has a power level twice that of another, one can say that the higher is $10 \cdot \log_{10} 2/1 = 3.01$ dB (usually rounded to 3 dB). If a signal is 1,000,000 times the power of another, it can be said to be 60 dB higher, and so on. Even when satellites operate in frequency bands that are free of interference from other users of the spectrum, the fact that satellite links must traverse the atmosphere can give rise to degradation in signal quality or even to a total loss of signal, due to a number of factors. The effect of the ionosphere in reflecting/refracting RF signals at lower frequencies has already been discussed, but as that effect is reduced at higher frequencies this is generally quite small at the frequencies in the microwave region normally used by satellites. However, the presence of oxygen and water vapor in the lower level of the atmosphere known as the troposphere, even on clear, cloudless days, attenuates RF signals passing through it. This attenuation also varies with the frequency of the RF waves, but it generally increases as one goes higher in frequency.

Figure 9-8 shows the approximate levels of attenuation caused by oxygen and water vapor in the atmosphere, as a function of frequency (Fortescue and Stark 1995). The figure is based on an assumed relative humidity of 40% at 20°C, and shows that, for a vertical penetration of the atmosphere (which would imply an earth station at the equator with a satellite directly overhead), the attenuation caused by water vapor at around 20 GHz, for which the effective vertical path length is about two kilometers, would be of the order of 0.2 dB, which would cause a reduction in signal level of about 5%. If an earth station were located in the higher latitudes,

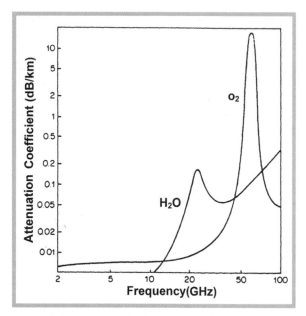

Figure 9-8: RF attenuation by atmospheric components (Fortescue and Stark, 1995)

the signal received at C-band (4 GHz) from a satellite, which could be compensated for by built in margins in the transmission, at Ka-band (20GHz), the attenuation could be as large as 25-50 dB, effectively wiping out the signal completely. Fortunately, however, except in very wet and rainy areas, the occurrence of the rainfall intensity that would cause such high attenuation levels is relatively rare, fairly well localized, and if it does occur, normally lasts only a short time.

Therefore, while these and other types of propagation problems are unavoidable with satellites, with proper system design and with suitable margins being included, such problems can be overcome except for a very small percentage of the time in any particular area.

9.3
Satellite Communication Links

The process described earlier, explaining what happens when two earth stations establish a communications link, is very similar to what happens when communications satellites are also employed. The satellites are seldom the initiators of messages. They merely act as intermediate stations to receive a message sent up from an earth station at one location on earth and to retransmit it down to one or more other earth stations at perhaps vastly different locations. The payload on a satellite that perform this function is known as a *repeater,* and the specific part of the

and/or if the satellite weren't directly overhead, the effective path length through the atmosphere could be significantly greater. With a ten-degree elevation angle, the effective path length for water vapor would be some 11.5 kilometers, and the attenuation would be 2.3 dB, causing a signal level reduction of over 40%.

Such attenuation problems can be catered for when designing a satellite communications system by incorporating sufficient compensating power margins in the satellites and ground stations. There is another propagation problem, however, caused by rainfall, for which compensation is not so easy.

Rainfall can also attenuate RF signals, as shown in figure 9-9 (Fortescue and Stark 1995). The rain attenuation increases as a function of frequency, so L-band and C-band transmissions are, for the most part, unaffected by rain. At Ku-band, the attenuation when there is heavy rainfall can be significant, however, and at Ka-band it can be severe. The attenuation is a complex function of raindrop size, rainfall rate, elevation angle, and other factors, but is roughly proportional to the square of the frequency of operation (WTEC 1993).. Thus, while a heavy rain squall might cause an attenuation of 1-2 dB in

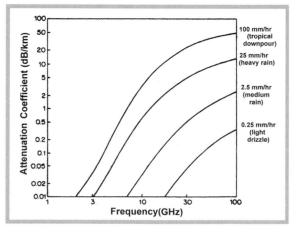

Figure 9-9: RF attenuation by rainfall (Fortescue and Stark, 1995)

payload that accomplishes it for any particular message, channel, or group of these, is known as a *transponder* (transmitter and responder) (web: *transponder*).

For many years, satellites provided simply this function, merely repeating the signal that it received from the ground. As illustrated in figure 9-10, the repeater on the satellite would receive a (very low level) signal sent up from an earth station, amplify it in an LNA, change the frequency of the modulated wave in order to avoid interference between reception and transmission, power amplify the resulting wave, and radiate it back down to earth. Repeaters operating in such a fashion are referred to as "bent pipe repeaters," in analogy to a water pipe or garden hose that can make a stream of liquid "turn a corner" without otherwise affecting it.

In recent years, advances in technology have made it feasible to put more intelligence in the satellite, so it can do more than simply blindly repeat what it receives. One such technology improvement has been the ability to provide more and smaller spot-beams on a satellite. Early satellites had beams that covered the entire portion of the earth visible from the satellite. Later, narrower beams could be generated, but

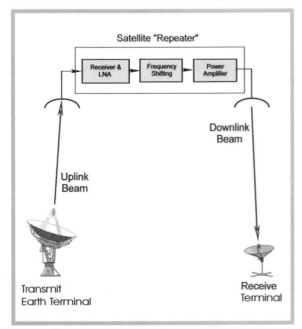

Figure 9-10: A satellite acting as a simple "repeater"

these still covered vast areas on the ground. More recently, technology developments have made possible satellites that can generate even hundreds of very narrow beams. This requires a very large antenna reflector on the satellite (somewhat counterintuitively perhaps, to produce a very narrow antenna beam, a reflector with a large cross sectional area is required), but this is now possible. The technology for manufacturing antennas that can be stowed inside a relatively small volume and then deployed once in orbit is now a reality.

This is a very major step in satellite communications, because it allows a satellite to reuse the same limited frequency spectrum many times over. If a frequency is used in one spot beam, it can be used in other such beams as well, provided that they are aimed sufficiently far away (typically, more than one beam width) from the first beam. This is illustrated in figure 9-11, which shows the spot beam pattern when the available frequency spectrum has been subdivided into four smaller bands (thus often referred to as a "four-color" pattern). Such a pattern can be replicated as many times as needed to cover the entire geographical area of interest, and the same frequency sub-band (e.g. f_1) then can be used many times over to send different information to different locations on the ground.

When it becomes possible to generate many spot beams, the question arises as to how to make best use of them. If a satellite can receive messages in many different spot beams, and can transmit in spot beams as well, what is the best way for the satellite to route signals coming up in one spot beam down again to a terminal located in another spot beam?

One way would be for the satellite to divide up its frequency spectrum between the various spot beams with each being assigned a discrete fraction of the total allocation. A user located in the coverage of one spot beam who wanted to send a message to a terminal in another spot beam could select just the right narrow frequency band in which to transmit the signal up to the satellite which would result in it being retransmitted down again in the proper spot beam.

Such a scheme would work, of course, but, particularly if the number of spot beams is large, it is a highly inefficient way to use the limited frequency

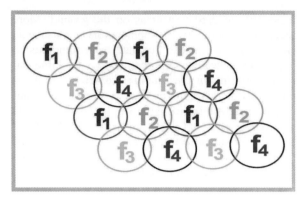

Figure 9-11: Frequency re-use in a four-color pattern of spot beams

resources of the satellite. The alternative would be to send the "address" of the intended receiving terminal along with the message, have this composite signal demodulated on board the satellite, and then have the payload on the satellite itself route the message to the proper downlink spot beam to reach the intended recipient. This process, termed "On Board Processing – OBP" requires a rather complicated payload which incorporates computer elements that can examine the address linked with each message and route it properly in real time. OBP processing lends itself well to reusing the same frequencies in (nonadjacent) spot beams, making it much more efficient than a method that relies on choosing specific uplink and downlink frequencies to route messages.

There is also another big advantage of using OBP technology. Both the uplink and downlink beams are susceptible to interference and corruption due to electrical noise and to the propagation effects mentioned earlier. When a satellite is used as a simple bent-pipe repeater, the interference on the uplink adds directly to that on the downlink, resulting in a received signal that might be relatively highly corrupted. Of course, in most cases, sophisticated FEC coding can be used to extract the uncorrupted signal, but this introduces an "overhead" that reduces the actual amount of information that can be sent over the satellite.

When onboard processing is used, the FEC imbedded in the modulated wave can be used on the satellite itself to correct for any errors encountered on the uplink. This process, known as "onboard regenera-

tion," means that what is then retransmitted down to earth will contain a signal without errors. The corrections that must be made at the receiving earth terminal are therefore only for any errors encountered on the downlink. By separating the errors on the uplink from those on the downlink in such a manner, the extent and complexity of the FEC that must be used to ensure error free end-to-end transmission can be reduced from that required in a bent-pipe repeater. The consequent reduction in overhead results in the link's being able to carry more information per unit time.

Finally, another advantage of OBP is that it can provide "bandwidth-on-demand," assigning capacity to individual spot beams (both on the uplink and downlink sides) as needed to accommodate the instantaneous traffic demand. This allows each spot beam to be used for different data rates, while still allowing full beam-to-beam interconnectivity. This is illustrated in figure 9-12.

9.4
The Present Satellite Services and Systems

Given all the communications techniques and technologies described above, what has been done with them? What has been done is to use them to develop and market various types of satellite communications *services*, as we will explain with examples below.

The discussion presented earlier concerning the wide bandwidth available with fiber-optic cables should not be construed to indicate that satellites have become, or are likely to become, superseded by such cable networks. Quite the contrary; when satellites began to be surpassed in capacity by cables, the roles in which they were dominant merely changed. To understand this, let us look at the various acronyms categorizing the various types of communications satellites. Communications satellites have historically (albeit with a rather short history) been classified as falling into one of several categories, depending upon the type of service they were developed to provide and for whom the service was intended. The traditional categories have acronyms depending upon the planned service, FSS (Fixed

Satellite Service), MSS (Mobile Satellite Service), BSS (Broadcast Satellite Service), and DRS (Data Relay Service). The FSS satellites were designed to interconnect, in a point-to-point type of connection, two (or sometime more) earth stations or earth terminals, both of whose positions were fixed. MSS satellites were intended to connect users moving around a region with fixed location communications "hubs," and vice versa. Later, this was extended to include direct mobile point-to-mobile point connections. BSS satellites were, as the name implies, intended for broadcasting radio or television to a large number of users in a region, and hence were designed for fixed point-to-multipoint connections. Finally, several governments have deployed or are considering GEO DRS satellite systems to relay data from low earth orbiting satellites to data reception and processing stations on earth.

Fixed Services

The early transoceanic satellites, of which those of Intelsat are perhaps the best example, were predominately FSS satellites. They largely acted as "trunking links" carrying large numbers of telephony channels from one large earth station on one side of an ocean to another large earth station on the other. When the transoceanic fiber cables began to provide more and more capacity between points which had prior to that been predominantly served by FSS satellites, the satellites did not become redundant. They were simply used more for point-to-point connections between points that did not (yet)

have such fiber cable connections, while still retaining capacity to allow them to be used as a means to provide at least a partial backup to fiber cables, should those cables suffer an outage. Nevertheless, the outgrowth of fiber cables has eroded a lot of what had previously been FSS satellite revenues, and in order to remain competitive, FSS satellites have had to evolve and reinvent themselves. Traditional FSS services are evolving into "multimedia services," a process that is still ongoing. FSS satellites are also widely used for such tasks as providing multi-establishment companies with private VSAT (Very Small Aperture Terminal) networks. These are two-way communications networks used by businesses, for example, to connect their remotely located establishments to their headquarters (which could be the site of the VSAT "hub" indicated in figure 9-13). Another common use is to connect remote point-of-sale locations, such as automotive filling stations or automated teller machines, to a central location to verify credit card sales or cash withdrawals.

FSS satellites are also used to distribute Internet data to a great number of Internet Service Providers (ISPs). The Internet relies to a great extent upon there being a high degree of redundancy of the information it contains, with many ISP sites having the same information stored on their servers. Such "mirror sites" often use satellites to synchronize and update the information which they contain, since the mirroring process is in fact just the type of data

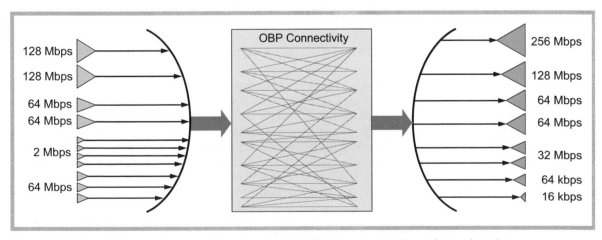

Figure 9-12: Beam-to beam interconnectivity via on board processing - regeneration and rate adaptation

broadcasting at which satellites excel. Also, since many ISPs are located in spots where there is no convenient fiber optic cable network access point, the use of satellites is often the only way they can obtain the broadband connectivity they require.

Mobile Services

As the name implies, mobile services concern communications with earth terminals that are mobile. That is, they may be terminals mounted on cars, trucks, boats, ships, or airplanes, or may be terminals that are not much bigger than conventional cell phones. Figure 9-14 illustrates such services. The frequency band traditionally allocated to satellites for the provision of mobile services has been in the L-band region of the spectrum, which is heavily crowded with other users, and was limited to a band some 35 MHz wide. Initially, the only mobile service provider was Inmarsat, which was granted a monopoly position and was expected, at its inception, to provide only telephony services to a relatively small number of ships at sea. Thus, the limited spectrum allocation was thought to be adequate. In the meantime, Inmarsat has expanded its service portfolio to include aircraft and terrestrial mobile terminals; its service to ships has expanded to include many thousands of even small sailing and power boats, and it has added data transfer and even limited video transmissions to the services it provides. In addition, other satellite mobile service providers have come into being, competing for the same limited spectrum. In order to cope with such competition in a narrow frequency spectrum, Inmarsat and others are moving to advanced technology using hundreds of narrow spot beams to allow their spectrum allocation to be reused many times.

Another effect of the competition is that, as has been the case for other types of satellite services, there has been a gradual move for MSS services into higher frequency bands. Non-GEO satellite systems such as Iridium and Globalstar, for example, in addition to L-band, also use a portion of the S-band spectrum for their services, and some operators, particularly for two-way communications with aircraft, are moving into the Ku-band and above. MSS satellites are able to provide a service that is not easily provided by fiber cables. Since the users of the terminals using MSS satellites are constantly on

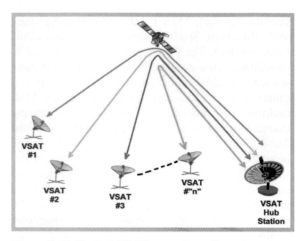

Figure 9-13: Typical VSAT network configuration

the move, only a competing means of wireless communications can erode the business of an MSS satellite. This has, however, happened in some areas and with some types of service. The economic failures of the Iridium and Globalstar endeavors were due in great part to the unexpectedly rapid outgrowth of cellular telephone services, and in particular to the near ubiquity of the GSM system, which took much of the global roaming business traffic which those two satellite systems had anticipated would be their main sources of revenue. Only one major global MSS system has been developed that continues to hold a significant subscriber base, and that is the aforementioned International Maritime Satellite Organization, Inmarsat.

Broadcast Services

Of all the uses to which satellites have been put, the most profitable has been their use for Direct to Home (DTH) broadcasting of television programs to small dish antenna located at people's homes or apartments. BSS satellites have, by far, produced the greatest revenues in the space field. The ability of a single GEO satellite to provide hundreds of television channels over a vast area has allowed some such satellites to generate annual revenues of hundreds of million of dollars. At that rate, a satellite operator can recover the initial outlay for the procurement and launching of a satellite in only a few years. Since these satellites are designed typically to operate for 15 years or more, the positive impact that

Figure 9-14: Mobile communications services - Connectivity for people on the move (web: *Mobile*)

the cash flow they can generate on an operator's balance sheet is considerable.

Satellite operators are able to earn such revenues because of their ability to provide services to a vast audience. In Europe, for example, SES-ASTRA claimed in 2003 some 36 million homes received DTH satellite and radio programming from its fleet of satellites (web: *SES-ASTRA*). Its major competitor in the region, Eutelsat, claimed a total of 110 million homes received its broadcast, either via DTH or cable (web: *Eutelsat*). In the USA, two companies, Hughes Network Systems with its DirecTV network and Echostar with its DISH network, dominate the Direct Broadcasting Satellite (DBS) market, together beaming their programs directly to more than 23 million homes (web: *SPCA*).

Not only do BSS satellites earn considerable sums of money, they have given FSS satellite operators a new source of income that has allowed them to continue to be profitable, despite the erosion of their previous telephony-centric market. FSS satellites are often used to bring television programs from several different content providers to a single location, from which they can then be multiplexed together and uplinked to a BSS satellite by its operator. FSS satellites are also widely used by the television cable industry to bring programming to its cable heads

from where it can be carried by cable to subscribers. Thus, a major market today for FSS satellites is for television distribution, as contrasted to the television broadcasting of BSS satellites.

Once a company has a fleet of satellites providing DTH services, however, the company can take advantage of its capacity in orbit to also provide other services. For this reason, BSS satellites often double as FSS satellites as well, being used to distribute television programming to, for example, TV cable head ends. BSS satellites also provide programming to SMATV (Satellite Master Antenna TV) antennas, which feed these signals into multiple family apartment buildings, hotels, and even office buildings (see figure 9-15).

There are signs that the DTH market is beginning to saturate in Western Europe, and while that point is still far away in the USA, it is bound to come eventually at the present growth rates for the service. For that reason, on both sides of the Atlantic, broadcast satellite companies are looking both for new services to market and for new geographical markets in which to provide services. Several big operators are dabbling (only dabbling so far, at least) with providing broadband Internet access and services to households. When this eventually comes to pass, they may find a considerable source of additional revenue.

245

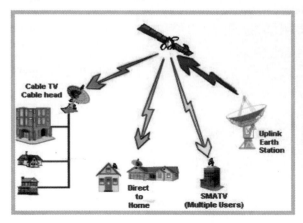

Figure 9-15: Broadcast satellite services - point to multi-point

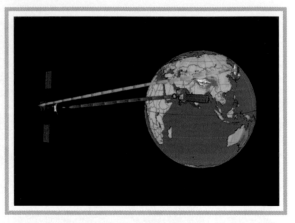

Figure 9-16: Japanese DRTS links to earth observation satellite, ADEOS (JAXA)

Data Relay Services

Perhaps the most notable of the DRS satellites is the Tracking and Data Relay Satellite Service (TDRSS) set up by the U.S. government. Its satellites can track and receive data from a number of LEO satellites simultaneously. In particular, the TDRS satellites are used to maintain contact between ground controllers and the astronauts/cosmonauts on board the International Space Station.

The TDRSS satellites put up by the USA have dual mechanically steerable antennas, each of which can exchange (both receive and transmit) data with satellites in lower orbits using either Ku-band or medium data rate S-band payloads. In addition, the TDRSS satellites each have a phased array S-band system that can dynamically form multiple beams and track and exchange lower bit rate data with several LEO satellites. A follow-on version, the Advanced TDRSS, adds Ka-band data relay services as well (ESA 2004).

In Europe, the European Space Agency had foreseen to set up its own autonomous data relay system, DRS, and as a predecessor, developed the Artemis satellite which has data relay payloads operating at S-band, at Ka-band, and at laser-optical bandwidths (web: *ARTEMIS*). After being put into the wrong and much lower orbit due to a launcher malfunction, Artemis seemed destined for a totally failed mission, but ESA managed to perform the "impossible": It used the onboard ion propulsion system designed to

provide North-South station keeping, to raise the orbit of Artemis up to geostationary altitude where it was then able to be put into service. Artemis is able to provide data relay services using both RF and optical communications links, and, indeed, is a pioneer in the civilian use of such optical communications technology.

Japan launched its Data Relay and Technology Satellite (DRTS – see figure 9-16) in September 2002 and will make good use of it to relay signals back to the Japanese mainland from their LEO earth observation satellites, as well as from the Japanese Experiment Module which forms part of the International Space Station.

9.5
What the Future Holds

So where do communication satellites go from here? Without a crystal ball (and perhaps even with one), it is difficult to predict very far into the future in what has already demonstrated itself to be a highly dynamic field. Nevertheless, there are sufficient trends and pointers to allow me to express my opinion on a number of things that seem highly likely to happen, at least in the reasonably near future, which include the following:

The worldwide demand for bandwidth, particularly for Internet bandwidth, is growing exponentially, and is expected to continue to do so for the foreseeable future. Figure 9-17 shows estimates made in

2001 for USA Internet traffic capacity needs alone, which predicted a Cumulative Annual Growth Rate (CAGR) near 100%. Actual growth in 2003 was closer to 75%, but predictions are that the CAGR will continue at around 60% through at least the year 2007, by which time US backbone traffic will exceed 1600 petabytes/month (web: *Techweb*). A relatively small, but still significant portion of this growth, as well as that in the rest of the world has been and will be captured by satellites, which will provide an expanding demand for satellite capacity for many years to come. This will be especially true in developing countries, where the rate of growth in demand will exceed the rate at which the terrestrial infrastructure can be put in place in these countries to service their capacity demand.

Another field in which the demand for satellite bandwidth is growing rapidly is that of military communications. The U.S. military was an early user of communications satellites, and has procured several generations of different satellite systems since the launch of its previously mentioned SCORE satellite. In Europe, Germany, France, the United Kingdom, Italy, and Spain have all developed and operated either their own military communication satellite systems, or have embarked dedicated military payloads on commercial systems.

The demand for capacity by the military is climbing exponentially. The U.S. satellite capacity demand had gone up by more than 650% since the Desert Storm Gulf war, and the increase is expected to continue as the U.S. military implements what it terms its Transformational Communications plans.

The projected military capacity demands of the NATO countries and particularly those of the U.S. military, far exceed what will be made available by the dedicated military satellite procurements understood to be underway or in a planning stage. The shortfall will have to be provided through military leases of commercial civilian satellite capacity, which should form an increasingly significant part of the market for commercial satellites in the future.

The attenuation of RF signals by the atmosphere and by weather conditions increases greatly as one goes up in frequency. This, at first sight, makes it difficult to see how frequency bands much above Ka-band can be usefully employed. Since increased demands

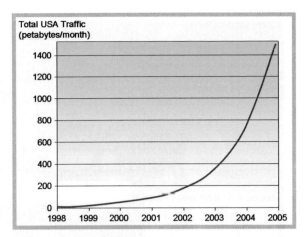

Figure 9-17: Growth of the internet communications traffic in the US (Thompson 2001)

for capacity can be expected, which will progressively "fill up" the frequency bands now in use, one might conclude that the use of satellites for communications will eventually reach some saturation limit. This is undeniably true, but a number of factors will make this occur much further in the future than one might expect.

The first factor is compression. The state of the art in compression technology is still being advanced, and there will therefore be a continuing increase in the bandwidth efficiency with which future services will come into being in the frequency bands now commonly used. Second, there are continuing improvements in the technologies involved in frequency reuse, which will enable more efficient use of the available spectrum.

Yet a third way that higher frequency bands may be exploited, and this includes both higher (above 70 GHz) RF bands and optical communications, is to use these to interconnect satellites in orbit. Since such interconnection does not involve transiting through the atmosphere, the problems of rain and atmospheric attenuation do not occur. Furthermore, by doing this, one can avoid having to use frequency bands for the feeder link connections needed for satellite double hops. These Inter-Satellite Links (ISL) are already used commercially in, for example, the Iridium mobile communications systems, and are also used in existing military communications satellites.

Figure 9-18: SAT-MODE – a low cost two-way interactive service via satellite (SES)

Laser optical communications technology, in particular, could come into extensive use if any of the several HAP (High Altitude Platforms) concepts now being studied ever come into commercial operation. These platforms are large unmanned aircraft (probably solar-electric powered) or lighter than air helium filled dirigibles or "blimps" that remain at a semi fixed location "hovering" some 20 km above the earth's surface. From such a location they can act as relays to and from the ground for terminals located within a radius of some 400 kilometers. Because a HAP would be much closer to the ground than a geostationary satellite, the propagation signal losses to and from it would be far less (by a factor of about 4,000,000, or 66 dBW!) than those from a GEO satellite, and it could then afford to carry the excess power needed to compensate for and "punch through" even the large values of atmospheric and rain attenuation that can be encountered in bands above Ka-band. Furthermore, since a HAP will be flying above the greatest portion of the atmosphere and weather, it could even be used as a relay to and from satellites, using optical communications.

Terrestrial fiber networks will continue to grow and expand, particularly in the developed world, pushing satellites out of some markets there. This will affect the point-to-point FSS markets most heavily, but

BSS markets will also feel the effect. The competition will force some operators to find new markets for satellite capacity already in orbit. These new markets will include in particular, as indicated in the previous point, the growing demand for Internet capacity, particularly in developing countries.

BSS satellite will see increasing competition, not only from the terrestrial cable industry, but from conventional telecommunications companies as well, who may introduce competitive "video by ADSL" services. The growth in High Definition television (HDTV) will tend to demand more satellite capacity, but this will at least partially be counterbalanced by improvements in compression technology that will tend to reduce the demands for bandwidth. The traditional distinction between the three commercial services, FSS, BSS, and MSS will continue to blur even more than it has to date. The operators of each type of service will begin to make inroads into the markets of the others.

Two-way interactive services using a narrow band return channel via satellite are now under development. Figure 9-18 depicts one such system, the SATMODE system, now under development by SES-ASTRA in Europe, which will provide real-time interactivity via satellite to customers receiving the ASTRA broadcast services. Such systems, which promise interactivity at a very minimal additional cost, will experience significant growth. This will gradually evolve, as more performing terminal prices drop in price to affordable levels, into asymmetric two-way broadband services for Internet access that will increasingly be offered by both FSS and BSS service providers.

A significant new growth industry will develop to take advantage of the inherent synergy that exists between communications and navigations by offering hybrid navigation and communications services to vehicles worldwide. Two-way satellite communications links combined with the position determination capabilities provided by GPS and Galileo navigation will allow a wide range of new services to be developed aiding travelers wherever they may be.

The cost per gigabyte of nonvolatile data storage will continue to drop, and the amount that can be stored with a given storage media will continue to rise. This has already made it possible for cable and satellite TV providers to provide "Personal Video Recorders" to subscribers at affordable rates, making it possible for them to record upward of 40 or more hours of programs on the internal hard disk of the PVR. This will lead ultimately to residential users themselves becoming "mirror sites" for many types of Internet data, and satellites will play a large role in feeding data to such residential "servers."

The synergy between communications and earth observation satellites will result in closer coordination among those offering each of these types of service, resulting in more efficient disaster monitoring, mitigation and relief. A global disaster monitoring, management, and mitigation organization should eventually come into being under the auspices of the United Nations, combining the services of earth observation, meteorological, search and rescue, navigation, and communications satellite systems into a composite capable of providing rapid detection and response to both natural and man-made disasters.

The idea of using GEO satellites in a DRS role in earth orbit will be extended to employing DRS satellites in orbit around a number of planets and/or their moons to monitor and at least semi-continuously relay information back to earth from scientific sensors placed on their surfaces. The first such interplanetary data relay satellite will be the planned Mars Telecom Orbiter (see figure 9-19), scheduled for launch in 2009, to relay data from future robotic landers, and eventually even from manned Mars landings, back to earth.

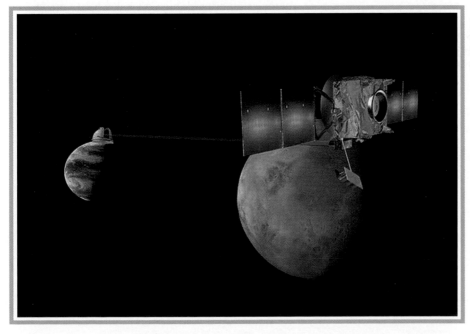

Figure 9-19: Mars Telecom Orbiter planned for 2009 (NASA/JPL, web: *fourth millennium*)

Very large "antenna farms" providing hundreds and perhaps even thousands of spot beams will be deployed in GEO, making it possible to reuse frequencies hundreds or thousands of times. This will allow satellite system data throughputs to rise from the present low-gigabit/second range up to many terabits/second. When this occurs, satellites may again become the medium of choice for provision of broadband services in many parts of the world, as they will once again become cost and performance competitive with terrestrial fiber networks.

The above are relatively short range predictions which depict what the future could hold in perhaps the next decade. If one wants to predict still further in the future, there is the risk of venturing out of the sphere of science and into that of science fiction. Nevertheless, many of the science fiction themes and gadgets of fifty years ago have now become reality, so perhaps at least some of science fiction's longer range predictions may do so as well. What is certain is that communications via satellite will continue to play a large role in disseminating information globally, eventually perhaps leading to truly total global interconnectivity, where everyone will be able to be in contact with anyone else. Only time will tell!

References

Ashford EW (1986) "Satellite Parameters Dossier," Space Communications and Broadcasting 4(4) Elsevier Science Publishers B.V.

ESA Publications Division (Feb 2004), "Artemis – Paving the Way for Europe's Future Data-Relay, Land Mobile and Navigation Services"

Fortescue P, and Stark J (1995) "Spacecraft Systems Engineering," John Wiley & Sons

Iida I, Pelton J, Ashford E (2003) "Satellite Communications in the 21st Century – Trends and Technologies," the American Institute of Aeronautics and Astronautics

Martin D (2000), "Communication Satellites" The Aerospace Press and The American Institute of Aeronautics and Astronautics, Inc.

Richharia M (1999) "Satellite Communications Systems" McGraw Hill Telecommunications series

Thompson C (2001) "Supply and Demand Analysis in Convergent Networks," MBA Thesis to Sloan School of Management, MIT

U.S. Patent 2292387 (August 11, 1942), entitled "Secret Communication System"

World Technology (WTEC) Division of Loyola University (1993), "Satellite Communications Systems and Technology," study report

Web references

ARTEMIS: http://www.space-technology.com

Digital cameras: http://www.photo.net/ equiment/digital/cameras/basics/ "Digital Cameras - A beginner's guide" by Atkins, R.

Eutelsat: http://www.ses-astra.com/

Capacity: http://www.friends-partners. org

ITU: http://www.itu.int

Jpeg: http://www.faqs.org/faqs/jpeg-faq/

Fourth millennium: http://www.fourth-millennium. net/mission-artwork/mission-index.html drawn by Coby Waste

Mobile: http://www2.crl.go.jp/ka/sat-com/images/ concept.gif, by permission of Mr. Noriaki Obara, CRL

SBCA: http://www.sbca.com

SES-ASTRA: http://www.ses-astra.com

TDRSS: http://leonardo.jpl.nasa.gov/msl/ programs/ tdrss.html

Techweb: http://www.techweb.com/wire/26803952, see also Techweb News, Feb 10, 2004

10 Satellite Navigation

by Günter W. Hein

The 21st century is often also considered the *"century of information."* However, what would information be without a date, time and location? It needs a *"position and time stamp"* in order to create value-added services and is now becoming part of a more general concept, that of "informability" where users receive information tailored to their needs and referenced to location and time.

Satellite navigation delivers positioning and time, independent of weather, around the globe and in space near the earth, 24 hours a day. It is thereby more efficient in terms of accuracy, availability[1],

---

[1] *Service Availability* represents the percentage of time averaged over the design lifetime (20 years) when the service is within the specified performance (accuracy, integrity and continuity) for any point within the service volume. It is derived from the availability of each operational configuration (nominal, without failures, or non-nominal, with one or more failures), weighted by its probability of occurrence, averaged over the design lifetime.

[2] *Integrity* is defined by the following parameters:

- Alert Limit: the maximum allowable error in the user position solution before an alarm is to be raised within the specific Time-to-Alert.

- Time-to-Alert: the time from when an alarm condition occurs until the alarm is received at the user level (including the time to detect the alarm condition).

- Integrity Risk: the probability, during any continuous period of operation, that the computed vertical or horizontal positioning error exceeds the corresponding Alert Limit, and the user is not informed within the specified Time-To-Alert.

[3] *Continuity Risk:* the probability that the specified performance (accuracy and integrity) is supported by the system over the time interval applicable and within the coverage area, given that it is supported at the beginning of the operation and predicted to be supported throughout the operation duration.

integrity[2] and continuity[3] than any other terrestrial sensor or method.

10.1
Satellite Navigation in the Context of Space Sciences and Applications

Satellite systems presently available and used for positioning, navigation and timing are: the *U.S. Global Positioning System (GPS)* and to a lesser extent the incomplete *Russian Global Navigation Satellite System (GLONASS)*. The European satellite navigation system *Galileo* is under development and might be completed around 2010 (see boxes later in this chapter).

Satellite navigation technology is increasingly used in almost all sectors of activity. Its high performance standards already make it an essential tool for very demanding professional, commercial and scientific applications. Converging factors have favored this remarkable expansion. The availability of communication networks and Geographical Information Systems (GIS), together with the overall decrease in the cost, size and power consumption of satellite navigation receivers have driven the market towards high-volume consumer applications. The public sector, for example the European Commission, but also worldwide bodies in different fields of application are setting up more and more regulatory frameworks which promote the use of satellite navigation services to improve the safety and efficiency of all types of transport modes (rail and road, maritime, aviation).

Moreover, nowadays applications of very precise satellite positioning can be found in surveying, geodesy and other fields of earth science (geodynamics, weather and climate), whereas very accurate timing is needed for telecommunications, energy, finance, banking and insurance.

No doubt that a significant military use of GPS also takes place for the guidance of modern weapons. This so-called NAVWAR (Navigational Warfare) application is out of the scope of this chapter on civil applications.

Most people have not yet realized that for the so-called critical infrastructure of our countries we rely on a continuous availability of high-precision time provided by a satellite navigation system at present only GPS. Examples are synchronization in telecommunication base stations, and the supply of energy for networks, banking and financial transactions. For these aspects of modern society, an uninterrupted service is vital for government and economic life. A disruption of satellite navigation services would be a threat to economic, safety and security-related applications (Volpe 2001). A misuse is a threat to national security. Therefore, precautions are taken in each global satellite navigation system to protect services that must not be interrupted; they require spectral frequency separation and secured controlled access. In other words, civil services must be "jammable" without affecting the military and/or security signals of a global satellite system. This terrestrial or airborne jamming is always done locally. Because of the high importance of satellite-based applications mentioned before, any manipulation in the satellites themselves would affect and penalize services and economy outside the

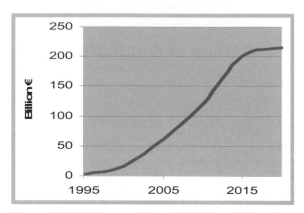

Figure 10-1: Market evolution of global satellite navigation: total annual global turnover (Source: Business in Satellite Navigation, Booklet of the European Commission, European Space Agency and the Galileo Joint Undertaking, March 5, 2003)

theater of operations, and is and will be therefore never done.

Coming back to the growing market in satellite navigation, a few graphs should make this fact more transparent. Figure 10-1 shows the total estimated annual global turnover for satellite navigation over the next decade.

Figure 10-2 illustrates the breakdown in satellite navigation applications. The automotive sector (car navigation) as well as integrated personal communications (satellite navigation chip in a mobile phone) are predicted to be the most dominant application areas.

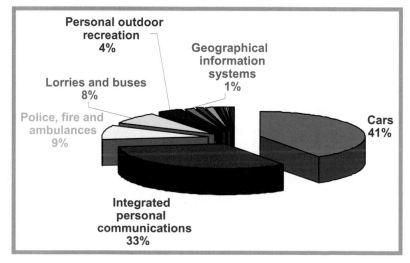

Figure 10-2: Application breakdown of satellite navigation (Source: CBA, European Commission, Nov. 2000)

Considering the information in the figures 10-1 and 10-2, we can conclude that satellite navigation really marks a general breakthrough in the utilization of space: Never before has a space project had so much impact on the daily life of each citizen on earth!

However, when trying to establish a business plan for satellite navigation, one is faced with the problem that the immediate products of satellite navigation like position coordinates and velocities cannot themselves generate considerable revenue. Only when building up so-called "value-added services" by combining position and time with communication and geographical information, a successful outcome is guaranteed and any applications can be found.

10.2
Principles of Operations – a Brief Outline

The heart of a satellite navigation payload is a very accurate atomic clock (rubidium, cesium, H-maser) in redundancy generating time signals, or more precisely, *carrier phases* (sinusoidal radio frequency signal) of a certain fundamental frequency (present satellite navigation systems are working in the L-band). On the carrier there is a binary random sequence, a so-called *ranging code* (pseudo-random noise -PRN- codes), modulated with properties carefully selected in order to mainly minimize inter- and intra-interferences. A binary-coded *broadcast message* consisting of data on the satellite orbit, the satellite health status, satellite clock bias parameters and an almanac (approximate positions of the other satellites in that constellation) modulated also on the carrier is transmitted with a low number of bits per second (e.g., 50 bps for GPS) to the user. Satellite clock time is usually aligned to an international time standard, like a real-time prediction of the *Universal Time Coordinated* (UTC) modulus, an integer number of seconds determined by one or several time laboratories of the satellite ground segment.

In order to differentiate between signals from different satellites, access could be arranged in one of the following ways. Code Division Multiple Access (CDMA) assumes that all signals are transmitted on the same frequency, but have different codes (example: GPS, Galileo). Frequency Division Multiple Access (FDMA) uses a separate frequency for each satellite (or at least for those satellites whose signals cannot be received at the same time), but the same code for all satellites (example: GLONASS). Time-multiplexed signals widely used in communication are not of interest in satellite navigation.

When these signals are received by the antenna of a satellite receiver, converted from analog to digital, and demodulated, two types of measurements are possible.

Pseudorange or code phase measurements (figure 10-3). Since the codes are known, replica are generated in the receiver, however, referring to the receiver time scale which is generated by a less accurate and, therefore, inexpensive clock (quartz).

By correlation the code phase shift is measured, which relates to the signal travel time from the satellite to the receiver antenna. Multiplying it with the velocity of light results in the pseudoranges, which still have to be corrected for various error sources like atmospheric refraction (ionosphere and troposphere), orbit errors, etc.

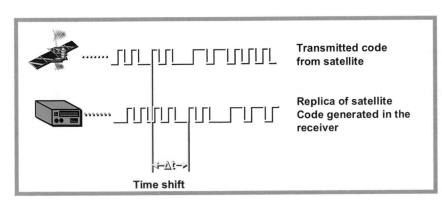

Transmitted code from satellite

Replica of satellite Code generated in the receiver

Time shift

Figure 10-3: Principle of pseudorange measurements: the time Δt between transmission and reception of the signal is measured, using the knowledge of the code structure. The pseudorange value derived in this way equals $\Delta t \cdot c$, referring to the less accurate receiver time scale

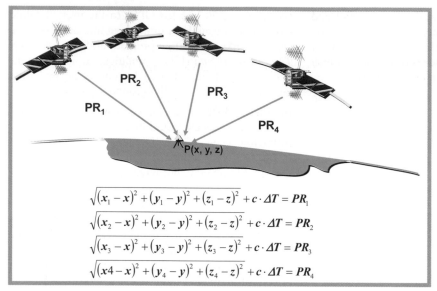

Figure 10-4: Principle of absolute positioning using measurements PR_i (pseudo- or code ranges). Using the orbital positions (x_i, y_i, z_i) of the satellites i (i=1,...4) the three-dimensional position (x,y,z) and the clock offset ΔT can be estimated

$$\sqrt{(x_1 - x)^2 + (y_1 - y)^2 + (z_1 - z)^2} + c \cdot \Delta T = PR_1$$

$$\sqrt{(x_2 - x)^2 + (y_2 - y)^2 + (z_2 - z)^2} + c \cdot \Delta T = PR_2$$

$$\sqrt{(x_3 - x)^2 + (y_3 - y)^2 + (z_3 - z)^2} + c \cdot \Delta T = PR_3$$

$$\sqrt{(x4 - x)^2 + (y_4 - y)^2 + (z_4 - z)^2} + c \cdot \Delta T = PR_4$$

Note that the pseudoranges are measured in two different time scales. In particular, the low accuracy of the receiver time scale has a certain bias to the satellite clock scale. In principle, for a three-dimensional position on earth, three ranges to three satellites are sufficient (three observations and three unknowns). However, because of the unknown time bias a fourth unknown has to be solved and therefore one more observation has to be carried out. This is also the reason that these measurements do not only give a position (and its derivative, the velocity, in a processing filter) - see figure 10-4 - but can also determine time very precise (time distribution, synchronization and timekeeping). Insofar, a satellite navigation system is also a very precise global, all-weather timing system.

Carrier phase measurements (figure 10-5). Analogously to pseudorange measurements, a carrier phase measurement is the difference between the Doppler-shifted transmitted carrier and the carrier generated in the receiver. Carrier phase measurements are ambiguous, because the receiver can

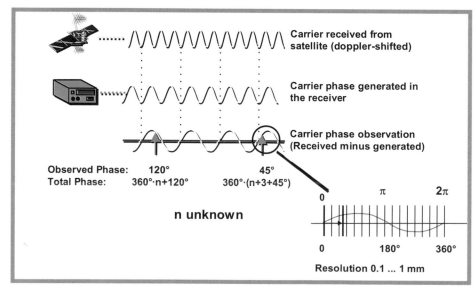

Figure 10-5: Principle of carrier phase measurements: the phase difference between the Doppler-shifted carrier from the spacecraft and the carrier generated in the receiver is measured. Ambiguities arise from integer multiples of the full wavelength, which cannot be determined

measure only fractions of the received wavelength (cycle). This, however, can be very accurately done in millimeters or even sub-millimeters (wavelengths are of the order of two decimeters in the L-band). The unknown cycles, the so-called carrier phase ambiguity, an integer number, has to be determined using least-squares techniques and redundant satellite observations and/or additional information and/or other sensor data. When adding the integer number of cycles, a phase range is derived which, again, has to be corrected for the same error sources as for the code ranges. The determination of the position (and the velocity) of the user is equivalent to that using pseudoranges.

The precise carrier phase data can also be used to smooth the less accurate pseudorange observations in a filter.

Instead of determining the carrier phase *integer* ambiguity separately, pseudorange and carrier phase data can be processed in a combined (Kalman) filter considering the integer ambiguity as an unknown of *real variable* type and solving for it in each observation epoch together with all other unknowns (coordinates, parameter of error sources, etc.). This approach is called "*float solution,*" which leads to an ambiguity which is numerically different and "floats" in each observation epoch (which is physically wrong). Nevertheless, this numerical trick

leads to two-decimeter-accurate solutions, which avoids the troublesome determination of the integer carrier phase ambiguity.

In order to eliminate further error sources, a differential measurement approach is used; this is a common practice in engineering sciences in order to cancel out (constant) errors by building up differences.

In satellite navigation a receiver is placed at a station with known coordinates. Comparing the observations with the real values derived from that position and the known orbits of the satellites leads to corrections (pseudorange or phase corrections) which the user applies to his observations. This is the key to millimeter accuracies in satellite navigation. Figure 10-6 illustrates the principle of differential GNSS.

Figure 10-7 informs about the present accuracy capabilities of the GPS system down to the millimeter level under ideal static conditions. GLONASS accuracies are similar; Galileo might show even a slight improvement. With respect to timekeeping, time transfer and synchronization with other sensors, accuracies in the nanoseconds level are possible.

For more details on observations and processing strategies the reader is referred to textbooks of satellite navigation (see the references to this chapter).

Standard products of satellite navigation are so-

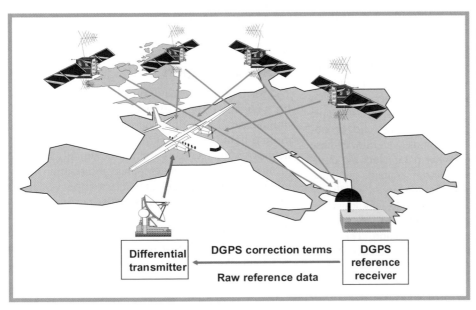

Figure 10-6: Principle of differential mode of GNSS. Here errors of the pseudorange or carrier phase measurements are eliminated by comparing the observed values with those derived from a station with accurately known coordinates and the broadcasted satellite or bits. This method allows for millimeter accuracies of the relative position

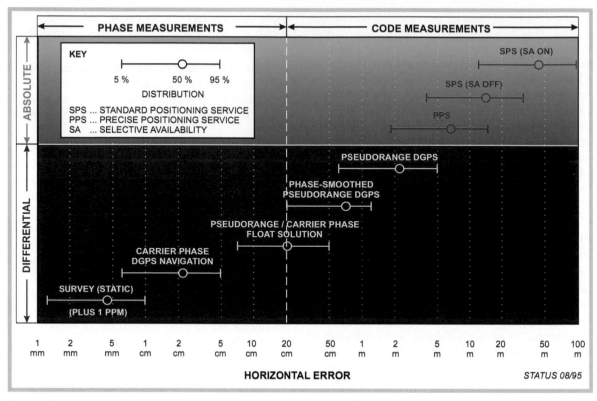

Figure 10-7: Horizontal errors of the Global Positioning System (GPS) in absolute and differential mode and type of processing (DGPS Differential GPS, PPM part per million of baseline length); vertical errors are approximately twice the horizontal errors

called PVT (position-velocity-time) results. The use of more than one antenna enables the user to make attitude determination (in general four antennas for three-dimensional attitude determination).

In recent years, however, many more nonstandard quantities have been derived. When deriving the pseudorange and carrier phase observations it is necessary to apply many corrections, for example atmospheric corrections (ionosphere and troposphere), precise orbit corrections, etc. Since these corrections rely again on other parameters like temperature, pressure, total electron content, etc., one could also consider those parameters - which are sometimes based on hypotheses.- as unknowns in the processing scheme. By doing so, many new products result from satellite navigation. These are described in chapter 10.3.11 under "Science Applications" of satellite navigation.

10.3
Applications

10.3.1
Road Transport

The road sector is a major potential market for satellite navigation. By 2010 there will be more than 670 million cars, 33 million buses and trucks and 200 million light commercial vehicles worldwide. Satellite navigation receivers are now commonly installed in new cars. Using geographical information, positioning, navigation and guidance of the vehicle is carried out with the *aid of satellite positioning and map matching*. Real-time traffic and travel information is provided via radio transmission in order to inform the driver about possible new routes due to incidents like congestion or accidents, for example (figure 10-8).

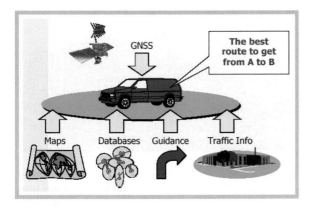

Figure 10-8: Positioning and guidance

Whereas in this first stage satellite navigation in conjunction with a geographical data base asks for no accuracy better than, approximately 30 m (positioning and guidance only), the next step in application aims to locate the vehicle in the right lane of the street (1 to 2 m accuracy through differential GPS), thereby enhancing the safety of the driver in traffic, possibly also by interacting with the car's advanced driving assistance system. In the near future the real-time availability of decimeter accuracies will ultimately lead to collision avoidance (location within the lane).

A very important application is managing emergency and rescue vehicles (figure 10-9). Combined with dynamic traffic information the vehicles can reach the required area much faster. Also, road tolling can use satellite navigation in order to determine distance-dependent charging (under an initiative of the German government a consortium is installing a

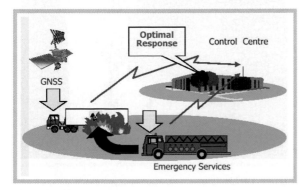

Figure 10-9: Satellite navigation and telecommunication functions for emergency services

highway toll system for trucks in Germany in 2004/5).

Taxi, bus and truck fleet management are crucial and complex tasks for operators. Companies have already equipped more than 500,000 vehicles in Europe with satellite navigation receivers to identify the location of each to a control center. Telemetry and *communication functions are added* for that purpose. Knowing the exact locations of the vehicles, operators can more effectively manage their maximum use and loading (figure 10-10).

10.3.2
Rail

Rail applications of satellite navigation can be grouped into the following areas: (i) train control, (ii) fleet management and goods tracking, (iii) con-

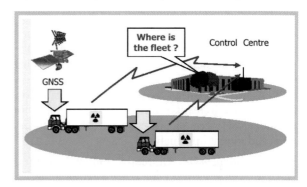

Figure 10-10: Fleet management

tribution to passenger information, (iv) energy optimization, and (v) track survey.

Train control. Very high integrity requirements are necessary for this safety-related application. The remaining integrity risk is in the same range (10^{-9}) as in aviation for a precision landing with almost no visibility (CAT III). A satellite navigation system alone cannot fulfill these high requirements; its integrity might be only of the order of 10^{-4} to 10^{-6}. The accuracy requirements are not so stringent. Differential satellite navigation (DGPS nowadays) can deliver the required positions in the 1 to 2 m range. Due to the integrity problem a hybridization of a satellite navigation receiver with other sensors, like odometers, balises and gyroscopes has to be considered. In addition, communication functions have to be added since a satellite navigation system is only a one-way system transmitting signals from the satellite to the user's antenna. All safety-related train control functions are standardized in the European Rail Traffic Management System (ERTMS).

Fleet management and goods tracking. As it is the case also for all other transport modes, fleet management is an important tool for improving the logistics and performance for both passengers and goods transportation. For freight traffic, knowing the location of goods in transit is important for customer confidence in timely delivery. The goods can be tracked if they are connected with the carriers (e.g. traction unit identification). Additionally, satellite navigation may help to organize rolling stock, improve rolling stock maintenance, enable effective goods tracking, simplify route pricing and supervise track usage. Accuracy requirements can be easily fulfilled with a single frequency receiver (5...10 m).

Contribution to passenger information. Information about train arrival and departure times, especially when there are delays, is important for maintaining good service. On-board passenger information is also essential. Knowing the position of the train (low accuracy requirements) can also be the basis for additional services to passengers, such as connections and tourist information.

Energy optimization. Currently, rail movement is generally not optimized for energy consumption. A driver normally controls the train according to a speed-profile table, which generally defines the allowed speed depending on track distance traveled. However, drivers often change speed without concern for saving energy. For example, they brake sharply before a tunnel instead of using regenerative braking at the appropriate distance before the tunnel. In order to save energy, the first question is to know the train's position with respect to its environment. Satellite navigation provides a cost-effective means of offering that information.

Track survey. Surveying and monitoring of tracks is an important task for ensuring safe passage for trains. A good survey needs accurate position determination in the millimeter level and synchronization between the positioning system and other testing/inspection systems. Traditional surveying techniques are replaced by an integration of satellite

Figure 10-11: Track irregularity measurement system investigating the derailment risk of high-speed trains using GPS, INS and Ultra-sonic Sensors (Lück et al. 2001); here: train engine with GPS satellite antenna

Figure 10-12: Inertial navigation system as part of a track irregularity measurement system investigating the derailment risk of high-speed trains (Lück et al. 2001)

Figure 10-13: Ultra-sonic measuring device as part of a track irregularity measurement system investigating the derailment risk of high-speed trains (Lück et al. 2001)

navigation receivers with inertial navigation systems of very high quality (see figures 10-11, 10-12 and 10-13).

10.3.3
Maritime Navigation

Satellite navigation is used in every phase of marine navigation: ocean, coastal, port approach and port maneuvers, under all weather conditions.

Offshore navigation. The International Maritime Organization (IMO) is implementing regulations on an Automatic Identification System (AIS) and vessel traffic management system in order to increase navigation safety and collision prevention. AIS is using a satellite navigation receiver (among other sensors).

Harbor operations. Approach and maneuvers in ports are critical operations, particularly under poor weather conditions. Locally-assisted satellite navigation is a fundamental tool for all kinds of harbor operations and precision docking. The increased availability of satellites will improve the economic sustainability of operations with satellite navigation in an environment where limited sky visibility might be an issue. Local components in the harbor and

communication links transmitting the exact position of vessels will encourage innovative and safe automated operations.

Inland waterways navigation. Satellites provide precise navigation along inland waterways, especially in critical geographical environments or meteorological conditions. This includes navigation on rivers and canals, where the accuracy and integrity of navigation data are essential to automate accurate maneuvers in narrow waterways.

Survey and marine engineering. Satellite navigation has revolutionized hydrographic surveying for dredging and maintenance of harbors and waterways, drilling operations, mapping the sea bottom and underwater obstacles (in combination with an echo-sounder type of sensor), pipe and cable laying, and oil and mineral exploration. Differential GPS in buoys is able to deliver instantaneous sea surface and wave heights in the millimeter level with a frequency of 10 Hz and higher (figure 10-14).

Science. The observation of tides and currents is possible using GPS in buoys or carrying out static surveys at tide gauge stations worldwide, thereby establishing a worldwide unified datum.

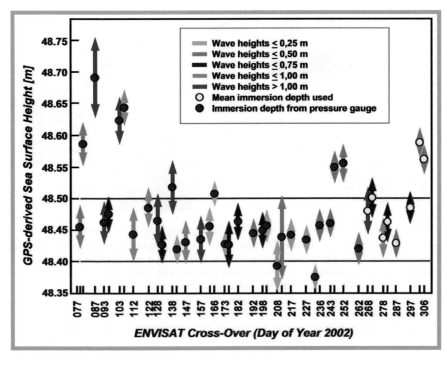

Figure 10-14: Results from the ENVISAT radar altimeter calibration campaign using moored high-sea GPS buoys (heights reduced by tidal and atmospheric pressure impact) (Schueler and Hein 2004)

Search and Rescue. The Galileo satellite navigation system (to be completed around 2010) will contribute to the international search and rescue service, enhancing the worldwide performance of the current COSPAS-SARSAT system (figure 10-15). The positioning accuracy of today's system is very poor (typically a few kilometers) and the alert is not always issued in real-time. The Galileo SAR service will drastically reduce the time to alert, and the position of the distress beacon will be determined to within a few meters. The Galileo SAR service foresees the rescue center's acknowledging the distress message. This will increase victim survivability and reduce the number of false alarms that dog the current system.

Commercial maritime operations. Commercial maritime activities will be assisted by satellite navigation. In fishing, it will help to locate traps and nets, manage fleets, and monitor cargo. Delivery and loading schedules can be optimized via GPS. Even the location of shipping containers can be facilitated, and satellite navigation could be used for automatic piloting of barges.

10.3.4
Aviation

Due to safety requirements commercial air transport has the most stringent requirements on satellite navigation. Table 10-1 informs about the detailed numbers for all phases of flight. It is integrity which

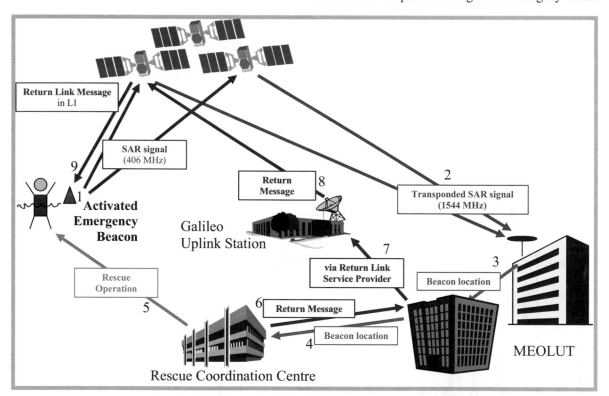

Figure 10-15: Planned search and rescue in the Galileo satellite navigation system. The approach is as follows: (1) The receiver of someone in emergency activates an emergency beacon which sends a search and rescue (SAR) signal in 406 MHz to the Galileo satellite. (2) The SAR signal is transponded by the Galileo satellite and sends the emergency position on 1544 MHz via the so-called MEOLUT center (3) to the SAR Mission control center (MCC). (4) SAR MCC reports beacon location to the Rescue Coordination Center. (5) The Rescue Coordination Center initiates the rescue operation. (6) The Rescue Coordination Center sends a return message via the SAR MCC, the Return Link Provider (7) to the Galileo Satellite (8). (9) The Galileo satellite sends a message to the receiver of the person in emergency in the standard broadcast message

Operation	Accuracy (horizontal 95%)	Accuracy (vertical 95%)	Integrity (probability of HMI)	Time to alert	Continuity Risk (1 - Continuity)	Availability
En Route	3.7 km	NA	10^{-7} / hr	1 min	10^{-4}/hr to 10^{-8} / hr	0.99 to 0.99999
Terminal	0.74 km	NA	10^{-7} / hr	15 sec	10^{-4}/hr to 10^{-8} / hr	0.999 to 0.99999
LNAV (NPA)	220 m	NA	10^{-7} / hr	10 sec	10^{-4}/hr to 10^{-8} / hr	0.999 to 0.99999
LNAV/VNAV	220 m	20 m	$2 \cdot 10^{-7}$/approach	10 sec	$8 \cdot 10^{-6}$ / 15 sec	0.99 to 0.999
LPV	16 m	20 m	$2 \cdot 10^{-7}$/approach	10 sec	$8 \cdot 10^{-6}$ / 15 sec	0.99 to 0.999
APV-II	16 m	8 m	$2 \cdot 10^{-7}$/approach	6 sec	$8 \cdot 10^{-6}$ / 15 sec	0.99 to 0.999
GLS/CAT I	16 m	6 m to 4 m	$2 \cdot 10^{-7}$/approach (150 sec)	6 sec	$8 \cdot 10^{-6}$ / 15 sec	0.99 to 0.99999
CAT II / IIIa	6.9 m	2 m	10^{-9}/15 sec	1 sec	$4 \cdot 10^{-6}$ / 15 sec	0.99 to 0.99999
CAT IIIb	6.2 m	2 m	10^{-9}/30 sec (lateral) 10^{-9}/30 sec (vertical)	1 sec	$2 \cdot 10^{-6}$/15 sec (lateral) $2 \cdot 10^{-6}$/15 sec (vertical)	0.99 to 0.99999

Table 10-1: Navigation requirements of the different phases of flight of commercial aviation (Source: Office of Architecture and Investment Analysis (ASD-1), Federal Aviation Administration, Washington, DC)

causes the biggest problem, the required accuracy is no problem. The integrity risk for precision landing with almost no visibility - so-called CAT III landing - has to be $< 10^{-9}$ with an alert time of 1 sec. A satellite navigation system itself without any augmentations can only deliver integrity of the order 10^{-4} to 10^{-6}. This gap can be overcome only by integration with other sensors, mainly using inertial navigation systems.

The high capabilities of satellite navigation over conventional techniques was expressed at the Tenth Air Navigation Conference in 1991 when the International Civil Aviation Organization (ICAO) endorsed recommendations of its Future Air Navigation Systems (FANS) committee and recognized that the primary stand-alone navigation in the 21st century will be provided by the Global Navigation Satellite System (GNSS). GNSS as a generic term for all kinds of satellite navigation systems was born. During the last decade it became clear that the use of satellite navigation would require more time due to regulatory and certification issues. Nevertheless, the idea is still to have a satellite-based gate-to-gate navigation concept.

Meanwhile GPS is certified for en-route flight and non-precision approach enabling aviation now to have more flexible routes, savings in fuel and time, and more aircraft airborne than ever before when using only terrestrial navaids. In the future, higher accuracy and service integrity will allow aircraft separation to be reduced in congested airspace, to cope with traffic growth (free flight concept).

For airports not equipped with instrument landing systems, so-called *wide-area differential* satellite navigation systems or *augmentation systems* were and are being built (2004) in order to allow a precision landing for CAT I (see table 10-1). Monitor stations at known locations are tracking the GPS satellites and send their information to control centers where (i) wide area differential corrections are determined, (ii) errors in the satellite broadcast data are detected and as a consequence so-called "unhealthy" satellites are flagged and integrity messages per satellite are formed (use/do not use/not monitored) and (iii) GPS-like navigation signals are generated. Geostationary satellites (INMARSAT) transmit these integrity messages and the GPS-like signal via the onboard transponder to the user where increased integrity (of the order of 10^{-6} to 10^{-7}) and availability (more GPS signals due to the geostationary satellites) can be obtained. Three of these wide area augmentation systems are established: in Europe EGNOS (European Geostationary Navigation Overlay System.-.see also figure 10-16), in North America WAAS (Wide Area Augmentation System), and in Japan MSAS (Multi-Transport Satellite Augmentation System).

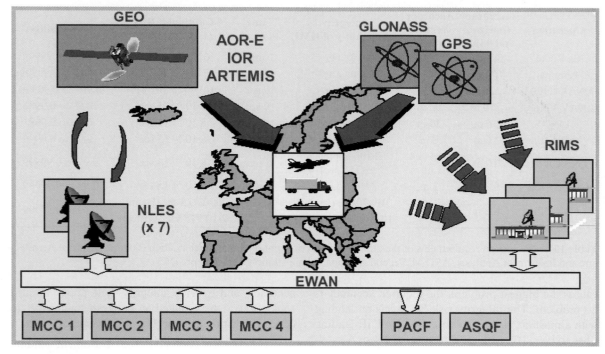

Figure 10-16: The European Geostationary Navigation Overlay System (EGNOS)

Efforts are on the way to apply satellite navigation (in combination with other sensors) for CAT III approaches using local area augmentation systems called LAAS (local differential GPS system at airports) and perhaps pseudolites (terrestrial signal generators transmitting a GPS-like signal).

Monitoring and Surveillance. Position, heading, speed and time information are needed by air traffic controllers for the continuous management of all aircraft. Some areas of the world lack the appropriate ground infrastructure, including secondary radar and communication links.

Taxiway guidance. Moving an aircraft on the ground requires assistance from the air traffic controllers. The airport may have surface radar, but sometimes the taxi movements are reported manually by the pilots and the aircraft is managed using visual aids only. Severe accidents have occurred during this supposedly safe phase. Satellite navigation together with its local elements and communication links (required accuracy 1 to 2 m) will improve the safety of these operations, creating the means for integrated surface movement guidance and control.

10.3.5
Space

Orbit determination of LEO (low-earth orbiting) satellites is usually carried out using dynamic (force) models. Mainly because of the limited knowledge of the near earth gravity field, but also due to other force model parameters, *dynamic* orbit determination is not highly accurate, in general. Therefore, nowadays all LEO satellites carry a GPS receiver making use of the *kinematic* determination of the orbit by satellite navigation. The results can even show centimeter accuracies when employing differential techniques with regard to ground stations. A well-known processing technique is also to smooth the single kinematic GPS solution by a dynamical model which is often called "reduced dynamic mode" (see for example Gill and Montenbruck, 2004, Martin-Mur, 1995, Parkinson and Spilker, 1996).

Using more than one antenna attitude measurements also become possible.

Such an approach is in general only possible if the altitude of the satellite is below the GPS satellites (<20,000 km above the earth). However, it was shown by Balbach et al (1998) with a GPS receiver on the Equator-S mission that even above those altitudes GPS signals can be received which are by-passing the earth from the other side. Insofar, an orbit determination in space was possible up to approximately 60,000 km altitude above Earth.

Future third generation satellites of GPS-III (after 2010) will also carry an additional antenna pointing into space in order to enable also more earth-distant orbit determination.

10.3.6
Telecommunication

The need to *locate callers* has two main drivers: (i) emergency calls (E-112 in Europe, E-911 in US) and (ii) new services based on the location. The first arises through new legislation in several countries aiming to offer efficient emergency services to their citizens by precise and fast response to distress calls. The second is more commercial and points to increasing traffic in the coming years. Technically, location can be achieved by integrating a satellite navigation receiver in the mobile phone (handheld solution) or by using the communication network itself. The last approach is not favored because it would require enormous additional costs for the infrastructure in the communication network. Once the caller's location is known, a great number of services can be offered. Some of these services are:

Location Based Services (LBS). At the beginning of this chapter on satellite navigation, it was already mentioned that information needs to be "time- und position-stamped" in order to create new value-added services in conjunction with geographical information and communication. A classical example is someone asking the way to the nearest hospital. The service provider compares the user's location with the hospital locations stored in a database,

The Global Positioning System (GPS) is a space-based, continuous, worldwide three-dimensional positioning and navigation, velocity and timing system operated by the United States Air Force which met its full operational capability in 1995.

The nominal configuration of the space segment consists of 21 satellites plus three active spares arranged in six 55-degree orbital planes approx. 20,190 km above the earth.

The main navigation payload consists of redundant atomic clocks (rubidium, cesium) transmitting very precise time signals right-handed circular polarized on the two frequencies L1 = 1575

1575.42 MHz and L2 = 1227.6 MHz modulated by a pseudo-random code (binary phase shift keying). For civil purposes a so-called C/A code or Standard Positioning Service (SPS) is provided. Access to the encrypted military P(W) code signals or Precise Positioning Service (PPS) requires authorization by the U.S. Department of Defense.

In order to determine a position, a receiver can measure the signal travel times from four or more satellites (code ranging). Multiplying it with the velocity of light results in the slant ranges to the satellites (pseudo-ranges) which have to be corrected for the influence of the ionosphere and troposphere. Measuring the instantaneous Doppler shift the receiver can also determine the velocity vector. Using the carrier phases itself in order to determine the highest accuracy in position requires solving the carrier phase ambiguities (cycles).

Among other activities, GPS modernization in the next years will offer new signals on a third frequency L5 = 1176.45 MHz and a civil code on L2 called L2C.

and then tells the user the nearest hospital and the fastest route. Service providers could also point customers to restaurants, movie theaters or parking lots. LBS will increase communication traffic significantly and generate important revenue to tele-communication operators and service providers.

Synchronization in communication networks. Network timing as well as synchronization of telecom base stations is usually done using atomic (rubidium) clocks. Satellite navigation also delivers as a by-product precise time and distributes it worldwide down to the nanosecond level. Precise time-synchronization of the different base stations can significantly increase the traffic capability of the telecommunication systems and the Internet.

10.3.7
Finance, Banking and Insurance

Secure electronic documents. The digital era has created electronic documentation as an effective alternative to paper. This means that new concepts for legal acceptance of electronic signatures and time-stamping must be developed. Satellite navigation provides certifiable and reliable data worldwide. For authentication and electronic-signing, the en-cryption system could be based on the trusted time signal and the integrity message of the global satellite navigation system, offering the additional value of traceability and liability for the time information.

Data encryption. The latest technologies for electronic encryption rely on highly precise time references - at performance levels obtainable only from atomic clocks - so they are not affordable to mass-market users. The timing service of satellite navigation, in addition to its certification, availability guarantee and integrity message, will enable secure transmission via inexpensive terminals, thus bringing data security within the reach of us all.

Insurance. The financial sector will certainly benefit from the innovations offered by satellite navigation in the field of insurance. It will open up new opportunities and bring about innovative prime and policy conditions. For example, the satellite navigation system could be an effective way of controlling and monitoring valuable goods, including the transportation of gold bullion between national banks, works of art, cash, and any insured risky item. Continuous tracking of these items would reduce risks, thereby creating financial benefits for insurance companies and their customers (e.g., distance-dependent insur-

The GLONASS System is managed for the Russian Federation Government by the Russian Space Forces. Like GPS it is a space-based, continuous, worldwide three-dimensional positioning and navigation, velocity and timing system. The fully deployed constellation is composed of 24 satellites in three orbital planes whose ascending nodes are 120 degrees apart. Eight satellites are equally spaced in each plane. Each GLONASS satellite operates in a circular orbit at an altitude of 19,100 km above the earth with an inclination angle of 64.8 degrees.

The satellites transmit two types of signal: standard precision (SP) and high precision (HP). SP

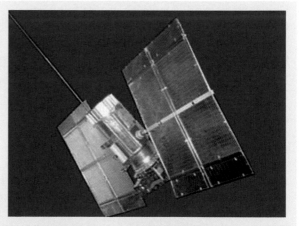

signals L1 have a frequency division multiple access L1 = 1602 MHz + n x 0.5625 MHz, where n is the frequency channel number (n = 0,1.2...). This means that each satellite transmits a signal on its own frequency, except satellites in antipodal slots of orbital planes which use the same frequency.

To determine position, the GLONASS receiver uses the same type of observations and principles as described for GPS. This would also mean that accuracies of GLONASS are in general similar to those of GPS. The GLONASS system is lacking the financial support of Russia and the satellites reveal a relatively short lifetime (2-4 years only). As of 20 July 2004 only 10 satellites were properly working in orbit.

ance charges). Certified satellite positioning would not only provide good legal support for the insurance sector, but also enable a great number of new services, including car and property insurance.

10.3.8
Precision Agriculture and Environment

Chemical spraying. More and more chemicals are being used to increase productivity by controlling pest and weed infestation of crops. Yet, as we become increasingly aware of the environmental impact, there is a need for better management of agricultural land. Besides, spraying chemicals where they are not needed is costly and bad for the environment. How can satellite navigation help? Precise aircraft positioning enables the pilot to spray the herbicides, insecticides or fertilizers in the right places and in the correct quantities. Automatic control also produces a more even distribution, reducing the quantity used. A positioning accuracy of better than 1 m is required. A satellite navigation receiver installed on the spraying vehicles will link the system to a database with other field information. For example, maps can then be generated to show where the spraying occurred.

Crop yield monitoring. Yield monitoring leads not only to effective resource management and consequently significant return, but also contributes to safeguarding the environment. Better control is becoming an issue. Farmers need to be able to map the high- and low-yield areas of fields so that a varying

application of chemicals can improve the yield with minimum environmental impact and cost. The yields of individual parcels of land can be monitored every season. Satellite navigation receivers on harvesters will lead to more automated systems and higher accuracy, drawing on the data stored in databases. By looking at yield maps, farmers can see where to take samples for analysis, with the satellite positioning system allowing specific areas to be targeted.

10.3.9
Surveying and Civil Engineering

The surveying community was the first to take advantage of GPS, in particular since many of their tasks have to be solved only post-mission. Nowadays (2004) the term "Real Time Kinematics (RTK)" is often used for offered hardware and software equipment, meaning nothing else than that measurements can be carried out in or near real-time. Insofar it is a special case of "high-precision navigation without guidance." Although most surveying tasks do not need real-time results it is very important to have a quality control in the field in order to avoid re-surveys and thus save money.

The surveying approach is characterized by the fact that mainly carrier phase observations in differential mode are used in order to achieve the ultimately highest accuracy. Carrier phases have the disadvantage that they are ambiguous with respect to the number of wavelengths (cycles). Therefore special algorithms and strategies have been developed in order to solve this "carrier phase ambiguity" in post-

GALILEO: The European Transport Council decided on 26 March 2002 to build up its own global satellite navigation system consisting of 30 satellites in a mean Earth orbit (at an altitude slightly higher than GPS and a similar inclination as GPS). It will transmit timing signals and a broadcast message on four frequencies in the L-band using Code Division Multiple Access (CDMA). It is envisaged to have all satellites working in orbit around 2010. The position determination follows the principle of GPS. Galileo and GPS will be interoperable for the user through an EU-US Agreement signed 26 June 2004.

mission and even in real-time (often also called "on-the-fly"). Due to the influence of the ionosphere dual-frequency solutions are preferred, but the carrier phase ambiguity resolution is still limited with respect to distances, approximately < 15-20 km, if a success rate near 100% is required (see e.g., Eissfeller et al. 2002; Misra and Enge 2001, pp. 209 ff).

The achieved accuracies in surveying are considered to be 5 mm ± 1 ppm (1 part per million of the measured baseline). However, under ideal conditions millimeter accuracies are possible in post-mission for small baselines, say centimeters, as well as for very long baselines (for example 3,000 km).

Structure Monitoring

Satellite navigation receivers in differential mode on and around bridges, barrages, dams, skyscrapers and historical monuments, for example, can provide important structural monitoring. Satellite techniques can also be used to predict natural events such as landslips, land settling, rising and rock falls, and measure the levels of rivers and lakes. A transmission link ensures the data reach a processing and monitoring center for real-time detection of any movement. Many bridges are carrying average loads higher than predicted during their design, so over the past few years there has been a significant increase in the need to monitor bridge performance. Some bridges are undergoing major repairs and retrofitting to fix critical deficiencies. Satellite receiver technology and data processing software are now cost-effective tools which can be integrated into an automated continuously operating system. Accuracies in the millimeter range are possible for these static networks since the major error source, which is the multipath, can be calibrated with time the system running (Hein and Riedl 2003)

Machinery Guidance

Civil engineers use heavy machinery in many types of construction. Satellite navigation receivers and real-time kinematic techniques can guide these machines precisely to perform their work. The same technique can be used for automated guidance of machines working in dangerous areas or simply to save manpower in repetitive work. The computer compares the actual satellite-derived position with the desired finished terrain, using grid files created

from topographic maps. A large, bright display provides visual guidance to the operator for maneuvering the vehicle and positioning the blade to achieve the cut- and fill values needed to match the computer model. A number of surface mines have recently installed satellite-based machine guidance systems with very positive results in productivity and costs.

Construction Site Management and Logistics

During the long construction periods for large structures, it is important to have efficient logistics as well as a coherent and common localization tool. The way the work area is accessed often changes as construction advances, and many vehicles are on the move at the same time. All these activities need efficient management to avoid confusion and wasted time.

10.3.10
Electricity Networks

The growing integration of networks for energy distribution and the emphasis on energy savings and efficiency require increasingly precise and accurate synchronization. Satellite navigation provides network synchronization for power generation and distribution in order to achieve efficient power flow. For that purpose precise timing of satellite navigation is used. Satellite navigation based mapping systems are also used to reduce power outage time by as much as 20%.

10.3.11
Science

Atmosphere-Related Research

Global navigation satellite systems allow for probing the atmosphere of the earth - both ionosphere and troposphere - by ground-based and space-based techniques. The first method requires networks of ground-stations, whereas the second option makes use of low-earth orbiting satellites that carry GNSS receivers.

Troposphere Monitoring (Ground-based)

Precise GNSS carrier phase measurements allow determination of the atmosphere's integrated water vapor content which plays an important role for the energy balance and the vertical stability of the atmosphere as well as for precipitation forecasts. Water vapor is responsible for about 62% of the natural greenhouse effect and as a consequence, GNSS meteorology has not only become a versatile tool for meteorologists during the past decade, but also for climate researchers.

The basic methodology can be summarized as follows: Carrier phase observations from permanent GNSS stations with precisely known positions are filtered for the tropospheric propagation delay which has the same size on all carrier frequencies used since the troposphere behaves as a non-dispersive medium for microwaves. However, it increases with decreasing elevation angle to the satellite due to the signals traversing longer fractions of the atmosphere. For this reason, the slant path delay is projected into zenith direction using a mapping function and the

zenith path delay is estimated in an adjustment process. Conversion into atmospheric water vapor requires knowledge of two important meteorological quantities, namely surface pressure (measured at the antenna site) and the mean vapor temperature. Pressure readings allow separation of the total zenith path delay into a hydrostatic and a wet component, where the hydrostatic delay can be precisely modeled and the wet delay can be directly related to integrated water vapor with the help of the weighted mean temperature. Since this value requires integration of vapor pressure and temperature profiles which are often unavailable, models have been developed to directly derive it from surface temperature measurements.

GNSS water vapor estimation benefits from a favorable error propagation: The conversion factor to transform zenith wet delays into integrated water

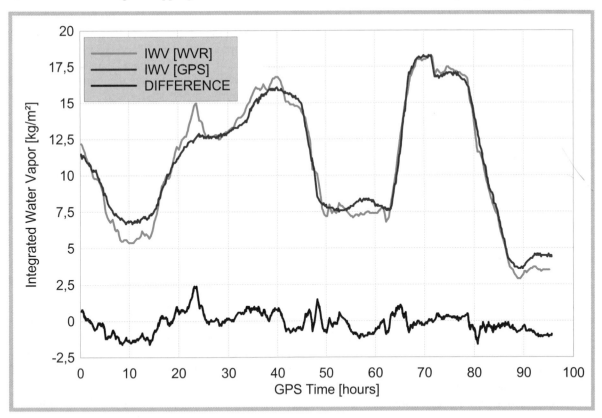

Figure 10-17: Water Vapour Radiometer (WVR) and GPS Signal Diagram, IGS Site Potsdam, start 19 January 1999. Comparison of integrated water vapour (IWV) estimates derived from GPS and measured by a ground-based radiometer near IGS Potsdam station

vapor is in the range of 1:6.3, i.e., if the tropospheric zenith path delay can be determined with an accuracy of better than 1 cm, the resulting error in the water vapor estimate is still in the range of better than 2 kg/m². In practice, an accuracy of 1 to 2 kg/m² can actually be reached, the RMS deduced from figure 10-12 is even slightly below 1 kg/m² (radiometer taken as reference). Ground-based microwave radiometers show a precision which is often slightly better, but their accuracy usually decreases due to calibration uncertainties. In principle, it would also be possible to perform tropospheric tomography yielding a 3-dimensional refractivity field, but this idea suffers from the very dense network of ground receivers required as well as from multipath and signal bending effects that can often not be fully mitigated.

Ionosphere Monitoring (Ground-based)

The idea to monitor the spatial and temporal behavior of the earth's ionosphere is even older than GNSS meteorology, since this part of the atmosphere behaves as a dispersive medium for the waves emitted by the satellites. For this reason, signals are broadcast on more than one frequency, allowing isolation of the ionospheric propagation delay and yielding more precise positioning results. Ionosphere researchers make use of this principle and form a so-called "geometry-free observation" from dual-frequency measurements that allows estimation of the first order ionospheric delay, the by-far dominant effect, whilst second order effects could be sensed by triple-frequency data in the near future. Many approaches generalize the ionosphere as a single-layer model with an effective height about 350 to 450 km. However, in contrast to the wet tropospheric refractivity which is concentrated in the lower few kilometers of the atmosphere, the ionosphere starts at an altitude of about 70 km and extends to an upper boundary layer of 1,000 km. This geometry also allows for ionospheric tomography with inter-station distances significantly smaller than those needed for the troposphere.

Space-Based Methods for Monitoring the Ionosphere and the Troposphere

Probing the ionosphere and troposphere can also be accomplished by means of space-based GNSS methods, particularly by the radio occultation technique. The basic idea dates back to the 1960s when it was practically applied to derive atmospheric properties of several planets of the solar system with the help of satellites that were orbiting those bodies. The Earth-related radio occultation method employs a LEO satellite orbiting at low altitude (700-1,200 km) and carrying a high-performance receiver that senses signals of GNSS satellites which are just about to rise or to eclipse behind the earth. Such a geometry implies that the signal will pass through the earth's atmosphere, where it is significantly bended. The bending angle can be related to the refractivity profile by means of an Abel integral transformation. Since the refractivity profile is a function of electron content, pressure, temperature and humidity (water vapor pressure), it is possible to derive profiles of these quantities. Pressure, temperature and humidity are not of concern when the signals are passing the ionosphere such that even a single-frequency receiver would be sufficient to probe this part of the atmosphere only. When the signals approach the boundary layer between ionosphere and neutral atmosphere, dual-frequency receivers are needed to separate the ionospheric refractivity properly. In the stratosphere, the water vapor term is negligible, whereas it is significant in the troposphere. Regarding the troposphere-related quantities, it should be stressed that temperature and humidity cannot be separated from GNSS measurements alone, i.e., external profile data are needed, such as humidity data from weather models when temperature profile are to be derived. Latest successful missions of this kind were SAC-C and CHAMP, for instance.

Geodynamics

The fact that GNSS satellite orbits are determined in a global terrestrial reference frame makes ground receivers a versatile tool for measuring global plate tectonics and monitoring regional geodynamic processes. Albeit this application sounds simple as it only implies to continuously determine the position of a very slowly moving receiver antenna, it still requires some skills to filter the time series appropriately by removing disturbing effects, since the target is to measure very slow processes with drift rates in the range of a very few centimeters per year, often also at the sub-centimeter level. Nevertheless,

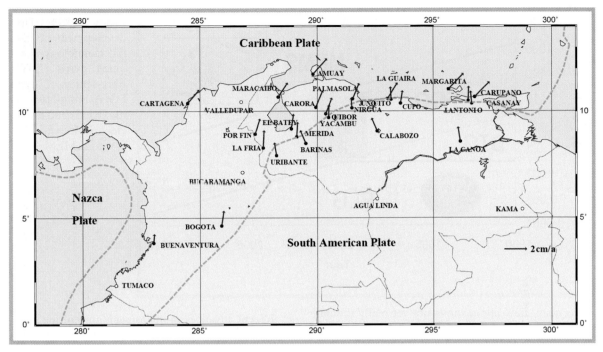

Figure 10-18: Regional geodynamics at the boundary of the South American and Caribbean plates revealed by GPS measurements (DGFI Annual Report 2001/2002, Deutsches Geodätisches Forschungsinstitut, München)

results for global plate tectonics obtained from GNSS measurements agree rather well with geophysical models, except for the plate boundary zones which are difficult to model. Figure 10-18 shows average horizontal motion rates in the boundary zone of the South African and Caribbean plates with typical rates between 1 and 2 cm per year.

10.4
Future Trends

With the advent of GPS many disciplines have revolutionized their measurement principles in positioning and navigation as well as in timing. In particular, satellite techniques are now replacing to a wide extent the traditional terrestrial methods. New applications come up every day. Water vapor estimation for the improvement of local weather models and occultation of the atmosphere are examples of byproducts never planned in the design of GPS.

Looking to the future, it seems that we are only at the beginning with respect to the applications of satellite navigation, which in many cases take ad-

vantage also of geoinformation and telecommunication in order to create value-added services. In this fast growing field it is difficult to predict long-term trends exceeding, say, five years. Nevertheless, the following can be expected:

- With the build-up of the European satellite navigation system Galileo to be completed around 2010 a new area in satellite positioning and navigation will start. Based on the frequency and signal structure and its bandwidths (all of them larger than those of GPS) *accuracy will increase* considerably (down to sub-millimeter level under ideal conditions). Real-time kinematic capabilities will become *more robust*. Integrity will be globally available at any time.

- Due to the *interoperability* agreement *between GPS and Galileo* the user will benefit in terms of availability of twice the number of satellites he is able to track with just the same receiver. In general, the user can observe two times eight to ten satellites and needs only four in order to determine a three-dimensional position). This *enormous redundancy* leads to more robustness and

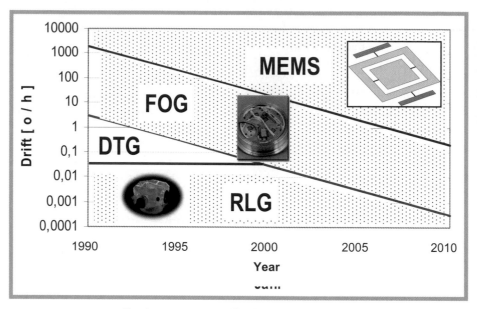

Figure 10-19: Drift development of inertial technology (Dry-tuned gyros (DTG), fibre-optics gyros FOG), ring-laser gyros (RLG) and micro-electro-mechanical systems (MEMS)

to many new applications not yet really thought of. The *increased availability* will make it happen that satellite navigation can be used also in cities, urban canyons and areas with limited sky coverage.

- Satellite positioning and navigation will be possible also *indoors* using new developments in signal processing to increase the sensitivity of tracking weak signals and to mitigate multipath through new methods. Data sent via mobile terrestrial telecommunication will enable the receivers to acquire satellites in a second or less (so-called "Assisted Global Navigation Satellite System").

- Satellite navigation receivers will be *integrated with various sensors* in the future, in order to overcome the deficiencies of satellite navigation in certain applications (interruption of signals, shadowing, autonomy, etc.).

- Considering complimentary properties, one prominent candidate is certainly the inertial navigation system (see e.g., Jekeli 2001). GNSS/INS integration can be carried out in different levels of closeness or tightness, both in the hardware and software, of the two sensors. Very important could be an ultra-tight coupling of a small INS (on a chip) with the tracking loops of a future GNSS receiver.

- Recent advances in micromachining of mechanical systems have led to the development and manufacture of very small, inexpensive inertial sensors (micro-electro-mechanical systems - MEMS, and micro-opto-electro-mechanical systems - MOEMS). Examples are safety and stability devices consisting of accelerometer(s) and/or gyro(s) in automobiles and other consumer products produced on a chip.

- Figure 10-19 outlines the expected drift behavior of different type of gyros like dry-tuned gyros (DTG), fiber-optics gyros FOG), ring-laser gyros (RLG) and MEMS devices over the next few years. Whereas no further development of ring-laser gyros is predicted and dynamically-tuned gyros will practically more or less vanish, a tremendous improvement of fiber-optical gyros can be foreseen (2005 drift $0.01...0.001^0$/hr; 2010 drift $0.001...0.0001^0$/hr). Also very interesting is that the MEMS technology may reach the 1^0/hr in 2008 to 2010.

- But also the integration of satellite navigation receivers with all kind of different sensors and techniques (odometer, UMTS/GSM, Bluetooth, WLAN, digital map, terrain-based information, etc.) will increase over the next years.

- Software receivers. Two decades have passed since the first commercial GPS receiver appeared on the market. Whereas the first ones were based on analog signal processing using large components demanding a lot of power, modern standard receivers are commonly based on ASICs (application-specific integrated circuits) for signal processing and fast microprocessors for position and application processing. ASICs guarantee on the one hand effective signal processing; however, every small change results in a costly-redesign of the system because it cannot be re programmed as easily as can a microprocessor.

- Considering the advances in the performance of microprocessors, all operations of an ordinary GNSS receiver can be carried out in future using a programmable microprocessor, making such a system highly flexible (changing just the source code). The software receiver consists then simply of the antenna, a front-end and A/D converter as well as a microprocessor. First PC-based versions are being developed which are able to process one frequency with 1 Hz data in real-time (Pany et al., 2004). In the near future, more sophisticated multiple frequency solutions allowing also higher update rates will be possible depending on progress in microprocessor performance.

- Whereas it is expected that the first use of software receivers as defined above will concern a development platform for receiver manufacturers, one can also think of a very personalized receiver (integrated with personal application-oriented modules containing information and digital maps) in a decade from now (after 2010) when already the core receiver might be even available in public domain as a software module.

Acknowledgements

Some text and figures on Galileo provided by the European Commission in public domain are used (Business in Satellite Navigation. An overview of market applications. Brochure dated March 5, 2003). I acknowledge also the contribution of Dr Torben Schüler, Institute of Geodesy and Navigation, University FAF Munich on science applications.

References

Balbach O, Eissfeller B, Hein GW, Enderle W, Schmidhuber M and Lemke N (1998) Tracking GPS above satellite altitude: First results of the GPS experiment on the HEO mission Equator-S. Proc. PLANS'98, Palm Springs, CA (USA), 20-23 April 1998, pp. 243-249

Eissfeller B, Tiberius C, Pany T and Heinrichs G (2002) Real-time kinematic in the light of GPS modernisation and Galileo. Galileo's World, Autumn 2002, pp. 28-34

Gill E and Montenbruck O (2004) Comparison of GPS-based orbit determination strategies. Proc. 18th Int. Symposium on Space Flight Dynamics, 11-15 Oct. 2004

Hein G and Riedl B (2003) Real-time monitoring of highway bridges using DREAMS. Proc. International Symposium on Deformation Measurements, Santorini Island, Greece, 25-28 May 2003

Jekeli Ch (2001) Inertial Navigation Systems with Geodetic Applications. Walter de Gruyter, Berlin New York

Lück T, Eissfeller B, Kreye Ch, and Meinke P (2001) Measurement of line characteristics and of track irregularities by DGPS and INS. In: Proc. Int. Symp. On Kinematic Systems in Geodesy, Geomatics and Navigation, Banff, Alberta, Canada, University of Calgary, 5-8 June 2001 pp 34-41

Martin-Mur T, Dow J, Bondarenco N, Casotto S, Feltens J, and Martinez CG (1995) Use of GPS for precise and operational orbit determination at ESOC, Proceedings of the ION GPS-95 pp 619-626

Pany T, Eissfeller B, Hein G, Moon SW, and Sanroma D (2004) IPEXSR: A PC based software GNSS receiver completely developed in Europe. Proceedings GNSS 2004 The European Navigation Conference, 16-19 May 2004, Rotterdam

Schüler T (2001) On Ground-Based GPS Tropospheric Delay Estimation, Ph.D. thesis, Universität der Bundeswehr München, Studiengang Geodäsie und Geoinformation No. 73

Schüler T and Hein GW (2004) ENVISAT Radar Altimeter Calibration with High-Sea GPS Buoys, ENVISAT/ERS Symposium, Salzburg, Sept. 2004

Volpe JA (2001) Vulnerability Assessment of the Transportation Infrastructure Relying on the Global Positioning System. Final Report. John A. Volpe National Transportation Systems Centre

Selected Textbooks in Satellite Navigation for Further Technical Information:

Hofmann-Wellenhof B, Legat K, and Wieser M (2003) Navigation. Principles of Positioning and Guidance. Springer, Wien / New York

Misra, P and Enge P (2001) Global Positioning System. Signals, Measurements, and Performance. Ganga-Jamuna Press Lincoln, Mass.

Parkinson BW and Spilker JJ (1996) eds: Global Positioning System: Theory and Applications. Vol. I and II, American Institute of Aeronautics and Astronautics, Inc., Washington, DC

Space as a Laboratory

Overleaf image: Astronaut Ulrich Walter during the D-2 Spacelab mission (DLR)

11 Fundamental Physics

by Hansjörg Dittus

High precision experiments and astronomical observations of increasing sensitivity offer a chance to learn about the microscopic and macroscopic structure of our universe. Physics, therefore, faces new challenges, which might result in a completely new understanding of our picture of the world. Space is an ideal laboratory enabling new fundamental experiments, orders of magnitude more precise than on earth.

11.1
Fundamental Physics - Definition

Our modern physical conception of the world is based on a number of universal theories which deal with four interactions. We distinguish the electromagnetic, the weak, the strong, and the gravitational interaction, which have completely different strength and interaction ranges. As the strong and weak interactions act in the range of only 10^{-14} m but are dominant on this scale, gravitation and electromagnetism act infinitely and determine the physics on the cosmological scale. Three of the interactions, the electromagnetic, the weak, and the strong force, are described by theories (the so-called standard model), which need exchange quanta to describe the physical processes: the massless photon within quantum electrodynamics, three vector bosons for the weak interaction (e.g. for the description of spontaneous radioactive β-decay), and the gluons for the strong interaction (describing the processes inside the atom nucleus). In contrast to this, all processes with gravitational coupling are described by General Relativity, a strongly deterministic theory based on geometrical constructions. A description of gravity with quantum exchange processes seems to be impossible, because the integration of gravitation into the standard model leads to higher order symmetry models and requires the violation of fundamental hypotheses of General

Relativity, as for example the violation of the universality of free fall, for which we have no experimental evidence up to now. Even more enhanced theoretical concepts to solve the conflict, such as string theory, miss any experimental hint. It is the intention of fundamental physics to find solutions to solve the unification puzzle and to prove and to test the theoretical and axiomatic foundations of the universal theories. In particular, these universal theories are:

- Quantum theory, characterized by the Planck constant $\hbar = 1.05457266(63) \cdot 10^{-34}$ Js, which is a measure of the quantization of a physical process.

- Special Relativity, characterized by the speed of light $c = 299.792458$ ms^{-1}, which is a constant in space and time.

- General Relativity, characterized by the gravitational constant G = $6.67259(85) \cdot 10^{-11}$ m^{3}kg^{-1}s^{-2}, which is a measure for the gravitational coupling between two masses.

- Statistical physics, characterized by Boltzmann's constant $k_B = 1.380658(12) \cdot 10^{-23}$ JK^{-1}, which is a measure for the statistical uncertainty.

As statistical physics dealing with many particle interactions is a special case, at least for the other three universal theories direct relations exist, reflected by figure 11-1, where each colored segment is assigned to a specific theory in which one or two of the constants $\hbar$, c, or G are set to zero or infinity. A consistent theory in which each of these constants has a finite value should accept a quantization of gravity, but does not exist up to now. Figure 11-1 demonstrates the inconsistency of the present universal theories and underlines the need to come to a more comprehensive description of nature beyond the standard physics.

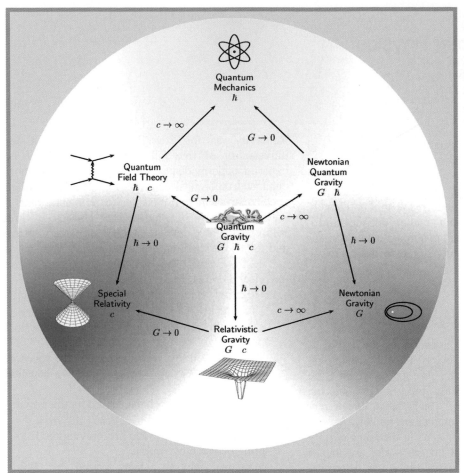

Figure 11-1: Schematics of universal theories. Each segment is assigned to a specific theory in which one or two of the universal constants ħ, c, or G is set to zero or infinity

Figure 11-1 also makes clear phenomenologically that all new theoretical approaches of "new physics" require violations of fundamental principles. It is the intention of fundamental physics to test standard physics and to search for new fields, interactions, and particles. On the theoretical side, the search for a theory combining quantum theory and gravitation is still the most challenging task of modern physics, whilst on the experimental side new methods of carrying out better high precision measurements may in the future enable discriminating tests of the current physical theories to be made. A significant aspect of these high precision tests of the fundamental principles underlying physics is metrology. This encompasses the very important task of preparing, reproducing, and transporting the fundamental units. All efforts to redefine basic units in terms

of uniquely reproducible quantum effects are only effective if the underlying non-gravitational physical laws are locally valid everywhere. Fundamental physics plays a guiding role, and it is obvious that our ability to carry out high precision experiments in space becomes an increasingly pressing issue (Lämmerzahl und Dittus 2002).

11.2
Space – A Unique Laboratory

Fundamental physics experiments usually have to be performed with very high accuracy. It is evident that the sensitivity of measuring devices and the sensing resolution increase if the experiments can be performed under conditions of free fall, that is, under conditions of weightlessness. Although in many precision experiments on earth gravitation might have a negligible influence and seismic noise is either irrelevant or at least can be shielded sufficiently enough, the more accurate the measuring devices are, the more terrestrial influences become the dominant disturbing effect. More than that, many experiments need an interaction-free environment and the absence of any linear or rotational acceleration. This indicates that it might be of great advantage to perform precision experiments in space. In particular, one can take advantage of the following:

- *Infinitely long, and periodic free fall*: As an example, long free fall conditions enable high precision tests of the universality of free fall. Free fall experiments on earth last only several seconds, whereas space conditions enable periodic tests, e.g. on the MICROSCOPE and STEP satellites.

- *Long interaction times*: These are, for example, advantageous in atomic interferometers, where the atoms may interact with external fields for a long time and do not fall down and touch container walls. The precision (peak width) of an interference measurement is inversely proportional to the interaction time.

- *High potential differences*: In a large class of experiments (e.g. experiments to test the universality of the gravitational redshift by clock comparison), one needs periodic differences in the gravitational potential. It is obvious that this can be achieved best on highly eccentric orbits. The maximum height difference on earth attainable with aircraft or balloons is about 40 km. Much larger values can be obtained in space.

- *Long-distance measurements*: In space, huge distances are available, which is essential for the study of low frequency gravitational waves by means of interferometric techniques with arm lengths of up to $5 \cdot 10^9$ m and for long-way satellite ranging to map gravitational fields and to search for post-Newtonian effects. For example, precise lunar laser ranging with an accuracy in the sub-cm range became possible with retroreflectors placed on the moon surface within the Apollo and Luna programs. Precise satellite ranging over long distances to measure the light deviation in the gravitational field of the sun was

carried out successfully for the first time within the Viking program. It is worth noting that a careful analysis of the radiometric tracking data from the Pioneer 10/11 spacecraft (launched 1972/73) at distances between 20-70 astronomical units (AU) from the sun has consistently indicated the presence of an anomalous, sun-directed acceleration of $(8.74 \pm 1.33) \cdot 10^{-10}$ m/s², which seems to be unexplained within the framework of General Relativity (Anderson et al. 1998).

- *Large changes in velocity*: By choosing appropriate orbits, the satellite velocity, i.e. its velocity with respect to the cosmic microwave background, can be extensively varied along its orbit. This can be used to test the constancy of light speed with respect to the laboratory velocity (e.g. Kennedy-Thorndike experiments). On earth, macroscopic bodies can experience velocity differences of the order of 1 km/s maximum. Along earthbound orbits, the velocity difference can be 1 to 2 orders of magnitude larger.

- *Low noise or low vibration environment*: Seismic noise is a limiting factor for many experiments on earth in the frequency range below 10 Hz. Although damping systems can reduce the mechanical disturbances and seismic vibrations effectively on earth, the cut-off frequency for a laser interferometer on earth is about 10 Hz. Nevertheless, measurements on satellites do not guarantee a complete reduction of the noise bandwidth, because systematic variations arise with orbital frequencies. But these effects can be reduced by a proper choice of spin rate and orbital path as well as by high precision attitude and orbit control (i.e. drag-free control).

11.3
Testing Fundamental Principles and Predictions in Space

By far the most proposals for fundamental physics tests on satellites are experiments to test the physical laws of gravitation. The reason is that one can use the global gravitational field of a huge massive planetary body, like the earth, much better for a satellite experiment than for a laboratory experiment on the surface of the earth. Satellites equipped with precise clocks and accelerometers can move on ideal free fall paths (geodetics) and can serve as gravitational probes. Modern gravitational physics is based on General Relativity which itself is based on Special Relativity. Therefore, the chapter contains two sections to describe tests of Special Relativity and General Relativity, respectively. Nevertheless, also quantum physics and experiments in statistical physics play a more and more important role which is discussed in the later sections.

Special Relativity Tests

Special Relativity is based on two fundamental and axiomatic principles:
- constancy of light speed c and
- validity of the Relativity principle.

The first principle was introduced by A. Einstein to give a consistent explanation of the interferometer experiments of A. Michelson and E. Morley, who measured already in 1881 the speed of light independently of the velocity of the laboratory and the direction. The Relativity principle states that the laws of physics are the same in all inertial systems, which includes the fact that the constant speed of light is a maximum, because a different speed of a specific particle would single out a preferred frame, thus violating the Relativity principle. A direct consequence of Special Relativity is the velocity dependent dilation of time and distance by the factor $1/(1-v^2/c^2)$ and $(1-v^2c^2)$, respectively.

Experiments of Special Relativity are usually described within the framework of kinematic test theories (Lämmerzahl et al. 2002). In a simplified way, one can write c as a function of the laboratory velocity v and the laboratory orientation θ with respect to the cosmic background:

$$c(v,\theta)=c\left(1+A\frac{v^2}{c^2}\sin^2(\theta)+B\frac{v^2}{c^2}+O\left(\frac{v^4}{c^4}\right)\right), \quad (1)$$

where A and B disappear if the metric tensor is symmetric and c is a constant and isotropic. Therefore, typical satellite tests use interferometers. The velocity varies along the orbit and the orientation changes with respect to the spin rate and

Optical Test of the Isotropy of Space (OPTIS)

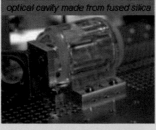
optical cavity made from fused silica

Scientific Objectives: Improved tests of the constancy and isotropy of light speed, universality of gravitational redshift, Doppler effect, Lense-Thirring effect, and absolute gravitational redshift.

Experiment: Three crossed optical resonators (cavities) to which 3 Nd:YAG lasers are locked; in addition an atomic clock and an optical comb generator.

Mission outline: The satellite moves on an high elliptical orbit. It is drag-free controlled and its spin rate can be varied. For the Michelson-Morley test, the changing orientation with respect to the cosmic microwave background (CMB) is used; for the Kennedy-Thorndike test, the large change of velocity with respect to the CMB is of advantage. The periodic changes of the gravitational potential can be used to test the gravitational redshift. Combined two-way laser ranging with drag-free attitude and orbit control enables all other experimental test (Dittus et al. 2003).

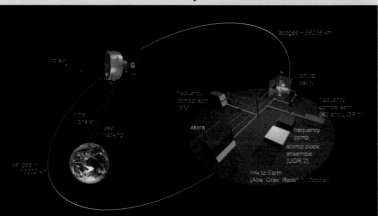

the trajectory, e.g. OPTIS (see box) and SUMO (Superconducting Microwave Oscillator). For $B \neq 0$, an eventual frequency shift in an electromagnetic resonator must appear within orbit frequency (Kennedy-Thorndike experiment). For frequency comparison, one needs a very precise clock on board. A will be determined by a comparison of two crossed resonators; if the satellite rotates, $A \neq 0$ would be measured as a relative frequency shift appearing periodically with spin rate. SUMO is an experiment proposed to be performed on board the International Space Station (ISS) inside the Low Temperature Microgravity Physics Facility (LMTPF), a liquid helium cryostat allowing access to temperatures down to 1.4 K for up to several months. (Buchman et al. 2000, web: *bigben*). The experiment consists of two cylindrical super-conducting microwave cavities mounted with their axes orthogonal. The cavity design reduces the sensitivity to disturbance accelerations caused by residual drag on ISS and vibrations, by a factor of about 1,000. The reference clock for the Kennedy-Thorndike test will be the Primary Atomic Reference Clock in Space (PARCS, see below). On ISS with its orbital height of approximately 350 km and its inclination of 50°, the velocity change during one orbit is 15 times larger than on the earth surface. The expected limit for the confirmation of the isotropy of space (constancy of light speed) is $\delta c(\theta)/c(\theta) \leq 10^{-18}$ and of the independence of velocity $\delta c(v)/c(v) \leq 3 \cdot 10^{-18}$

General Relativity Tests

General Relativity tests can be categorized in
- (1) tests to explore the structure of the coupling of matter (bulk or quantum) with the gravitational field; in Einstein's theory there is only one quantity describing the gravitational field, namely the metric $g_{\mu\nu}$, in other theories, additional couplings may appear;
- (2) tests to verify predictions of General Relativity.

Within the first category, one summarizes tests (or search of violations) of:
- the universality of free fall (the principle of equivalence), which states that all point-like par-

ticles fall along the same path (geodetic)[1]. The inertial mass is equivalent to the gravitational mass.
- the Local Position Invariance (LPI) which states that physics does not depend on the laboratory position in the gravitational field. It implies the universality of the gravitational redshift.
- the constancy of the gravitational interaction (G = *const.*) in space and time.

The second category (predictions) covers tests like:
- the search for gravitomagnetic effects, like the Lense-Thirring effect describing the dragging of an inertial frame due to rotation of a nearby massive body. This non-Newtonian effect results in a precession of a gyroscope moving in the gravitational field of a rotating planet or star.
- the gravitational time delay due to the light deflection in strong gravitational fields.
- the search and observation of gravitational waves.
- the search for deviations from or modifications of Newtonian physics.

In the following, the tests and the individual experimental space missions will be discussed in detail:

The Weak Principle of Equivalence (WEP): The basic hypothesis of General Relativity is the WEP. It states that inertial mass m_i and gravitational mass m_g are equivalent and cannot be distinguished. In Newton's law

$$m_i a = G \frac{M_{Earth} \cdot m_g}{r^2} \qquad (2)$$

where a is the gravitational acceleration of the earth and M_{Earth} its mass, we usually set $m = m_i = m_g$, where a is the same for all bodies falling at the same place. A measure for the validity of the WEP is the Eötvös parameter

$$\eta_E = 2 \frac{a_1 - a_2}{a_1 + a_2} = 2 \frac{m_{g1} / m_{i1} - m_{g2} / m_{i2}}{m_{g1} / m_{i1} + m_{g2} / m_{i2}} \qquad (3)$$

---

[1] In General Relativity, the Equivalence Principle also holds for massive bodies, like planets and stars. Considering the self-gravitation of a massive body, the theory assumes that self-gravitation contributes to inertial and gravitational mass in the same way.

where the indices 1 and 2 characterize two different test masses made from different materials. If the WEP holds, $\eta_E = 0$. The WEP has been proven in laboratory experiments by means of torsion balances to $\eta_E \leq 10^{-12}$. Although one would expect to attain better results with free fall experiments, torsion balance experiments are more accurate up to now, because they allow observation of a periodic signal.

Equivalence Principle tests have also been done with Lunar Laser Ranging (LLR) using retro-reflectors which had been placed on the lunar surface during the Apollo and Luna missions at five different locations between 1969 and 1973. Aligned perfectly to earth, the retro reflectors reflect pulsed laser light sent from earth. Since only one single photon of the 10^{19} photons sent from earth to the moon returns, a special data analysis had to be developed. It needed enormous efforts to improve the LLR resolution by a factor of 1,000 during the last 30 years. A single photon can only be detected by comparing its actual arrival time with its pre-calculated one. The resolution for earth-moon distance measurements attainable today is in the range of 1 cm. If one considers the iron-dominated earth and the silica-dominated moon as two test masses freely falling in the solar gravitational field, the distance measurements can be used to determine the Eötvös factor to about $\eta_E \leq 10^{-13}$ (Nordtvedt 2001) [2].

A violation of the WEP is predicted in the low energy limit of string theory at an order of $\eta_E < 10^{-15}$ (Damour and Polyakov 1996) and in cosmodynamical theories at an order of $\eta_E < 10^{-14}$ (Wetterich 2003). Therefore improved WEP tests are necessary.

Satellite-based experiments to test the WEP yield to measure $\eta_E \leq 10^{-15}$ on MICROSCOPE (Micro-satellite à trainée Componsée pour l'Observation du Principe d'Équivalence) and $\eta_E \leq 10^{-19}$ on STEP (Satellite Test of Equivalence Principle). MICROSCOPE (see box) (Touboul 2001a, web: *onera*) is an approved CNES mission. STEP (Lockerbie et al.

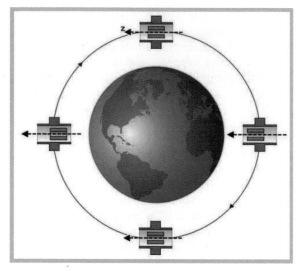

Figure 11-2: Two test masses would fall along a different path if the WEP would be violated. Their relative distance changes periodically if their centres of mass coincide at the beginning of the experiment. As one test mass falls along a circular orbit, the other falls along an elliptical path (N. Lockerbie, Univ. Strathclyde)

2001, web: *einstein*) is a NASA-ESA study. Free fall experiments in space take advantage of an extremely low level of residual acceleration acting on the test masses and enable long term signal integration [3]. If a WEP violation occurs, two test masses of different composition would fly on different orbits (see figure 11-2). As one test mass moves on a circular orbit, the other one moves on a slightly elliptical one, and the differential position with time is a sinusoidal signal. Two test masses arranged in a way that their centers of mass coincide form a differential accelerometer. As at STEP the relative motion of the test masses will be observed and measured by a SQUID (Superconducting Quantum Interference Device)-based sensing system, the WEP test on MICROSCOPE is performed by controlling the relative motion of both test masses at null with capacitors sensing and electrostatic actuators so that any WEP violation appears through the measured forces necessary to nullify this relative motion. Both satellites have drag-free control. Drag-free control guarantees that the satellite moves on a free fall

---

[2] The WEP test with LRR is only valid if the Equivalence Principle also holds for planetary bodies with self-gravitation. An eventual violation of this strong Equivalence Principle with gravitation is called the Nordtvedt effect, which can be rejected by LLR to about 0.001.

---

[3] It is a periodic signal. The free fall is repeated with orbit frequency.

MICROSCOPE
(Micro-satellite à trainee Componsée pour l'Obser-vation du Principe d'Équivalence)
Scientific Objectives: Test of the Weak Principle of Equivalence to an accuracy of $\eta = 10^{-16}$.
Experiment 2 differential accelerometers; each with cylin-

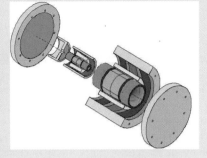

drical test masses with coinciding centres of mass; capacitive sensing.
Mission outline: The satellite is on a circular polar orbit; orbit height is 700 km. The orbit is controlled with drag-free attitude and orbit control. The spin rate can be varied and modulates the signal frequency in order to be off-resonance with orbit frequency (Touboul 2001a). MICROSCOPE is approved to be launched 2008.

trajectory unless disturbances like air drag, solar radiation pressure, etc. are acting. The satellite spin (ω_{spin}) modulates the WEP violation (ω_{WEP}) signal and avoids couplings from disturbances appearing with orbital frequency (ω_{orbit}): $\omega_{WEP} = \omega_{spin} + \omega_{orbit}$. The acceleration acting on the two test masses in each accelerometer can be split in two parts: the differential mode signal (i. e. the WEP violation signal) and the common mode signal which is used to control an ensemble of µN-thrusters of the satellite. MICROSCOPE is a "room" temperature experiment, whereas the STEP accelerometers have to be mounted inside a liquid helium Dewar to cool them down to a nominal temperature of about 2 K, which guarantees superconductivity. The test masses are superconductor (niobium)-coated and face pick-up coils on each side along their symmetry axis. Any movement towards a pick-up-coil changes the inductance value, which can be measured by the SQUIDs with very high resolution.

The WEP must hold for any kind of matter. STEP and MICROSCOPE are experiments to test the WEP only for neutral bulk matter. Therefore, it is of great importance to carry out experiments for other kinds of matter. Space tests would allow to attain improvements in the results for WEP tests of antimatter (Walz and Hänsch 2004), charged matter (Lämmerzahl et al. 2004a), and quantum matter.

Also spin coupling experiments[4] could be carried out.

Space experiments to test Newton's inverse square law (equation (2)) are closely related to WEP tests and have been proposed recently (Speake et al. 2004). These tests will prove theoretical predictions of the existence of additional (Yukawa-like) terms to the Newtonian potential U which would then be modified to $U = GM / r (1 + \alpha_5 \cdot \exp(\lambda_5/r))$. The additive term can be treated as a hypothetical "fifth force", an unknown new interaction with strength α_5 and range λ_5. Space experiments under microgravity are ideal tests for short distance measurements (test for $\lambda < 0.1$ mm) on one side, because of the possibility to depress disturbing effects by test mass levitation. On the other side, deep space tests with satellite radio and laser ranging gives the opportunity to test Newton's law on distances up to more than 100 AU.

Universality of Gravitational Redshift (UGR): UGR states that all clocks, independent of the working mechanism or physical principle they are based on, experience the same redshift (change of frequency) in a gravitational field with the gravitational potential U between positions x_1 and x_2:

$$f(x_1) = \left(1 - \frac{U(x_1) - U(x_2)}{c^2}\right) f(x_2). \quad (4)$$

---

[4] Experiments with "spin matter" (magnets)

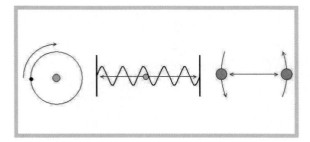

Figure 11-3: Schemes of different clock types. The oscillation of an atomic clock (left) is characterized by the interaction of photons with electrons in an atom. The oscillation in a light clock (mid) is characterized by the wavelength stabilized in a cavity, and a molecular clock (right) is defined by molecular vibrations. The physical principles of the clocks are completely different

Although an atomic clock using the motion of the electron in the Coulomb potential of the nucleus and a light clock defined by the bouncing of the photons at the mirrors at both ends of an optical cavity are based on completely different physical processes, they experience the same retardation in gravitational fields (see figure 11-3).

Different clocks based on different physical processes are defined by different physical constants, e.g. the fine structure constant, or the mass ratio of electron and proton. Therefore, UGR tests are even tests of the constancy of those constants. Standard UGR tests carried out on satellites compare different types of clocks in a varying gravitational potential. With equation (4), the ratio of the frequencies of two clocks is given for first order potential difference by

$$\frac{f_{clock1}(x_1)}{f_{clock2}(x_1)} \approx \left(1 - \left(\alpha_{clock1} - \alpha_{clock2}\right)\frac{U(x_1) - U(x_2)}{c^2}\right)\frac{f_{clock1}(x_2)}{f_{clock2}(x_2)} \quad , \quad (5)$$

where the parameters α_{clock1}, $\alpha_{clock2} \neq 0$ indicate an eventual deviation from General Relativity. These tests were carried out on Gravity Probe A (GP-A), a ballistic rocket flight (Vessot and Levine 1979) the first time. Improved tests will be carried out on the ISS with SUMO and PHARAO/ACES (web: *pharao*), and are proposed to be carried out with high accuracy on satellite missions with large potential variations like OPTIS and the deep space mission SpaceTime (STM, Maleki and Prestage 2001). GP-A confirmed $\alpha \leq 10^{-4}$ by comparing a

hydrogen maser clock on the rocket with one other on ground.

Gravitomagnetic effects: General Relativity predicts many important but weak effects which cannot be proven on earth experimentally. Those effects are precessions of a gyroscope in the gravitational field of a massive self-rotating body (like the earth). Because of their analogy to the spin-spin-coupling or spin-orbit-coupling in atom physics (a rotating charge induces an electromagnetic field), these non-Newtonian effects are called misleadingly gravitomagnetic effects. The total precession of the angular momentum of the gyroscope is given by

$$\frac{d\mathbf{S}}{dt} = \mathbf{\Omega} \times \mathbf{S} = \left(\mathbf{v} \times \left(-\frac{1}{2}\mathbf{a}\right) + \mathbf{v} \times \frac{3}{2}\nabla U + \nabla \times \mathbf{h}\right) \times \mathbf{S} \quad (6)$$

where the first term of the right hand side describes the precession of the spin due to inertial acceleration **a** (Thomas precession). The second term describes the geodetic precession due to the gravitational acceleration ∇U. The last term is related to the Lense-Thirring (or frame dragging) effect due to spin-spin-coupling between the gyroscope and the gravitating body rotating nearby (e.g. the earth) (Lämmerzahl and Neugebauer 2001). In order to verify gravitomagnetic effects, one can observe an inertial gyroscope on a satellite in a polar orbit around the earth or track a satellite precisely along its orbit with an eccentricity greater than 0. The latter has been done with the LAGEOS I and II satellites (Ciufolini 2000, web: *lageos*). For an orbital height of 5,900 km, and an eccentricity of 0.0045, as well as an inclination of 110° (LAGEOS I) and 152° (LAGEOS II), General Relativity requires a rotation of the knots of the ecliptic and the orbital plane resulting in a perigee advance of 0.033 arcsec per year (related to a shift of 1.9 m) due to the Lense-Thirring effect. Tracking of the satellites, which are metallic spheres covered by a large number of retroreflectors, confirms the Lense-Thirring effect with an error of about 20%. The limiting disturbance parameters are the poorly known quadrupole and higher order moments of the gravitational field of the Earth. Satellite laser ranging can be improved by additional precision attitude and orbit control as proposed for the OPTIS mission.

Gravity Probe B (GP-B)

Scientific Objectives: Highly precise tests of gravitomagnetic effects, precession of gyroscope in the gravitational field of the rotating earth:
(1) Geodetic effect (6.6 arcsec per year) due to the curvature of space-time by the earth to an accuracy of $2\cdot10^{-5}$,
(2) Frame dragging (Lense-Thirring) effect (0.042 arcsec per year) to an accuracy of 0.3.

Experiment: 4 superconducting gyroscopes (superconducting spheres spinning with 100 Hz); SQUID-based read out of the direction of the magnetic (London) moment

Mission outline: The satellite is on a circular polar orbit; orbit height is 640 km. The orbit is controlled with drag-free attitude. Direction is precisely controlled by a telescope designed as a folded Cassegrain system. The telescope is attached to a quartz block housing the gyroscopes. The quartz block is mounted inside a 2,300 l helium cryostat (Everitt et al. 2001). GP-B has been launched 2004.

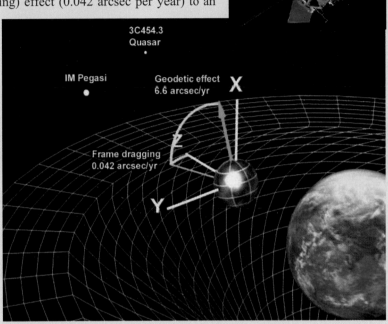

Much more precise measurements will be made with Gravity Probe B (GP-B), a satellite launched in April 2004 into a circular polar orbit with a height of 640 km. The satellite carries four inertial high precision gyroscopes with SQUID-based sensing, wherefore influences of Earth's gravitational field irregularities are eliminated. The scientific goal of GP-B (see box) is to observe the geodetic precession of 6.6 arcsec per year with an accuracy of two parts in 10^5 and the Lense-Thirring precession of only 0.041 arcsec per year with an accuracy of three parts in 10^3 (Everitt et al. 2001).

Experiments to test gravitomagnetic effects on the highest precision level could be done by using atom-interferometry. A mission proposal for HYPER (Hyper-precision Atom Interferometry in Space) has been studied intensively by ESA (Ertmer et al. 2004).

Gravitational time delay: Light is deflected in gravitational fields due to the space-time curvature. This deflection can be measured as a time delay of an incoming electromagnetic signal. Signals sent from earth to a planet or satellite on the opposite side of the sun and reflected on the planet's surface (radar ranging) or from retroreflectors on the satellite (laser ranging) are delayed twice as they pass the solar gravitational field. In a first order approximation (Schwarzschild solution), the time delay δt is given by

$$\delta t = \frac{2R_S}{c} \cdot \frac{1+\gamma}{2} \cdot \ln\left(\frac{r_{ES} + r_{RS} + r_{ER}}{r_{ES} + r_{RS} - r_{ER}}\right) \qquad (7)$$

R_S is the Schwarzschild radius, which is a measure for the strength of the solar gravitational field and the space-time curvature due to it. r_{ES} and r_{RS} are the distances between earth and sun and between reflector and sun, respectively, whereas r_{ER} is the

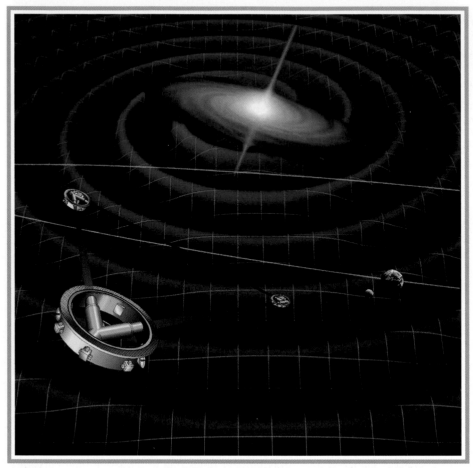

Figure 11-4: LISA (Laser Interferometer Space Antenna) on its Earth trailing heliocentric orbit. 3 spacecraft form an equidistant triangular Michelson arm interferometer. The distance between the spacecrafts is ca. $5 \cdot 10^6$ km. The triangle is inclined by 60° against the eclipse whereas not only the amplitude but also the direction of the incoming gravitational waves can determined during one year of operation (MPG)

distance between the earth and the target reflector. The Eddington parameter γ is a measure for eventual deviations from the predictions of General Relativity; it is identical to 1 if General Relativity holds. Careful space experiments need an analysis of the ephemerides of all greater objects in the solar system, wherefore equation (7) has to be modified[5]. First delay measurements have been carried out with the Mars probe Viking. These observations have been in agreement with relativistic predictions to 0.2% (Reasenberg et al. 1979). Much more precise measurements have been proposed to be carried out

on missions like STM (SpaceTime Mission) which should pass the sun as close as four solar radii, LATOR (Laser Ranging Test of Relativity), or ASTROD (Astrodynamical Space Test of Relativity using Optical Devices), which would allow laser ranging over a distance of 2 AU.

Gravitational waves: One of the most important predictions of General Relativity is the radiation of space-time deformations (gravitational waves) from masses accelerated in space-time. Direct observations of gravitational waves enable the study of e.g. the dynamics of black holes, the merging of galaxies, as well as the bursts of supernovae and open a new window for astronomy. Ground-based gravitational wave detectors are large scale Michelson-Morley type interferometers with arm lengths up to 4 km. These detectors are under construction at

---

[5] The appropriate equation is the Einstein-Infeld-Hoffmann equation. In this equation not only the eventual deviations from curvature (γ) are considered, but also the complete set of 10 post-Newtonian parameters, indicating all other possible deviations from General Relativity.

several sites worldwide. Earth-bound detection is limited to the frequency range above 0.1 Hz due to terrestrial seismic noise. For detection in the low frequency range down to the μHz-range, space-borne interferometers are needed. The ESA-NASA satellite project LISA (figure 11-4) (Laser Interferometer Space Antenna) is planned to be launched prior to 2012 and to measure gravitational waves in the frequency range between 10^{-1} and 10^{-4} Hz. The strain sensitivity of the detector will be $\delta L/L = h/2 \leq 4\cdot10^{-21}$ $Hz^{-1/2}$ (L is the detector arm length, h the relative amplitude, Danzmann and Rüdeger 2003, web: *interferometry*). With this sensitivity, the two main categories of gravitational waves observable by LISA are binaries in our galaxy and massive black holes existing in the centers of most galaxies. Galactic sources of interest are close white dwarfs as well as normal stars with white dwarf companions (cataclysmic binaries), neutron star binaries, neutron stars with black hole companions of up to 20 solar masses, and black hole pairs which can only be observed with gravitational wave astronomy. The main objective of LISA, however, is to observe the formation and growth of massive black holes with masses up to 10^8 solar masses. Most powerful sources detectable are the merges of massive black holes in distant galaxies.

The space-borne interferometer will be built by three spacecraft arranged in an equilateral triangle formation and separated by about $5\cdot10^6$ km. The center of the spacecraft triangle moves on a heliocentric orbit trailing the earth by 20° and is inclined by 60° with respect to the ecliptic. Therefore, the triangle as a whole rotates around its center of mass. Each spacecraft carries two phase-locked laser systems and two mirrors. Two sides of the triangle arrangement are the arms of the interferometer, as the third arm is used for a redundant measurement of wave polarization. The mirrors are actively driven by phase locking and the phase information of detected incoming photons is used to control the phase of the outgoing laser light. The mirrors are 40 x 40 x 40 mm^3 Au-Pt alloy cubes controlled by a system of electrodes and serve as reference masses for the high-precision drag-free attitude and orbit control system which guarantees a residual acceleration of only $3\cdot10^{-15}$ m/s^2 in the signal frequency band. The pointing performance by means of the star trackers is less than 10 nrad $/Hz^{1/2}$. The interferometer fringe resolution is $4\cdot10^{-5}$ $\lambda/Hz^{1/2}$ (λ = 1,064 μm is the wavelength of the Nd:YAG laser used). To test the relevant technology, in particular the interferometer system and the drag-free AOCS concept, the LISA Pathfinder mission is under construction. Approved by ESA and NASA, it will be launched in 2008.

Tests of Quantum Physics

Quantum physics is based on two fundamental principles: the superposition principle and the uncertainty relation. Due to the tremendous progress in the manipulation of single quantum systems it is possible with today's experimental techniques to prepare and to isolate macroscopic quantum systems (e.g. Bose-Einstein condensates) consisting of an atom cloud with very low velocities in an extremely low temperature regime. These quantum systems have lifetimes of several seconds and enable interference experiments. On earth, however, the quantum system is influenced by gravity, which limits the experimental time. Simply spoken, the ensemble of cold atoms falls out of the interferometer. A weightlessness environment guarantees a long free evolution time of the quantum states and improves the experimental accuracy by orders of magnitude. Although quantum physics experiments have not been carried out in space so far, there is a large potential to test quantum mechanics on satellites. In particular, the development of atom laser interferometry and the preparation of cold quantum gases are powerful experimental tools and will enable a study of quantum phenomena with very high accuracy.

By theory, quantum systems suffer decoherence in vacuum which might be observable in atom interferometers by searching for a decrease of the visibility of interference fringes as a function of the free evolution before recombination of the coherently split wave function. There are hypotheses that decoherence is a quantum gravity effect due to space-time fluctuations (Amelino-Camelia 2004).

Interference of quantum systems is a strange phenomena from the standpoint of classical mechanics, but has been experimentally confirmed. The best results for tests of potential deviations from linearity have been attained with neutron interferometry. A much longer free evolution time using space-born atom interferometers would improve those tests considerably.

Ultra-cold quantum gases, in particular Bose-Einstein condensates, are systems of identical boson states (typically 10^6 to 10^8 atoms). Cooled down to the nK-range, the wave-functions of the individual atoms overlap forming a macroscopic quantum system where all atoms are coherently linked. Therefore, Bose-Einstein condensates offer a unique insight into the "quantum world." Tests in laboratories on earth are limited by gravity, because the condensate falls down and hits the container wall after several seconds. Weightlessness offers the possibility to push Bose-Einstein condensates into an ultimate model system for fundamental quantum tests. Microgravity permits extending the free time of evolution by one to three orders of magnitude and enables an unperturbed evolution unbiased by gravity and without need of levitation. One can expect that Bose-Einstein condensates will be cooled down in the pK- and fK-range under weightlessness conditions, far beyond the present terrestrial temperature limits. This gives the chance to observe phase transitions never observed so far, e.g. dipolar magnetism in the quantum domain where ultra-weak forces govern the kinetics of quantum gases.

Bose-Einstein condensates offer the possibility to study entanglement, too. Many quantum phenomena are related to the entanglement of states, e.g. teleportation, the Einstein-Podolski-Rosen paradoxa, and the Greenberger-Horne-Zeilinger states. Recently the feasibility of adopting quantum information techniques to a space infrastructure has been studied (Aspelmeyer et al. 2003). Technological potential is obvious with respect to quantum computing.

Bose-Einstein condensates may also be the source for a coherent atomic beam serving as an atom laser. Due to the small de Broglie wavelength of atoms compared to electromagnetic waves, the sensitivity of interference experiments can be increased by orders of magnitude. Weightlessness would improve the experimental resolution with respect to spatial, energy, and time scales.

Experiments in Statistical Physics

Systems consisting of many particles are described by statistical methods serving as theoretical tools to derive averaged quantities like temperature, pressure, specific heat, and density. The underlying organizing principles for condensed matter systems, like the spontaneous formation of long-range order below a critical temperature, can be explained by renormalization group theory. The main result of this theory is that critical phenomena (phase transitions) appearing through manipulation of parameters show universality and a certain scaling behavior. For example, the critical temperature between the normal and the superfluid state of a system depends on the pressure. However, as we approach the critical temperature, other thermodynamic quantities like the specific heat, the magnetization, or the density behave universally, independent of the pressure. This means that near the critical temperature certain quantities describing the many particle system depend only on the temperature. Therefore, all systems must behave in the same way, which seems a very surprising prediction.

The importance of renormalization group theory for many fields of theoretical physics makes it worth proving its validity experimentally. Critical points are found in many systems. Beside the already mentioned transition from normal fluidity to superfluidity (λ-point), there are most prominent the critical pT-value in thermodynamics, or the critical temperature at which the transition from normal conductivity to superconductivity occurs. The prediction from renormalization group theory states that the exponents describing the phase transitions have exact and universal values. In general, critical points are characterized by a divergent correlation length ξ and by singularities of the temperature dependence of thermodynamic quantities, like the specific heat C for which the phase transition is characterized by the critical exponents ν and α:

$$\xi = \frac{\xi_0^+}{\xi_0^-} \left| \frac{T - T_c}{T_c} \right|^{-\nu} \quad \text{and} \quad C = \frac{C_0^+}{C_0^-} \left| \frac{T - T_c}{T_c} \right|^{-\alpha} \quad (8)$$

+ and − mark the approach from the different temperature domains $T > T_c$ and $T < T_c$, respectively (T_c is the temperature at the critical point.). Between the critical exponents exact relations exist, e.g. $2 - \alpha = 3\nu$.

A variety of fundamental microgravity experiments to test the predictions of renormalization group theory with respect to scaling and universality have been proposed and developed to be carried out on the International Space Station (ISS). See the review by Lämmerzahl et al. (2004 web: *fundamental*). Microgravity is needed to avoid density variations induced by gravity. Three examples will be discussed in the following:

Superfluid Universality Experiment (SUE): This experiment is designed to study the second order phase transition occurring when helium is supercooled into a liquid, and further cooled into a superfluid (Greywall and Ahlers 1973). The transition to superfluidity at the λ-point of helium is characterized by the exponent of the superfluid density, which is predicted to be a universal constant calculated exactly with a value close to 2/3. The superfluid density is related to the velocity of second sound (i.e., temperature waves) and the heat capacity which will be measured in the superfluid universality experiment along various isobars near the λ-line of helium. Detected pressure dependence would violate the theory. The experiment consists of a thermally insulating cylindrical sample chamber mounted in a helium cryostat on an orbital platform. The sample chamber forms a second sound resonator equipped with a highly sensitive detector and a second sound generator which is a thin film heater to be used also for heat capacity measurements. Temperature can be measured with a resolution of 10^{-10} K. A superconducting pressure gauge attached to the sample chamber guarantees constant pressure conditions and control. The temperature is raised incrementally closer to the λ-point and the fundamental resonant frequency of the cavity as well as the pressure, the heat capacity, and the temperature are measured. These measurements will be repeated until the transition to superfluidity is reached. The measurements will then be repeated along different isobars in order to measure the entire λ-line. A complete experimental cycle needs 50 days (web: *chex*).

Boundary Effects Near Superfluid Transitions (BEST): The scientific objective of this experiment

also to be carried out at liquid helium temperature on an orbital platform is the study of molecular-level boundary issues of liquid helium during the transition in the superfluid state. The experiment will measure the thermal conductivity in a three-dimensional ^{4}He sample along the λ-line over three orders of magnitude and in one- and two-dimensional confinements of various sizes. In addition, the cross-over behavior from three-dimensional to the fundamentally different two-dimensional superfluid transitions will be studied. In a confined geometry of characteristic length L, finite-size scaling for the specific heat C is predicted (see also equation (8)):

$$C\left(\frac{T_c - T}{T_c}, L\right) - C\left(\frac{T_c - T}{T_c}, \infty\right) = L^{\alpha/\nu} f\left(L / \xi\right) \quad (9)$$

The scaling function f is predicted to be universal for a given geometry (Dohm and Haussmann 1994). In the experiment, the transition to the superfluid state is observed along different isobars and in different confinements. The confinements are cylindrical pores with diameters between 1 and 100 μm and a length between 1 to 5 mm as well as rectangular channels of sizes varied within the same range (web: *best*).

Microgravity Scaling Theory Experiment (MISTE): has been proposed to measure thermodynamic quantities in the ^{3}He liquid-gas region. Ground-

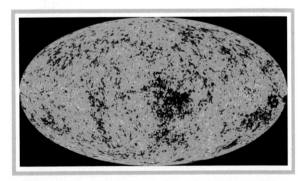

Figure 11-5: The spatial anisotropy from the close-to-perfect blackbody radiation of the cosmic microwave background from WMAP. As the mean temperature is about 2.7 K according to theory; the fluctuations are in the range of μK. An analysis of the anisotropies confirms the standard cosmology model and inflation to first order approximation and rules out a flat universe (NASA)

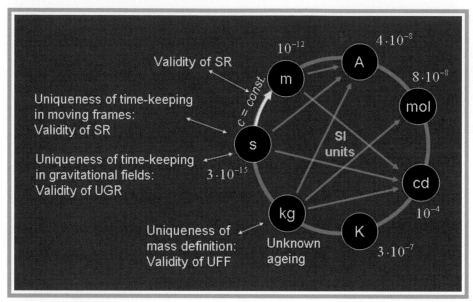

Figure 11-6: The international standard unit system is based on the validity of universal theories. Violations of the underlying hypotheses would have direct consequences for the unit definition

based critical point experiments show a gravity-induced density stratification because of the diverging isothermal susceptibility. According to renormalization group theory the isothermal susceptibility near the critical pT-value is a function of two critical exponents: the leading critical exponent and a correction-to-scaling exponent. The experiment is performed along various paths in the pT-diagram, like the critical isochore, the critical isotherm, and the coexistence curve, where the critical exponents of the thermodynamic quantities will be determined. It consists of a cylindrical sample cell positioned with its symmetry axis along the residual gravitational field. The sample density can be varied. High precision measurements of temperature, pressure, and density will be made (web: *miste*).

Cosmology and High Energy Physics

The observation of cosmic radiation and the cosmic microwave background is a major task of fundamental physics. It is obvious that space offers a unique environment for unperturbed observations. The standard model includes the Big Bang scenario and is based on General Relativity and the cosmological principle. The cosmology principle states that the matter in our universe is homogeneous and isotropic. The model fits the observations quite satisfyingly and the expansion of the universe can be

described by solutions of the Einstein field equations. Nevertheless, the standard model does not answer questions about the creation of the universe and how structures of matter came to exist, because nothing is known about the initial conditions. These missing conditions and the causality problem require inflationary models of the cosmic expansion. Cosmological models are strictly related to General Relativity and the standard model of quantum physics. Therefore, observation and investigation of cosmic phenomena like the cosmic microwave background radiation and the high energy particle flux of unknown source may help to adjust fundamental theories on the large scale.

Observations (figure 11-5) of the microwave background radiation with the NASA satellite Wilkinson Microwave Anisotropy Probe (WMAP) in 2003 confirmed the predictions from the cosmic standard model and inflation theory quite impressively (web: *wmap*), but raised new questions: only less than 5% of the matter in our universe is baryonic matter, as 25% dark matter and about 70% dark energy are not identified up to now. To study these questions, new satellite mission scenarios have been developed, like the ESA mission Planck to observe the spatial distribution of microwave background radiation with much higher resolution (web: *planck*) and the NASA/DOE Joint Dark

SQUID (Superconducting Quantum Interference Device)-based sensing relies on two phenomena:
- flux quantization in superconducting loops,
- Josephson effect

A SQUID-based accelerometer consists of a pair of two superconducting pick-up coils with inductance L_0 and a superconducting diaphragm covering the test mass. The moving test mass changes the inductance of both pick-up coils. Because of flux conservation in the superconducting loop results in a small current inducing an inductance in the SQUID-coupling coil (L_i).

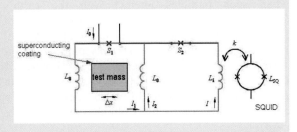

The sensitivity is given by: $\delta x = \dfrac{2L_i + L_0}{2k\sqrt{L_i L_{SQUID}}} \dfrac{1}{I_0} \dfrac{1}{\alpha} \delta\Phi_{SQUID}$,

(I_0 is the current in the loop, α describes the sensor characteristic, L_{SQUID} is the SQUID inductance, k a SQUID-specific coupling factor, and $\delta\Phi_{SQUID}$ is the flux resolution of the SQUID.) For a given flux resolution, a sensor accuracy of $2\cdot10^{-14}$ m has been realized (Vodel 2001).

Energy Mission (JDEM), a space-based telescope to search for farthest supernovae (web: *jdem*).

Another task of interest is the observation of high energy particles in space. Earth is bombarded permanently by particles like electrons, positrons, protons, antiprotons, and nuclei of different energy. Because many particles are absorbed or annihilated through collisions in the atmosphere of the earth, they have to be detected in space. On ISS, for the first time a detector for long term observation of particles will be placed in space. The Alpha Magnetic Spectrometer (AMS) has been designed to observe high energy radiation from space. Those observations may answer important questions about the Big Bang, e.g. the problem of the inequality and asymmetry between existing matter and antimatter, and the question how much matter and of what type exists in the universe. AMS consists of a transition radiator detector for measuring the velocities of the highest energy particles and a silicon tracker for following the particle paths through the instrument. A superconducting magnet produces the high magnetic field for the detector with 14 coils arranged in order to give a dipole field with minimal external stray field. The AMS magnet will be operated at a temperature of 1.8 K, cooled by 360 kg of superfluid helium. The curvature of particle paths through the magnetic field gives information about the momentum-to-charge ratio and about the sign of the charge. In addition, AMS has a time-of-flight counter, a ring-imaging Cherenkov radiation detector, an electromagnetic calorimeter, and an anticoincidence counter to detect spurious particles passing sideways through the apparatus (web: *ams*). A successful predecessor of AMS was already flown on the Space Shuttle in 1998.

The Extreme Universe Space Observatory (EUSO) is an instrument placed also on ISS to investigate cosmic rays and neutrinos of extreme energy (> $5\cdot10^{19}$ eV) using the atmosphere of the earth as a giant detector. The highest energy event ever measured was observed at $3\cdot10^{20}$ eV. At such high energies particles are no longer confined by the magnetic field of our galaxy. They must come from areas further away. Nevertheless, any particle should interact with photons of the cosmic microwave background which induces a general cut off, the GZK[6] cut off, at $5\cdot10^{19}$ eV. This means that one has to find possible sources, and to identify in the decay model a long-lived particle that exists as a residual from the Big Bang in sufficiently large number density and lifetime. Therefore, one can expect to learn new physics by observing extreme high energy particles. The detection will be performed by looking at the streak of fluorescent light produced

---

[6] Named after K. Greisen, G.T. Zatsepin, and V.A. Kuzmin

when such particles interact with the earth's atmosphere. The experiment will look downward from space into the dark atmosphere. Fluorescent light is imaged by a Fresnel lens onto a finely-segmented focal surface detector. EUSO is expected to detect on the order of 1,000 events per year with $E > 10^{20}$ eV. The instrument will be mounted as an external payload of the ISS Columbus module (web: *euso*).

11.4
Metrology and Fundamental Units

A significant aspect of high precision fundamental physics experiments is metrology. It encompasses the very important task of preparing, reproducing, and transporting from place to place the fundamental physical units like the second, the meter, the kilogram, and the ampere. All efforts to redefine units in terms of uniquely reproducible quantum effects are only effective if the underlying non-gravitational physical laws are locally valid everywhere. The procedure of basing the second on a certain atomic transition is possible because time keeping in a gravitational field is universal, although corrections due to General Relativity must be made to account for different gravitational potentials. The definition of the meter in terms of the second (measuring the frequency of a fixed electron state transfer of cesium) is only valid if the speed of light is constant, the key hypothesis of Special Relativity. Modern developments like the reduction of electrical units using the quantum Hall effect and the Josephson effect (these effects are most appropriate because of their close-to-ideal reproducibility) rely on the validity of quantum theory and Maxwell's theory[7] (figure 11-6). In many cases, the conditions in space (large distances, nearly weightlessness, changes of the gravitational potential along the orbit, large velocity variations, low-noise environment) lead to improvements in experimental sensitivity. Consequently, fundamental physics plays an important role, and it is obvious that our ability to carry out high precision experiments in space becomes an increasingly pressing task, in order to

test the predictions of the current universal theories for all four physical interactions.

11.5
Technology
Sensors

Many experimental techniques, recently developed for high precision experiments in the laboratory, need to be adapted to space platforms. The relevant technological developments can be categorized in order to measure:
- time
- frequency
- distance
- acceleration.

Time measurement in space as well as on earth can be done with atomic clocks, light clocks, or molecular clocks.

Standard atomic clocks for space application, like the H-maser, attain accuracies of $> 10^{-15}$. Recent developments of atomic fountain clocks, like PHARAO (Projet d'Horloge Atomique par Refroidissement d'Atomes en Orbite, figure 11-7), PARCS (Primary Atomic Reference Clock in Space), or RACE (Rubidium Atomic Clock Experiment) to be set up on ISS, will improve present standards up to 10^{-17} (Dittus et al. 2003). These clocks are based on the laser cooling technique; atoms are captured in magneto-optical traps and slowed and cooled down by lasers to about 1 μK. From the optical molasses, the atoms are released with a velocity of only about 10 cm/s. They enter a cavity in which they experience two successive Ramsey interactions with a microwave field tuned near a frequency defined with respect to a certain atomic transition. Atoms excited are detected downstream by fluorescence. The resonance signal is used to lock the central frequency of the oscillator to the atomic transition. The interaction time in weightlessness is much longer than on earth. Therefore, the typical resonance width is only 0.1 Hz under weightlessness conditions, 10 times narrower than in atomic fountain clocks on earth. PHARAO (web: *cnes*) and PARCS (*web: parcs*) uses cesium atoms, RACE is a rubidium clock with

---

[7] A non-zero photon mass or a non-constant speed of light would require modifications of Maxwell's theory.

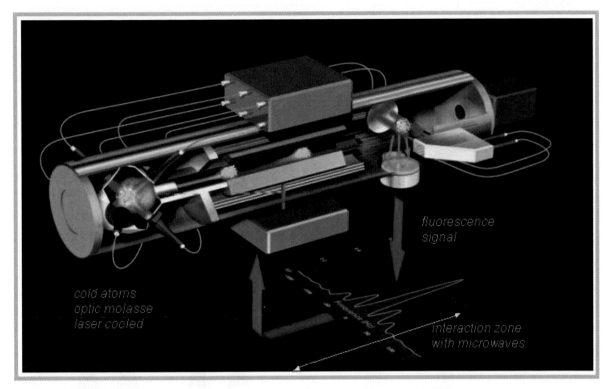

Figure 11-7: Atomic fountain clock PHARAO to be installed on ISS (Salomon et al. 2001)

much fewer frequency shifts by atomic collisions due to quantum mechanical effects.

Light clocks are based on ultra stable monolithic resonators with length stability of $\delta L/L < 10^{-15}$ $Hz^{-1/2}$ as well as highly precise lasers. The resonators are optical cavities made from fused silica or sapphire. Standard lasers are diode-pumped Nd:YAG lasers which guarantee an extremely narrow line width, a low intensity noise, and a very low frequency noise. The lasers are locked to the stable resonators and the arrangement forms an extremely precise resonator. Due to the required length stability, the cavities need a precise temperature control of 10^{-5} $K \cdot Hz^{-1/2}$, typically, and have to be operated in a weightlessness environment with drag-free control.

Frequency can be measured very precisely by means of optical or atomic clocks, because the wavelengths of emission lines are well known. Nevertheless, comparison of different clocks working in completely different frequency ranges is a problem in experimental physics. Optical combs enable the

coherent referencing of the frequency of an optical oscillator with a frequency of 10^{15} Hz to a hydrogen maser with a frequency of 10^{10} Hz only. The frequency comb is generated by a mode-locked fs-laser emitting a series of very short laser pulses with a well defined repetition rate. The comb enables a quasi-digitizing of the oscillator spectrum and can be stabilized to about $5 \cdot 10^{-16}$.

Distance must be measured over very long ranges in space as well as over ultra-short intervals in space experiments. Long range measurements can be done by means of laser ranging from earth to a satellite or between satellites, as well as with radio signal travel time analysis of deep space probes. Laser ranging measurements are precise on the sub-cm level, as deep space observations are strongly dependent on clock accuracy and modeling. The most precise distance measurements will be done with laser interferometry for LISA over a distance of about $5 \cdot 10^{6}$ km with an accuracy of 1 nm.

Atom Interferometry consists of an optical bench, lasers, one (or two) magneto-optical traps, and a detection unit. The magneto-optical trap consists of three orthogonal pairs of circularly polarized counter-propagating laser beams. Typically, up to 10^8 atoms are cooled down by exchange of momentum with photons. The lasers are tuned just below the resonance frequency. In an interferometer the atom beam is divided, deflected, and finally recombined with a sequence of three laser light pulses resulting in $\pi/2$-,π-,$\pi/2$- phase shifts. Any interference is detected by measuring the ground state populations. A rotation ω of the interferometer whose two atomic beam paths enclose an area A results in a phase shift

$$\delta\Phi = \frac{4\pi}{\lambda v}\vec{\omega}\cdot\vec{A}$$

Therefore, the resolution is strongly dependent on the de Broglie wavelength λ and the velocity v of the atoms. The most prominent examples of atom interferometers are the symmetric type in Mach-Zehnder configuration (left) and the asymmetric Ramsey-Bordé type (right). The vertical arrows indicate the beam splitting, re-directing and recombining lasers, respectively (*Interferometry*).

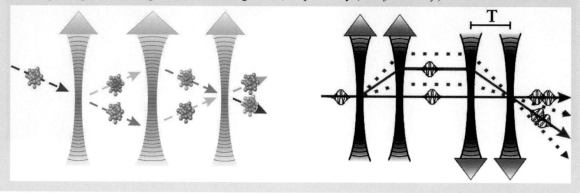

High precision short range distance measurements can be realized with capacitive sensing (see the description of MICROSCOPE) or with SQUID-based sensing (see box SQUID). Depending on the SQUID-noise, a resolution of $2\cdot10^{-14}$ m has been realized (Vodel 2001).

Acceleration is measured very precisely by means of reference sensors with electrostatic (capacitive) sensing successfully flown already on the CHAMP and GRACE missions (Touboul 2001b). Differential accelerometers (gradiometers) as they are being developed for the MICROSCOPE and STEP missions enable much higher precision measurements. These reference sensors are based on free-falling test masses which are levitated by electric or magnetic forces. The outer caging and the test masses form a spring-mass system with very weak spring constants.

Higher precision can be attained by using atom interferometers based on the Sagnac effect. In an Atomic Sagnac Unit (ASU), an atom beam interacts with light in order to split and recombine the beam by three counter-propagating lasers. The linear and rotational accelerations are sensed as an interference pattern of the split beam along the different (see box Atom Interferometry). Usually, high precision accelerometers need drag-free attitude and orbit control of the satellite they are mounted on.

Attitude and Orbit Control

In addition to sensors, also precise actuators have to be developed for fundamental physics space missions. An important technique is drag-free attitude and orbit control in order to move a reference test mass on an ideal free-fall path (geodetic) with an accuracy of at least 10^{-10} m/($s^2\cdot Hz^{1/2}$). Many satellites approved or planned to fly in the future, like GP-B, MICROSCOPE, LISA-Pathfinder, LISA, STEP, HYPER, ASTROD need drag-free control (see box Drag-free Attitude and Orbit Control). Drag-free orbit control is based on two principles:

Drag-free Attitude and Orbit Control

The general principle of drag free control is to ensure the trajectory of a satellite's centre of mass as close as possible to a free fall path. A gravity reference sensor has to be used in which the movement of a test mass relative to the satellite is observed with respect to all six degrees of freedom. A common concept is to control the movement capacitively. Electrodes for sensing and active servo control are surrounding the free falling test mass. The signal is used to control the satellite's attitude and orbit by a set of thrusters. The thrust force has to be controlled with an accuracy of 0.1 µN. Depending on the orbital height and the cross section of the satellite, the aerody-

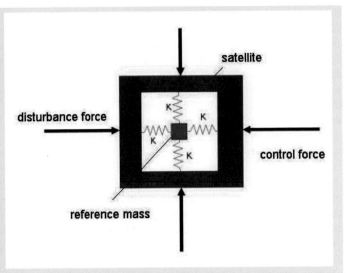

namic drag is between 10 and 150 mN, the resulting torques are between 10 and 100 µN. Dominant disturbing forces are the air drag, interactions with electromagnetic fields, tidal forces, coupling between test mass and satellite via feed-back reactions from the detection system, and electrostatic forces caused by charging through cosmic rays.

- precise acceleration measurement and thrust control by state estimation,
- modelling and simulation of the entire spacecraft.

The control concept has to minimize the non gravitational accelerations inside the satellite. The displacement of the satellite with respect to the reference mass is observed to keep the satellite following. The acceleration must be minimized at the test mass location which is denoted as the drag-free reference point. Usually, the states (position and velocity) of the satellite and reference mass are measured directly or determined by an observer model based on real observations. The estimated states, the output of the observer model, are fed into the controller which then commands the thrusters. The estimated values are directly fed back to cancel out the disturbances, and the residuals are compensated by the controller (Theil 2002). Thrust control is a major challenge for drag-free orbit maintenance. Different concepts have been developed which allow the thrust to be controlled to 0.1 µN. Field emission electrical propulsion or colloidal thrusters are based on the acceleration of ions or charged atom clouds in a high voltage electric field.

11.6
Summary and Outlook

Fundamental Physics stimulates research with respect to the universal laws of physics, special and General Relativity, quantum mechanics, many-particle physics, high energy physics, and cosmology. Today, some of our basic theories of physics are questioned. They cannot be correct, because it is not possible to unify them along the lines given by conventional quantum theory. As all approaches to attack this problem predict violations of the underlying assumptions and hypotheses of the basic theories, it is of great interest to improve the experimental search for New Physics according to deviations from standard theoretical models. It requires an increasing experimental accuracy which can be achieved only under well defined conditions. Space, however, is a unique environment and will be used more and more to carry out high precision experiments in fundamental physics.

It is obvious that the search for new physics is a key issue for finding new methods and concepts for technological application.

Space is a unique laboratory for gravitational physics. Nevertheless, it can be shown that all other fields of fundamental physics can benefit from research in the space environment. The missions approved and planned up to now are usually based on state-of-the-art technology. Therefore, experimental accuracy is limited by space technology mainly. As the laboratory techniques have been developed for decades, their adaptation to space conditions is a challenging and time-consuming task. Along many studies, various proposals for dedicated missions and experiments on platforms like the ISS have been developed. In any case, technology development for sensors and actuators is the most important issue. Experimental gravitational physics is based on the measurement of distance, acceleration, time, and frequency. Modern physics enables the use of quantum effects for highly precise sensing. Therefore, cold atom physics is of highest interest and needs to be developed for space application.

Beside tests of special and General Relativity, the investigation of the long list of open questions in cosmology, such as the search for dark matter and dark energy as well as the investigation of eventual anomalies of the Newtonian potential on long and short distances give rise to plans for future missions and single experiments which can be mounted on any deep space probe. Challenging experiments set the goals, and the development of completely new technology for these experiments and missions is useful and promising, anyway for our future.

References

Amelino-Camelia G (2004) Fundamental Physics in Space: A Quantum–Gravity Perspective. Gen. Rel. Grav 36, p. 539

Anderson JD, Laing PA, Lau EL, Liu AS, Nieto MM, Turyshev SG (1998) Indication, from Pioneer 10/11, Galileo, and Ulysses Data, of an Apparent Anomalous, Weak, Long-range Acceleration. Phys. Rev. Lett. 81, 2858, e-print gr-qc/9808081

Aspelmeyer M, Jennewein T, Böhm HR, Brukner C, Kaltenbaeck R, Lindenthal M, Molina-Terriza G, Petschinka J, Ursin R, Walther P, Zeilinger A, Pfennigbauer M, Leeb WR (2003) Space, Quantum Communications in Space, Exec. Summary Rep. ESA/ESTEC 16358/02/NL/SFe

Buchman S, Dong M, Wang S, Lipa JA, Turneaure JP (2000) A Space-based Superconducting Microwave Oscillator Clock. Adv. Space Res. 25, p. 1251

Ciufolini I (2000) The 1995-99 Measurements of the Lense-Thirring Effect with Using Laser-ranged Satellites. Class. Quantum Grav. 17, p. 2369

Damour T, Polyakov AM (1996) String Theory and Gravity, Gen. Rel. Grav. 12, p. 1171

Danzmann C, Rüdeger A (2003) LISA Technology – Concept, Status, Prospects. Class. Quantum Grav. 20, p. 1

Dittus H, Lämmerzahl C, Selig H (2004) Testing the Universality of Free Fall for Charged Particles in Space, Gen.Rel.Grav.

Dohm V, Haussmann R (1994) The Superfluid Transition of ^{4}He in the Presence of a Heat Current: Renormalization-group Theory Physica B 197, p. 215

Everitt CWF, Buchman S, DeBra DB, Keiser GM, Lockhart JM, Muhlfelder B, Parkinson BW, Turneaure JP, and other members of the Gravity Probe B team (2001) Gravity Probe B: Countdown to Launch. In: Lämmerzahl C, Everitt CWF, Hehl FW eds: Gyroscopes, Clocks, Interferometers,...: Testing Relativistic Gravity in Space. Lecture Notes in Physics 562, Springer, Berlin, p. 52

Greywall DS, Ahlers G (1973) Second-sound Velocity and Superfluid Density in ^{4}He Under Pressure Near T_{lambda}. Phys. Rev. A, 7, p.2145

Jentsch C, Müller T, Rasel EM, Ertmer W (2004) HYPER – A Satellite Mission in Fundamental Physics Based on High Precision Atom Interferometry: Gen. Rel: Grav. 36, p. 2197

Lämmerzahl C, Ahlers G, Ashby N, Barmatz M, Biermann PL, Dittus H, Dohm V, Duncan R, Gibble K, Lipa J, Lockerbie N, Mulders N, Salomon C (2004) Experiments in Fundamental Physics Scheduled and in Development for the ISS, Gen. Rel. Grav. 36, p. 615

Lämmerzahl C, Neugebauer G (2001) The Lense-Thirring Effect: From the Basic Notions to the Observed Effects. In: Lämmerzahl C, Everitt CWF, Hehl FW eds: Gyroscopes, Clocks, Interferometers,...: Testing Relativistic Gravity in Space. Lecture Notes in Physics 562, Springer, Berlin, p. 31

Lämmerzahl C, Dittus H (2002) Fundamental Physics in Space: A Guide to Present Projects. Ann. Phys. 11, p. 95

Lämmerzahl C, Braxmaier C, Dittus H, Müller H, Peters A, Schiller S (2002) Kinematical Test Theories for Special Relativity: A Comparison, J. Mod. Phys. D, p. 282

Lämmerzahl C, Ciufolini I, Dittus H, Iorio L,Müller H, Peters A, Samain E, Scheithauer S, Schiller S (2004a) OPTIS – An Einstein Mission for Improved Tests of Special and General Relativity Gen. Rel. Grav. 36, pp 2373-2416

Lockerbie N, Mester J, Torii R, Vitale S, Worden PW (2001) STEP: A Status Report. In: Lämmerzahl C, Everitt CWF, Hehl FW eds: Gyroscopes, Clocks, Interferometers,...: Testing Relativistic Gravity in Space. Lecture Notes in Physics 562, Springer, Berlin, p. 213

Maleki L, Prestage J (2001) SpaceTime Mission: Clock Test of Relativity at Four Solar Radii. In: Lämmerzahl C, Everitt CWF, Hehl FW eds: Gyroscopes, Clocks, Interferometers,...: Testing Relativistic Gravity in Space. Lecture Notes in Physics 562, Springer, Berlin, p. 369

Nordtvedt K (2001) Lunar Laser Ranging – A Comprehensive Probe of the Post-Newtonian Long Range Interaction. In: Lämmerzahl C, Everitt CWF, Hehl FW eds: Gyroscopes, Clocks, Interferometers,...: Testing Relativistic Gravity in Space. Lecture Notes in Physics 562, Springer, Berlin, p.317

Reasenberg RD, Shapiro II, MacNeil PE, Goldstein RB, Breidenthal JC, Brenkle JP, Cain DL, Kaufmann TM, Komarek TA, Zygielbaum AI (1979) Viking Relativity Experiment: Verification, of Signal Retardation by Solar Gravity. Astrophys. J. Lett. 234, p. L219

Salomon C, Dimarcq N, Abgrall M, Clairon A, Laurent P, Lemonde P, Santarelli G, Uhrich P, Bernier LG, Busca G, Jornod A, Thomann P, Samain E, Wolf P, Gonzalez F, Guillemot P, Leon S, Nouel F, Sirmain C, Feltham S (2001) Cold Atoms in Space and Atomic Clocks: ACES, C. R. Acad. Sci. Paris 4, p. 1313

Speake CC, Hammond GD, Trenkel C (2004) The Feasibility of Testing the Inverse Square Law of Gravitation at Newtonian Strength and at Mass Separation of 1 μm. Gen. Rel. Grav. 36, p. 503

Theil S (2002) Satellite and Test Mass Dynamics Modeling and Observation for Drag-free Satellite Control of the STEP Mission, Doctoral Thesis, Fac. of Engineering, University of Bremen

Touboul P (2001a) MICROSCOPE, Testing the Equivalence Principle in Space. Comptes Rendus de l'Acad. Sci. Série IV: Physique Astrophysique 2, p. 1271

Touboul P (2001b): Space Accelerometers: Present Status. In: Lämmerzahl C, Everitt CWF, Hehl FW eds: Gyroscopes, Clocks, Interferometers,...: Testing Relativistic Gravity in Space. Lecture Notes in Physics 562, Springer, Berlin, p. 274

Vessot RFC, Levine MV (1979) A Test of the Equivalence Principle using a Space-borne Clock. Gen. Rel. Grav. 10, p. 181

Vodel W, Dittus H, Nietzsche S, Koch H, von Zameck Glysinski J, Neubert R, Lochmann S, Mehls C (2001) High Sensitivity DC SQUID Based Position Detectors for Application in Gravitational Experiments at the Drop Tower Bremen. In: Lämmerzahl C, Everitt CWF, Hehl FW eds: Gyroscopes, Clocks, Interferometers,...: Testing Relativistic Gravity in Space. Lecture Notes in Physics 562, Springer, Berlin, p. 248

Walz J, Hänsch TW (2004) A Proposal to Measure Antimatter Gravity Using Ultracold Antihydrogen Atoms, Gen.Rel.Grav. 36, p.561

Wetterich C (2003) Probing Quintessence with Time Variation of Couplings, arXiv:hep-ph/0203266

Web References
ams: www.nasa.gov
best: www.nls.physics.ucsb.edu/best/
bigben: www.bigben.stanford.edu/sumo
chex: http://chex.stanford.edu/sue/
cnes: www.cnes.fr/activities/connaissance/physique/1index.htm
einstein: www.einstein.stanford.edu
euso: www.euso-mission.org

fundamental: www.funphysics.jpl.nasa.gov/technical
/ltcmp
interferometry: www.iqo.uni-hannover/ertmer
jdem: www.doe.gov/engine
lageos: www.galileo.crl.go.jp/ilrs/lageos.html
miste: www.miste.jpl.nasa.gov/
onera: www.onera.fr

parcs: www.boulder.nist.gov/timefreq/cesium/parcs.
htm
pharao: www.opdaf1.obspm.fr/www/pharao.html
planck: www.esa.int/science/planck
wmap: http://map.gsfc.nasa.gov/index.html

12 Materials Sciences

by Lorenz Ratke

Since the early days of Skylab and the Apollo-Soyuz mission, research on metallic materials and crystal growth has been an inherent part of any microgravity research program (Naumann 1980). This chapter is devoted to materials processing and aims to describe challenges and goals of materials science in space in a way that enables a nonspecialist to understand the objectives of materials research in space, the principles of materials engineering, and the major achievements made in various microgravity missions in the past two decades. At the end of this chapter a prospective view of research to be performed on the International Space Station (ISS) is given.

12.1
Objectives of Materials Sciences in Space

Materials of various origins accompany us in our daily life. We live in houses made from stone, sand, concrete, glass; we use porcelain, enjoy diamonds or sapphires, or use them for grinding, polishing and cutting; we use polymers like polycarbonate (CDs) and polyethylene (bags); metals are used in our cars, in trams, trains, window frames, knives, scissors and cooking pots; silicon is used in our computers, digital cameras, television; graphite is used in electrical engines and pencils; hard metals are used in drilling tools, copper is used in electrical wires and so on. Even ice can be thought of as a structural material. Billions of tons of water crystallize and melt back every year. It is used as a structural material in the northern hemisphere and as refreshing agent in southern areas.

In the past, materials characterized cultural and historical ages, such as the stone, bronze and iron ages; today's era could be called the silicon age. Progress in the production, handling, and machining of materials defined these ages of mankind, since it marked major past technological breakthroughs. Without being able to produce metals and machine them into the desired shape with the desired properties, we still would live like our ancestors ten thousand years ago. The ability to handle the various kinds of materials mentioned above and to artificially produce materials with desired properties is a prerequisite for further technological progress.

Nevertheless, typically we do not perceive the materials surrounding us nor do we think about the complicated processes required to make them or the details of physics and chemistry underlying the process steps from production to final product. An example is the processor inside a computer, which is going to determine and change dramatically our lifestyle: who notes that approximately 50 years of technological development in crystal growth is behind a silicon chip, during which thousands of researchers investigated the details of the crystal growth of silicon, producing crystals of ever increasing size, free of defects that could disturb their utilization as a processor chip?

Materials are produced by various routines or processes from the molten or liquid state: metals and alloys are cast, semiconductors like silicon are grown from the melt or by deposition from the vapor phase; porcelain or glass is made by mixing various oxides in suitable proportions, heating them such that at least a part melts. Polymers are often made from solutions of monomeric molecules. A fluid phase, which can be a gas or a liquid, is most often a necessary step in the sequence from raw materials to final product (casting, welding, spraying, sintering, etc.). Fluids are, however, strongly affected by gravity. It is such a part of our daily experience and so strongly anchored into our thinking that we do not really notice it unless gravity is absent, as in space laboratories, and we are faced with problems never experienced on earth for hundreds of thousands of years.

Gravity induces in fluids a pressure called hydrostatic pressure in all kinds of water engineering, metallostatic pressure in metals processing, barometric pressure in atmospheric physics (weather forecast). This pressure helps, for instance, to empty all types of containers upon opening them (water taps). Density differences between fluids (oil with vinegar) or mixtures of fluids with solid materials (water with sand) lead to the well-known effect of sedimentation. The heavier material sinks to the bottom, the lighter to the top. Hot air rises, cold air sinks. Without this effect, no oven would work, no heating of rooms. The same holds for all fluids and occurs in every process having a hot liquid phase, like a metallic melt or a glass. These simple statements neglect an important fact: the rising of hot air induces a flow acting on the whole fluid. Circulation heats up the whole room, not just the air above the heater. Fluid flow lines are always closed loops. Fluid flow induced by gravity is of utmost importance in nearly all technological processes where materials are fabricated. Flow or convection transports not only heat, but also species or mass.

Preparing and designing materials is, however, not just controlling fluid flow. During cooling of a liquid below its freezing point, an important process occurs: crystallization. Whereas a liquid can be thought of as a material in which all atoms or molecules are in random motion at random places, they rearrange themselves during crystallization into regular positions forming a lattice, as shown in figure 12-1. It is the same difference as between the spatial arrangement of people in a rock concert in New York City's Central Park and a military parade on the Champs-Elysees on July 14th. How do atoms or molecules know where their place is? How does this arrangement occur on the microscopic scale of atoms? Is there any way to influence the crystal arrangement? Are the atoms perfectly arranged on the lattice places? If there are defects, how do they affect the crystal's properties? If there are impurities (foreign atoms), where are they built in? Do they change properties? If there is a mixture of molecules like in glass or metallic alloys, what microstructures evolve on crystallization? Gravity plays an important role in crystallization or solidification, since flow or convection induced in the liquid will change the heat and mass transport (stirring sugar in a cup of hot tea accelerates its dissolution).

Looking microscopically into a material it turns out that things are even more complicated. Not only do crystals appear on cooling, but they form structures at a scale that can be revealed in a light microscope (see figure 12-2). This is especially the case in all kinds of metals and their alloys. It also turns out that these microstructures determine essential properties.

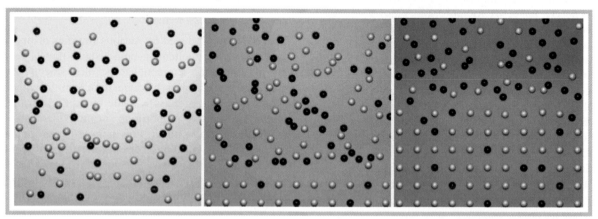

Figure 12-1: Crystallization of a two component melt (black and white atoms, binary alloy). The left picture shows a snapshot of the atomic arrangement in the melt. In the picture in the middle two rows of atoms have already crystallized into a regular lattice. In the right picture five lattice planes are finished. Notice that the melt above the crystal contains more black atoms than does the solid. This enrichment of the second component in the melt is typical for metallic alloys. The pictures are taken from an animation developed by ACCESS eV, Aachen

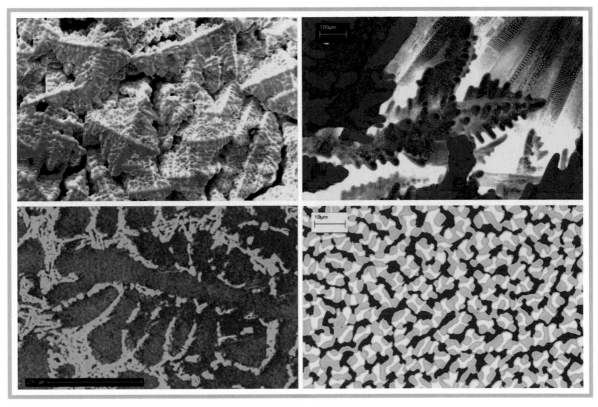

Figure 12-2: Microstructures observable in the interior of metallic alloys. The top left picture shows dendrites in an Al-alloy. Dendrites look like trees. The picture at the top right also shows a dendrite structure, but here in a ternary alloy of Al, Cu, and Ag. This yields not only a tree like shape of the crystals; they have in addition a finer interior structure of two different phases or crystals. The bottom left picture shows a dendrite in a conventional Al-cast alloy (AlSiMg). The blue dendrites are imbedded in a eutectic matrix of Al with Si platelets. The red and yellow inclusions are intermetallic phases. They appear due to impurities in technical alloys, like iron and manganese. The bottom right picture shows a ternary eutectic of AlCuAg. Directly from the melt three different types of crystals appear at the same temperature growing cooperatively

Materials science is simply the science of understanding the relation between the processing conditions during preparation of a material, the microstructure evolving, and the properties attained.

Classically, the whole area of engineering science (mechanical, iron and steel metallurgy, foundry engineering, metal physics, glass and ceramics engineering, mineralogy, crystallography) was and is devoted to gaining experience and understanding in the preparation, handling, machining, and designing of materials, and to developing models to understand the processes and the properties. Within the past hundred years much progress has been made in understanding materials from the atomic scale to the

scale of a product which can be measured in meters or kilometers (iceberg). At least empirical or phenomenological models are available that can be combined with modern computer simulations of heat and mass transport. The results of these combinations are often intriguing and give the impression that one can model the complex processes from raw material to final product already in a computer (see figures 12-3 and 12-4).

Looking into the details it becomes obvious, however, that even with the computers which will be available in 20 years, it will be impossible to design a material from the atomic scale to the meter scale, since most of the processes important in the trans-

Figure 12-3: Numerical simulation of the cooling process of a motor block casting. The blue areas show cold regions, the yellow areas are very hot regions. Numerical modeling of heat transport in complexly shaped castings is today a standard in foundry shops. This type of simulation is a pre-requisite for further modeling of the microstructure developing during cooling in the different locations of the part cast (Magma Gießereitechnologie GmbH, Aachen)

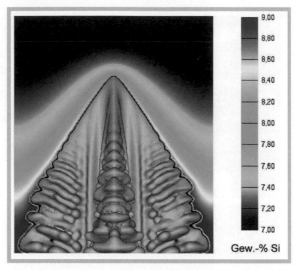

Figure 12-4: Three dimensional modeling of the growth of a single dendrite in an AlSi alloy. A nucleus was set at the bottom center and then the temperature was continuously decreased. The dendrite grew first as paraboloid of revolution (cigar shape) and later developed side branches (secondary arms). Around the dendrite, which mainly is Al, silicon enriches in the melt surrounding the growing dendrite. This kind of modeling is nowadays performed with a so-called phase field approach (ACCESS eV, Aachen)

formation from a liquid to a solid are not understood in sufficient detail to allow the set-up of suitable physical and mathematical models. Especially the connections between the processing and the microstructure of a material, and between the microstructure and the properties, are far from being understood.

The challenge of materials research in space is that only the space environment, the absence of gravity, allows the study of crystallization or solidification of fluids not disturbed by gravity induced fluid flow or, alternatively, allows the study of the effect of flow under control of the scientist, not disturbed by "natural" or "gravity dependent" convection.

But even if full understanding of all processes occurring on the scale of atoms or molecules would be available, on the mesoscopic scale of the microstructure and on the macroscale of the product, and even if we would be able to set up computer models that can connect all scales and simulate the fabrication

process and evolution inside the material we wish to prepare, we would still need data on the thermophysical properties of the materials involved.

Thermophysical properties are, for instance, the heat conductivity, the latent heat of melting, the viscosity, the heat capacity, the surface tension and the electrical conductivity. How do we obtain reliable, accurate data on these properties? Without them the best computer model cannot predict anything. Unfortunately, certain thermophysical properties are not available in the earth laboratory with sufficient precision. Measuring, for instance, the viscosity of a melt in different laboratories with different experimental devices led to values scattering by a factor of ten, although all laboratories performed the experiments to the best of their engineering knowledge. The reason is gravity or gravity induced fluid flow, some sort of natural convection that cannot be avoided on earth, even with the most sophisticated set-up. The only way to determine these properties

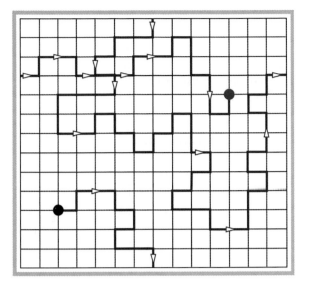

Figure 12-5: Diffusion of species can be modeled as a random walk. Suppose someone starts out from a bar in New York City (black dot) and wishes to walk to his hotel (red dot). At each crossing he throws dice, which gives him four choices of the way to go. Will he ever reach the hotel? What is the average distance he moves in time t? It turns out that the average distance he moves is proportional to the square root of time. There is no guarantee that he reaches his hotel in a finite time. In real diffusion in liquids and solids there is another complication: if an atom would like to move to a particular place, this place must be empty. If its occupied by another atom no movement is possible, although the thermal energy to jump to the next crossing would be available. Diffusion in materials depends, therefore, on the number of vacant sites, which is controlled by the temperature, but also by all the defects a microstructure has obtained during processing

with high precision is in space. The same holds not only for viscosity, but also for transport of mass or species by diffusion (see figures 12-5 and 12-6). If there is any kind of even the smallest fluid flow acting, the transport of atoms by diffusion is disturbed in an unpredictable way. Ever since the early days of experimentation in space (Skylab), diffusion has been one of the evergreens in microgravity research. The microgravity research performed in this area has led to a complete revival of investigations on diffusion in liquids and it became clear in the past few decades that more is unknown than understood.

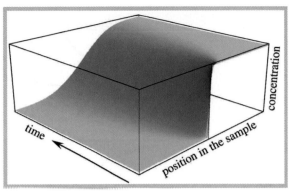

Figure 12-6: Assume that initially a sample is made of two different materials (A and B) joined at one face. Then heat the sample to a high temperature and wait. Measuring after a certain time of anneal the concentration of B atoms along the sample length, B atoms have moved from the right side to the left. The longer the anneal takes, the more B has moved to the other material. This is the effect of diffusion: diminishing of concentration differences

12.2
Principles of Materials Engineering

Modeling of materials processing from the liquid state always starts with the thermodynamics of the system of interest and tries to elaborate what type of phase transformations occur in the system if it would always be at equilibrium, a state that usually is not achieved in nature and generally not preferred in materials processing. Thermodynamics describes the energetics of a system as it depends on pressure, temperature and concentration of species (Hillert 1998). Consider a single component liquid, for instance pure water or an aluminum melt. On heating such a liquid the amount of liquid that vaporizes increases until, at a well defined temperature, the liquid cannot exist anymore and is completely converted into the gaseous phase. For water this so-called boiling point is at 100°C (at normal pressure), for aluminum at 2220°C. On cooling a liquid the amount of thermal energy in the system decreases, the vibration amplitudes of atoms become smaller and at a well defined temperature, the melting point, it crystallizes. For water it is 0°C, for aluminum 660°C. Below the melting point the solid phase is

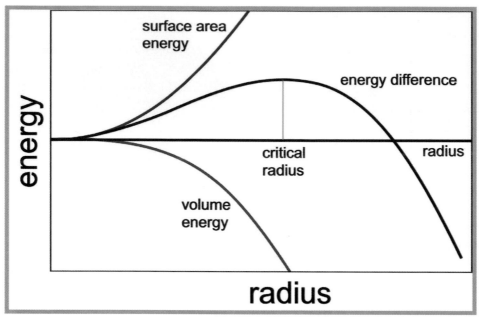

surface area
energy

energy difference

energy

critical
radius

radius

volume
energy

radius

Figure 12-7: The nucleation of a solid from a liquid requires interfacial energy which grows proportionally to the surface area. The solid has a lower energy than the liquid and thus on forming a nucleus energy is gained proportional to the volume (counted negative). Adding both shows that below a certain size the energy spent to build the surface is higher then the energy gained. Above the critical size nuclei can grow

the equilibrium phase. On boiling as well as on melting an additional amount of heat is necessary for the conversion of a solid to the liquid or a liquid to the vapor state (latent heat of melting or vaporization). This energy is gained back in the reverse process. These phase transitions can be expressed in terms of energy. The stable phase at given external conditions always is that state having the lower energy.

The transformation from the liquid to the solid state does not occur exactly at the transformation temperature. Although thermodynamics says that below the melting point the solid state is the stable state of the material, the appearance of the solid state needs additional energy, which is provided by undercooling below the equilibrium melting temperature. Multiplying the undercooling with the entropy yields an energy. The larger the undercooling the more energy is contained in the undercooled melt. At a certain undercooling, nuclei of the solid phase can form inside the melt (homogeneous nucleation), but mostly they will appear at container walls (heterogeneous nucleation), since the energy to form them there or at foreign particles inside the melt is smaller (Hillert 1998, Hurle 1994, Kurz and Fisher 1989, Porter and Easterling 1992). The extra energy is needed to create the interface between the solid and the liquid. Any interface between different phases or

materials contains energy. Therefore at the melting point there is no energy in the system that could be spent to establish the solid-liquid interface. The undercooling provides this energy.

Once a nucleus has formed it can grow, if the gain in energy per volume is larger than the interface energy needed to increase the nucleus size (see figure 12-7). The statistical nature of nucleation processes was discovered and models developed for it in the 1930s. Despite thousands of scientific papers on the mathematical modeling of nucleation, the process of nucleation is not really understood, since the models developed so far are difficult to validate against experimental results, at least in metals and their alloys. In condensation of drops from a vapor (mist, fog, rain) the understanding and modeling of nucleation is much more advanced. The essential difference to metallic materials or to the liquid-solid transformation is the different nature of metallic binding in the molten and the solid state compared to binding forces in water. Quantum mechanical effects can be important and are considered nowadays to calculate the interfacial energy and the gain in energy during growth. Molecular dynamics are used to directly calculate the atomic arrangements of the atoms at the interface between solid and liquid and the interfacial energy.

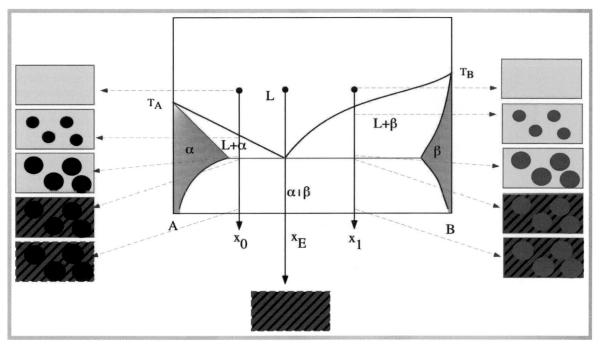

Figure 12-8: The diagram in the center shows a phase diagram of a binary alloy. It depicts the existence range of various phases. Above the blue "liquidus" line the alloy is liquid, independent of the composition. Below the horizontal red line two solid phases co-exist. In the area bounded on the left by the straight blue and red "solidus" line, liquid L and solid α coexist. In the area bounded to the right by the curved blue line and the horizontal line, a liquid L and solid β co-exist. Cooling of an alloy with two components A and B (for example Ag and Cu) leads to different microstructures depending on the composition. An alloy of the "eutectic" composition x_E crystallizes at one temperature simultaneously into two solid phases (α and β). A liquid of composition x_0 nucleates below the blue liquidus line solid α, in an amount increasing on cooling (black circles). Once the temperature of the red line is reached, the remaining liquid between the solid α phase transforms as the eutectic

The simple statement "on cooling" made above has to be refined. Cooling means that heat is extracted from the melt or liquid. In typical castings or crystal growth facilities the melt is within a suitable container. The heat is extracted by transport of heat from the melt volume towards the container walls or generally towards the surface of the melt. Heat may be transported in the volume either by diffusion or by convection. Cooling through the surface inevitably means that the surface has a lower temperature than the interior. Thus temperature gradients are established in the melt. This means that there are regions of different densities leading to natural convection. In a gravity field, fluid flow enhances the heat extraction but also tends to eliminate the temperature differences (Kurz and Fisher 1989, Flemings 1974, Khou 1996, Hurle 1994).

Besides heat transport by diffusion or natural convection, there are other kinds of convection, also known from daily life. Take a thin liquid layer in, for example, a cooking pan (margarine) and heat it from below. Heat enters the liquid from the bottom and a certain amount is extracted at the top surface by the surrounding air. Thus a temperature difference between top and bottom establishes. This gives rise to a special convection pattern if the temperature difference exceeds a special critical value, the so called Rayleigh or Rayleigh-Benard pattern, often having a honeycomb structure.

However, the surface of a melt is, like in weld pools, prone to another type of heat transport, essentially discovered and becoming a topic of fluid flow research through intensive experimentation in microgravity in the last decades. If the surface of the melt

Figure 12-9: Attachment places of atoms at a flat crystal interface. At various positions the gain in energy differs. The red place has a gain of three bonds, the blue one two bonds. In the volume each atom has six bonds. The green square indicates that at a flat surface a new layer can be introduced by nucleation on the surface. Growth instead occurs by attachment to steps (blue and red places). Growth is much more rapid than nucleation

pool does not have the same temperature everywhere, but temperature differences along the surface, the surface tension induces a fluid flow called Marangoni or thermocapillary convection. Origin of this flow is the temperature dependence of interfacial tension. Typically, the interfacial tension decreases with increasing temperature. The fluid reacts on surface tension gradients with a flow aimed at reducing the tension gradient. Especially in welding this effect is of dramatic importance and was not taken into account in the years before microgravity research on soldering, brazing and welding revealed its importance (Khou 1996, Porter and Easterling 1992).

In systems with more than a single component the situation is more complicated. For metallic melts, where the pressure is not an important variable (in contrast to many organic and inorganic liquids), one can express the areas of stability in a so-called phase diagram, with temperature and concentration of a second species as coordinates (Hillert 1998). In certain regions a single phase is stable, but in other regions two phases coexist in equilibrium. In addition, these diagrams always show that generally the solid state dissolves a lower amount of second species than does the liquid state (the gaseous state would dissolve all species in arbitrary amounts). The existence of two or multiphase regions and the dif-

ference in solubility between solid and liquid state is of utmost importance in alloy solidification, since this is the basis of the large variety of microstructures that evolve from the molten state in passing such an area on cooling.

In order to understand the intricate nature of the liquid to solid phase transformation of alloys, consider an AlSi alloy of, e.g., seven weight percent Si. If a melt of this alloy is cooled from approximately 700°C it passes the liquidus line at about 620°C. Below this temperature a solid of Al with less than 1 wt.% Si coexists with an AlSi melt which enriches in Si inasmuch as the temperature decreases. Depending on the cooling rate, the melt enriches Si up to amounts of 12.6 wt.%. At that temperature the melt decomposes simultaneously into silicon platelets embedded into an Al-rich matrix. This special structure is called a eutectic structure and quite common (Kurz and Fisher 1989, Flemings 1974, Kassner 1996, Hurle 1994).

Whereas in the case of a single component melt essentially heat transport determines the crystallization, in multicomponent melts mass transport is generally more important than heat transport, since it is typically slower by a factor of 100 to 1000 in metallic alloys. The slowest step in any kind of phase transformation determines its speed. In a binary mixture of A and B atoms species transport is characterized under diffusive conditions achievable in microgravity by the diffusion coefficient. In three or more component mixtures several diffusion coefficients determine the flux through the material and these diffusion coefficients are not independent but influence each other (an effect well known in the solid state, but rather unexplored in liquid metals due to the overwhelming effect of gravity-induced convection, Rosenberger 1979).

Once a nucleus has formed and is able to grow the resulting morphology depends on the crystallography of the phase growing, the solid-liquid interfacial energy, the attachment kinetics of atoms at the interface and the transport of atoms through the melt. Whereas in semiconductors, especially in compounds like GaAs, the attachment kinetics at the interface are important, solidification of alloys usually is determined by the species transport through the bulk volume. The origin of this difference is the surface roughness of the solid-liquid interface. In semiconductors or compounds the interfaces are usually atomically flat, while metallic interfaces are rough. Thus an atom easily finds a place to join the solid. Transport in the bulk volume can occur by diffusion of species or convection (Tiller 1991).

The solid-liquid interfacial tension usually depends on crystallographic orientation. In cubic crystals a fourfold symmetry axis exists. Thus, starting from a spherical nucleus this sphere becomes unstable during growth and develops six needle-like spikes. During further growth each needle develops four side branches and thus gives the appearance of tree-like morphologies, the dendrites.

Although crystal growth and solidification of alloys have in common the phase transition from the liquid to the solid, there are essential differences. In crystal growth either nearly pure materials or compounds of close to stochiometric composition (GaAs, InP, Al_2O_3) are grown from the melt. "Nearly pure" in crystal growth means that well defined impurities are added (doping), for example in Si very small amounts of Sb or As are added to change the electrical conductivity (p- or n-doped Si).

Semiconductors are usually covalently bonded and brittle. This means that thermal stress originating during growth may lead to severe deformations or

Figure 12-10: Screw dislocations in a SiC crystal. These dislocations lead to a spiral growth pattern. On the terraces of the spiral smaller steps are visible. The big and small steps are the preferred attachment points of atoms during growth

even cracks in the crystal. Deformation leads to dislocations just behind the solidification front, where the crystal still is at a high temperature and the flow stress is low. Dislocations have harmful effects on the physical properties of semiconductors. A high density of them may make the whole crystal useless for any application (see figure 12-10). In contrast to solidification, crystal growth does commonly not start with a melt, but single crystalline seeds are used (and thus nucleation is avoided) onto which atoms from the melt may attach and grow during controlled cooling. Large single crystals can be grown by carefully controlling the cooling process (Hurle 1994).

Whereas in alloy solidification polycrystallinity is a method to improve the mechanical properties of a product, in crystal growth this has to be avoided, since all boundaries between the crystals (grain boundaries) act in the same way as dislocations; they are traps for electrons and holes. Therefore the techniques of crystal growth and solidification are quite different. In solidification sophisticated techniques to cast metals and alloys have been developed for a few thousand years and are still being improved. Typically a mould is made into which the liquid metal is simply poured. Moulds may consist of a metal (heat resistant steel), compacted sand bonded with organic resins, or a ceramic shell. In crystal

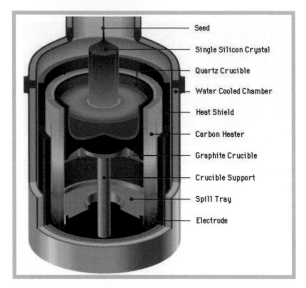

Figure 12-11: In the Czochralski method of crystal growth a single crystal is dipped into a melt and slowly withdrawn while rotating. This method is the standard to produce single crystalline semiconductors (like Si) for electronic applications

growth a classical method was developed in the early 1920s, the Czochralski method (see figure 12-11). Other techniques, like the Bridgman technique, are also used, especially in scientific investigations. This technique will also be available on the International Space Station and be used for solidification and crystal growth (see figure 12-12). Another important difference is given by the fact that in solidification and casting alloys are commonly used, meaning a base metal is chosen to which other metals or semimetals are added (alloyed) not in vanishingly small quantities as in crystal growth, but in the order of a few to a few tens of weight percent.

This difference leads to completely different microstructures, and the techniques for solidification of metals therefore differ from those for crystal growth. Assume that a cylindrical crystal is moved through a furnace (see figure 12-12) such that the solid-liquid interface is always located at the same position. Ahead of the interface the temperature profile ensures a temperature increase towards the melt and vice versa in the solid below the interface. In crystal growth the aim is to have the temperature gradient at the interface in the molten part small and the absolute temperature in the solid part as high as possible (below the melting point) such that the overall temperature difference along the whole sample is small. This helps to avoid deformation and dislocation generation during growth. In addition the growth velocities are small, so that an almost perfect crystal can establish and defects heal during growth.

In solidification processing the opposite approach has to be used. If the temperature gradient towards the liquid is small, a dendritic interface inevitably develops. High temperature gradients can avoid this. The solidification velocity also is large, ranging from tens of microns per second to m/s. The interface typically is dendritic instead of planar or flat. This is a consequence of the difference in solubility between solid and liquid (constitutional supercooling). The region between the fully liquid state and

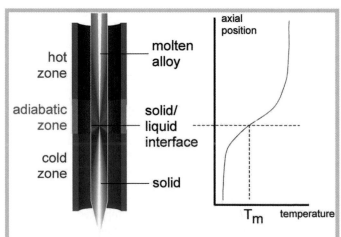

Figure 12-12: In the Bridgman method a long cylindrical sample is pulled with constant velocity through a furnace consisting of three zones. A hot zone in which the alloy is molten, a cold one acting as a heat sink (for instance a water bath) and an adiabatic zone made from isolating materials to avoid radial heat extraction. Under perfect experimental conditions the solid-liquid interface stays at a fixed position and the temperature profile is constant during the pulling process

the fully solid state is called a mushy zone (see figure 12-13).

Solidification engineering is the science and art of controlling the casting process in such a way that microstructures establish with properties tailored for applications. Mechanical properties like yield strength, fracture strength and toughness, and the ductility of a material depend strongly on details of the microstructure. The general rule is that the finer the microstructural features like primary dendrite stem, secondary arm and eutectic lamellae spacing, the higher the strength values and the ductility. The microstructure also determines the physical properties, like electrical and thermal conductivity, magnetic properties in the case of iron, cobalt or nickel based alloys, and the corrosion susceptibility.

A broad spectrum of casting technologies is available adapted to the alloy classes and the product usage (Kurz and Fisher 1989, Flemings 1974). Shape casting is done by lost mould or permanent mould techniques, the first one includes for instance sand and investment casting, the latter one chill casting. These techniques are usually only used for cast alloys. Continuous casting is a method applied for semifinished products, especially wrought alloys. The production of metallic powders is economically done by melt or gas atomization and spray compaction of semisolid droplets. Advanced castings technologies emerged within the last decade like casting of microparts, micron thick structures with extremely high cooling rates, microstructure control

during casting by rotating or alternating magnetic fields, nano- to micron-structured surfaces of castings adapted to their usage, extreme rapid solidification from deeply undercooled melts, etc. These advanced techniques are partly unexplored and partly need to be further developed to become mature enough to be transferred to industrial practice. Today all classical casting technologies are mainly handled within the cast shops by experienced engineers and workers. Since most of the foundries in Europe are small or medium size enterprises, progress in replacing traditional casting techniques with new techniques supported for instance by numerical simulation is very slow. Traditional methods and thinking predominate over advanced technologies.

The long term aim of solidification research is to design a material, its proper fabrication process, in the computer given a list of specifications or requirements for the properties of a product. This material design from the melt is the challenge for which solidification research in space is a necessity, since only there does a convection free environment offer the chance to understand the physical processes during solidification in such detail that mathematical modeling becomes feasible (figure 12-14) (Beckermann 2002).

12.3
State-of-the-Art Applications

In this section an overview of achievements made in

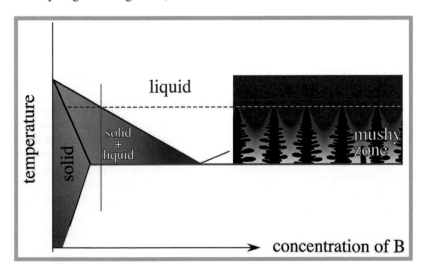

Figure 12-13: Solidification of alloys leads to the formation of a transition zone between the fully liquid and the fully solid state, called the mushy zone. The typical microstructural feature existing in this zone is an array of dendrites (black). The dendrites show a stem and branches. The stem is called a primary dendrite and the branches are called secondary arms

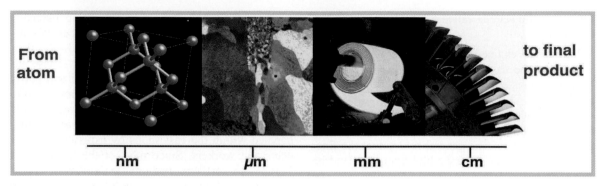

From atom

to final product

nm   µm   mm   cm

Figure 12-14: The challenge for solidification research is to understand all processes occurring in a material from the molten state on an atomic scale to the final solid state of a product on a centimeter to meter scale. Thus materials design from the melt requires intensive research in space to understand the physics of liquid metals and the physical processes of solidification not disturbed by gravity induced fluid flow

the last two decades by research under microgravity is given, where emphasis is put on a few well performed experiments that were able to answer some important question on microstructure formation in a quantitative manner. A large number of experiments on materials processing have been performed since the mid-1970s, mainly of an engineering and explorative character, since the tool of "microgravity" was brand new and a lot of experience had to be collected both on the experimental side (since experiments in space differ drastically from experiments performed in the laboratory) and scientifically, since a lot of experimental and scientific ideas turned out to be in reality much more complicated, so classical engineering approaches were not as successful as they would have been on ground. The "microgravity" environment was and is a challenge for the highest standards in experimentation and re-thinking of the experience in experimentation made in daily work in the laboratory on the ground (Seibert 2001, Feuerbacher 1984, Hurle 1989, Legros 1992, Walter 1987, Roosz and Rettenmayr 1996, Roosz et al. 2000).

12.3.1
Dendrites

The fundamental microstructural feature observed in metallic alloys are dendrites (figure 12-15). Although these structures have been observed and described in the metallurgical literature for more than 200 years, first attempts at a theoretical descrip-

tion date back just 50 year. The problem of growth morphologies developed quite recently in the 1960s and 1970s. Modern theories of dendrite growth originated in the past ten years. Theoretical investigations of the stability of a planar crystallization interface performed by Tiller and Chalmers 50 years ago showed that a planar front is unstable against perturbations if it is constitutionally supercooled (for a detailed discussion and a reprint of all fundamental papers on cells, dendrites and morphological stability, see Pelce 1988).

To understand the basic idea of "constitutional supercooling" consider a binary alloy (A with a minority component B) and let it solidify at a constant velocity. The difference in solubility between solid and liquid induces a pile-up of solute at the solid-liquid interface. Whereas the advancing solidification front rejects continuously species (B) into the liquid, the liquid tries to equilibrate the concentration by diffusion of atoms. This gives rise to an exponential decay of the solute concentration ahead of the interface (the decay length is roughly given by D/v, with D the diffusion coefficient and v the solidification velocity). The concentration profile can be converted into a virtual temperature profile indicating a position ahead of the interface where the concentration there would be in equilibrium. The real temperature is, however, determined by the heat transport, which is orders of magnitude faster than species transport. The difference between the real temperature and the virtual, equilibrium phase diagram temperature defines a region in which two

Figure 12-15: Dendrites in an aluminum-copper alloy (P.W.Voorhees, Northwestern University, Evanston)

ute ahead of a curved, wavy interface. A perturbation induces a perturbed solute or thermal field and either heat or solute (or both) have to be transported away from the tips by diffusion. The longer the wavelength of the perturbation the larger the diffusion distance to equilibrate the concentration profile. Diffusion of heat or solute would prefer a sharply curved interface. The competition between transport and curvature determines the wavelength that is selected. This Mullins-Sekerka instability leads to a first description why planar or spherical fronts are not stable during growth, and what type of microstructures can be observed. In follow-on theoretical analysis it was shown that there always is a unique sequence of microstructures in constrained growth. At a given temperature gradient a low solidification velocity will yield a planar interface, which becomes cellular at higher velocities. These cells become unstable and develop into dendrites. At even larger velocities a planar solidification front can be achieved again. The theory does, however, not explain the appearance of dendrites and the relation between their growth velocity and shape.

conditions can prevail: the real temperature always is much larger than the virtual temperature or vice versa. The first case means whenever a perturbation grows in advance of the interface, it sees a region which is hotter and thus melts back. If the virtual temperature is larger than the real one, a perturbation sees a region which is in reality colder, and thus it can grow (Kurz and Fisher 1989, Flemings 1974).

The concept of constitutional supercooling only states when a front becomes unstable; it does not predict the most stable growth mode, its shape, etc. This problem was solved later by Mullins and Sekerka, who introduced another important aspect: the curvature of a perturbation (Pelce 1988). The melting point of a planar front differs from that of a curved interface. Convex interfaces have a lower melting point than concave shaped ones. This means, if a perturbation is highly curved, it melts back easier than would a shallow curved interface. Capillarity therefore tries to avoid sharp and highly curved interfaces. This is counteracted by the diffusion, or more generally the transport, of heat or sol-

Here one has to notice that there are two different kinds of growth conditions, so-called free growth and constrained growth. The above mentioned situation is one of constrained growth; the heat is withdrawn from the sample via the already solidified material. In free growth the melt is colder than the crystal and the heat is withdrawn via the liquid. This is for instance the case for ice crystals (snow) or in the central region of cast ingots. Constrained growth is usually achieved in directional solidification using for instance a Bridgman-type furnace.

In microgravity research in the last decades free dendritic growth was investigated as well as constrained growth. From the large number of successful experimental investigations a few important results will be described briefly.

A number of theories of free dendritic crystal growth based on various transport mechanisms, physical assumptions, and mathematical approximations have been developed over the last forty years. These theories attempt to predict a dendrite's tip velocity, V, and radius of curvature, R, as a function of the supercooling, ΔT. The growth of dendrites in pure melts is known to be controlled by the transport of latent heat from the moving crystal-melt interface as it advances into its supercooled melt. Ivantsov, in 1947, provided the first mathematical solution to the dendritic heat conduction problem, and modeled the steady-state dendrite as a paraboloidal body of revolution, growing at a constant velocity, V. The original paper of Ivantsov can be found for instance in (Pelce 1988). The temperature field solution is known as the Ivantsov, or "diffusion-limited" transport solution. This solution is, however, incomplete insofar as it only specifies the dendritic tip growth Peclet number, $Pe=VR/2\kappa$, (here Pe is the growth Peclet number, and κ is the thermal diffusivity of the molten phase) as a function of the initial supercooling, and not the unique dynamic operating state, V and R, which is observed in experimental investigations. The Peclet number obtained from the Ivantsov solution for each supercooling yields instead an infinite range of V and R values that satisfy the diffusion-limited solution at that particular value of ΔT. In the early 70's, succinonitrile (SCN), an organic plastic crystal, was developed as a model metal analog system for studying dendritic growth. SCN solidifies like the cubic metals, i.e., with an atomically "rough" solid-liquid interface, yet retains advantages because SCN displays convenient properties for solidification experiments, such as a low melting temperature, optical transparency, and accurate characterization of its thermophysical properties. The use of SCN greatly facilitated dendritic

Figure 12-16: Dendrite of succinonitrile grown under microgravity conditions in the Isothermal Dendritic Growth Experiment (IDGE) facility (Glicksman et al. 1994)

growth studies over the past thirty years, where because of its use, dendritic tip velocities could be accurately measured and used as a critical test of theory.

Theoretical efforts have concentrated in the last two decades on trying to discover an additional equation or length scale which, when combined with the Ivantsov conduction solution, "selects" the observed operating states. Although the underlying physical mechanisms for these "theories of the second length scale" are quite different, their results are invariably expressed through a scaling constant, $\alpha^* = 2\kappa d_0/(VR^2)$, where d_0 is the capillary length scale, a materials parameter defined from the equilibrium temperature of the crystal-melt interface, the solid-liquid interface energy, and the specific and latent heats. Although some theories predict a numerical value, in practice the scaling constant is used as an adjustable parameter to describe dendritic growth data in various materials.

Numerous experiments with SCN performed in laboratories worldwide showed that gravity-induced convection dominates dendritic growth in the lower supercooling range typical of metal alloy castings. Convection, unfortunately, confounds any straightforward analysis of dendritic solidification based on conductive heat transfer. This also led to the situation that modern theories of dendritic crystallization could not be compared quantitatively with experimental results. There have been attempts to estimate the effects of natural or forced convection on dendritic growth, but these calculations need to be compared with experiments in which convection can be controlled with respect to the flow pattern and velocity field. The experimental situation prior to the microgravity experiments performed by Glicksman and co-workers was that there appeared to be too narrow a range of supercooling in any crystal-melt system studied terrestrially that remains both free of convection effects and permits an accurate determination of the dendrite tip radius of curvature.

Research under microgravity conditions was the only chance to establish conditions that allowed for a quantitative comparison between theoretical predictions and experiments. Such experiments were performed by Glicksman and co-workers in the Isothermal Dendritic Growth Experiment (IDGE)

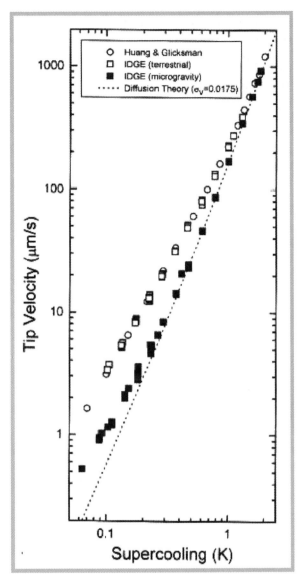

Figure 12-17: Dendrite tip velocity of SCN as measured under terrestrial conditions and under microgravity. The difference shows clearly the effect of fluid flow in earthbound experimentation, and comparison with the theoretical prediction of pure diffusive transport of heat shows that at least for supercooling larger than 0.3 K the agreement is satisfactory. The differences at lower supercooling could be attributed to finite size effects (the extension of the temperature field becomes comparable to that of the sample container) (Glicksman et al. 1994)

facility on the space shuttle using SCN and another organic material pivalic acid (PVA) (Ratke et al.

1996). Extremely carefully performed experiments allowed for the first time the results of growth velocity, tip undercooling and Peclet number to be compared quantitatively with theoretical predictions (figures 12-16 and 12-17). One result of the investigations is that the theory of microscopic solvability of dendritic growth is now widely accepted as the theory to describe free dendritic growth. Meanwhile, theoretical investigations using especially phase field methods include fluid flow to study its effects, compare it with analytical theories on the influence of flow (figure 12-18), and prepare new experiments on the ground and in space.

12.3.2
Columnar to Equiaxed Transition

In constrained growth of alloys a large number of experiments have been performed in the last decades using single phase alloys like AlSi, AlCu, AlNi, CuMn, steels, Ni-base superalloys and transparent alloys like the SCN described above. Investigations have been performed to study the transition from planar to a cellular and from cellular to a dendritic microstructure, trying to clarify the effect of fluid flow prevailing on earth compared to diffusional conditions in space. One beautiful investigative series was performed by the groups of Billia, Camel, Dupouy and co-workers. They studied dilute AlNi and AlCu alloys and investigated under what conditions columnar dendrites appear and how these can be changed to equiaxed dendrites.

In industrial practice equiaxed grain structures are induced and forced by the addition of so-called grain refiners, which are suitable intermetallic compounds like TiB_2. They promote dendritic growth at their interface to the melt. If a suitable amount of grain refiners is added to a melt and dispersed evenly, a cast ingot may develop which is polycrystalline and has a large number of dendritic grains evenly distributed in the ingot. The situation near each grain refiner is comparable to free dendritic growth except that the artificial refiner particle acts as a center of nucleation. In castings, however, the heat is always extracted via the mould and thus via the already solidified material. Therefore, typically a columnar growth mode is first observed. Only in the center of a big ingot can equiaxed globular structures es-

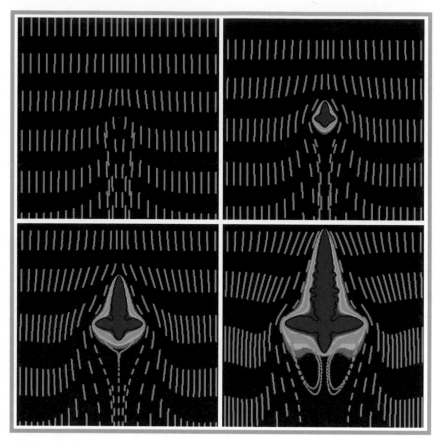

Figure 12-18: Free growth of a dendrite when it is imposed on a fluid flow coming from above. Starting from a sphere placed in the center of the computation cell, a dendrite develops in the plane with four branches. The flow accelerates growth in the flow direction since the concentration profile is steepened by it (large gradients mean large transport currents and large velocities). Note the swirl flow behind the dendrite. (Tong and Beckermann, University of Iowa)

tablish. The transition from columnar to equiaxed is of tremendous importance for industrial castings, since typically a fine globular microstructure has better and more uniform properties compared to a columnar structure. There are numerous models for this transition in the literature. Probably an effect called fragmentation of dendrites plays an important role. Fragmentation means that parts of a dendrite break away and the fragments act as centers for further growth. Such fragments could be, for instance, the cause of the globular zone in big ingots. Fluid flow is important in the fragmentation process itself and decisive in the distribution of the fragments in the melt. Since typically the dendrites have a density different from the parent melt they either rise to the top or settle to the bottom. Their spatial distribution, possible re-establishment of a network with large open space, will affect the solidification process and the microstructure evolving. Instead of fragments emerging from the columnar dendritic

network, grain refiners act in a similar way. The details of fragmentation and the columnar-to-equiaxed transition are not sufficiently understood.

The influence of natural convection on the columnar-to-equiaxed transition during directional solidification was analyzed by the group of Camel and co-workers using a refined Al-3.5%wt. Ni alloy. A series of comparative ground and microgravity experiments was performed in order to discriminate between the relative influence of solidification rate and natural convection on the columnar-to-equiaxed transition during the solidification: experiments AGHF 6 in STS-95 (1998), and TITUS 1 and 2 in MIR-PERSEUS (1999). After partial melting, several successive solidification steps with different solidification velocities were applied and the resultant microstructures analyzed in two and three dimensions.

Figure 12-19: Transverse section of a refined Al-3.5wt% Ni solidified at V = 8.2 μm/s in microgravity (TITUS 2). Successive observations of the same region by conventional optical microscopy, SEM, and polarized light microscopy after anodic oxidation of the surface (M.D. Dupouy, EPM Grenoble)

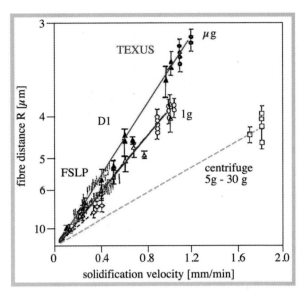

Figure 12-20: Fiber spacing in the eutectic InSb-NiSb. The results under microgravity, 1g and hyper-g conditions clearly show that there is an effect of natural convection on the structural arrangement of the fibers

Under microgravity conditions, a continuous microstructural transition is observed for decreasing velocities where crystal size increases and the morphology becomes more dendritic with branches growing preferentially in the direction of the temperature gradient. But the grain structure remains equiaxed without any selection mechanism. On the ground, the transition from columnar to equiaxed structures is observed under the same conditions. It could be determined that solutal convection and settling of refining particles and the growing crystals themselves induce that transition. For instance, free floating equiaxed crystals settle and are thus overgrown by the columnar grains.

12.3.3
Eutectics

The solidification of all multicomponent alloys usually ends at the eutectic. In a eutectic reaction a melt decomposes simultaneously in two or more solid phases at a given temperature. In a binary system of A and B the two solid phases consist of an A-rich phase α and a B-rich phase β (see figure 12-8). The formation of these phases often leads to lamellar or fibrous microstructures, meaning the two solid phases alternate as thin slabs, or one solid phase is a cylindrical fiber surrounded by the other solid phase. Both structures are well described by a theory of

Jackson and Hunt developed in the late 1960s. For a very detailed description see the book by Kassner (Kassner 1996). For an overview see the article by Liu and Hunt in (Hurle 1994) and for an introduction see (Kurz and Fisher 1989, Flemings 1974). Both structures are a result of the coupled growth of the two phases at the solid-liquid interface. The A-rich phases rejects B atoms, the B-rich phase rejects A atoms. By diffusion parallel to the interface these pile-ups of solute atoms are diminished and an oscillatory concentration profile is established. The competition between interface energy necessary to establish the two-phase solid towards the single phase liquid interface and the diffusion distance determines the spacing. The classical result of the Jackson and Hunt analysis of this problem gave a unique relation between fiber spacing R and growth velocity v, namely $R^2 v$=const., where the constant is determined by the thermophysical properties of the system and the phase diagram. This relation has been verified for various systems. In industrial castings there are two eutectics of extreme importance: the iron-carbon and the aluminum–silicon eutectic.

It is unclear if fluid flow would change the microstructure of eutectics, since in contrast to the above-

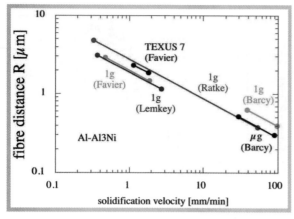

Figure 12-21: Fiber spacing in the eutectic Al-Al3Ni. The results of various authors show that there is no clear trend on the effect of solutal convection on the microstructure; even opposite effects are obtained (Alkemper and Ratke 2000)

mentioned single phase alloys, the diffusion parallel to the interface reduces drastically the extension of the concentration pile-up perpendicular to the interface. Instead of having a decay length of D/v, the decay length is only of the order of the lamellae or fiber spacing and thus considerably smaller. It is not clear if fluid flow is able to change the growth modes or the fiber spacing. Experiments performed in sounding rockets, Spacelab, drop towers and centrifuges with eutectics establishing lamellae structures like Al-Cu and fibrous structures like InSb-NiSb and AlNi did not gave a clear indication. Whereas in InSb-NiSb the results suggest a clear dependence on fluid flow, since the fiber spacing in microgravity, 1g and hyper-g conditions clearly showed remarkable differences (figure 12-20), the results on AlCu and AlNi were not conclusive. There are experiments showing a decrease in spacing, whereas other researchers found an increase with the same alloy system. This problem is still to be solved under microgravity conditions with improved facilities (figure 12-21).

In the last decade the investigation of eutectics had a revival, since new types of microstructures were observed (Kassner 1996). Fibers were observed varying the diameter along their length periodically, lamellae tilting towards the growth direction with a fixed angle, etc. The large variety of possible microstructures was analyzed theoretically, especially

with phase field methods. Although the topic of eutectics seemed to be understood since the theory of Jackson and Hunt, it turned out that this is just the starting point to a rich family of structures and that fluid flow induced by, e.g., gravity can change the local constitutional supercooling and affect the structure selection during growth. There is much to be done in the future.

12.3.4
Coarsening

Whenever a single phase system is brought into a two-phase metastable state, for example by quenching from a high temperature or changing pressure, a second phase nucleates, grows, and coarsens. Nucleation of the second phase occurs since the energy of the single phase system can be reduced by forming regions of the second phase, as explained above. Nanometer-sized precipitates of the new phase appear. Growth of nuclei then proceeds due to heat diffusion away from the nucleus into the matrix or mass diffusion from the matrix towards the nuclei. This results in a dispersion of second phase particles in a matrix. Coarsening, also called Ostwald ripening or competitive growth, then occurs and large particles grow at the expense of small particles. The process sequence nucleation, growth, and coarsening is very common, occurring in systems ranging from solid alloys to the precipitation of drops from clouds (Ratke and Voorhees 2002, Porter and Easterling 1992, Koster and Sani 1990).

Provided there is sufficient atomic mobility, a two-phase dispersion will coarsen by transfer of matter from small to large particles, thus reducing the overall free energy associated with the particle-matrix interfacial area (figure 12-22). In 1900, Ostwald reported the first systematic study noting the dependence of the solubility of small HgO particles on their radii. The phenomenon has come to bear his name. The basic theory of particle coarsening, however, was developed sixty years later by Lifshitz, Slyozov and Wagner.

Given that a two-phase mixture always possesses excess energy due to the presence of significant interfacial area, this Ostwald ripening process is observed in the late stages of nearly all phase trans-

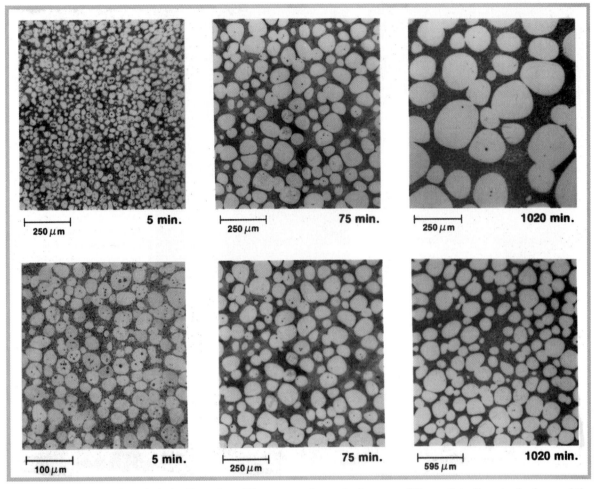

Figure 12-22: Coarsening of Pb-Sn dispersions. The top row shows the increase of the particle diameter in time and the reduction of particle number. In the bottom row the same pictures are rescaled with the average particle radius, demonstrating that Ostwald ripening is a self-similar process (after Ratke and Voorhees 2002)

formation processes. There are some technically important situations in which nucleation, growth, and coarsening play an essential role in microstructural evolution. Quenching a multicomponent alloy from a temperature in a single phase field of the phase diagram rapidly to room temperature generally produces a supersaturated matrix. If such a supersaturated solid solution alloy is then annealed at an elevated temperature, precipitates nucleate and grow by diffusion of solute from the matrix towards the precipitates.

After long annealing times the initially large supersaturation of solute has decreased sufficiently for interfacial curvature effects to play an important role: the spectrum of precipitate sizes leads to curvature-induced solubility differences in the matrix such that the dispersed two-phase system (precipitates and matrix) coarsens, i.e., the big particles grow at the expense of the smaller ones. This is the basis of all age-hardening alloys. During the nucleation and growth stages the yield strength and hardness generally increase, whereas coarsening usually leads to overaging and a degradation of the mechanical properties. The processes of nucleation, growth, and coarsening do generally not occur in a strict sequence but rather overlap.

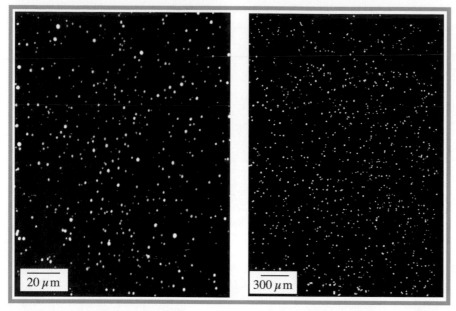

Figure 12-23: Zn and Pb exhibit a miscibility gap in the liquid state. Annealing them at high temperatures when both components are miscible and quenching them rapidly into the solid state leads to the microstructure shown in the left figure. Annealing them in the miscibility gap for 1h under reduced gravity conditions leads to a considerably coarser microstructure (right figure) as can be seen from the ruler at the bottom of both figures (Kneissl et al. 1983)

Similarly one can quench an alloy from a single phase liquid state into a two phase liquid state if the phase diagram exhibits a miscibility gap in the liquid state. Then droplets of the second phase will nucleate, grow, and coarsen as in solid solutions, but significantly faster since mass transport in liquids is orders of magnitude faster. However, liquids of different composition usually have different densities. Therefore they show a rapid spatial phase separation: the denser liquid settles to the bottom of a crucible, similar to the separation of oil and water.

Under reduced gravity conditions this problem can be avoided and the coarsening of liquid-liquid dispersions can be investigated similar to that of precipitates in solid solutions.

As explained above the solidification of alloys frequently results in the formation of solid dendrites. During growth the dendrites develop side branches and sufficiently far from the dendrite tips they exhibit coarsening due to variations in curvatures along the dendrite surface. Thus the morphology of a cast microstructure depends on growth and coarsening of

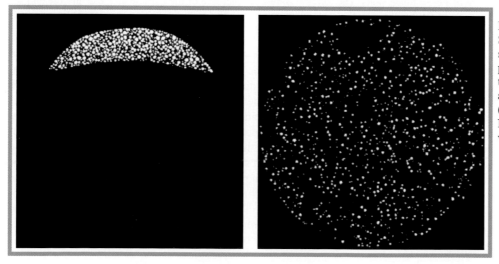

Figure 12-24: Section of Pb-Sn solid solution sample coarsened on the ground (left) and in space (right) (P.W. Voorhees, Northwestern University, Evanston)

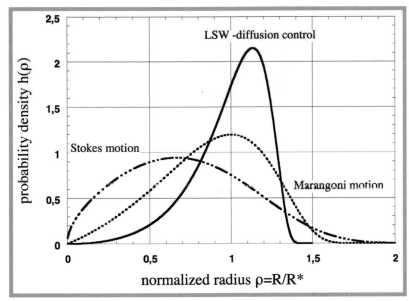

Figure 12-25: Probability distribution of particle or drop sizes in coarsening processes after very long times. If transport of solute or heat between the particles occurs by diffusion (LSW), the particle size distribution has am maximum at 1.2 times the average size and particles bigger than 1.5 should not exist. If transport of species is dominated by interfacial (Marangoni) convection, the shape of the distribution changes drastically and the largest size observable is two times the average. If drops settle by Stokes motion, the size spectrum is even more different. The particle or drop radius also changes in various ways with time, depending on the dominating transport mechanism (Ratke and Voorhees 2002)

the network of primary dendrite stems and their secondary arms. It is also typical in metallurgical processing to heat treat cast ingots in order to reduce micro- and macro-segregation of alloying elements, which are a natural result of solidification, and to achieve morphological changes of the as-cast microstructure. For example, in cast aluminum-silicon alloys the eutectic is of flake-like type. The silicon platelets act as nuclei for internal cracks; they reduce especially the fatigue strength, etc. Heat treating the as-cast alloy at high temperature, just below the eutectic temperature, leads to a morphological change. Flakes are transformed into spherical particles and grow in size due to the reduction of the total interfacial energy.

The process of Ostwald ripening is of fundamental importance to all microstructures in any kind of first order transformation. The theory of Ostwald ripening (Lifshitz-Slyozov-Wagner) predicts that a unique length scale exists in dispersions, a so-called critical radius, which, on scaling all radii in a dispersion with this length, yields at all times a unique distribution of particle sizes independent of the actual coarsening or annealing time. The Lifshitz-Slyozov-Wagner theory, however, deals with a physically unrealistic situation. The dispersion must be infinitely dilute, i.e., the volume fraction of particles or droplets dispersed should be close to zero. Also the interface tension should be isotropic and the trans-

port must occur by diffusion of solute. The theoretical assumptions are difficult to establish on earth. Firstly the volume fraction is always finite. Secondly there are inevitably density differences between

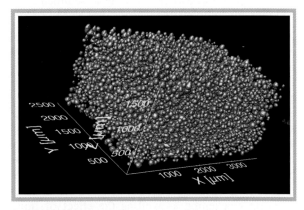

Figure 12-26: 20 vol% Sn particles in Pb-Sn matrix after 9500s of coarsening in microgravity. This picture is a three dimensional reconstruction of a space processed sample. The sample was cut micron by micron in an ultra-milling machine and immediately after cutting the surface was photographed with a light microscope. From the two dimensional serial sections this three dimensional reconstruction was calculated. It shows that even at a relatively low volume fraction of 20% the particles already stick very close together such that interparticle diffusion and correlation effects change the ripening kinetics (P.W. Voorhees, Northwestern University, Evanston, Illinois)

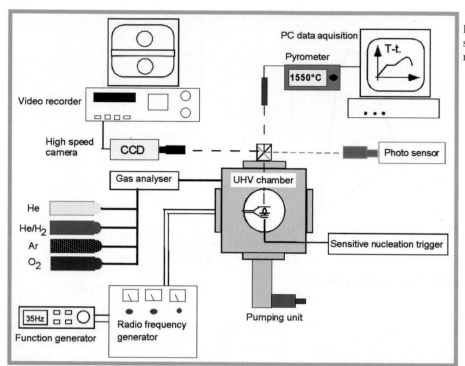

matrix liquid and particles leading to sedimentation and thus spatially inhomogeneous dispersions.

Therefore, theories have been developed by several authors in the last two decades to handle the problem of finite volume fraction and to address effects of fluid flow induced by, for instance, Stokes motion of particles or droplets (sedimentation) or Lifshitz-Slyozov-Wagner theory for diffusional transport of solute, which can only be tested in microgravity. The group of Voorhees at Northwestern University investigated carefully the coarsening in Pb-Sn dispersions on several shuttle flights. The experimental results revealed the influence of the starting conditions on coarsening as well as the effect of finite volume fraction and spatial correlations.

12.3.5
Thermophysical Properties

Precise knowledge of the thermophysical properties of high temperature melts, in particular metallic alloys, is indispensable for the prediction and simulation of industrial processes like casting, welding, spray forming or surface laser treatment. Yet reliable

data are scarce due to the experimental difficulties in obtaining them. While characterization of the solid phase with respect to properties like thermal conduc-

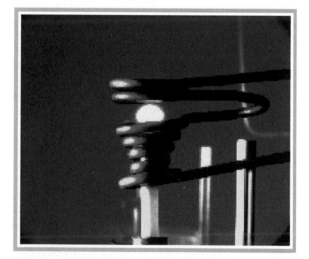

Figure 12-28: Levitated metallic drop in an earthbound EML. Notice the toroidal shape of the lower part of the coil, which helps to establish a gradient in the magnetic field, and the upper two windings of the coil, which build a loop of opposite sense with respect to the lower coil and thus induce a magnetic dipole fixing the drop in space

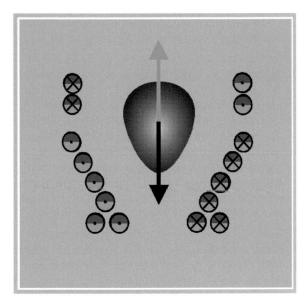

Figure 12-29: Schematic of the shape of a levitated droplet in earthbound EM

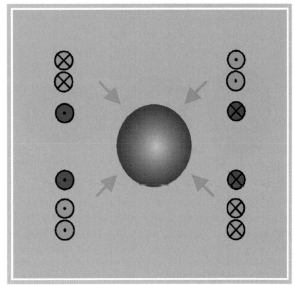

Figure 12-30: Schematic of the shape of a levitated droplet in spacebound EML. The green coils are used to fix the samples location in space. The red coil system is used to heat the sample

tivity, species diffusivity, thermal expansion, magnetic propertics, electrical conductivity, specific heat, etc. is nowadays standard and readily available, the measurement of thermophysical properties of the liquid phase is still a challenge. See the review article on experiments performed in the International Microgravity Laboratory (IML2) by the TEMPUS team in (Ratke et al. 1996).

The experimental difficulties have several origins. Experimentation with metallic melts usually means working at high temperatures up to 2000°C. At temperatures above 1000°C melts are often very reactive either with the surrounding atmosphere or with many crucible, mould or container materials. The measurement of thermophysical properties, especially transport properties like viscosity or diffusivity, can be influenced by the existence of parasitic fluid flow, especially natural or thermocapillary convection. Whereas diffusion in solids can easily be measured with high accuracy using the above mentioned diffusion couples, the diffusion in liquids requires the suppression of any kind of fluid motion, as otherwise data are obtained which yield something like an effective diffusion coefficient altered to an unknown extent by mass transfer due to convection. The same is true for viscosity measurements of liquids in which a well defined motion should exist

not disturbed by any parasitic flow outside the control of the experimenter.

Thermophysical properties of interest are properties like enthalpy, heat of fusion, heat of mixing, specific heat, the melting range, the fraction solid, the density and thermal expansion, the surface tension of the melt, and interface tension with solids like the container wall, intermetallic compounds, and the melt with its own solid, or the different phases of the solid. Other properties like thermal and electrical conductivity, viscosity and the diffusions coefficient matrix in multicomponent alloys are important for process simulation (Egry 1998, 2003, Egry et al. 1999, 2003). All the properties mentioned depend on temperature and on concentration of alloying elements.

The reactivity of molten alloys can be bypassed by utilizing containerless processing techniques like electromagnetic or electrostatic levitation (EML, ESL). Using EML in the space environment also allows one to bypass problems with parasitic fluid flow.

An electromagnetic levitator consists of a UHV vacuum chamber housing the levitation coil, a high

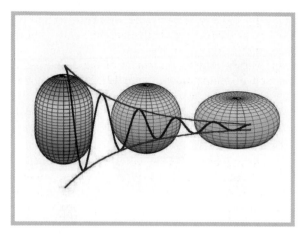

Figure 12-31: Schematic of drop oscillations with an overlay of a damped harmonic sketching the time dependence of the radius of the levitated sphere after an oscillation is induced. The amplitude decreases with time due to the internal friction. This damping yields the viscosity of the fluid

frequency power supply, video cameras for observation and measurement, pyrometric temperature measurement, a nucleation trigger device, a pumping unit, and a gas supply (figure 12-27). For semiconducting samples, a preheating device can be added or electrostatic levitation has to be used. A splat cooler and liquid metal bath can be integrated for rapid quenching of levitated samples. A sample of a few grams in weight is placed onto a sample holder and is positioned inside the levitation coil. The chamber is evacuated and subsequently backfilled with inert gas. The generator is turned on and the sample is levitated.

Levitation of a sample relies on the fact that a high-frequency current in the coil system induces an alternating magnetic field which induces in electrically conducting materials a current in the opposite direction. Manufacturing the coil such that a magnetic field gradient exists, a volumetric Lorentz force is established proportional to the field strength and its gradient. This force can be adjusted to compensate for gravity, leading to a free floating drop. Heating of the sample always accompanies levitation, since Joule heat losses of the induced current are proportional to the square of the magnetic field strength. Thus heating and positioning against gravity is always coupled on earth (figure 12-28 and 12-

29). For an overview of magnetohydrodynamics and electromagnetic levitation see the book by Davidson (2001).

In laboratory levitation facilities typically a sample is not processed in an ultra-high vacuum environment but a cooling gas is used in order to cool the sample while it is being levitated. The heat exchange of the sample with the gas atmosphere reduces its temperature, although it is still heated by the electrical current losses. In this way a molten alloy can be undercooled far below the equilibrium melting temperature since there are no container walls which induce heterogeneous nucleation of crystallization.

Solidification can be initiated at a preselected undercooling level using a suitable nucleation trigger needle. Dendrite growth velocities are obtained by means of the signals from the trigger device and photo sensor. Levitated samples can be quenched by a splat cooling device or in a liquid metal bath to freeze solidified phases for subsequent metallographic analysis. Levitated drops can also be imaged from various sides with video cameras to observe their shape, from which their volume and density can be calculated as a function of temperature. A levitated sample usually performs oscillations. These can be analyzed and the surface tension derived as well as the viscosity, because the internal friction damps the oscillations. On earth there is, however, a limitation to this technique, since the shape of the levitated sample is egg-like rather than spherical (figure 12-29, 12-30). This means firstly that there is no unique oscillation frequency but several eigenstates possible, and secondly that fluid flow inside the molten alloy leads to additional damping and thus the viscosity measured is a mixture of the true diffusional viscosity and convective damping (figure 12-31). The sample and the coil system are a free oscillation circuit. Measuring the impedance of this circuit allows extraction of the electrical conductivity. The specific heat can be measured by modulating the coil current with a suitable sinusoidal wave and observing the temperature reaction of the sample.

Performing EML in space has been done with the TEMPUS facility in the IML2 mission and the Materials Science Laboratory Missions MSL-1 and MSL-1R (Egry 1998, 2003, Egry et al. 1999, 2003).

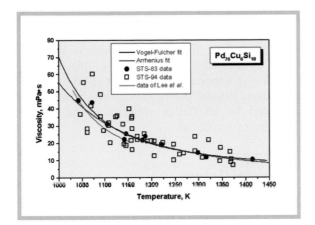

Figure 12-32: Viscosity of PdCuSi, measured in the MSL-1 R mission in microgravity. Several possible theoretical models are fitted to the data points

As stated above it solved a series of problems EML has on earth and therefore is of great interest for experimentation on the ISS. The drop shape is spherical; heating and positioning can be done independently, since there are no fields necessary to levitate the sample in space but only to fix its position. Due to this decoupling also the fluid flow in-

duced by the electric currents inside the sample is reduced to a large extent, changing the flow pattern from turbulent to slowly laminar. The spherical shape allows easy measurement of the density as a function of temperature (figure 12-33). Crystallization is not influenced by fluid flow and thus the formation of stable and metastable phases from undercooled melts can be investigated in a detail, which is not possible on earth.

In recent years an advanced version of the TEMPUS facility has been developed suitable for experimental studies in an ISS facility or for investigations in parabolic aircraft flights. Parabolic flights campaigns offer only twenty seconds of microgravity, but this can be sufficient for some thermophysical property measurements. In recent years dozens of thermophysical measurements on technical casting alloys were made for the first time in parabolic flight campaigns. These investigations are part of a European project called "Thermolab," funded by ESA and national agencies, in which many European laboratories and industrial companies collaborate to obtain from space research reliable thermophysical data for technical alloys.

As explained above, on cooling a molten alloy below its melting point a crystalline phase will appear whose lattice structure is well defined. The way this

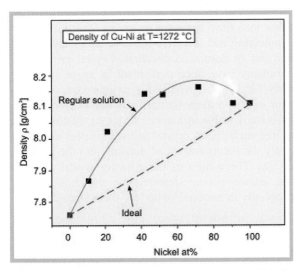

Figure 12-33: Density of Cu-Ni melts as it varies with the nickel content. The data were collected in earthbound EML processing. The experimental results clearly show that a Cu-Ni alloy is not a mixture of atoms without any interaction between the Cu and Ni atoms ("Ideal" line), but that the heat of mixing changes the density considerably with alloy content

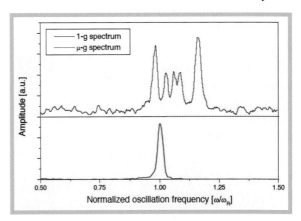

Figure 12-34: Fourier analysis of drop spectra taken on earth (top) and in space (bottom). The spectrum on earth shows five eigenfrequencies which are due to the non-spherical shape of the drop, whereas the space levitated sample shows a single eigenfrequency as expected for a perfect sphere. The eigenfrequency allows direct calculation of the surface tension

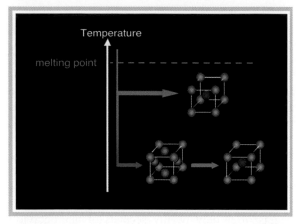

Figure 12-35: Crystallization sequence of an alloy. In equilibrium the melt should follow the blue line directly to a body-centered cubic lattice. A possible other mode of crystallization would be that first a metastable face centered cubic phase appears (red line), which later transforms into the bcc phase

structure is achieved is, however, not really fixed. Frequently metastable structures are easier to nucleate and appear first, while the stable structure develops later (figure 12-35). In solid state processing this is the basis of age-hardening of alloys. In molten alloys there are hints in the solidified microstructure that this sequence also occurs, but it is difficult to verify as the lifetime of the metastable phase may be too short, or fluid flow changes the lifetime. Per-

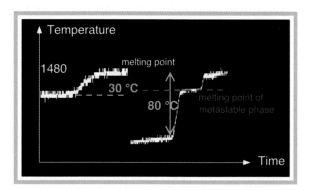

Figure 12-36: The experimental result obtained with the advanced TEMPUS facility in a parabolic flight confirms that first a metastable phase is formed, which has a melting temperature distinctly different from the equilibrium phase. The experiment also shows that under reduced flow conditions in microgravity the lifetime of the metastable phase increases

forming such experiments in microgravity could reveal that indeed after sufficient undercooling a metastable phase first appears, which is later followed by the stable phase (figure 12-36). This observation is important with respect to so-called eigenstresses induced during crystallization, which are important for all further mechanical and thermal treatment of a material.

12.3.6
Crystal Growth

The continuously increasing demands for a higher packaging density of integrated circuits require semiconducting materials with a higher homogeneity of doping elements and larger crystal size. In silicon, the most widely used semiconducting material, as well as in other industrially relevant materials like GaAs and InP dopant inhomogeneities can be found in the form of striations having a spacing of a few microns. They are caused by time dependent fluid flow in the melt. They are due to gravity dependent flows, thermocapillary flows and forced flows (crucible rotation, crystal rotation) that change heat and mass transport and induce striations (Hurle 1994).

The advantages of crystal growth experiments in space thus have the same origin as for solidification engineering and research: the absence of any kind of buoyant or natural convection. Crystal growth experiments have been performed in space since the early Skylab and Apollo-Soyuz missions. Suppression of the gravity-induced natural convection allows study of the influence of Marangoni convection at free surfaces on crystal growth kinetics and especially the incorporation of dopants into the growing crystal. The weight of the growing crystal on earth can induce plastic deformation and thus defects, especially in materials with low shear strength.

Crystal growth in space also allows the use of growth techniques that are difficult to handle on earth but may be the only suitable ones for certain semiconductor types, such as the floating zone method utilizing a mirror furnace, or the traveling heater method (figure 12-37).

The principle of this technique is quite simple: in one of the two foci of an ellipsoid a halogen lamp is placed which leads to a zone of high temperature in

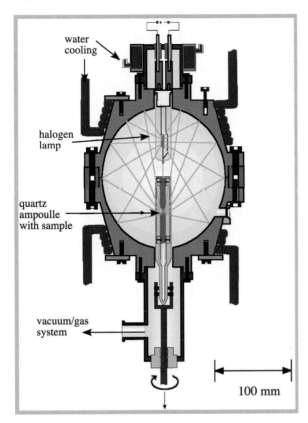

Figure 12-37: Monoellipsoidal mirror furnace ELLI developed at the Institute of Crystallography at the University of Freiburg (Germany)

the other focus. If a sample (held in a transparent crucible) is moved in one direction, the liquid zone

moves through it and a crystal grows. Most of the crystal growth experiments in space have used monoellipsoid or double ellipsoid mirror furnaces (see figure 12-37). The mirror furnaces exhibit a high temperature symmetry around the central axis such that frequently a rotation of the sample can be avoided. Defocusing the lamp makes it possible to modify the maximum temperature in the sample and to vary the temperature gradient at the solid-liquid interface. The absence of hydrostatic pressure allows liquid zones to be established with much larger diameter than on earth without the utilization of strong magnetic fields (figures 12-38 and 12-39).

From the very beginning of crystal growth in space unexpected results have been obtained (Feuerbacher 1984). This induced a complete rethinking of crystal growth processes and led to many industrial improvements indispensable for today's large single crystal growth of silicon for processor chips or solar power plants. One of the earliest achievements was made already 20 years ago, when it was clearly shown that convection induced by surface tension variations along a free surface due to temperature gradients (Marangoni or thermocapillary convection) lead to dopant inhomogeneities, called striations.

The assumption that Marangoni convection was responsible for the dopant inhomogeneities could be proven by covering the surface with a layer of silicon dioxide. Samples processed with such a cover layer did not exhibit dopant striations. This technique is used today in Czochralski growth, in which the free liquid surface is covered with a highly vis-

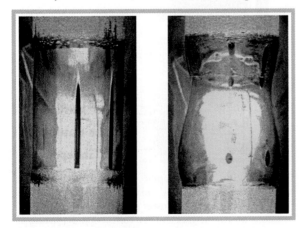

Figure 12-38: Silicon floating zone under microgravity (left, TEXUS) and on earth (right)

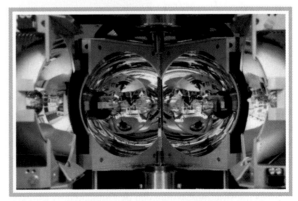

Figure 12-39: Mirrors in an ellipsoidal furnace

cous melt. The importance of Marangoni convection and fluid flow in general was not accepted until that time and not even mentioned in the crystal growth literature. Since the successes of space experiments it is meanwhile standard to analyze growth processes with respect to all kind of fluid flow and to develop methods to control the flow (for example by static or rotating magnetic fields).

The scientific objectives of crystal growth in space are to measure the thermophysical properties of molten semiconductor materials, to study the influence of fluid flow on the crystal growth, the crystal quality, the incorporation of dopants, the generation of defects and the detailed microscopic growth ki-

netics. The material studied was and is especially silicon, but interest moved in the recent decade to complex compound semiconductors like GaAs, GaSb, HgCdTe or HgCdZnTe.

A comparison between floating zone grown crystals in space and on earth yielded the following results. Dopant striations in the crystals are produced by time dependent Marangoni flows. The onset of these flows is determined by the so-called Marangoni number, which is the ratio of thermocapillary to viscous stress. If a critical value is exceeded Marangoni flows establish (figure 12-40 and 12-41). The critical Marangoni number determining the onset of these flows could be determined in space and be compared with three dimensional numerical simulations (figure 12-42). Experimentally it is important that in space the liquid zones are much larger compared to those that can be realized on earth by at least a factor of four in diameter. This allows detailed parametric studies to be performed on the onset Marangoni flows, the fluid flow structure and the fine structure of striations. Covering the surface with a quartz film suppressed the dopant striations. The natural convection on earth gives an additional flow but does not change the general flow pattern.

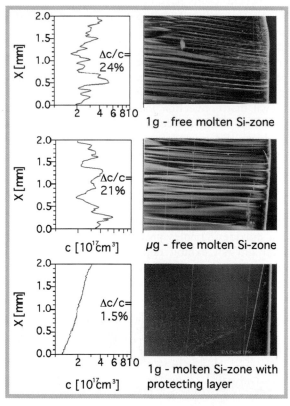

Figure 12-40: Doping striations in Si crystals grown with the floating zone method. The right side shows the striations after etching a cut through the sample. The left side shows a resistance measurement along the sample center that can be converted into a relative variation of the dopant concentration. Time dependent fluid flow (Marangoni) induces the striations, which can be eliminated by a silicon dioxide surface layer

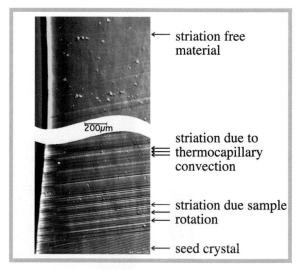

Figure 12-41: Metallographic section through a GaSb-crystal processed with the floating zone method in the Spacehab 4 mission. The lower part shows time dependent striations due to thermocapillary convection. Reducing the zone height reduces these striations since only laminar flow is left

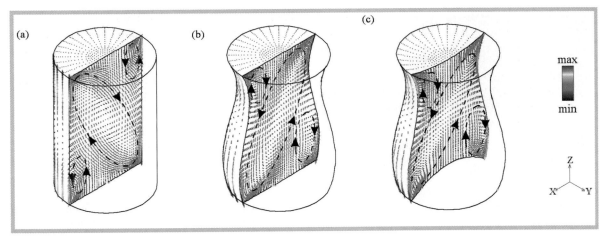

Figure 12-42: Fluid flow field as calculated for floating zone growth of silicon in the case of a flow which exceeds the critical Marangoni number such that time dependent thermocapillary convection is established. Picture (a) shows a molten zone with a flat surface and crystal-melt interface, (b) shows the typical surface shape evolving under gravity conditions and (c) adds to it a convex melt-crystal interface. The flow is a single primary roll with two counteracting rolls in the corners

Comparing 1g and μg crystals shows that the natural convection on earth adds an additional concentration variation to that induced by Marangoni convection by about a factor of two. This can be suppressed on earth by the application of high static magnetic fields (Croell et al. 1991, 1996, 1998).

Using a solution growth process in the mirror furnace called the traveling heater method, the dopant distribution in GaSb: Te, (Ga, Al) Sb and GaAs crystals grown from GaP or InP from indium solution was investigated. This technique is very similar to the floating zone method, but instead of a molten zone a molten solution is used to supply the growth front with suitable species from a solid feed material. All experiments performed with the traveling heater method in the Automatic Mirror Facility (AMF) of the European Retrievable Carrier (EURECA) exhibit only a special type of striations in contrast to 1g reference samples. Detailed analysis of the dopant distributions in the space grown samples and reference samples grown on earth could clarify the kinetics of these striations and revealed that different striation types exist. On ground so-called type I striations are dominant, whereas in microgravity type II striations exits. These cannot be observed in ground experimentation. The analysis revealed that these striations can be suppressed if the temperature gradient in the molten solution and the crystallization speed are below a critical value (Ratke and Benz 1998).

During the D2 and Spacehab 4 mission GaAs and GaSb crystals were grown with free molten zones

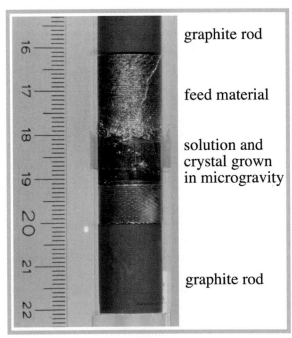

Figure 12-43: Growth ampoule of a (Ga,Al)Sb material after processing in space during the EURECA mission

325

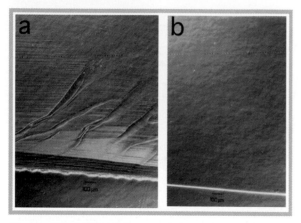

Figure 12-44: (Ga,Al)Sb crystal grown by the traveling heater method. Doping striations of type I and II are apparent in the sample grown on earth (a) and absent in the space grown crystal (b)

(floating zone method). Compared to earth grown crystals the diameter was much larger, since the instability limit of a liquid bridge on earth is different from that in space. Striations induced by time dependent Marangoni convection could be observed in all space grown crystals. The dislocation density was reduced by almost a factor of 10 and growth speeds of around 60 to 120 mm/h could be used without compromising crystal quality (figures 12-43 and 12-44) (Ratke and Benz 1998).

Within the last decade it became clear that the control of fluid flow is essential to controlling the crystal quality. Therefore, methods have been investigated to add to either buoyant or Marangoni convection artificially induced flow fields that could either counteract and probably cancel these flows or that are strong enough to dominate the flow and thus the heat and mass transport. Interesting ideas came up, like vibrating a sample by for instance ultrasonic transducers, or using alternating crucible rotations or axial vibrations with various amplitudes and frequencies. Research along this line both in crystal growth and solidification engineering is still ongoing. An alternative to mechanically induced flows are magnetic fields (figure 12-45 and 12-46). In conducting liquids an alternating magnetic field induces currents and thus, together with the magnetic field, a Lorentz force. This force in turn induces a fluid flow. Of the various possible alternat-

ing magnetic field configurations (traveling or rotating magnetic fields) the rotating field has been investigated on the ground and in space first in crystal growth facilities and is now used for solidification purposes. It will be available on the ISS in ESA's Material Science Laboratory (MSL) in both configurations of the Bridgman furnace (Kaiser et al. 1996, Davidson 2001).

Crystal growth experiments with a rotating magnetic field using the Bridgman or the floating zone method were performed in the past decade on earth and in space to change the interface curvature and the macrosegregation of dopants. Superimposing time-dependent buoyancy convection with rotating magnetic field driven forced convection leads to two different regimes that have to be distinguished. First the flow is dominated by the natural convection, showing low frequency temperature fluctuations with high amplitudes in the crystal melt. Second, the regime is governed by the magnetic field induced convection, and the temperature fluctuations are reduced by more than an order of magnitude. Typically field strength of less than 5 mT and frequencies

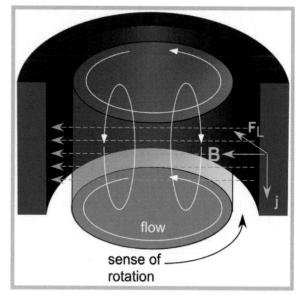

Figure 12-45: Rotating magnetic fields induce a three dimensional fluid flow, consisting of an azimuthal flow and two secondary flows in radial and axial directions. The azimuthal flow velocity depends on the field strength squared, linearly on the rotation frequency and to the fourth power on the sample radius

Figure 12-46: Experimental set-up of a rotating magnetic field device. Three pairs of Helmholtz coils are driven by a three phase current similar to an asynchronous motor

below 100 Hz are used (Croell et al. 1991, 1996, 1998).

The main results obtained so far with different types of crystals (Ge, Si, compound semiconductors) are that dopant inhomogeneities due to time-dependent buoyancy or thermocapillary convection can be reduced to a high degree by adequate fields. The interface curvature can be modulated by rotating magnetic fields; concave shapes are flattened. Radial dopant distribution can be improved in Bridgman growth as well as in floating zone growth. Compared to static magnetic fields the reduced power consumption is remarkable (the utilization of static fields requires typically a field strength of around 1T and more).

12.4 Future Research on the ISS

The International Space Station (ISS) is becoming a large space-based reality and offers unprecedented access to space conditions. All ISS partners have invested heavily in the development of this endeavor and it is necessary to perform fundamental and application oriented research in this laboratory to get a return on investment. The international materials science community is well prepared to perform excellent scientific research on the ISS (not forgetting other microgravity platforms like sounding rockets or parabolic flights). In the past five years hundreds

of research program proposals in material sciences have been submitted to the relevant space agencies, were peer evaluated by international committees, and the best of them selected for funding if industrial support for them is substantial. It is anticipated that these research programs will have a typical duration of 5-10 years. A few of the projects will be briefly described here with their perspective and horizon of research.

A research program called MICAST, MIcrostructure Formation in CASTing of Technical Alloys under Diffusive and Magnetically Controlled Convective Conditions, was initiated in 1998. It is intended to utilize the ISS with industrially relevant topics in the field of materials science, identified the high scientific and industrial importance of a detailed knowledge and a quantitative theoretical description of the influence of convective heat and mass transport on microstructure formation and segregation in technical alloys. Although numerous papers have been published investigating the effect of fluid flow on dendrite formation, coarsening of secondary arms, precipitation of intermetallics, micro- and macro-segregation and the effect of magnetic fields (static, alternating, rotating) on the cast microstructure, understanding of fluid flow in the liquid to solid transformation is rather in its infancy. Research under microgravity conditions in the past decades has shed light on the importance of fluid flow and it

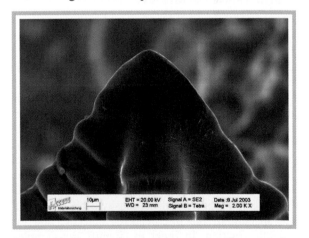

Figure 12-47: Tip of a dendrite taken with a scanning electron microscope, an Al-Si alloy whose solidification process was interrupted and the residual melt rapidly removed by decanting (ACCESS eV, Aachen)

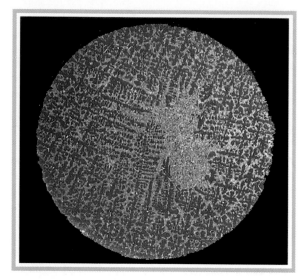

Figure 12-48: Cross section through an AlSIMg alloy solidified with artificial fluid flow induced by a rotating magnetic field device. Whereas in most areas a typical dendritic microstructure can be seen, there is slightly off center an area free of dendrites. This zone exhibits a eutectic microstructure only. Thus the fluid flow induced a macrosegregation not observed if fluid flow is suppressed

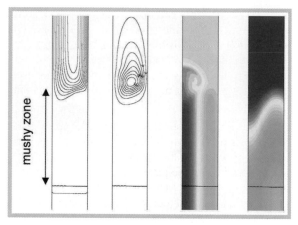

Figure 12-49: Interaction of a flow with the mushy zone. Shown are (from left to right): isolines of the azimuthal velocity, streamlines in the meridional plane, the mixture concentration, and the liquid volume fraction (the axis of symmetry is in each case on the left side). This numerical simulation performed at the Crystal Growth Laboratory in Erlangen predicts the experimental result shown in figure 12-48

became clear that even in the best experimental setups used on earth residual flows can change the microstructure significantly and thus prevent a detailed quantitative comparison with theoretical predictions (Ratke and MICAST team 2005).

The MICAST research program focuses on a systematic analysis of flow effects on microstructure. Questions are, for example, how the intensity of convection and the direction of the flow act on the evolution of the mushy zone, on macro- and microsegregation, on the morphology of dendrites, and on the growth mode and spatial arrangement of intermetallic precipitates. The microstructure is characterized by the dendrite tip radius (figure 12-47), the dendrite envelope, the primary dendrite stem spacing, the secondary arm spacing, the distribution and morphology of intermetallic phases, the three-dimensional topology of the mushy zone, the lamellae spacing of the interdendritic eutectic, and the eutectic fraction. In order to simplify the complex interactions between heat and mass transport and microstructure evolution, the experiments performed by the MICAST team are carried out under well

defined thermal and magnetically controlled convective boundary conditions using directional solidification and advanced in-situ diagnostics and metallographic techniques.

The experiments should provide benchmark data to validate theoretical models developed by the MICAST team. The theoretical modeling involves phase field simulation and micro-modeling as well as global simulation of the heat and mass transport and the magnetically induced fluid flow in the experimental set-ups used by the team. The MICAST project investigates conventional aluminum based cast alloys used for example in automotive applications (cylinder heads, motor blocks). The project presently uses sounding rocket flights (TEXUS 39, TEXUS 41) for the preparation of the ISS experiments and for first results. In the ISS the Materials Science Laboratory (MSL) will be used, which offers especially a rotating magnetic field device (figure 12-50) in both Bridgman type furnaces. On the ground the team uses conventional Bridgman furnaces equipped with rotating and traveling magnetic fields and a new type of furnace, based on aerogel technology, which also is used in the TEXUS experiments. The overall goal of the project, to quantitatively understand the effect of flow on alloys solidification, can only be achieved if a series of experiments in microgravity is performed with various alloy compositions and various fluid flow configurations. If the goal is reached, the result will lead to a significantly improved numerical modeling of castings and better performance of today's casting processes.

Another ESA application oriented project on monotectic alloys (MONOPHAS) aims to develop a technically usable, aluminum-based, lead-free bearing material with adequate hardness, wear and friction properties and good corrosion resistance. Engine bearings are subject to high pulsating load and high surface friction simultaneously. Bearing materials therefore generally consist of a solid solution or precipitation hardened nonferrous matrix in which hard and soft phases are incorporated to provide good tribological properties. Measures to enhance tribological performance often adversely affect mechanical strength and vice versa (Ratke 1993).

Recent developments to reduce fuel consumption, emission and air pollution, the size and weight of engines for automotive, and truck, ship propulsion and for electrical power generation have lead to significant increases in crankshaft bearing loads Direct fuel injection, for instance, raises ignition pressures up to 30% and the engine operating temperature up to 20°C over current levels. Compact engine dimensions require reduced bearing size.

The standard bearing material is of the bronze-lead type with lead content up to 20 wt.%. The most often used production method is that of direct casting onto a steel support in a continuous way. Generally the bronze-lead type bearings cannot fulfill future requirements concerning, for example, strength and friction properties. Since the early 1960s, materials from the Al-Sn alloy system have been used increasingly frequently in the sliding bearing industry and comprise in the meantime an entire product spectrum. Attempts to utilize alloys exhibiting a miscibility gap in their phase diagram, like Al-Pb, Al-Bi, Al-In, etc. all failed in the past 80 years. The reason was identified as the gravity induced sedimentation of the dense lead phase precipitating during the cooling through the liquid miscibility gap existing in the Al-Pb system. Unfortunately, early attempts to overcome this problem in micro-

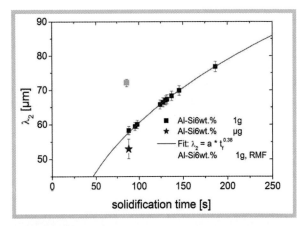

Figure 12-50: Secondary dendrite arm spacing in a AlSi alloys as measured in the laboratory (black squares), in reduced gravity in TEXUS 39 (blue star) and in the laboratory using the same solidification conditions but now with an applied rotating magnetic field leading to artificial fluid flow (grey square), see figures 12-48 and 12-49

gravity also failed, but these investigations led to a better understanding of the complicated problem of phase separation in immiscible systems (figure 12-51). They pointed especially to the central role of Marangoni motion of drops. These investigations led to the development of a new vertical strip casting process which for the first time in decades is able to produce two phase microstructures from immiscibles that seem to be suitable for bearings. The vertical strip casting process, however, is far from being understood in all details of processing and microstructure evolution, especially with respect to microstructure optimization by appropriate changes of the process parameters.

The requirements set by the microstructure of bearings implies that immiscible alloys must be investigated and developed in this project. Space experimentation plays a crucial role in optimizing a microstructure suitable for the applications intended. In order to achieve its goal the team will utilize advanced techniques of numerical modeling and thermodynamics, solidification experiments on earth and in space with in-situ X-ray studies of the microstructure evolution, and a pilot plant-scaled vertical strip casting process. The project concentrates on the family of AlZnBi alloys with a series of additional elements to suitably modify both mechanical and corrosion properties. Alloys from the AlZnBi system seem to be the only choice left to prepare advanced bearings for engine applications, since all other possible immiscible alloy systems have either metallurgical or ecological disadvantages.

Intermetallic materials possess remarkable mechanical, physical, chemical and biological properties which make them very attractive for extreme applications where new functionalities and higher performance are urgently needed, in particular for turbine blades and catalytic devices. The EU project IMPRESS (Industrial Materials Processing in Relation to Earth and Space Solidification) will study, in a systematic across-the-board manner, the intimate relationship between materials processing, structure, and the mechanical, physical, chemical and biological properties of Ti-Al-Ni-based intermetallics (figure 12-52). Quantitative understanding of the complex coupling between the many physical phenomena that occur during intermetallic materials processing will be sought in both ground and space experimentation. Experiments on board the International Space Station will yield critical information about microstructure formation as a function of heat

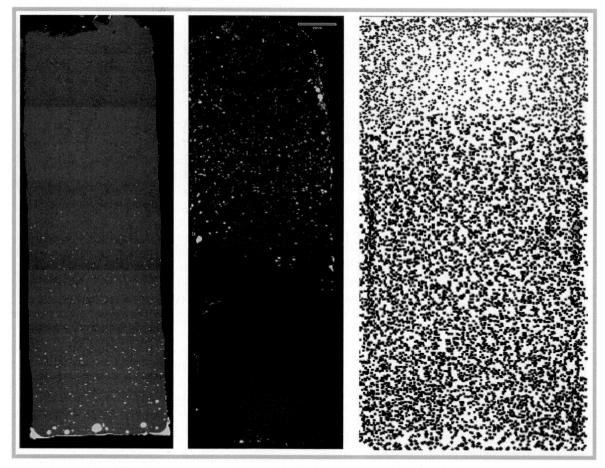

Figure 12-51: Left, AlPb alloy cast on earth. Most of the lead droplets are enriched at the bottom due to sedimentation when passing the miscibility gap (Ratke and Korekt 2000). Middle, overview of an AlPb alloy composition cast and solidified during a parabolic flight. The Marangoni motion has moved the droplets towards the hot zone of the sample, that zone solidifying last (Ratke and Korekt 2000). Right, numerical modeling of the phase separation in Al-Pb alloys under earth gravity conditions in a cylindrical cavity comparable to the experiments shown in figures 12-52 and 12-53. The black dots do not directly represent droplets but show an index composed of droplet number, drop radius and the Pb concentration in a given area (Wu et al. 2003)

Figure 12-52: Rotor turbine for which the IMPRESS project develops new advanced intermetallic blades replacing the Ni-base superalloys used in to day's engines

and mass transport conditions. The measurement of thermophysical properties of intermetallic materials in space will also deliver critical data unattainable on the ground. These data are urgently required by industry for the validation of computer modeling of material processes.

The integration of all these results will provide industry with the predictive capability to select the intermetallic alloy composition, process type and process parameters which should yield the set of properties required for a given application. Improved gas turbine technology will increase the efficiency of power generation; better catalytic devices will provide energy saving and reduce pollution.

Compared to conventional cast and wrought alloys the unexplored alloy area in materials sciences are the so-called multicomponent, multiphase alloys. In alloys with three or more components and three or more phases the process of microstructure formation during solidification is barely understood. These types of alloys cannot be treated by experience or classical methods, but require sophisticated theoreti-

cal and numerical methods. They are a challenging long-term research task not yet covered systematically in research activities due to inherent difficulties associated with them. Although all industrial alloys used are multicomponent in nature, many are designed to give single-phase microstructures during sol during solidification, e.g. grey iron with graphite, steels with Cr carbides, Al alloys with silicon, etc. Depending on the design task, knowledge of the formation of all the phases (solid solutions, intermetallics, carbides, nitrides, oxides, etc) in a given alloy system is crucial, either to avoid their formation or to control their size, morphology and dispersion. Here the investigation and exploitation of techniques to design the microstructure by grain refiners and modifiers as well as techniques to control fluid flow play an important role; thus microgravity investigations are crucial (Hecht et al. 2004).

A special class of multiphase alloys are the so-called in situ composites, based on the concept of generating several system-inherent but quite different phases in a well defined, periodic pattern. By appropriate solidification processes such patterns can potentially be tailored to yield dedicated properties. Some pioneering research has been performed so far on aligned lamellar patterns of ternary eutectics like Al-Ni-Ta, Cr-Co-C, Ni-Al-Cr, Al-Cu-Ni, Al-Nb-Ni and a few peritectic alloys like Ti-Al-Cr, but this work just started to exploit the available potential. This is especially true when considering the potential of obtaining multiphase pattern gradients that fit to the functionality of defined products, e.g., bio-implants. In addition multiphase microstructures can be processed such that a submicron structure is achieved with superior mechanical properties compared with conventional alloys. By intelligent solidification processing the microstructure can be varied inside a product to adapt for instance to the local mechanical loads. Multiphase and multicomponent alloys could be designed in the future in the lab and then brought to industrial practice. This is possible to date, since numerical modeling using complex thermodynamics and phase-field simulation has advanced much in the past decade. Multiphase alloys are per se knowledge-based alloys that cannot be handled in foundry practice by experience alone. A European project Solidification Along a Eutectic Path in Ternary Alloys (SETA) has already worked in the last years on multicomponent, multiphase alloys and is devoted to multicomponent multiphase systems.

Outstanding problems are the morphological stability in such systems and the nature and dynamic be-havior of the multiphase patterns forming in coupled growth (figure 12-53 and 12-54). This is an especially challenging field, as the high number of degrees of freedom and the multiplicity of phases lead to phenomena specific only to such complex systems. In this regard, ternary alloys of the SETA project serve as an example. Growth of two solid phases from the liquid, e.g., in eutectic reactions, is no longer nonvariant as in binary alloys but univariant. Accordingly, the two-phase solid/liquid interface can exhibit morphological transitions to cellular or dendrite-like patterns. Nonvariant growth in a ternary alloy involves three solid phases growing simultaneously from the liquid, thus the respective patterns are more complex than in binary (nonvariant) eutectics. The SETA team plans experiments with transparent analog materials and metallic system in the ISS. Without pure diffusive conditions for microstructure formation the multicomponent multiphase alloys will not be understandable.

The ESA project METCOMP (Metallic Composites) investigates two different types of composite materials: (i) multiphase materials consisting of at least two different phases produced in situ during a peritectic reaction and (ii) two materials composites (metal matrix composites, MMCs), which contain a metallic element as a matrix material with dispersed ceramic particles. The production of both types of composites from the liquid state of matter underlies some serious restrictions in the gravitational field of earth (Herlach 1994).

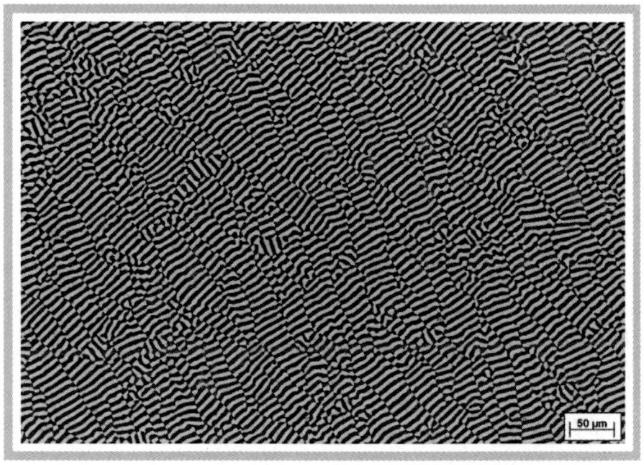

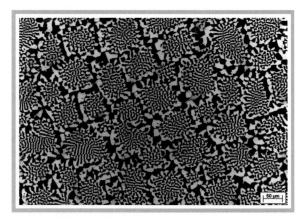

Figure 12-53: Univariant eutectic growth in Al-Cu 16.5 at%-Ag 5at% exhibits the well-known lamellar pattern consisting of α(Al)–black and Al$_2$Cu–white lamellae

Figure 12-54: The morphological transition to eutectic cells is governed by topologic and crystalline anisotropy. Cellular patterns shown here can be quadratic or elongated

Peritectic alloys, which can produce such composites, are of utmost importance for industry. Steels and copper alloys, to mention just the economically most important, fall in this category. Peritectic reactions are in general sluggish phase transformations, since diffusion in solids is required to transform a solid phase β in contact to a liquid phase L into a solid β+L → α. As a direct consequence, peritectic reactions are in most cases far out of thermodynamic equilibrium and are very much influenced by changes in heat and mass transport in liquid due to convection phenomena. On the other hand, dispersing of ceramic particles of mass density different from the metallic liquid suffers from sedimentation processes. In both cases the use of the beneficial conditions of microgravity in space may lead to a better understanding of the influences of gravitational related phenomena such as natural convection and sedimentation on the solidification route if ex-

perimental results of solidification studies on earth are compared with those obtained in space. Such comparative studies could lead to the construction of microstructural selection maps with convection and sedimentation as "process parameters" that may serve industry as guides in improving materials quality and making production routes more efficient by avoiding energy and time consuming post solidification treatment.

Transparent model systems of peritectics are quite attractive for deepening our understanding of solidification of metastable in situ composites. Such systems offer the advantage that both the morphology and the dynamics of solidification can be investigated by optical diagnostic means. Comparison of experiments both on earth and in space makes it possible to determine directly the effect of gravitational phenomena such as natural convection and

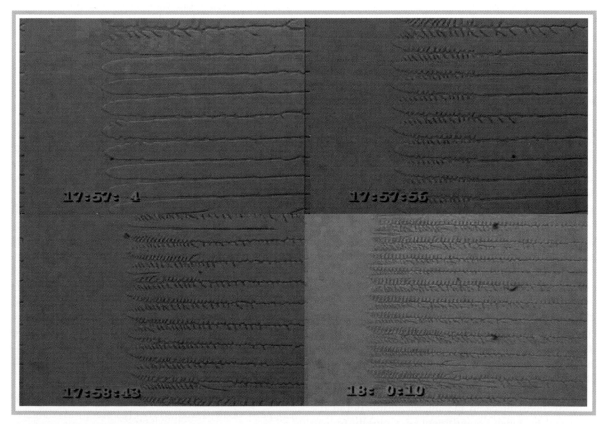

Figure 12-55: Morphological pattern formation observed during solidification of Neopentylglycol, Tris(hydroxymethyl) Aminoethan, and Pentaerythritol. The transition from a cellular to a dendritic array is shown

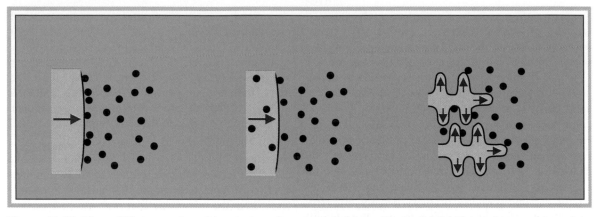

Figure 12-56: Three different modes of interaction of ceramic particles with a solid-liquid interface: pushing of the particles of a planar interface propagating slowly in the liquid (left), engulfment of particles of a rapidly propagating planar interface (middle) and trapping of particles in the interdendritic regions during dendritic solidification (right)

sedimentation on the growth morphology and dynamics of peritectic solidification. A Bridgman furnace will be used to observe the peritectic solidification of transparent model systems of such peritectics as Neopentylglycol, Tris(hydroxymethyl) Aminoethan, and Pentaerythritol (figure 12-55). On board the ISS the DECLIC facility is especially useful for analyzing transparent materials.

MMCs consisting of various phases and substances are of high interest since they combine advantageous properties of different phases within one material. Ceramic particles are often added to metallic alloys to reinforce metallic alloys and make them usable at high temperatures. Production of high quality composite materials requires homogeneous distribution of composite material within the matrix. In most cases such conditions are hindered by the pushing of particles by an advancing solid-liquid interface and by sedimentation of particles due to buoyancy effects in liquids. The interaction between a solid-liquid interface and ceramic particles has been investigated during directional solidification of a planar interface both under terrestrial and reduced gravity conditions. But for dendritic solidification, which is of much more practical importance in casting processes, a description of the conditions of homogeneous distribution of particles in a metallic matrix is still lacking. In contrast to a planar interface, particle trapping takes place in the interdendritic liquid regime and by kinetic engulfment of a rapidly propagating dendrite. This process is controlled by the

growth dynamics of the interface, transport phenomena in liquids, and particle size and morphology. On the other hand, industrial production processes require particle pushing during casting processes in order to remove foreign phases and purify the cast material. In general, embedding of particles is favored for large particles and rapid growth, while pushing is expected at small particle size and small growth velocity of the crystal (figure 12-56).

After finding suitable metal-ceramic particle composite systems, the interaction between solid/liquid interface and ceramic particles will be investigated. The current experiments will be extended to binary alloys of Cu-Ni, Fe-Ni and industrial steel alloys. The critical velocities for particle entrapment both for the dendrite stem and the dendrite branches will be measured. For comparison a furnace for directional solidification of undercooled melts will be set up to solidify metal-ceramic particle composites without inductive stirring of the melt as present in electromagnetically processed samples. Experiments will be prepared for studies of particle pushing and engulfment in melts processed under reduced gravity conditions. Testing experiments will be performed during parabolic flights to define the critical experimental parameters for the corresponding experiments to be conducted on board the ISS.

The ESA project Undercooling and Demixing of Copper-Cobalt Based Alloys (Coolcop) investigates so-called metastable miscibility gap alloys, Cu-Co and Cu-Fe (Egry 2003). The copper-cobalt and cop-

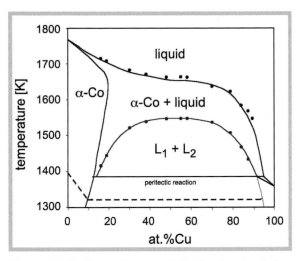

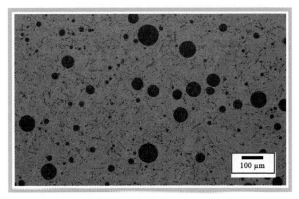

Figure 12-57: Phase diagram of Cu-Co showing below the stable liquidus line a metastable miscibility gap (blue) which can be reached in containerless processing using for instance electromagnetic levitation. Avoiding by this

Figure 12-58: Cu-Co sample (composition indicated in figure 12-58) as processed in a parabolic flight under microgravity conditions. The microstructure shows a dispersion of Co-rich droplets (dark) embedded in the Cu-rich phase (bright). The demixing indicates that the sample was solidified below the binodal

per iron alloy systems are fascinating model systems because both have a peritectic phase diagram with a metastable miscibility gap in the undercooled region (figure 12-57). Therefore, rapid solidification and quenching of the early phases of phase separation can be controlled by the amount of undercooling. While copper is one of the best electrical conductors, cobalt and iron are ferromagnetic elements. Therefore, both alloy systems permit study of the interplay among three different phenomena: undercooling, demixing, and magnetic ordering. The new aspect of this project lies in the metastable nature of the miscibility gap, allowing, on the one hand, study of the interplay between two different phase transitions, namely solidification and phase separation, but requiring, on the other hand, a containerless technique as a prerequisite for accessing the undercooled state. For metallic systems, this is accomplished by electromagnetic levitation. Therefore, the objective of the project is to undercool alloys of various compositions into the miscibility gap by electromagnetic levitation in order to study liquid-liquid phase separation and coarsening, to study correlations between undercooling and microstructure evolution, to measure surface and interfacial tensions before and after demixing, to observe a wetting and prewetting transition, and to study the onset of magnetic ordering at deep undercooling.

Electromagnetic levitation experiments benefit generally from being carried out in microgravity, as explained above. This is because the electromagnetic fields necessary to provide stable positioning of the sample are reduced by at least a factor of 1000. Consequently, the detrimental side effects of these fields are also reduced (figure 12-58). The CoolCop project will use the MSL-EML (Materials Science Lab, Electromagnetic Levitator) facility.

These alloy systems have some potential applications. Being a monotectic system, Cu-Co or Cu-Fe could be used as a high-temperature bearing material. The most striking property of Cu-Co is, however, its giant magnetoresistance (GMR). Most GMR materials consist of thin film multilayers with interchanging magnetic and nonmagnetic layers. Recent experiments have shown, however, that high performance magnetic materials can also be successfully produced by solidification from undercooled melts. Cu-Co is already in use as catalyst for synthesizing alcohols. However, by producing the appropriate microstructure through undercooling, porous granular material can be produced with a large surface-to-volume ratio, thereby improving catalytic efficiency. Finally, in the field of recycling of scrap from automobiles, the main problem is contamination of the steel to be recovered by copper, stemming from the car's electrical components. Already small amounts of copper lead to detrimental effects in processing the steel. Therefore, the separation of copper from iron is a central technological issue.

The ESA-MAP project "Non-Equilibrium Solidification, Modeling for Microstructure Engineering of Industrial Alloys (NEQUISOL)" analyzes the effect of fluid flow and thus gravitational effects on the formation of crystalline phases, stable and metastable ones, from the liquid state utilizing electromagnetic levitation in order to control the undercooling of the melt below the equilibrium melting temperature. The ultimate goal of the project is to develop with the help of earth and space experimentation so-called microstructure selection maps (figure 12-59) which show what type of microstructure an engineer can expect, depending on the processing parameters used in processes like powder atomization, drop tubes, and spray deposition (Eckler et al. 1997 and Herlach et al. 2004).

In crystal growth the future challenges are to understand the interaction of fluid flow without the disturbing effects of natural convection. Of special interest in recent years are compound semiconductors. There are several projects in Europe, Canada and Japan to investigate the growth of compounds like CdZnTe (CZT) by several methods. There are many examples of the use of CZT detectors in medical imaging and diagnostics, ranging from simple X-rays carried out in a dentist's office to cardiac angiography (imaging "slices" through the heart), bone densitometry measurements, and the use of nuclear medicine to pinpoint areas of activity within the brain to help characterize conditions such as epilepsy. Astronomers use CZT arrays to study the origin of high energy gamma-ray bursts. CZT is also suitable for high resolution measurements and isotope identification in the nuclear industry and for X-ray radiography applications.

One of the new techniques to be used under microgravity is the so-called liquid phase electroepitaxy (LPEE). This is a solution growth technique that allows the production of high quality semiconducting single crystals. It is one of the techniques with proven potential for growth of bulk ternary crystals with the desired material and structural properties. Although the effect of natural convection occurring in the solution due to the gravitational field of earth is small in LPEE growth compared with other crystal growth techniques, the natural convection is one of the most important phenomena adversely affecting the quality of grown crystals. Therefore its effect must be reduced to a minimum. There are two ways to achieve this reduction. One uses very high static magnetic fields and the other is to perform experiments on the ISS and also to perform experiments in which fluid flow is artificially induced by rotating magnetic fields or acoustic waves such that the flow becomes a parameter under the control of the experimenter.

In one research program the Traveling Heater Method (THM) already used in EURECA and Spacelab will be employed. Mathematical simulations will be used to define optimum growth parameters, under the effect of a steady, vertical magnetic field, to simulate low gravity conditions. Controlled, forced mixing, induced within the solvent zone by the action of an applied rotating magnetic field, and by acoustic waves, will be investigated by both mathematical modeling and experimentation, with the objectives of improved crystalline quality and increased growth rate, consistent with quality. The results obtained will be used to improve earth-

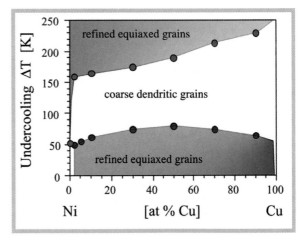

Figure 12-59: Microstructure selection map for binary Ni-Cu alloys. Two different types of microstructures can be observed. At low undercooling (red shaded area) fine polyedric grains are observed. The same is true for large undercooling, but with coarse dendritic grains in between. Especially the lower critical undercooling depends very much on experimental conditions and fluid flow. In the NEQUISOL project the effect of fluid flow in microstructure selection will be investigated

based processes for improved yield and material characteristics.

A classical material of interest are alloys of silicon-germanium, which play an important role in the electronic and optoelectronic industries. Although silicon and germanium can be mixed in the alloy for any amount of silicon in germanium, the silicon-germanium system suffers from strong segregation and, together with the large density difference between molten silicon and germanium, leads to macroscopic and microscopic segregation in the alloy system.

One Canadian research project of Dalhousie University intends to grow silicon-germanium alloy using a crystal growth method, namely, the traveling solvent method, which was recently shown to produce silicon-germanium crystals with the least amount of segregation compared to other crystal growth methods. An intensive theoretical and experimental study is planned to solve after many years of research worldwide the problem of growing homogeneous Si-Ge semiconductors. Si-Ge are extreme fast transistors and an interesting material for substrates since the lattice distance can be varied by alloying silicon with germanium.

As already mentioned above one important topic in the area of thermophysical properties is the precise measurement of diffusion in molten alloys, which should eventually lead to a theoretical understanding comparable to that for the solid state. Especially the Canadian Space Agency and Europe have performed a large series of experiments with various alloys in the past to understand diffusion in liquids. For the study of this complex problem the Canadian Space Agency (CSA) has developed a special new furnace, the so-called ATEN furnace (Advanced Thermal ENvironment). A specialty of this furnace is that it fits into the so-called MIM device (Microgravity vibration Isolation Mount base unit), a Canadian development of the past decade to eliminate the residual disturbances of the microgravity level on board spacecraft by active and passive damping measures. In the past, as mentioned above, it was shown that the isolation of facilities against disturbances of the microgravity level by, e.g., spacecraft maneuvers or crew exercises, is especially important if one studies diffusion in molten alloys, as planned

by Queens University and the University of Manitoba. Both research projects will investigate the atomic diffusion in molten alloys and should improve the accuracy of the data such that comparisons with theoretical models can be performed and the long-standing question of temperature dependence be answered for various types of alloys. There are also Japanese experiments planned to study diffusion in liquids. They concentrate, however, on diffusion in transparent materials since there interferometric methods can be used that allow in-situ analysis of the process of diffusion by directly measuring the concentration field as it develops over

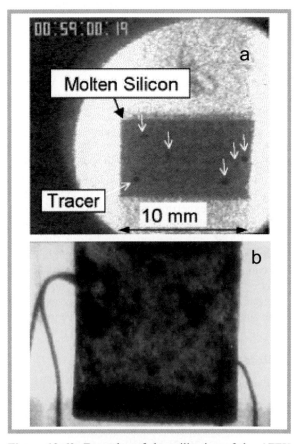

Figure 12-60: Examples of the utilization of the AFEX furnace. Figure a shows tracer particles inserted in molten silicon. The analysis of their motion allows study for example of Marangoni motion in molten silicon. Figure b shows the phase separation in two immiscible liquids. The X-ray absorption contrast allows study of the growth and coagulation of droplets

time.

AFEX is a material experiment facility with the X-ray visualization system for in situ observation of materials processing to as high as 1600°C. The proposed experiments to be performed on AFEX are: 1) crystal growth and in-situ observation of Marangoni convection of molten silicon under microgravity, 2) in-situ observation of the solidification phenomena of immiscible monotectic alloys. The group of Motegi at Chiba Institute of Technology aims to investigate the separation and the solidification process of the hypermonotectic alloys which become separated in liquid phases because of the density difference on the ground. The hypermonotectic alloys with solubility gap (Al-Pb) are heated with an AFEX and the liquid separation and the solidification processes are observed directly by using an X-ray radiography system under microgravity conditions (figure 12-60). The results obtained will support research and development of homogeneously solidified functional materials.

In contrast to the research possible 20 years ago, all experimental research activities in space are now always accompanied by theoretical analysis using the most advanced computer models, and the experiments are guided by the results of these modeling. Also, the experimental facilities will be fully modeled numerically such that the space experiments will be simulated on earth before flight and optimized with respect to the scientific output. Thus one can expect that once the international materials science facilities on the ISS are working nominally, research on solidification and crystal growth will make substantial progress in the next 10 years.

References

J.Alkemper J, Ratke L (2000) Ordering of the Fibrous Microstructure of Al-Al3Ni due to Accelerated Solidification Conditions. Acta Materialia 48, pp. 1939 – 1948

Beckermann C (2002) Modelling of Macrosegregation: applications and future needs. Int. Mat. Rev. 47, p. 243

Croell A, Kaiser T, Schweizer M, Danilewsky AN, Benz KW (1998) Flowting zone and floating so-lution zone growth of GaSb under microgravity. Journal of Crystal Growth 191, p. 365

Croell A, Müller-Sebert W, Benz KW, Nitsche R (1991) Natural and thermocapillary convection in partially confined silicon melt zones. Microgravity Sci. Technol. III/4, p. 204

Croell A, Schweizer M, Tegetmeier A, Benz KW (1996) Floating zone growth of GaAs. Journal Crystal Growth 166, p. 239

Davidson PA (2001) *An Introduction to Magnetohydrodynamics*, Cambridge University Press, Cambridge

Eckler K, Gärtner F, Greer AL, Herlach DM, Moir SA, Norman AF, Ramous E, Zambon A (1997) Microstructure-Selection Maps applied to droplet solidification In: Beech J, and Jones H eds: Proc. 4th Decennial Int. Conf. on Solidification Processing (Sheffield, UK), pp. 577-581

Egry I (2003) Thermophysical property measurements in microgravity: chances and challenges. Int. J. Thermophysics 24, pp. 1313-1324

Egry I (1998) Properties, nucleation and growth of undercooled liquid metals: results of the TEMPUS-MSL-1 mission. J. Jpn. Soc. Microgravity Appl., 15, pp. 215-224

Egry I, Herlach D, Kolbe M, Ratke L, Reutzel S, Perrin C (2003) Surface tension, phase separation, and solidification of undercooled Cu-Co, D. Chatain Advanced Eng. Materials 5, pp. 819 – 823

Egry I, Lohöfer G, Seyhan I, Schneider S, Feuerbacher B (1999) Viscosity and surface tension measurements in microgravity. Int. J. Thermophys. 20, pp. 1005-1015

Flemings MC (1974) Solidification Processing, MacGrawHill, New York

Feuerbacher B (1984) ed: Proc. 5th European Symposium Material Sciences under Microgravity, Schloß Elmau, Germany, ESA SP 222, ESA Publications Division, Noordwijk

Glicksman ME, Koss MB, Winsa EA (1994) Dendrtitic Growth Velocities in Microgravity, Phys. Rev. Letters 73, 573-576

Hecht U, Granasy L, Pusztai T, Böttger B, Apel M, Witusiewicz V, Ratke L, De Wilde J, Froyen L, Camel D, Drevet B, Faivre G, Fries SG, Legendre B, Rex S (2004) Multiphase solidification in multicomponent alloys. Mat.Sci.Eng. R 46, pp. 1–49

Herlach DM (1994) Non-equilibrium solidification of undercooled metallic melts. Materials Science and Engineering Reports R12, pp. 177-288

Herlach DM, Funke O, Phanikumar G, Galenko P (2004) Free dentrite growth in undercooled melts: experiments and modelling. In: Rappaz M, Beckermann C, and Trivedi R eds: Solidification Processes and Microstructures: A Syposium in Honor of Prof. Wilfried Kurz. Proceedings of the TMS Annual Meeting, March 14-18, 2004, Charlotte, North Carolina. (TMS, Warrendale, Pennsylvania, 2004) p. 277

Hillert M (1998) Phase Equilibria, Phase Diagrams and Phase Transformations, Cambridge University Press, Cambridge

Hurle DTJ (1998) ed: Proc. 7th European Symposium Material Sciences under Microgravity, Oxford, UK, 1989, ESA SP 295, ESA Publications Division, Noordwijk

Hurle DTJ (1994) Handbook of Crystal Growth, Elsevier Science B.V., Amsterdam

Kaiser T, Croell A, Dold P, Benz KW (1996) Numerical simulation of heat and mass transfer in silicon crystal growth by the floating zone method. Proc. 2nd European Symp. Fluids in Space, p. 245

Kassner K (1996) Pattern Formation in Diffusion-Limited Crystal Growth, World Scientific, Singapore

Khou S (1996) Transport Phenomena and Materials Processing, Wiley Interscience, New York

Kolbe M, Reutzel S, Patti A, Egry I, Ratke L, Herlach DM (2003) Undercooling and demixing of Cu-Co melts in the TEMPUS facility during parabolic flights. In: Li BQ ed: TMS Multiphase Phenomena and CFD Modeling and Simulation in Materials Processes (The Minerals, Metals and Materials Society), 2004, pp. 55-63

Koster J, Sani RL (1990) eds: Progress in Low-Gravity Fluid Dynamics and Transport Phenomena, AIAA Washington D.C.

Kurz W, Fisher DJ (1989) Fundamentals of Solidification, TransTechPubl, Ackermannsdorf

Kurz W, Sahm PR (1975) Gerichtet erstarrte eutektische Werkstoffe, Springer-Verlag, Berlin, Heidelberg, New York, Band 25

Legros JC (1992) Proc. 8th European Symposium Material Sciences under Microgravity, Universite Libre de Bruxelles, Belgium, ESA SP 333, ESA Publications Division, Noordwijk

Wu M, Ludwig A, Ratke L (2003) Modelling Simul. Mater. Sci. Eng. 11, pp. 755 - 769

Müller G, Kyr P (1986) Directional solidification of InSb-NiSb eutectic. In: Sahm PR, Keller MH, and Schiewe B eds: Proc. Norderney Symp. 1986, Scientific Results of the German Spacelab Mission D1, DLR, Köln

Naumann RJ (1980) Material Processing in Space: Early Experiments, NASA, Washington D.C., SP-443

Pelce P (1988) Dynamics of Curved Fronts, Academic Press, Boston

Porter DA, Easterling KE (1992) Phase Transformations in Metals and Alloys, 2nd Edition, Chapmann & Hall, London

Ratke L (1993) ed: Immiscible Liquids and Organics, DGM Informationsgesellschaft, Oberursel

Ratke L, Benz KW (1998), Erstarrung und Kristallzüchtung: Vom Spacelab zur Raumstation. In: Keller MH, Sahm PR (eds): Bilanzsymposium Forschung unter Weltraumbedingungen, Norderney 1998, WPF RWTH Aachen, 2000, pp. 198 222

Ratke L, Korekt G (2000) Solidification of Al-Pb base alloys in low gravity. Z. Metallkde. 91, pp. 919 – 927

Ratke L and MICAST team (2005) MICAST – The effect of magnetically controlled fluid flow on microstructure evolution in cast technical Al-alloys. In: Proc. 2nd International C Conference on Physical Sciences in Space, Spacebound 2004, Toronto. J. Microgavity Science and Technology, March 2005

Ratke L, Voorhees PW (2002) Growth And Coarsening – Ripening in Materials Processing. Springer Verlag, Heidelberg

Ratke L, Walter H, Feuerbacher B (1996) eds: Materials and Fluids under Low Gravity. Lecture Notes in Physics 464, Springer Verlag, Heidelberg

Roosz A, Rettenmayr M (1996) eds: Solidification and Gravity. Materials Science Forum Vols. 215 – 216, Trans Tech Publications, Switzerland

Roosz A, Rettenmayr M, Watring D (2000) eds: Solidification and Gravity 2000, Materials Science Forum Vols 329 – 330, Trans Tech Publications, Switzerland

Rosenberger F (1979) *Fundamentals of Crystal Growth* I. Springer Series in Solid-State Sciences 5, Springer Verlag, Berlin

Sahm PR, Hansen PN, Conley JG (2000) eds: Modeling of Casting, Welding and Advanced Solidification Processes – IX, Shaker Verlag, Aachen

Sahm PR, Keller MH, Schiewe B (1995) eds: Scientific Results of the German Spacelab Mission D-2, DLR Köln

Seibert G (2001) *A World without Gravity*, ESA SP 1251, ESA Publications Division, Noordwijk

Tiller WA (1991) *The Science of Crystallization*, Cambridge University Press, Cambridge

Walter HU (1987) ed: Fluid Sciences and Materials Science in Space. Springer Verlag, Berlin

13 Life Sciences

by Rupert Gerzer, Ruth Hemmersbach, and Gerda Horneck

Since the advent of space flight, the scope of life sciences has greatly expanded, adding new facets to existing disciplines, such as to gravitational biology, or even creating new disciplines, such as astrobiology. Exposing living systems to an environment which they have not yet experienced in their lifetime and also not their ancestors during evolution brings new exciting experimental approaches and results. Excluding the important environmental stimulus gravity provides a new experimental tool towards our understanding of the physiology and fundamental mechanisms of life. It requires an integrative experimental approach at all levels of organization —from molecules, single cells, tissue, and organs, up to the whole organism. This field is covered by the expanding discipline of gravitational biology using the unique possibilities to perform experiments in space.

Sending humans into space also requires understanding the effects of potential hazards of space and space flight, such as microgravity, radiation, temperature, atmospheric quality, noise, vibration and confinement, in order to safeguard the mission and the space travelers against adverse health effects. Human physiology research in space aims at understanding the physiological changes occurring in humans in space, information required to develop efficient countermeasures for long-term space exposure of human beings. These studies also bear potential spin-on effects for the living conditions and health of human beings on Earth.

Space exploration has extended the boundaries of biological investigations beyond the earth and even beyond earth orbit, to other planets, moons, comets, meteorites, and space at large (see also Chapters 7 and 8 under "Looking up: stars and planets"). This research in the multidisciplinary field of astrobiology has focused on the different steps of evolutionary pathways through cosmic history that may be related to the origin, evolution and distribution of life on Earth, or elsewhere in the universe. The overriding objective of astrobiological research has been to attain a better understanding of the principles leading to the emergence of life from inanimate matter, its evolution, and its distribution, thereby building the foundations for the construction and testing of meaningful axioms to support a theory of life. The space environment offers an ideal platform for testing some of the hypotheses of astrobiology, such as the import of precursors of life by meteorites, the transport of life between the planets, or the limits of life at all.

13.1
The Space Environment

Arriving in space without any protection, living beings are confronted with an extremely hostile environment, characterized by a high vacuum, an intense radiation field of solar and galactic origin and extreme temperatures (table 13-1). This environment or selected parameters of it are the test-bed for astrobiological investigations, thereby exposing chemical or biological systems to selected parameters of outer space or defined combinations of them.

In practical cases, the biological systems including humans are protected from most of these hostile parameters of space, either by containment within a space capsule, in other words a pressurized module and an efficient life support system (LSS), or at least by a space suit during extravehicular activity (EVA). In this case, mainly microgravity and radiation are the parameters of interest or concern, respectively.

Microgravity

As gravity is always present on Earth, scientists developed methods in order to vary the influence of gravity and made efforts to achieve the physiological status of functional microgravity. Today different

Space parameter	Interplanetary space	Low Earth orbit (≤ 500 km)
Space vacuum		
Pressure (Pa)	10^{-14}	10^{-6}-$10^{-4(a)}$
Residual gas (part/cm⁻³)	1 H	1×10^5 H 2×10^6 He 1×10^5 N 3×10^7 O spacecraft atmosphere[b]
Solar electromagnetic radiation		
Irradiance (W/m²)	Various values[c]	1380
Spectral range (nm)	Continuum	Continuum
Cosmic ionizing radiation		
Dose (Gy/a)	$\leq 0.1^{(d)}$	$\leq 0.4^{(e)}$
Temperature (K)	$> 4^{(c)}$	wide range[c]
Microgravity (g)	$< 10^{-6}$	10^{-3}-10^{-6}

Table 13-1: Physical Conditions Prevailing in Interplanetary Space and in Low Earth Orbit.
[a] Depending on outgassing of the spacecraft; [b] sources of contamination: waste dumping (H_2O, organics); thruster firing (H_2O, N_2O, NO); [c] depending on orientation and distance to sun; [d] depending on shielding, highest values at mass shielding of 0.15 g/cm²; [e] dose measured within the spacecraft, value depending on altitude, orbit and shielding, highest values at high altitudes and mass shielding of 0.15 g/cm²

experimental and technical designs exist in order to study the effect of acceleration on the physiology of living systems and thus identify potential problems during long-term space flight. While the influence of increased gravitational stimulation can be generated by means of centrifuges, the condition of functional microgravity is achieved, for example, by clinostats for small biological systems (cell cultures, small plants and animals) (for review see Häder et al. 2005) and for humans by means of -6° head down tilt positioning (simulating fluid shift as in microgravity) or for small animals (rats) by hind limb suspension, meaning that the hind legs can no longer support the body weight or touch any surface. Real free fall conditions and their impact on living systems can be studied by means of different experimental approaches. Short-term microgravity is provided in drop towers or drop shafts (2.1 to 10 s), balloons (30-60 s) and parabolic flights of aircraft (20-25 s) or sounding rockets (up to 15 min). These methods are suitable for fast reacting systems and for studies of the first responses to microgravity. In order to study long-term effects of microgravity, automatic satellites, platforms (e.g., EURECA) and

human-tended space laboratories such as space shuttles, Spacelab, Spacehab, Mir and ISS have been developed. One important step to space was the space station Mir as it offered the long-desired challenge of a long term stay of humans in space. For 15 years Mir circulated at a height of 300-400 km above ground and was visited by more than 100 astronauts and cosmonauts. Mir provided essential experience with respect to the stay and experimentation in microgravity. Since 1998 the International Space Station (ISS) is being assembled in space 400 km above the ground and should be completed before 2010. Pressurized modules and external platforms provide living and working accommodations for up to seven astronauts. Specific equipment offers laboratory conditions for dedicated and systematic studies in microgravity.

Radiation

During space missions living systems are exposed not only to microgravity, but also to a radiation environment which they normally do not experience in our biosphere where the surface is largely protected from the cosmic radiation source because of

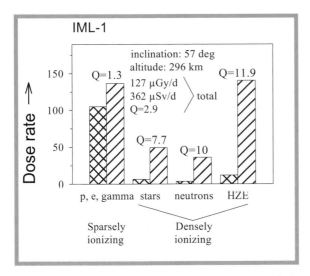

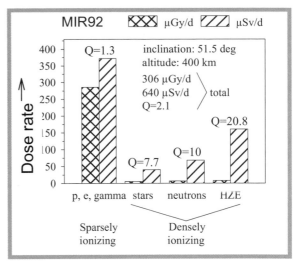

Figure 13-1: Absorbed dose rates (μGy/d) and equivalent dose rates (μSv/d) of the sparsely ionizing and the three densely ionizing components of the radiation field in LEO, determined from dosimetric measurements during the Spacelab mission IML-1 and the Mir 92 mission (data from Reitz et al. 1993)

the deflecting effect of the geomagnetic field and the huge shield of 1000 g/cm² provided by the atmosphere. At the orbit of the ISS, the radiation field is composed mainly of: (i) solar cosmic radiation (SCR), (ii) galactic cosmic radiation (GCR), and (iii) the radiation trapped by the earth's magnetosphere, the so-called radiation belts (see chapter 1).

The different types of radiation in space cause different amounts and kinds of biological damage per unit of absorbed dose. Therefore, the different types of radiation received in space have to be known and weighted using the quality factor Q. Figure 13-1 shows the contribution to the absorbed dose rate (Gy/d) and dose equivalent rate (Sv/d) attributed to the various components of cosmic radiation. It is clearly seen that the total dose rate is about twice as high for Mir 92 as for IML-1 and the contributions of the different components of radiation to the total dose are quite different for the two missions. The reason is the higher altitude of the Mir orbit compared to that of the IML-1 mission and a different period in the solar cycle.

For radiation protection issues, it is important to know the radiation dose received at the different organs of the human body. The human phantom Matroshka, mounted at the outer wall of the ISS,

measures the depth dose distribution which will be experienced by astronauts during EVA (figure 13-2).

Human Factors

Future human exploratory missions to the moon and Mars, including long term habitats on the lunar surface, will extend the distances traveled, the dose allotted to the radiation environment, the gravity levels, the duration of the mission, and the levels of confinement and isolation to which the crew will be

Figure 13-2: Matroshka, a human phantom, mounted at the outer wall of the ISS to measure the depth dose distribution of cosmic radiation in a human body

exposed. This will raise several health issues which may be limiting factors during these missions, in particular radiation health, gravity related effects and psychological issues. For missions to the moon or other planets the safety measures employed so far in LEO (Low Earth Orbit) have to be developed much further. Crew health and performance have to be ensured during transfer flights and planetary surface exploration, including EVAs, and upon return to Earth, within the constraints of safety objectives and mass restrictions (Horneck et al. 2003). Thus, prior to the design of an exploratory type mission (hardware and operations), numerous issues in life sciences need to be addressed.

In addition, before sending humans to Mars, an intense preparatory robotic research program is required in order to serve the following purposes:

- to search for signatures of past or present Martian life. This requires protecting the planet from the unintended import of terrestrial microorganisms, which can only be accomplished with robotic missions to Mars. Special planetary protection requirements are laid down in COSPAR guidelines (Rummel 2001). The scenario will change when humans are involved in a mission. Since humans carry vast amounts of microbes required to sustain important body functions, Mars will become inevitably contaminated with terrestrial microorganisms as soon as humans arrive on its surface. Although the surface of Mars seems to be globally very hostile to microbial life, it cannot be excluded that some microorganisms accidentally imported may find protective ecological niches where they could survive or even metabolize, grow, and eventually propagate. Having this in mind, the ESA Exobiology Team stressed the vital importance of a substantial series of robotic missions to precede the human mission to Mars in order to carry out the essential exploratory search for life by in situ measurements at selected sites (Brack et al. 1999, see also chapter 8).

- to determine the potential surface hazards to humans. Knowledge of the radiation climate, especially the ionizing, neutron and UV components, the composition and chemical reactivity of the dust and soil (regolith), also under wet conditions, as well the existence and nature of a potential

Martian biota are a must before human missions to Mars take place.

13.2
Astrobiology

Astrobiology—formerly called exobiology—includes the study of the origin, evolution and distribution of life in the universe. Its central focus is directed towards questions that have intrigued humans for a long time: What is life? Where do we come from? Are we alone in the universe? They are jointly tackled by scientists converging from widely different fields, reaching for example from astrophysics to molecular biology and from planetology to ecology (summarized in Horneck and Baumstark-Khan 2002). This spilling beyond the boundaries of classical sciences opens completely new opportunities for research, a state described by some contemporaries as the "Astrobiology Revolution of the Sciences."

Astronomers explore the vast realms of the universe for signatures of life beyond the earth. In the interstellar medium, as well as in comets and meteorites, complex organics teem in huge reservoirs that eventually may provide the chemical ingredients for life. More and more, other planetary systems are found in our galaxy, which supports the assumption that habitable zones are frequent and not restricted to our own solar system. Within such a habitable zone, life may not be confined to its planet of origin: Martian meteorites, detected in Antarctica, provide the evidence that material can be expelled from a planet by natural mechanisms of expulsion caused by a large meteorite impact. Astrobiology tackles the question whether this impact scenario may provide the vehicle to transport microbial communities through space.

Astrobiology also provides clues to reconstruct the history of life on Earth (figure 13-3). Geologists study the few places on Earth exposing sedimentary rocks older than 3.3 billion years where they detect traces of the oldest morphological fossils of microbial communities. The isotopic signatures of the organic carbon of the Greenland metasediments provide indirect evidence that life may be even 3.85 billion years old.

In the endeavor to search for life on other celestial bodies of our solar system, our neighbor planet Mars and Jupiter's moon Europa have gained increased interest. However, in order to design the right "search for life" experiments, it is prerequisite to first understand the limits of life on Earth, and to gain more data on the geology and climate of the celestial body under investigation, present state as well as past evolution (see also chapter 8). Microorganisms have invented strategies to cope with and adapt to environments of a wide range of physical and chemical parameters (Horneck and Baumstark-Khan 2002). Examples are microbial ecosystems in the deep subsurface down to a depth of several thousand meters, in the interior of rocks in cold and hot deserts, and in crystalline salts from evaporite deposits. Microorganisms have been isolated from extremely cold environments, such as the Antarctic soils, as well as from hot environments at temperatures in the range of 80°C to 113°C which are usually associated with active volcanism ("black smokers"). These examples demonstrate that nearly all niches on Earth seem to be inhabited by microbial communities where an energy source is available and which are compatible with the chemistry of carbon-carbon bonds. They provide ideal test-beds for designing the right "in-situ search for life" experiments, e.g., for Mars.

13.2.1
Astrobiology Studies in Earth Orbit

Since the eighties, with the accessibility of Spacelab, the free flying satellite EURECA and Russian Cosmos or Foton satellites, the European Space Agency ESA has provided opportunities for astrobiology experiments in earth orbit. Utilizing outer space or special components of this environment as a tool, specific questions of astrobiology were tackled. Some examples are described in the following.

Relevance of Extraterrestrial Organic Molecules for the Emergence of Life

During the EURECA mission of ESA, in the Exobiology Radiation Assembly ERA, complex organic residues produced in the laboratory were exposed for six months to the full spectrum of solar UV and vacuum-UV radiation in order to simulate the photolytic processes assumed to occur during chemical evolution of interstellar grains (figure 13-4). Infrared

Figure 13-3: A reconstruction of the different evolutionary steps during the history of life on Earth (Artist view, NASA)

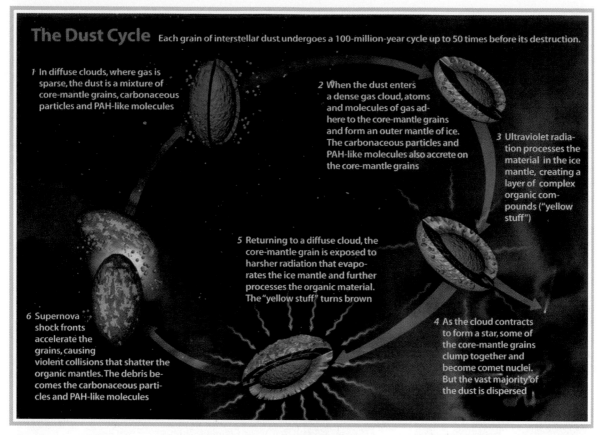

The Dust Cycle Each grain of interstellar dust undergoes a 100-million-year cycle up to 50 times before its destruction.

1 In diffuse clouds, where gas is sparse, the dust is a mixture of core-mantle grains, carbonaceous particles and PAH-like molecules

2 When the dust enters a dense gas cloud, atoms and molecules of gas adhere to the core-mantle grains and form an outer mantle of ice. The carbonaceous particles and PAH-like molecules also accrete on the core-mantle grains

3 Ultraviolet radiation processes the material in the ice mantle, creating a layer of complex organic compounds ("yellow stuff")

5 Returning to a diffuse cloud, the core-mantle grain is exposed to harsher radiation that evaporates the ice mantle and further processes the organic material. The "yellow stuff" turns brown

6 Supernova shock fronts accelerate the grains, causing violent collisions that shatter the organic mantles. The debris becomes the carbonaceous particles and PAH-like molecules

4 As the cloud contracts to form a star, some of the core-mantle grains clump together and become comet nuclei. But the vast majority of the dust is dispersed

Figure 13-4: Cyclic evolutionary model for interstellar dust (Greenberg 1995)

spectra obtained after the retrieval of the samples matched quite well with the spectra of the interstellar medium, thereby supporting the cyclic evolutionary model for interstellar dust (Greenberg 1995).

Studies in earth orbit have also contributed to answering the question whether the precursors of life were produced on the primitive earth or transported to the early earth via meteorites. Support for the latter supposition comes from an analysis of micrometeorites collected in the Greenland and Antarctic ice sheets. These grains contain 2% carbon, e.g., in the form of amino acids, as well as a high proportion of metallic sulfides, oxides, and clay minerals that belong to various classes of catalysts. On the early earth such micrometeorites may have functioned as tiny chondritic chemical reactors when reaching oceanic water. In recent years, passive aerogel collectors have been used to collect these particles

directly in space, e.g., on NASA's Long Duration Exposure Facility LDEF (figure 13-5) and on the Mir station. The advantage of this type of collection—compared to sampling on Earth—is that the samples are not altered by atmospheric entry effects. Aerogel collectors are used on the NASA Stardust mission to collect and return to Earth grains from the interplanetary medium and from the coma of comet Wild 2. Studies on board the Russian Foton satellite and the Mir station showed that amino acids which naturally occur in carbonaceous meteorites were protected against the destructive effects of outer space, if embedded in clays.

Solar UV Radiation in Evolutionary Processes Related to Life

Space experiments using the solar UV radiation as authentic energy source represent new approaches

Figure 13-5: The Long Duration Exposure Facility (LDEF) of NASA accommodated instruments to collect interplanetary dust and to expose microorganisms to the space environment (NASA)

towards our understanding of the evolution of life and potential evolutionary steps on other celestial bodies. They can especially contribute to tackle the following issues: the formation and stability of organic molecules in the prebiotic environment of the early earth, the role of the ozone layer in protecting life on earth, and the role of solar UV radiation in planets and moons of astrobiological interest, such as Mars, Saturn's moon Titan, and Jupiter's moon Europa.

So far, our understanding of the consequences for the biosphere of increasing environmental UV-B irradiation due to decreasing concentration of stratospheric ozone was merely based on model calculations. Using extraterrestrial sunlight as natural radiation source and optical filters in an experiment on board the German Spacelab D2 mission, the terrestrial UV radiation climate was simulated at different ozone concentrations down to very low ozone values. Biologically effective irradiances as a function of a simulated ozone column thickness were measured with bacterial spores (*Bacillus subtilis*) immobilized in a biofilm and compared to expected irradiances, using radiative transfer calculations and the biofilm action spectrum (figure 13-6). Analyses after the mission showed that with decreasing simulated terrestrial ozone concentrations, the biologically effective solar UV irradiance strongly increased (figure 13-7). Compared to the biologically effective UV irradiance at the surface of the earth (at an aver-

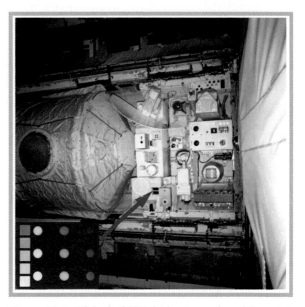

Figure 13-6: Biological UV dosimetry on board Spacelab D-2 to determine the biological consequences of a shrinking stratospheric ozone layer. Insert: Biofilm with bacterial spores as UV dosimeter

age total ozone column), the biologically effective UV irradiance in space (full spectrum of extraterrestrial solar UV radiation) was increased by nearly three orders of magnitude, thereby confirming model calculations (Horneck et al. 1996).

Transport of Life in our Solar System

The idea of interplanetary transfer of life was originally suggested at the turn of the 19th century as the theory of "panspermia." However, it has been subjected to criticism with arguments such as: it cannot be experimentally tested; it leaves aside the question of the origin of life, and living organisms will not survive long-time exposure to the hostile environment of space, especially vacuum and radiation. Recent experimental evidence, summarized below, has led to a reexamination of the feasibility of the notion of interplanetary transfer of living material, particularly microorganisms.

The scenario of interplanetary transfer of life (figure 13-8) involves three basic hypothetical steps: (i) the escape process, i.e., removal of biological material which has survived being ejected from the surface into space; (ii) an interim state in space, i.e., survival

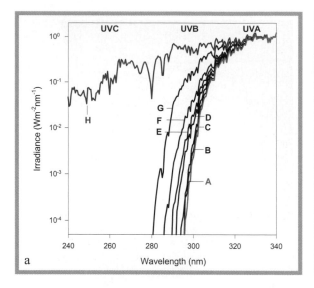
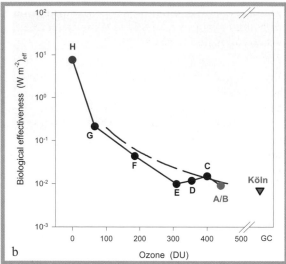

Figure 13-7: Increase in biologically effective UV irradiance with decreasing simulated ozone column thickness as measured by the spore biofilm technique in a space experiment. Extraterrestrial solar spectral irradiances filtered through a quartz (7 mm) plate (H) or additionally through cut-off filters simulating different ozone column thicknesses with a progressive depletion from curve A to G (a); biofilm data of biologically effective solar irradiance for the different ozone column thicknesses A to G, the extraterrestrial UV spectrum (H) and ground control (GC, Köln) (b). The dashed line shows the corresponding curve for calculated DNA damage (data are modified from Horneck et al. 1996)

of the biological material over time scales comparable with the interplanetary or interstellar passage; (iii) the entry process, i.e., nondestructive deposition of the biological material on another planet. We do not know with certainty whether panspermia is likely to have occurred in the history of the solar system, or whether it is feasible at all, but additional support for the idea of panspermia is given by a variety of recent discoveries (summarized in Mileikowsky et al. 2000; Nicholson et al. 2000).

It now seems almost certain that the only natural process capable of ejecting surface material from a planet or moon into space is the impact of a large object such as a meteorite, asteroid, or comet. Thus, two of the central questions of microbial interplanetary transfer theory are: "What physical conditions in the vicinity of an impact are sufficiently energetic to accelerate ejecta to planetary escape velocities?" and "What are the chances of spores or other microorganisms to survive these 'launch' conditions?" Because impacts are highly energetic processes, the major factors predicted to influence bacterial survival during an impact-generated launch are extremes of heat, pressure, and acceleration resulting

from transfer of energy during the shock of collision (Mileikowsky et al. 2000). However, many of the meteorites originating on Mars or the moon showed evidence of little or no shock-induced internal heating. The explanation is given by the so-called spall zone surrounding an impact, where pressure is zero at the surface and increases with depth. Thus, only the top few meters of surface crust would be ejected in an unshocked or lightly-shocked state. Laboratory

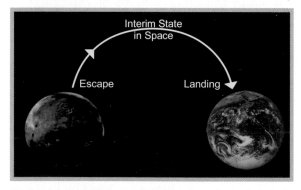

Figure 13-8: Different steps of the hypothetical scenario of interplanetary transfer of life (lithopanspermia) (modified from Horneck and Baumstark-Khan 2002)

experiments with bacterial spores simulating such a meteorite impact showed that an appreciable fraction of the population survived this simulated meteorite impact.

Once rocks have been ejected from the surface of their home planet, the microbial passengers then face an entirely new set of problems affecting their survival, namely exposure to the space environment (table 13-1). In several space experiments, the responses of resistant microorganisms (physiological, genetic and biochemical changes) to selected factors of space, applied separately or in combination, were determined.

Effect of space vacuum. Because of its extremely dehydrating nature, space vacuum has been considered to be one of the factors that may prevent interplanetary transfer of life. However, space experiments have shown that up to 70% of bacterial and fungal spores survived short-term (e.g., 10 days) exposure to space vacuum, even without any protection. The chances of survival in space were increased when the spores were embedded in chemical protectants such as sugars or salt crystals, or when they were exposed in thick layers. This was demonstrated during the six year LDEF mission (figure 13-5), where up to 80% of the bacterial spores survived exposure to space vacuum. The high resistance of bacterial spores to desiccation is mainly due to the specific structure of spores: a dehydrated core enclosed in a thick protective envelope, the cortex and the spore coat layers (figure 13-9), and a chemical protection of their DNA by small proteins whose binding greatly alters the chemical and enzymatic reactivity of the DNA (Nicholson et al. 2000).

Effects of extraterrestrial solar UV radiation. Solar UV radiation has been found to be the most deleterious factor of space, as tested with dried preparations of viruses, and of bacterial and fungal spores. The reason for this is the highly energetic UV-C and vacuum UV radiation that is directly absorbed by the DNA. The full spectrum of extraterrestrial UV radiation killed unprotected spores of *Bacillus subtilis* within seconds. These highly damaging effects of extraterrestrial UV radiation are based on specific photoproducts in the DNA that are highly mutagenic and lethal. Fortunately, these harmful UV ranges do not reach the surface of the earth because they are

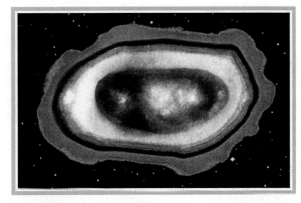

Figure 13-9: Cross-section of a spore of *Bacillus subtilis*. The spore core is surrounded by a protective envelope of the cortex and coat layers. The long axis of the whole spore is 1.2 μm, the core area is 0.25 μm² (S. Pankratz, Michigan State University, Lansing, USA)

effectively absorbed by the earth's atmosphere (figure 13-7).

This damaging effect of extraterrestrial solar UV radiation was even aggravated when the spores were simultaneously exposed to both solar UV radiation and space vacuum. Through the vacuum-induced conversion in the physical structure of the DNA, an altered DNA photochemistry occurred, resulting in a tenfold increase in UV sensitivity of the spores in space vacuum as compared to spores irradiated at

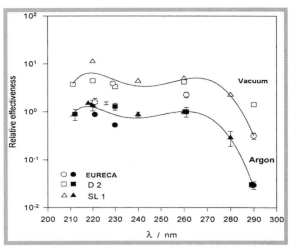

Figure 13-10: Spectral effectiveness of extraterrestrial solar UV radiation in killing bacterial spores, applied singly or in combination with space vacuum (data from several space missions)

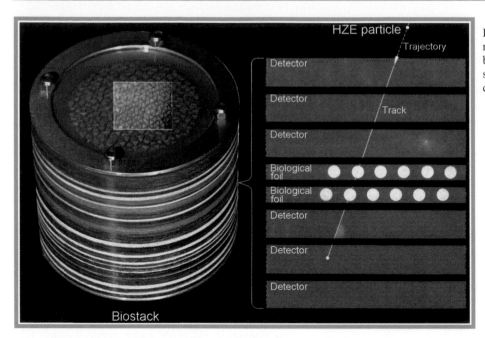

Figure 13-11: Biostack method to determine the biological effects of single HZE particles of cosmic radiation

atmospheric pressure (figure 13-10).

Effects of cosmic ionizing radiation. Among the ionizing components of radiation in space, the heavy primaries, the so-called HZE particles, are the most biologically effective species (reviewed in Kiefer et al. 1996). Because of their low flux (they contribute approximately 1% of the flux of particulate radiation in space), methods have been developed to localize precisely the trajectory of an HZE particle relative to the biological object and to correlate the physical data of the particle relative to the observed biological effects along its path. In the biostack method (figure 13-11) visual track detectors are sandwiched between layers of biological objects in a resting state, e.g., seeds, animal cysts, and bacterial spores (Bücker and Horneck 1975). This method allows (i) localization of each HZE particle's trajectory in relation to the biological specimens; (ii) investigation of the responses of each biological individual hit separately, in regard to its radiation effects; (iii) measurement of the impact parameter (i.e., the distance between the particle track and the sensitive target); (iv) determination of the physical parameters [charge (Z), energy (E) and linear energy transfer (LET)]; and finally (v) correlation of the biological effect with each HZE particle parameter.

Taking the results from several experiments in space as well as those obtained at accelerators, it was concluded that the inactivation probability for spores, centrally hit, is always substantially less than unity, and that the effective radial range of inactivation around each HZE particle, the so-called impact parameter, extends far beyond the range where inactivation of spores by secondary electrons (δ-rays) can be expected (figure 13-12). The dependence of inactivated spores on impact parameter points to a superposition of two different inactivation mechanisms: a short ranged component reaching up to about 1 μm may be traced back to the δ-ray dose, and a long-ranged one that extends at least to somewhere between 4 and 5 μm off the particle's trajectory, for which additional mechanisms are conjectured, such as shock waves, UV radiation, or thermophysical events. These data show that bacterial spores can survive even a central hit of an HZE particle of cosmic radiation. Finally, these HZE particles of cosmic radiation are conjectured to set the ultimate limit on the survival of spores in space because they penetrate even thick shielding. Because they interact with the shielding material by creating secondary radiation, with increasing shielding thickness the dose rates go through a maximum. Calculations based on space experiments have shown that, if shielded by 2 to 3 m of meteorite material, a substantial fraction of a spore population (10^{-6}) would survive exposure to cosmic radiation, even after 25 million years in space (figure 13-13).

These studies on the biological effects of the HZE fraction of cosmic radiation are not only of interest for answering astrobiological questions, they also provide important basic information in the attempts to assess the radiation risks for humans in space (see part 13.4).

Combined effects of the complex matrix of space parameters. With the LDEF mission, for the first time bacterial spores were exposed to the full environment of space for an extended period of time (nearly six years), and their survival was determined after retrieval. The samples were separated from space by a perforated aluminum dome only (figure 13-5), which allowed access of space vacuum, solar UV radiation and most of the components of cosmic radiation. It was found that even in the unprotected samples thousands of spores survived the space journey (from an initial sample size of 10^8 spores). All spores were exposed in multilayers and predried in the presence of chemical protectors. Probably, the

spores in the upper layers were completely inactivated by the high flux of solar UV radiation, but formed a protective crust which considerably attenuated the solar UV radiation for the spores located beneath this layer.

Time scales of interplanetary or interstellar transport of life. To travel from one planet of our solar system to another, such as from Mars to Earth by random motion, a mean time of several hundred thousands to millions of years has been estimated. Therefore, can the LDEF experiment be considered a reasonable simulation of an interplanetary transfer event? Because LDEF was in a near-equatorial low-altitude earth orbit, during each orbit the satellite traveled about 40,650 km every 90 minutes. Thus over the nearly six-year duration of the LDEF experiment the satellite traveled over 1×10^9 km, the approximate equivalent of the distance from Earth to Saturn if it had been flying on a straight course rather than in an elliptical orbit. The likelihood of spore survival over the time spans involved in interplanetary transfer is therefore quite high (Mileikowsky et al. 2000). More data on long-term effects are required. Exposure platforms such as the Expose on the International Space Station (ISS) (figure 13-14) are especially suited for this kind of research. Expose will support long-term in situ studies of microbes in artificial meteorites, as well as of microbial communities from special ecological niches, such as endolithic and endoevaporitic eco-

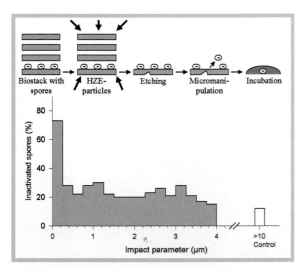

Figure 13-12: Biostack method to localize the effect of single particles (HZE particles) of cosmic radiation (biological layers are sandwiched between track detectors) and results on the inactivation probability of spores of *Bacillus subtilis* as a function of their distance from the particle's trajectory (data from Facius et al. 1994)

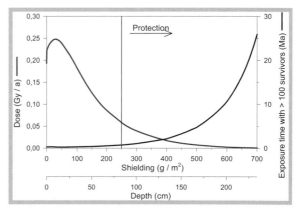

Figure 13-13: Shielding of bacterial spores against galactic cosmic radiation (GCR) by meteorite material and survival times ($\geq 10^{-6}$ survivors) at different depths of the meteorite (data from Mileikowsky et al. 2000)

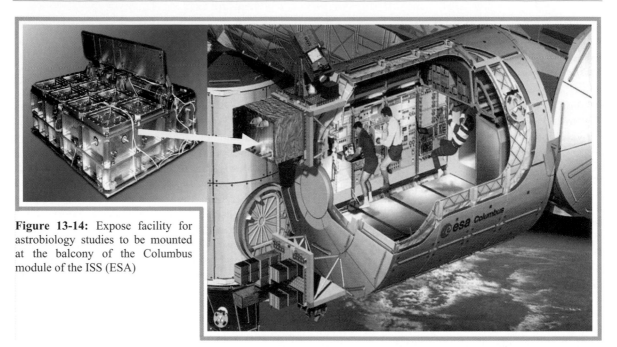

Figure 13-14: Expose facility for astrobiology studies to be mounted at the balcony of the Columbus module of the ISS (ESA)

systems. Such studies will eventually allow extrapolation to the time spans required for interplanetary or interstellar transport of life. Expose experiments will finally provide a better understanding of the processes regulating the interactions of life with its environment.

Step (iii) of the hypothetical scenario of interplanetary transfer of life described above concerns the deposition of spores from space onto a target planet. When captured by a planet with an atmosphere, most meteorites are subject to very high temperatures during entry. However, because the fall through the atmosphere takes a few seconds only, the outermost layers form a kind of heat shield and the heat does not reach the inner parts of the meteorite. During entry, the fate of the meteorite strongly depends on its size: large meteorites may break into pieces, however these may be still large enough to remain cool inside until hitting the surface of the planet; medium sized meteorites may obtain a melted crust, whereas the inner part still remains cool; micrometeorites of a few μm in size may tumble through the atmosphere without being heated at all above 100 C. Therefore, it is quite feasible that a substantial number of microbes can survive the landing process on a planet. A first attempt to experimentally study the

fate of microorganisms embedded inside a small meteorite is scheduled for the flight of the FOTON satellite in 2005 in the stone experiment (figure 13-15).

13.2.2
Perspectives of Astrobiology

Considering the upcoming opportunities to explore

Figure 13-15: Russian Foton satellite after landing: it carries the ESA facility Biopan and three Stone experiments (ESA)

the planets, moons and other bodies of our solar system by orbiters and robotic landing missions, as well as the perspectives of future human exploratory missions to the moon and to Mars, astrobiology is anticipated to provide exciting inputs for our understanding of the nature of life. The search for habitability and hence for life beyond the earth has been considered one of the intellectual driving forces in the endeavor to explore our solar system. Among the planets and moons of our solar system, our neighbor planet Mars, the Jovian moon Europa, and Saturn's satellite Titan are of prime interest to astrobiology. Here, scientists converging from widely different fields of research will jointly tackle the question of the general rules underlying the origin, evolution and distribution of life, on Earth and beyond.

13.3
Gravitational Biology

Through millions of years of evolution, life forms on earth have developed under the continuously acting force of gravity and they have taken advantage of this influence. As a consequence, the absence or increase of this force might induce both acute and chronic changes in biological systems. Investigation of the responses of single cells up to complex organisms to gravity as well as to acceleration changes are central topics in gravitational biology on the ground and in space.

Gravitational biological studies started in 1880 when Charles Darwin and Julius Sachs demonstrated the role of the root cap in downward growing plants (gravitropism). Space experimentation started in 1948 by exposing dogs, apes, and monkeys as well as microorganisms to ballistic flight conditions. The dog Laika experienced microgravity on board the satellite Sputnik 2 for one week, and died thereafter due to the lack of a recovery system. Prolonged experimentation time and possibilities for dedicated studies were subsequently provided by numerous unmanned and manned missions and by space stations. Thus, a huge amount of data on very different species is available today. Some results are summarized within this chapter (for further details see Moore et al. 1996; Cogoli 2002; Marthy 2003; Häder et al. 2005).

13.3.1
Life Under Gravity Conditions

Before speculating about the impact of microgravity on living systems, a few aspects about our life under permanent 1g conditions should be considered. Looking for example at sessile plants, their response to gravity can be easily observed. While primary roots grow downwards (a behavior called positive gravitropism) for anchoring the plant in the soil and providing it with nutrients, primary shoots grow upwards (negative gravitropism) thus providing light and the distribution of propagating stages like spores, seeds and fruits. Likewise, motile systems use gravity for their spatial orientation and have developed mechanisms to use the ecological advantage of the unique and reliable stimulus gravity, thus achieving optimal living conditions with respect to light, food, oxygen, and temperature.

This capacity is already found in primitive, unicellular organisms (protists), such as the ciliate *Paramecium*, which manage to overcome their passive sedimentation by a gravity-induced upward movement (negative gravitaxis) and a gravity-dependent regulation of their swimming velocity (gravikinesis): speeding up during upward swimming and decelerating during downward swimming (compared to the swimming velocity of horizontally moving cells). Spatial orientation of multicellular animal systems requires complex processing of a great variety of sensoric inputs. Specific receptors provide them with information about their orientation relative to the gravity vector in order to control their position and activities in three-dimensional space.

Leaving the aquatic environment and exploring land brought for biological systems a lot of problems which had to be overcome, such as body stabilization by means of a musculo-skeletal system or fluid distribution, etc. Thus it seems likely that switching off the stimulus gravity will affect the physiology of living systems.

13.3.2
Gravisensors

Living systems have developed sensors for the perception of environmental stimuli. Although the so far known gravisensors in living systems differ in their morphology and structure, there are general principles concerning how the gravity stimulus is perceived, transduced and finally used for distinct responses. Some structural descriptions of sensors are quite old; however, considerable progress in knowledge about their mechanisms resulted from experiments under altered gravitational stimulation.

Gravity only interacts with masses. Thus, size and specific density difference matter, and a mass has to be coupled to sensitive structures and to a signal transduction cascade in order to transduce the physical information into a physiological or biochemical signal. In animal systems we find the general principle of statocysts. These are fluid-filled cysts containing a dense mass, either a single statolith (otolith) or multiple statoconia, which fall as their density is higher than the density of the surrounding medium and give a signal to ciliated mechanoreceptors. Deviations of the organisms from the vertical position lead to a bending of the cilia and thus the stimulation of mechanosensitive ion channels of the receptor cells. The subsequent steps are changes in membrane potential, signal transduction and computation in the nervous system (brain), finally resulting in motor responses. A fascinating analogy is already found in unicellular systems in the form of statocyst-like organelles in the ciliated protist *Loxodes*. They possess so called Müller organelles, which are vacuoles containing a heavy body of barium sulphate connected to a stick of microtubules (figure 13-16).

When *Loxodes* deviates from the vertical, displacement of the barium granulum provokes changes in membrane potential, which finally controls the functioning of the body cilia and thus the orientation of the whole cell. Disruption of the connection between the stick and the heavy body by means of a laser beam leads to disorientation of the cell, demonstrating the function of this structure as a gravisensor. The statocyst-like organelle of *Loxodes* seems to be an acquisition due to adaptation to special living conditions as the cell often creeps on surfaces of detritus or in narrow gaps of the benthos, an environment which might favor the development of an intracellular gravisensor (Häder et al. 2005; Bräucker et al. 2002). For other systems, such as slime molds or fungi, ubiquitous cell organelles, for example nuclei with only a marginally higher den-

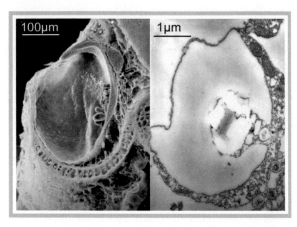

Figure 13-16: Electron micrographs of a multicellular gravisensor (vertebrates; frog *Xenopus laevis*) (left) and a cellular gravisensory organelle (protist; ciliate *Loxodes striatus*) (right). (*Xenopus*; J. Neubert, DLR, Cologne

sity as compared to the surrounding cytoplasm, have been proposed to be gravireceptor candidates. A further and more common mechanism and maybe early invention in evolution seems to be the ability of unicellular systems to recognize the direction of gravity by using their whole cell mass as statolith (statocyst hypothesis). Electrophysiological studies revealed that—at least in ciliates—two kinds of mechanosensitive ion channels are distributed in a polar manner in the cell membrane, which is ideally suited for perception of a unidirectional stimulus such as gravity (figure 13-17). Mechanical stimulation of these channels induces distinct changes of the membrane potential (de- or hyperpolarisation) and, in turn, as the membrane potential controls the beating pattern of the cilia, distinct behavioral responses (faster or slower swimming and reorientation). Based on this structural prerequisite, it was proposed that the higher density of the cytoplasm of the cell (*Paramecium:* 1.04 g/cm^3) compared to the density of the surrounding medium (1.00 g/cm^3) leads to an outward deformation of the cell membrane and therefore stimulation of the mechanosensitive ion channels in the lower region of the membrane. As a consequence, an upward swimming cell should increase its forward swimming velocity due to a stimulation of mechanosensitive potassium channels, thus intensifying the beating frequency of the cilia, while downward swimming cells reduce their swimming velocity due to a stimulation of mech-

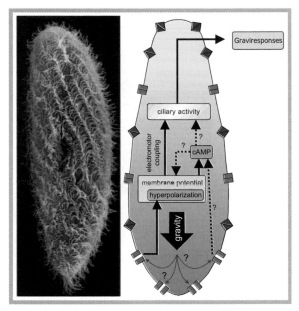

Figure 13-17: Proposed model of graviperception in the ciliate *Paramecium*. Mechanoreceptor channels are located in the cell membrane in a bipolar manner: potassium (blue, posterior) or calcium channels (red, anterior). They are activated by the mechanical load of the cytoplasm. Depending on the cell's orientation different kinds of channels are activated leading to distinct changes in behavior, which result in the graviresponses. Left: scanning electron micrograph of a *Paramecium* (W. Foissner, Salzburg); right: scheme (modified after Bräucker et al. 2002)

anosensitive calcium channels—and this is exactly what has been measured (for review Häder et al. 2005; Bräucker et al. 2002).

Regarding plant systems we find again the principle of a heavy mass in combination with a sensing apparatus. The multicellular green alga *Chara* produces unicellular rhizoids which grow downwards (positive gravitropism) (figure 13-18). The region where graviperception takes place is restricted to the apex of the rhizoid where 50 to 60 vesicles filled with dense crystals of barium sulphate (the same material which is also used by the already mentioned animal system *Loxodes*) serve are sedimentable masses. Interestingly, these statoliths do not sediment all the way into the tip of the rhizoid. They are kept in a dynamically stable position 10-30 μm above the apex by cytoskeletal forces, namely, the acto-myosin system (Braun et al. 2002). When rhizoids are placed horizontally, the statoliths sediment within a few minutes onto specific sensitive plasma membrane areas, where according to a current hypothesis gravisensor molecules are located. These molecules are activated by the sedimenting statoliths and play a pivotal role in the initiation of the signal transduction pathway of gravitropism in *Chara* rhizoids. Activation of the gravireceptors was shown to result in differential growth of the opposite cell flanks by a local reduction of cytoplasmic free calcium and thus a reduction of elongation growth of the lower cell flank until the positively gravitropic tip-downward orientation is re-established.

While in *Chara* rhizoids graviperception and response are located within the same cell, the sites of gravity perception and response are spatially separated in higher plants. Here, graviperception occurs in specialized cells (statocytes) in the root cap columella cells and in the shoot endodermal cells adjacent to the vasculature, while growth response occurs in the elongation zone and in the epidermis, where differential growth of cells on the opposite flanks of the organ generates the gravity-guided curvature. At the perception site, statoliths in the form of starch-filled amyloplasts represent the heavy masses, which sediment towards the lower cell flank and rest on a cushion of endoplasmatic reticulum vesicles (statolith sedimentation hypothesis) (for review see Hemmersbach et al. 1999; Blancaflor and Masson 2003). There is clear evidence from experiments using a high-gradient magnetic field to displace amyloplasts in statocytes of nongravistimulated roots that amyloplast displacement in the statocytes is sufficient to induce the gravitropic response (Kuznetsov et al. 1999).

Alternatively it was postulated that not only sedimenting amyloplasts but also the entire protoplast might serve as a statolith, thus activating putative gravireceptor molecules and initiating the gravitropic response (protoplast pressure hypothesis). Gravistimulation, or positioning a plant horizontally, causes rapid changes in the membrane potential and the cytosolic pH in the specialized columella cells of the root cap in *Arabidopsis*, but not in other cells. It was also proposed that sedimenting statoliths might stimulate mechanosensitive ion channels by affecting actin as a transducer of the gravity signal. Modu-

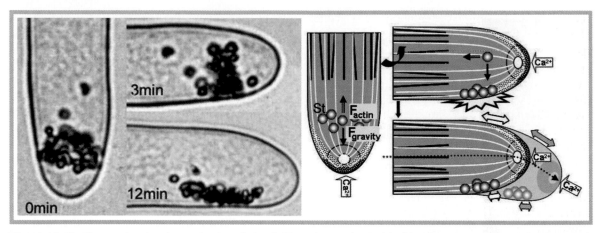

Figure 13-18: The gravity-signal transduction chain of *Chara* (green algae) rhizoids (diameter = 30 μm, left microscopical picture, right scheme). In tip-downward growing cells, statoliths (St) are positioned by two compensating forces: actomyosin forces (Factin) and gravity (Fgravity). Upon gravistimulation (horizontal positioning), statoliths sediment onto the lower cell flank where activation of gravireceptor molecules causes a local reduction in the concentration of cytosolic free calcium leading to inhibition of exocytosis and, thus, differential extension of the opposite cell flanks (double-headed arrows). The white arrows point to the area of maximal calcium influx that defines the cell tip. Microtubules, black lines; actin microfilaments, white lines (M. Braun, University of Bonn, Germany)

lation of ion channel activity might create a signal that is transmitted to the site of differential growth and this transmission step involves the phytohormone auxin, both in the shoot and the root (for more information see Blancaflor and Masson 2003). Understanding the signal transduction chain from perception to gravitropic bending provides benefits for agriculture on earth. Wind and rain have a devastating effect on crop production if they occur late in the life cycle of a plant shortly before harvesting. While young plants have the capacity to straighten up and resume upward growth, this capacity decreases with age (Blancaflor and Masson 2003). Understanding the mechanisms can provide the basis for developing counterbalancing mechanisms to prevent the decreasing gravity responses of old plants. Furthermore, understanding the multifaceted mechanisms plants use to cope with manifold environmental stimuli and their potentials to adapt to altered gravitational fields is essential for building life support systems to be used during prolonged space flights, where plants play an essential role.

There is increasing evidence that multiple mechanisms exist in different organisms, and maybe even parallel mechanisms in the same organism which would warrant a response to the stimulus, if one of

the mechanisms was disabled by mutation. While it seems fairly clear that at least the major players involved in graviperception are identified, their interaction remains less certain.

13.3.3
Experiments in Microgravity

Impact on Single Cells

The observation that free-swimming unicellular organisms tend to swim to the top of a tube was made more than 100 years ago. Scientists ever since have been eager to understand the underlying mechanism. The real breakthrough came with space technology, which has provided the tools to study these processes in the absence of the gravitational force (for reviews see Hemmersbach et al. 1999; Häder et al. 2005). It had to be shown that "gravitactic" organisms are really capable of detecting the gravitational field of the earth and thus the term "gravi" is justified for the observed orientational movement. A clear proof was derived from space experiments on sounding rockets with model organisms: the ciliated protist *Paramecium* and the unicellular green alga *Euglena*. Both species showed pronounced upward swimming before start of the rocket flight (figure 13-19).

However, after onset of microgravity the precision of orientation deteriorated and after about one minute swimming in random directions was observed (figure 13-19). Likewise, cress roots oriented in random directions in microgravity, compared to the downward growing (positive gravitropic) ground controls (figure 13-20).

The fact that the graviresponses do not immediately disappear after transition to microgravity but show relaxation times in the range of minutes indicates the involvement of elastic (cytoskeletal) elements in gravisensation. Unambiguous evidence for the predominant role of the cytoskeleton (acto-myosin system) in a cellular gravity signal transduction chain was provided by a number of experiments with the *Chara* rhizoid under altered gravitational stimulation (space shuttle, sounding rocket flights, clinostats and centrifuge) (Braun et al. 2002).

Recently a potential role of the ubiquitous second messenger cyclic adenosine monophospate (cAMP) in the gravity signal transduction chain was shown for protists (*Paramecium* and *Euglena*) and can be predicted also for other systems. Fixation of protozoan cultures at dedicated times during a microgravity experiment lasting seven minutes revealed gravity-dependent changes in the levels of cAMP, though the time course of events in the transduction chain remains so far open (for review Häder et al. 2005).

The answer to the question of a threshold for gravity sensitivity in a biological system can only be given in space. One result is mentioned here: during the IML-2 mission (1994) samples of different protozoan species were investigated on different mission days. Samples which had been cultivated for three days in space on a 1 g reference centrifuge were transferred to a centrifuge microscope ("NIZEMI") accelerated to 1 g, then stepwise down to 10^{-3} g (= stop of the centrifuge). Negative gravitaxis of the paramecia was measured with decreasing precision up to 0.32 g and achieved a random distribution at and below 0.16 g. Alternatively, cells which had been cultivated for 10 days in microgravity were exposed to an increasing acceleration profile. Now random distribution was measured up to 0.16 g

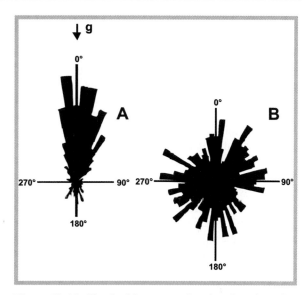

Figure 13-19: Circular histograms showing the orientation of *Paramecium* cells. While they show negative gravitaxis (precise upward swimming) under 1 g conditions (A), random distribution can be observed in microgravity (B) (modified after Hemmersbach-Krause et al. 1993)

while negative gravitaxis was induced at ≥ 0.32 g. Comparable experiments with *Euglena* algae revealed threshold values of ≤ 0.12 g for this species, and of ≤ 0.16 g for *Loxodes*. The data indicate that these small organisms are able to detect at least 10% of the normal gravitational field strength. The low threshold values also indicate that from an energetic

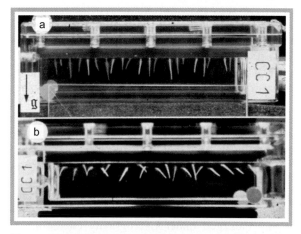

Figure 13-20: Cress roots grown on ground (a) (positive gravitropism), and in microgravity (D1 mission) (b) (random orientation) (D. Volkmann, University of Bonn)

point of view a system for signal amplification has to be postulated; otherwise, the signal to noise ratio would become critical. Signal enhancement and signal focusing might be due to the cytoskeleton and second messenger cascades (for review Häder et al. 2005). It can be speculated that gravity is not the limiting factor for perception and orientation capacities of unicellular organisms on other celestial bodies, e.g., the moon (0.166 g), Mars (0.380 g), Venus (0.879 g) and even on Jupiter (2.339 g), as centrifuge experiments on the ground revealed that protists are even able to sense accelerations higher than 1 g. Besides the effects on gravity orientation, other cellular and physiological effects in protists were altered in microgravity, such as an increased proliferation rate compared to 1 g conditions, which might indicate faster aging in space, though the general fitness and structure of the cells appeared unchanged.

Behavioral and biochemical studies with sperm cells, which are also free-swimming flagellated cells, revealed a gravity-dependent increase in sperm velocity and motility correlated with biochemical events, such as phosphorylation of specific proteins (Tash and Bracho 1999). It is important to study the long-term effect of microgravity on spermatogenesis, sperm function and fertilization, as they might have a direct impact on reproduction, e.g., of fish, which may provide an essential food source during space travel.

We have seen that unicellular systems respond to altered gravitational stimulation, thus the question arises about the impact of gravity (microgravity) on the cells in our body, that means is gravisensation a general ability of each kind of cell? Experiments in microgravity revealed that many functions in humans, such as bone development, heart function and immunocompetence are altered (see 13.4). One reason might be a direct effect of gravity on the cellular level. An experiment on Spacelab 1 (1983) revealed that mitogenic activation of T lymphocytes, key players of the immune system, was nearly completely inhibited in microgravity. Further experiments revealed, for example, that lymphocytes were highly damaged under microgravity conditions in comparison to the intact 1 g controls, leading to the hypothesis that programmed cell death (apoptosis) is

increased in microgravity. This hypothesis was later supported by microscopical and biochemical studies (for review see Lewis 2002; Cogoli, 2002).

Another system which has been studied in microgravity is the bone system. Bones are highly sensitive to mechanical stress. It was shown that microgravity affects mineral metabolism and cell differentiation in organ cultures of living bone rudiments and in osteoblastic cell lines, thus a direct effect on the cellular level is assumed (Hughes-Fulford 2002). Understanding the underlying mechanism might provide an explanation for the so far limited effectiveness of exercises during space flight. Recent experiments stress the cytoskeleton as an appropriate player in interacting with gravity as it controls fundamental cellular functions, such as maintenance of cell architecture, cell motility, cell division, as well as signal transduction steps with respect to transduction of mechanical stimuli into bone cells, regulation of enzymes, of ion channels and of gene expression. The spontaneous self-organization of in-vitro preparations of microtubules was studied in microgravity, revealing disturbances of these processes, thus showing on the other side that gravity triggers self-organization processes of microtubules under the specific experimental conditions. Such findings are of primary importance because they indicate a potential direct effect of gravity on intact cells. Cytoskeletal anomalies have been reported for many, but not all, space-flown cell cultures such as disturbances in the linearity of actin filaments or reduction in the number of stress fibers. Furthermore, the inhibition of the key enzymes such as protein kinase C (PKC) was stated. In addition, the genetic expression of the early oncogenes c-fos and c-jun was found to be significantly depressed in microgravity, and by using DNA microarrays it was shown that the expression of 1,632 genes in human renal cortical cells was altered after eight days of cultivation in space, though the final consequences for the organism are unclear (for review see Lewis 2002).

Future investigations and understanding of the microgravity-induced alterations in cellular mechanisms and structures in mammalian cells are essential, since, for example, suppression of the immune system or the induction of cancer may become a serious health issue on long-term space flights.

Development in Space – Multigeneration Experiments

The development of a complex organism follows a sequence of events during which the fertilized egg divides and gives rise to millions of cells. A complete development undergoes cleavage, gastrulation and organogenesis, all of which are under genetic control but can also be influenced by environmental factors. Research activities address aspects of embryonic development and thus behavioral, morphological and physiological studies in developing and growing organisms in microgravity. Topics are to elucidate the role of gravity on development and aging as well as long-term adaptation to a new environment. Can organisms undergo normal development in microgravity? How stable is the eucaryotic genome to environmental changes and thus to space conditions? If not, this might be a limitation with respect to long term exploration or colonization of other planets or celestial bodies by human beings. Are the space-induced changes reversible? Is there a gravity-dependent sensitive period in the development of organisms, especially in the development of the gravity sensory systems, as has been shown for other stimuli (e.g., light – visual system)?

So far, current knowledge from relatively short-term microgravity experiments reveals the following contradiction: while unicellular systems sense the absence of gravity as indicated by altered signal transduction pathways, the development of higher organisms appears to be unchanged in microgravity as the adult and complex organisms appear normal. This has been shown for a broad variety of model organisms, from low to highly evolved invertebrates and vertebrates (for review see Marthy 2003). This situation makes predictions about what will happen in microgravity rather difficult. Though there are a lot of hints about effects of microgravity on early developmental processes, regulatory capabilities are obviously able to correct and overcome some of these microgravity-induced modifications. But how stable are these changes? Though for example amphibian embryos grown in microgravity show a thicker blastocoel roof compared to 1 g controls, they nevertheless develop into normal tadpoles. Some species grown in microgravity showed an increase in statoliths in their statocysts, but what is the physiological consequence for them? Long-term multigeneration experiments and complete life cycles in dedicated hardware taking account of the lessons learned are a prerequisite for future successful developmental biological studies in a space laboratory. Knowledge of complete genomic sequences offers a new aspect in understanding how biological systems function, adapt and evolve. Ideal model organism such as the fruit fly *Drosophila*, with 300 generations in 15 years, or *Arabidopsis*, a small plant in the mustard family which is currently the focus of an international gene sequencing project analogous to the human genome project, or fish with extensive homology to humans, offer exciting and promising approaches for space experiments.

The maintenance of life and the capability to reproduce do not necessarily ensure that higher brain functions have developed normally and are functional in organisms raised in microgravity. The development of the brain and vestibular system in microgravity and the maintenance of the vestibular system's capacity for neuroplasticity are of high importance with respect to adaptation to an altered gravitational environment—transition to microgravity as well as readaptation to 1 g conditions after long-term space travel. Humans and other vertebrates experience dysfunction of the vestibular system due the lacking stimulus of gravity, resulting in transient behavioral disturbances (kinetoses, called "space motion sickness in the case of humans; see 13.4). After hours to days in microgravity these symptoms disappear and the systems have obviously adapted. In order to understand the phenomena, fish has shown to be an ideal model system (figure 13-21). In the case of fish, transition to microgravity alters their swimming behavior and "spinning" and "looping" responses have been described (for review see Anken 2003).

Further studies in altered gravity revealed that asymmetric otoliths are the main reason for the observed kinetoses. Furthermore, otolith development is obviously regulated by a feedback mechanism controlling the gravity-dependent mineralization processes and enzyme activities (carboanhydrase). In addition, it was stated that microgravity induces the development of additional synapses in the brain and increases metabolism in the vestibular brain

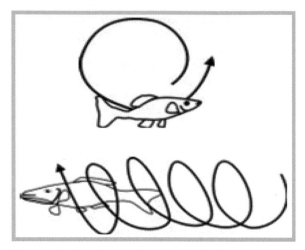

Figure 13-21: Abnormal transient behavioral responses (kinetoses) of fish after transition to microgravity (R. Anken, University of Hohenheim)

parts, indicating compensation processes via neuronal activities (Anken 2003).

With respect to plants and their reproductive development, the microgravity environment has a lot of scientific potential. Besides serving as test systems in order to understand the signal transduction leading to the well known gravitropic responses of plants, plants are an important source of fresh food for astronauts. Furthermore, plants contribute to a comfortable environment in space stations and thus the well-being of astronauts. Studies on how to grow plants successfully in space involve assessments of plants suited to grow in space, the conditions needed for optimal crop yield, and the problems in growing plants in a gravity-free closed system. One of the major outcomes from relatively short-term space experiments with higher plants performed so far, e.g., observation studies concerning growth and movement studies of the statoliths, clearly demonstrated that the actomyosin system mediates the movement of the amyloplasts in roots. Determination of perception and presentation time (parameters to estimate gravity sensitivity) occur within seconds, therefore the simple weight of the amyloplasts may be sufficient to alter the tension of actin filaments, to induce signal cascades and finally the gravitropic response. Future studies focus on identifying how the physical signal resulting from the sedimentation of amyloplasts is translated into a physiological,

biochemical signal (e.g., involvement of second messengers and regulation of lateral auxin transport) to initiate the response. Future experiments in prolonged microgravity in a space laboratory will also focus on growing plants for several generations in space (seed-to-seed experiment) (for review see Blancaflor and Masson, 2003).

Long-term experimentation in a space laboratory poses technical challenges with respect to the demands of the organism and automation. Problems of long-term storage and sample preparation in microgravity have to be overcome by suitable hardware. Control experiments have to be performed in a 1 g reference centrifuge, in order to discriminate unspecific space effects from the effects solely induced by microgravity. Experiment racks such as Biolab foresee the incubation and processing of samples under controlled conditions, in microgravity and simultaneously on a 1g centrifuge (figure 13-22).

For long-term cultivation the development of space bioreactors and artificial ecosystems are needed (Binot 2002). Regenerative life support systems are necessary for long-term manned space missions in order to recover edible biomass, water and oxygen from feces, urea, carbon dioxide and minerals. One example is C.E.B.A.S (Closed Equilibrated Biological Aquatic System) which allows cultivation of different aquatic species in order to study the influence of microgravity on the physiology, morphology, and biochemistry of the species and providing the astronaut with oxygen, water and food and the possibility to deposit waste (Slenzka 2002). Another ecosystem called MELiSSA (Micro-Ecological Life Support System Alternative) has been conceived as an ecosystem based on micro-organisms and higher plants. A functioning pilot plant for long-term validation of the system is planned for 2006/2007 (http://www.estec.esa.nl/ecls). Suitable life support systems will be the basis for exploration and for the success of human space flight, for example to Mars or a lunar base.

13.3.4
Perspective of Gravitational Biology

What will happen if living systems are exposed to prolonged microgravity conditions? How will this

Figure 13-22: Biolab – a multi-user experimental rack for biological experiments in space (D. Ducros, ESA)

change in environmental conditions influence behavior, vitality, proliferation, ageing, and differentiation processes? Multigeneration experiments in a space laboratory hold still unknown results. Excellent hardware and preparation set-ups are the prerequisite for successful experimentation in space. Knowing complete genome sequences and the finding that molecular mechanisms are conserved across phylogeny offer new approaches in understanding adaptation and evolutionary processes. Knowing how organisms manage the absence of gravity is mandatory for planning long-term, orbital and interplanetary missions.

13.4
Human Physiology

Since the first flight of Juri Gagarin on April 12, 1961, tremendous progress has been made in human space flight. Humans landed on the moon on July 20, 1969, space stations were developed, a reusable human space flight system, the space shuttle, has been operative for over two decades and as of today, three nations—the USA, Russia and China—have direct access to space for humans. We are presently in a transition phase in which the International Space Station helps us to prepare for further human exploration beyond earth orbit, and in which new space transportation system generations will be developed.

The role of medicine and psychology in health, performance and safety of crews has always been crucial for the success of missions and for the question of which next step in human exploration can be taken with acceptable risk and costs. Beyond this direct operational task of medicine and psychology, medicine makes use of the opportunity to eliminate the gravity factor when establishing and examining concepts of how the human body functions. If we knew how our body functions on earth, we could simply predict what will happen without gravity, which always influences our body functions on earth. Indeed, it is most fascinating to see that a multitude of body functions behaves in an unpredicted manner in microgravity, which points to obvious problems in our understanding of the human body and helps in the fascinating mission to explore it.

Since these investigations often require the development of new tools that by themselves have interesting applications and markets on earth, lay persons often assume these additional successful application aspects ("spin offs") for medicine to be a reason per se for human space flight. However, they are just welcome add-on benefits to the primary goal to understand our human body and to apply these results of basic research in medicine both on earth and in space (further reading, e.g., in Seibert 2001).

Medicine serves the purpose of expanding the possibilities of humankind and enabling human exploration of our solar system. It profits from human space flight by using the possibility to have an additional tool—the absence of gravity—for the equally fascinating challenge to explore the human body. It has the additional advantage over many other areas of basic research and exploration in that it develops knowledge and technologies that can be applied directly on earth.

13.4.1
Space-related Aspects of Human Health

Launch and Transport to the ISS

The risk of the severity of infectious diseases appears increased in microgravity and treatment is difficult. Astronauts and cosmonauts therefore undergo a phase of seven days of preflight quarantine. If the launch time is in the night or early morning, this quarantine phase can also be used for a sleep shifting program to achieve that the astronauts are at the "wake" phase in their circadian rhythm when they launch. While the shuttle system allows crews to float around freely soon after launch, the Soyuz capsule is very small. Crews are therefore forced to stay most of the time in their seats until docking two days after launch to the International Space Station (ISS). To avoid both systems (shuttle and Soyuz) from having to interrupt launch procedures due to the need of crew members to urinate while they wait in launch position in their launch and entry suit, they usually wear absorbable underwear ("diapers") during launch. In addition, cosmonauts launching in the Soyuz system can use a diuretic (furosemide) on the preflight evening and receive a rectal enema to decrease urination frequency and avoid defecation, respectively, during the two days before docking to the ISS (Ewald et al. 2000).

Acceleration during launch is moderate in the shuttle system (< 3 g), and in the Soyuz system (three short peaks up to 3.5 g). Especially the landing in the Soyuz system poses very high deceleration loads on the astronauts (up to 7 g). Therefore, preflight centrifuge training of the accelerations during launch and landing is mandatory in the Russian system; it is not done in the American system.

Space Adaptation Syndrome and Neuro-Vestibular Adaptation

A mismatch between the sensory systems that indicate the position and acceleration state of the body occurs immediately after entry to microgravity within eight minutes after launch. Especially otolith function, which is adjusted to the 1 g acceleration on earth, has to readapt dramatically to microgravity. In some individuals, otoliths apparently are of different mass, which is on earth compensated by neuronal plasticity. In microgravity, where the body is weightless and not moving, otoliths are not giving any signal on acceleration. However, neuronal compensation still exists in persons with different otolith sizes, indicating to the subject a tilted body position. Such astronauts may be very sensitive to space motion sickness. Once gravity-sensing systems are eliminated, the human brain may answer the question "where is up, where is down?" in different ways. For about 46% of astronauts, "up" seems to be where the head is (body axis), for the same percentage "up" is where the spaceship "ceiling" is, and for the rest, the brain switches between both principles (Reschke et al. 1998). Since the gravity sensors cannot indicate body position, movements of the head do not allow the human body to decide clearly whether the head or the environment (the astronaut or the spaceship) has moved. This and many other mismatches lead to many illusions.

Phylogenetically, the human body experiences a sensory mismatch between body systems (e.g., sensory illusions) only from intoxication. Therefore, such mismatches between visual and other inputs automatically lead to the decision of the brain "I must be intoxicated" and the brain orders the body to immediately get rid of the toxin. As a consequence, over 60% of astronauts suffer from symptoms of motion sickness and up to 30% have episodes of vomiting ("space motion sickness" or "SMS"). Fortunately, this episode is terminated after two to three days for most space travelers, when neuronal plasticity allows the body to adapt and learn that the new situation is not intoxication, but a normal situation.

A number of sensory inputs change during entry into microgravity. Proprioceptors that usually tell us whether and how we stand or sit give no signals; the arms and legs, whose weight we do not normally notice, suddenly feel floating. Rapid movements may be dangerous because they may lead to touching a wall, which may result in an unexpected movement of the whole body into a new and unexpected direction. Rapid head movements may lead to otolith and labyrinth organ stimulations that are perceived much more intensely than under 1 g conditions. Dropping something does not work anymore, since things float. All these changes lead to decrements in performance. This can be demonstrated impressively by difficulties in writing in microgravity. Eye-hand coordination and several aspects of single and multiple tasking deteriorate. Adaptation of these performances to preflight standards fortunately occurs, but this adaptation may require up to one month in microgravity. Thereafter, performance may return to and remain at preflight levels for very long stays in space (Manzey et al. 1998). Massive decrements in performance then will again occur after reentry to 1 g. Reentry makes readaptation necessary and may lead to motion sickness, illusions and to performance decrements.

Fluid Shift, Salt and Fluid Regulation, Hunger and Thirst

Immediately after entry into microgravity, body fluid is distributed freely in the body without the interfering influence of earth gravity. Surprisingly, there are not only about two liters of redistribution from the lower extremities to the upper parts of the body; fluid is especially redistributed in the subcutaneous skin areas of the upper body. As a result, the heads of astronauts appear swollen ("puffy face syndrome") and wrinkles disappear, while the legs decrease in volume ("bird legs" or "spider legs"). Astronauts sense the redistribution of fluid to the head and report on an unpleasant feeling of nose, mouth and throat congestions "as if I had a severe cold" (H. Schlegel, ESA astronaut, personal communication).

This upward fluid redistribution is accompanied by a rapid extravasation of 10 to 15% of plasma volume into the interstitial spaces. Due to this extravasation, blood cell concentration and the hematocrit increase by over 10% within hours of launch (Churchill and Bungo 1997). In contrast to predictions based on our knowledge of physiology on earth, the upward fluid shift has—to the knowledge of the authors—never induced a documented increase in fluid excretion. In contrast, so far all measurements of fluid excretion after entry into microgravity have found an unaltered or a decreased level (Norsk and Epstein 1991; Drummer et al. 2000), although the size of the heart is increased, demonstrating an increased return of blood into the heart. Moreover, measurements of central venous pressure on the first day in space demonstrate a decreased central venous pressure when the reference pressure determination is done

outside the body (Buckey et al. 1993). This finding supports the findings of decreased diuresis and natriuresis in space. It does, however, not explain the discrepancy to what was predicted from our knowledge of physiology.

The main reason for the discrepancies might simply be that a compression of the thorax is induced by gravity on earth, but it does not occur in microgravity. This missing compression would allow the heart in space to expand and even to manage an increased return of blood without an increased transmural pressure. In addition, massive stress on the first days of microgravity may activate fluid retaining mechanisms. So, the main reason for the unexpected changes would be an increased volume capacity in the thorax, combined with stress effects. All reflex mechanisms acting on earth, such as the "Henry Gauer Mechanism" (diuresis and natriuresis while pressure in the heart rises), would thus indeed work normally.

Interestingly, our knowledge of the effects of thorax pressure on normal physiology is extremely small, although such changes are regularly induced during artificial ventilation, during narcosis and in intensive care. Due to the possibly high relevance for intensive care medicine—and for understanding changes in microgravity—DLR and ESA have now initiated a study series to systematically clarify the influences of changes in intrathoracic pressures on a variety of body functions including natriuresis. Preliminary data suggest many such effects (Baisch and Gerzer 2004).

Most astronauts lose between 1 and 3 kg body weight during their first week in space. For many, but not all astronauts, this loss in body mass continues and adds up to several kilograms during a long flight. Fluid loss by increased diuresis cannot account for these losses, since it simply does not occur. This loss appears to result from decreased hunger and thirst sensation resulting in a negative body fluid and energy balance. The scientific reasons for this decreased thirst and hunger are still not clear, but may be related to the shift in body fluid and the concomitant extravasation in microgravity.

During experiments with only a single astronaut on the Mir 97 mission, we obtained a very unexpected result: When this astronaut did a two week diet monitoring experiment, he not only complained about having to eat too much food, he also had to eat without hunger (he ingested calories needed for "light work" and did not lose nor gain weight during this phase), but his body stored large amounts of sodium without storing water (figure 13-23; Drummer et al. 2000). Since this phenomenon had not been known before to occur on earth and is in complete disagreement with physiology textbook knowledge, we undertook several studies to find out whether an important mechanism for sodium handling could have been overlooked. It turned out that this was indeed the case (Heer et al. 2000).

The human body has a very high capacity sodium storage mechanism that had not been discovered before, and was discovered by carefully studying a single astronaut in space. Probably the threshold to activate this storage mechanism is decreased in microgravity and already operative in space at "normal" sodium (Na^+) ingestion of 200 meq Na^+ per day, while it requires high sodium ingestion on earth.

Since a variety of body functions and diseases are linked to sodium homeostasis in the body, we assume that these findings open not only a novel field for scientific research, but will have impact on the way in which several diseases, such as several forms of high blood pressure, will be treated in the future. Thus, recent data from Titze et al. (2004) not only confirm our data also in rats, but extend them to localization of the storage mechanism in the skin through exchange of sodium with protons. These findings also suggest that rats prone to salt-sensitive high blood pressure do not have the novel salt storage mechanism. We are therefore currently preparing to test whether this mechanism is also missing in humans with salt sensitive high blood pressure.

Stress and Hormonal Alterations

Entry into microgravity alters sensory inputs and reaction patterns of the whole body and induces stress, especially in space novices. This is increased by the huge workload usually imposed on astronauts, by their own motivation to perform optimally, by the burden to be "visible" to the whole world, and by the initial decrements in performance due to the space environment. This high stress is, naturally,

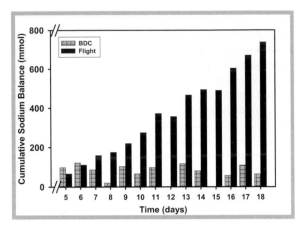

Figure 13-23: Sodium storage of one astronaut during a two week metabolic balance study in space (from Drummer et al., 2000). Daily sodium ingestion was 180 meq per day. The astronaut did not store fluid during the same study period. Daily sodium excretion in space (black bars; "flight") and during an identical study protocol on earth (grey bars; "BDC", Baseline Data Collection Control)

often accompanied by sleep problems leading to sleep deficits that increase performance problems and stress levels, respectively. Especially during the first week(s) in microgravity, physiological reaction patterns are often accompanied by high stress. Thus, scientific results obtained from investigations at least during the first weeks of microgravity must be considered very carefully and might be strongly influenced by stress effects.

Many hormonal patterns are strongly influenced by stress. The most well known "stress hormones" include catecholamines and cortisol. These hormones influence a myriad of body functions. Recent results suggest an upregulation of the sympathetic nervous system in microgravity (Ertl et al. 2002). So far, it cannot be decided whether microgravity per se increases the sympathetic nervous tone or whether these results are solely consequences of increased stress during short-term missions. Similar questions are open concerning cortisol and the immune system (see below), and concerning other hormones, such as fluid regulating hormones. One of the most stress-sensitive hormones is the antidiuretic hormone ADH. Its levels are increased tremendously under different stress forms, and high ADH decreases fluid excretion. Thus, when we investigate any body function such as fluid excretion, we have to be aware that all these functions may be influenced by stressful conditions. Taken together, it is not surprising that the levels of various hormones have been inconsistent (Leach Huntoon et al. 1994), since many hormones are very sensitive to stress situations, which can vary considerably for different space flights and different astronauts. One can therefore not be careful enough in taking possible stress effects into account when one makes conclusions about "space" effects on body functions.

Cardiovascular and Pulmonary Adaptation

Blood pressure and heart rate show only small changes in space. During the initial adaptation to space, there is a slight increase in heart volume during the first week in space followed by a decrease that probably occurs due to a continuous reduction in blood volume. Later on even the thickness of the heart muscle decreases in space (Charles et al. 1994). Direct measurements of sympathetic activity during the Neurolab space shuttle mission in 1999 indicated an increased sympathetic tone that may be responsible for a slightly increased diastolic blood pressure in space (Ertl et al. 2002). After return to earth, astronauts experience orthostatic intolerance that is most probably due to altered cardiovascular reflex mechanisms by a remodeling of sympathetic responses, especially in astronauts that have low alpha 1 adrenergic activity already preflight (Platts et al. 2004). An important underlying factor appears also to be a fluid deficit—not necessarily due to increased renal fluid loss, but rather due to decreased thirst and hunger during space flight. Figure 13-24 summarizes several components that may contribute to this intolerance. Recent observations suggest that anti-g suits or special boots worn during landing and/or medications (alpha 1 adrenergic agents like midodrine) might be promising approaches for developing countermeasures for orthostatic intolerance.

Lung function in space has been studied intensively during recent years. No major alterations in lung functions occur during microgravity. In general, lung function appears to improve slightly in space, while ventilation distribution changes in an unpredictable way and is by far not as homogeneous as

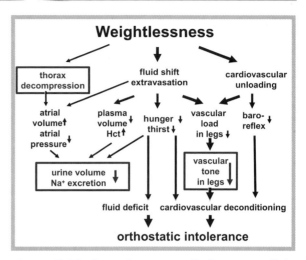

Figure 13-24: Some factors contributing to postflight orthostatic intolerance of astronauts. Factors that were only recently recognized are highlighted in red

expected from our terrestrial knowledge (West et al. 1997).

Blood Cell Formation, Immune System

Immediately after entry into microgravity, about 10% of the body's plasma shifts to the extracellular volume. This increases hematocrit and, in turn, induces the kidneys to reduce erythropoietin formation. Reduced plasma erythropoietin levels in turn induce a decrease in the formation of red blood cells. Newly formed erythrocytes are also reduced in size and oxygen transport capacity. However, due to the long half life of erythrocytes of about 100 days, a decrease in erythrocyte count and oxygen transport capacity is apparent only after months in space. In contrast to erythrocytes, white blood counts increase in space. This appears to correlate with high catecholamine levels (Mills et al. 2001). It is not completely clear whether the decreased capacity of T lymphocytes to produce cytokines (Cogoli et al. 1997) is a result of microgravity per se or also a consequence of increased stress levels of astronauts in flight. However, regardless of the reasons, reports on increased growth of bacteria and fungi in space and higher resistance to antibiotics also point to a decreased capability to fight infections in space. This is very important for long-term space flight. Incidents of fungal infections of the skin during the

Mir missions also provide evidence for the importance of this field.

The immune system is also involved in combating cancer. Thus, a stimulated and active immune system would also be essential to prevent possibly increased cancer incidents due to high radiation in microgravity and to prevent preexisting clones of cancer cells, such as malignant melanoma cells, which may be present in humans for decades after sunburns during childhood and might become activated when normal immune function becomes impaired, e.g., during long-term space flight.

Bone and Muscle

Especially those bones that are used to counteract gravity on earth are degraded in space. The calcaneus, pelvis and lumbar spine, followed by the legs and thoracic spine, are strongly affected in microgravity, while the ribs, arms and the head are only slightly affected (Cann 1997). Decrease in bone density can reach up to 2% per month. It has repeatedly been found that such decrements in bone density continue for one to two months after landing. Recovery may even take many months. With some Russian cosmonauts, lost bone mass seems not to recover at all, even after several years postflight (Kozlovskaya, personal communication). Bone loss also increases calcium excretion through the kidneys and leads to the risk of kidney stone formation.

While bone loss is focused on bone that counteracts gravity, muscle loss appears more generalized. This may be correlated with the need to move differently in space compared with on earth and, in general, not to make fast movements. In addition, things lose their weight, and therefore not only the legs and trunk, but also the arms and shoulders are unloaded in space. However, postural muscle that supports standing and walking on earth is degraded much more than nonpostural muscle. Under space conditions, muscle protein metabolism is decreased and muscle atrophy occurs. This appears to be due to a decrease in fiber size rather than to a change in fiber numbers. Extensor muscles are affected much more than are flexors (Di Prampero et al. 2001).

Psychology

Astronauts undergo several emotional phases during stays in space. These are highly variable and cannot be predicted for individual astronauts. They not only depend on the characteristics of individual astronauts such as their experience, specific personal traits etc., but also on the whole complex scenario and the interactions of all the players in the complex team in space and on earth.

Stays in space roughly dissociate into four to five phases (Kanas and Manzey 2003):

The first phase is an adaptation phase with problems of vestibular adaptation, fullness of the head, possibly back pain and decreased work capacity. Usually, the main problems of this phase are over within one week.

This is followed by the "phase of complete adaptation." This phase is, especially for newcomers, a phase of excitement and stress. Everything is new, microgravity is fascinating, and performance increases again after the first week of adaptation. Ambitions are very high and the astronaut feels the pressure to perform optimally in every respect. When problems arise he/she will work them through and sacrifice free time and sleep time to achieve the mission goals. All this leads to increased stress and decreased effectiveness of sleep—which are compensated by all the excitement of the dream that came true to be one of the privileged humans to be sent into space.

After six to eight weeks, reality strikes. One adapts to microgravity and finds it "normal"; one has seen space and the earth through the windows hundreds of times, and, although this view remains spectacular, one "knows" it. The astronaut needs to recover from too much stress and too little sleep. He/she needs time for privacy, such as time for contact with the family on earth, and one may become bored, since a space station is huge and fantastic from far off, but shrinks to a tiny tin can if one cannot escape it for months. And the people down in the control center may start to become bothersome, since they do not know the real situation, but still give order upon order. Thus, in this phase of "asthenia," astronauts suffer from decreasing motivation, increased tiredness, emotional lability, sleep problems and decreased appetite.

This intermediate phase may stretch into the "long-term phase," where astronauts may assume the properties of "hermits" (figure 13-25). This phase is characterized by disturbed communications between crew ("insiders") and ground ("outsiders") and by astronauts becoming more sensitive to depression and overexcitability. They may change personal tastes (like music and leisure time activities). Interestingly, astronaut teams that are more "allergic" to tasks given them from the ground and try to start telling the ground their rules may be more successful than crews who follow the rules from ground even during long term stays in space. Therefore, control room teams need to be flexible in adapting to new situations of collaboration with long-term crews.

The final phase is full of preparations for the return and full of emotion and excitement, but also lack of self control, and may therefore also be difficult to handle.

Radiation

The radiation doses the astronauts in space are exposed to have already been described in the introduction to this chapter. When in space, astronauts regularly experience "light flashes" after phases of dark adaptation. Even for the non-dark adapted eye these light flashes can sometimes be as strong as a

Figure 13-25: Psychology in space: an astronaut looking out of the window of the Mir space station during long-term space flight (IBMP)

flash from a camera or a beam of light. Experiments at ground-based heavy ion accelerators have supported the supposition that they are caused by heavy ions of cosmic radiation striking through the eyeball and affecting the retina. The frequency of the light flashes varies with the orbital parameters. When the space ship passes the South Atlantic Anomaly (an area off the coast of South America where the inner Van Allen radiation belt is closest to earth), about three flashes per minute have been reported (Pinsky et al. 1975). The fact that the human eye can act as a radiation biosensor indicates the severity of the radiation problem in space. Radiation will be most probably the biggest medical problem during human missions to Mars.

In astronauts, after long-term space flights an elevation of the frequencies of chromosomal aberrations in peripheral lymphocytes has been reported. Obe et al. (1997) investigated the lymphocytes of seven astronauts who had spent several months on board the Mir space station. They showed that the frequency of dicentric chromosomes increased by a factor of approximately 3.5 compared to preflight control and that the observed frequencies agreed quite well with the expected values based on the absorbed doses and particle fluxes encountered by individual astronauts during the mission. These data suggest the feasibility of using chromosomal aberrations as a biological dosimeter for monitoring the radiation exposure of astronauts.

In addition to chromosomal aberrations, other biomarkers for genetic or metabolic changes may be applicable, such as germ line minisatellite mutation rates or radiation induced apoptosis, metabolic changes in serum, plasma or urine (e.g., serum lipids, lipoproteins, ratio of HDL/LDL cholesterol, lipoprotein lipase activity, lipid peroxides, melatonin, or antibody titers), hair follicle changes and decrease in hair thickness, triacylglycerol-concentration in bone marrow and glycogen concentration in liver. Whereas the first three systems mentioned are noninvasive or require only blood samples for analysis, the latter systems are invasive and therefore appropriate for radiation monitoring in animals only. Dose response relationships have been described for most of the intrinsic dosimetry systems.

Extravehicular Activity (EVA)

During EVA, the crew members involved require a fully functioning life support system. Two such systems exist, the American "Extravehicular Mobility Unit" EMU and the Russian ORLAN suit (figure 13-26). Both designs have a liquid cooling garment for the astronaut and the possibility of telemonitoring vital and environmental functions. Since astronauts working during an EVA are exposed to the space vacuum, the pressure inside the suit has to be reduced, while astronauts breathe pure oxygen in the suit. Otherwise, the space suit would become blown up in vacuum like a balloon and working would become very difficult if not impossible. The normobaric pressure inside the ISS of 101.3 kPa is reduced in the Russian suit to 40.6 kPa and to 29.6 kPa in the American suit (Newman and Barratt 1997). Due to this reduction in pressure, astronauts—especially in the American suit—have to prebreathe oxygen under reduced pressure conditions in order to avoid symp-

Figure 13-26: Extravehicular activity (EVA): an astronaut during an EVA in the U.S. EMU (extravehicular mobility unit) suit (NASA)

toms of decompression sickness during their EVA (figure 13-26).

13.4.2
Countermeasures

The human body adapts rapidly to microgravity. This leads, e.g., to cardiovascular deconditioning, to bone degradation, to neurovestibular alterations and to muscle atrophy. All these factors may become health hazards after return to 1 g conditions on earth. Therefore, physical fitness programs are implemented during prolonged stays in space. These programs include ergometer and treadmill training as baseline physical training methods. Astronauts performing EVAs also have arm ergometer training to keep up their ability to do physical work with their arms when they have to fight the pressure difference between the space suit and the vacuum while they work physically with their hands and arms. Astronauts have to exercise a variety of muscle functions during their daily two hours of fitness training.

Additional methods in the Russian system may involve the "Penguin Suit" with elastic bands in the suit to simulate gravitational effects, or the CHIBIS suit, a "lower body negative pressure" (LBNP) device in which decreased pressure in the lower body pushes the blood to the legs and decreases the venous return to the heart. Thus, cardiovascular reflexes are stimulated that activate the heart and increase blood pressure to ensure sufficient blood flow to the brain. LBNP training is not done by American astronauts, since previous results with American or Western European LBNP devices have not proven sufficient.

Astronauts like to wear "bracelets," elastic bands worn on the thighs. Due to the increased pressure, venous blood and extracellular fluid return to the upper part of the body is restricted and the sensation of fullness of the head decreases (figure 13-27).

There are several additional countermeasures, such as the use of expandable cords or the ingestion of salt tablets plus water before reentry to try to decrease symptoms of orthostatic collapse after landing.

Recent results suggest that the problem of orthostatic intolerance can be reduced if astronauts have the anti-g suit function of the launch-and-entry suit activated or wear special boots during reentry. Both principles reduce the capacity of veins in the lower limbs to store blood. While these are countermeasures that give acute support after landing, they cannot substitute preventive countermeasures.

All present preventive countermeasures have not resolved the problems that (1) astronauts need to do physical training for two hours every day in long-term microgravity and (2) no countermeasure protocol can reliably counteract the detrimental effects of microgravity on the human body. Therefore, new forms of countermeasure are being developed. The greatest hopes come from two new approaches that may possibly be used in the future:

Both NASA and ESA are currently developing new

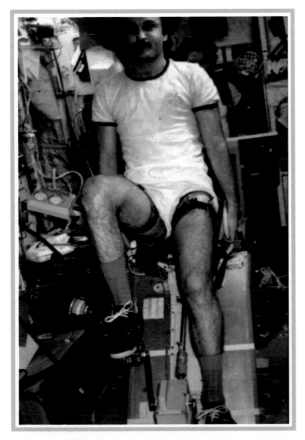

Figure 13-27: Countermeasure: a Russian cosmonaut during bicycle ergometry wearing bracelets to decrease fluid fullness of the head (IBMP)

short arm human centrifuges (figure 13-28) for research purposes (ESA) and countermeasure development (NASA). The NASA centrifuge with a radius of 3m will be built in triplicate and located in Galveston, Cologne and Moscow. The purpose is not only to rapidly develop a novel countermeasure program for use in space, but also to initiate a process of joint harmonized international studies to optimize outcome and avoid duplication of efforts. The hope is to establish a countermeasure program under 1 g conditions integrating other countermeasures such as ergometry and resistive exercise, which requires only little crew time. It is hoped that the other novel countermeasure principle, vibration, can also be directly integrated.

Recent results in space and during bed rest studies on earth suggest that vibrating platforms may be very efficient in decreasing bone loss during inactivity or in space. Presently, various groups worldwide try to establish vibrating platforms both for terrestrial application as a countermeasure to osteoporosis as well as for space applications.

Since bed rest studies in head down tilt position resemble many effects of microgravity, novel countermeasures are usually established first in bed rest studies before application in space. During the upcoming years, a multitude of such studies will be performed internationally. One must, however, be careful when conducting bed rest studies, since some aspects of microgravity, such as fluid homeostasis,

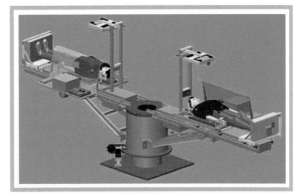

Figure 13-28: Design of a short arm centrifuge (radius 3m; Wyle Laboratories) for future countermeasure development during microgravity (see Web: *Centrifuge*)

are not simulated correctly. In addition, long-term bed rest studies may carry the risk of thromboembolism, which should be carefully avoided.

13.4.3
The Future of Human Space Flight

Human space flight is still in its pioneering phase. This is exciting and dangerous at the same time. Pioneers always have the same problems: complete failure is possible every day, success comes only after many failures. The "normal" world does not understand the pioneers. A pioneer strives for completely new things that nobody knows—and therefore nobody "needs." A pioneer will always be in the situation that finances are limited and other uses for available funds appear to be more promising and safer investments.

Humankind has always expanded its possibilities to act and it is extremely unrealistic to think that it will quit this drive to go beyond limits. And our present frontier is space. If we imagine that we would live in the year 3000, then there will for sure be space stations, research outposts on the moon and Mars, and since centuries tourist trips to space and to hotels on the moon and Mars. There will be tremendous financial returns from investments in space for those countries which had invested strategically in human space flight. And one will remember the pioneering times when this all happened: it will then be irrelevant whether the first human mission to Mars was in 2020 or in 2070. It will all have been in those exciting pioneer days when some governments had the foresight and courage to invest into widening the possibilities for human action.

The present International Space Station is an important test-bed for the next steps in expanding human presence in space. With the conditions aboard the ISS, we can learn much about the effects of long-term microgravity and exposure to the space radiation environment on the human body, and for medicine on earth.

With the plans to return to the moon and to go to Mars within the next decades, we also have the possibility to resolve many questions that we do not only need to answer in order to enable long-term human presence beyond earth, but which are at the same time important questions on earth today. Many of these questions have to do with bioregenerative life support systems for space stations, or lunar and Mars bases for humans. Subquestions like regenerating waste, keeping water clean and recycling it without waste, using solar energy as effectively as possible, understanding the interactions of the different components of a biosphere including humans, and understanding how to care for individuals who are isolated and need protection due to immobility and decreased load on the body are at present of very high importance for terrestrial life.

Focusing on these tasks with the vision of flights to the moon and Mars will help to answer many questions about our fragile biosphere, about how to keep our planet Earth habitable in the future and about how to individually support humans in our ageing population. For example, when we do bed rest studies and develop novel countermeasures in order to prevent degradation of body functions during long-term stays in space, we find at the same time novel applications for the health care of bedridden patients. Therefore, when we focus on tasks for human space flight that are at the same time of relevance on earth, we will be able to put up habitats on the moon and Mars when they are required, but we will—long before such applications—answer questions of high relevance for our life on earth.

Thus, the future of human space flight and the future possibilities to expand the human horizon are closely related with space medicine. Space medicine not only serves the purposes of human space flight, but also provides an opportunity to better understand human physiology and to profit from the concentrated and complex approach that must be taken to answer specific questions that arise when humans are in a habitat with limited resources, in a dangerous environment and in a condition where several body systems are degrading.

References

Anken RH (2003) Neurophysiology of Developing Fish at Altered Gravity: Background – Facts – Perspectives. In: Marthy HJ ed: "Developmental Biology Research in Space," Adv. Space

Biol. Med. Vol. 9. Elsevier, Amsterdam, pp 173-200

Baisch FJ, and Gerzer R (2004) Positive pressure breathing by total body negative pressure: a new simulation model for fluid balance in microgravity. Acta Astronaut. 54, pp. 649-655

Binot R A (2002) The Coming Space Science Programme in Animal Cell/Tissue Research. In: Cogoli A ed: "Cell Biology and Biotechnology in Space," Adv. Space Biol. Med. Vol. 8. Elsevier, Amsterdam, pp. 237-48

Blancaflor, EB and Masson PH (2003) Plant gravitropism. Unraveling the ups and downs of a complex process. Plant Physiology 133, pp. 1677-90

Bräucker R, Cogoli A, and Hemmersbach R (2002) Graviperception and Graviresponse at the Cellular Level. In: Horneck G, and Baumstark-Khan C eds: "Astrobiology, The Quest for the Conditions of Life," Springer, Heidelberg, pp. 284-97

Brack A, Fitton B, and Raulin F (1999) Exobiology in the Solar System & the Search for Life on Mars. ESA SP-1231, European Space Agency, ESTEC, Noordwijk

Braun M, Buchen B, and Sievers A (2002) Actomyosin-mediated statolith positioning in gravisensing plant cells studied in microgravity. J. Plant Growth Regul. 21, pp. 137-45

Buckey JC, Gaffney FA, Lane LD, Levine BD, Watenpaugh DE, and Blomqvist CG (1993) Central venous pressure in space. NEJM 328, pp. 1853-1854

Bücker H, and Horneck G (1975) Studies on the Effects of Cosmic HZE-Particles in Different Biological Systems in the Biostack Experiments I and II, Flown on Board of Apollo 16 and 17. In: Nygaard OF, Adler HI, and Sinclair WK eds: "Radiation Research." Academic Press, New York, pp. 1138-1151

Cann CE (1997) Response of the Skeletal System to Spaceflight. In: Churchill SE ed: "Fundamentals of Space Life Sciences." Vol. 1, Krieger Publishing Comp., Melbourne, Florida, USA, pp. 83-103

Charles JB, Bungo MW, and Fortner GW (1994) Cardiopulmonary Function. In: Nicogossian AE, Leach Huntoon CE, and Pool SL eds: "Space Physiology and Medicine." Lea & Febiger – a Waverly Company, Philadelphia, pp 286-304

Churchill SE, and Bungo M (1997) Response of the Cardiovascular System to Spaceflight. In: Churchill SE ed: "Fundamentals of Space Life Sciences" Vol. 1. Krieger Publishing Comp., Melbourne, Florida, USA, pp 41-64

Cogoli A (1997) Signal transduction in T lymphocytes in microgravity. Gravit. Space Biol. Bull. 10, pp. 5-16

Cogoli A (2002) ed: Cell Biology and Biotechnology in Space. Adv. Space Biol. Med. Vol. 8, Elsevier, Amsterdam

Di Prampero PE, Narici MV and Tesch PA (2001) Muscles in Space. In: Seibert G ed: "A World Without Gravity," ESA SP-1251, Noordwijk, The Netherlands, pp. 69-82

Drummer C, Hesse C, Baisch F, Norsk P, Elmann-Larsen B, Gerzer R and Heer M (2000) Water and sodium balances and their relation to body mass changes in microgravity. Eur. J. Clin. Invest. 30, pp. 1066-1075

Ertl AC, Diedrich A, Biaggioni I, Levine BD, Robertson RM, Cox JF, Zuckerman JH, Pawelczyk JA, Ray CA, Buckey JR Jr, Lane LD, Shiavi R, Gaffney FA, Costa F, Holt C, Blomqvist CG, Eckberg DL, Baisch FJ, and Robertson D (2002) Human muscle sympathetic nerve activity and plasma noradrenaline kinetics in space. J. Physiol. 538, pp. 321-329

Ewald R, Lohn K, and Gerzer R (2000) The space mission Mir'97: operational aspects. Eur. J. Clin. Invest. 30, pp. 1027-1033

Facius R, Reitz G, and Schäfer M (1994) Inactivation of individual Bacillus subtilis spores in dependence on their distance to single cosmic heavy ions. Adv. Space Res. 14 (10), pp. 1027-1038

Greenberg JM (1995) Approaching the interstellar grain organic refractory component. The Astrophys. J. 455, L177

Häder D.-P, Hemmersbach R, and Lebert M (2005) Gravity and the Behaviour of Unicellular Organisms. Cambridge University Press, Cambridge, New York

Heer M, Baisch F, Kropp J, Gerzer R, and Drummer C (2000) High dietary sodium chloride consumption may not induce body fluid retention in humans. Amer. J. Physiol. 278, F585-F595

Hemmersbach R, Volkmann D, and Häder D-P (1999) Graviorientation in protists and plants. J. Plant Physiol. 154, pp. 1-15

Hemmersbach-Krause R, Briegleb W, Häder D-P, Vogel K, Grothe D, and Meyer I (1993) Orientation of Paramecium under the conditions of microgravity. J. Euk. Microbiol. 40(4), pp. 439-446

Horneck G, and Baumstark-Khan C (2002) eds: Astrobiology, the Quest for the Conditions of Life, Springer, Heidelberg

Horneck G, Facius R, Reichert M, Rettberg P, Seboldt W, Manzey D, Comet B, Maillet A, Preiss H, Schauer L, Dussap CG, Poughon L, Belyavin A, Heer M, Reitz G, Baumstark-Khan C, and Gerzer R (2003) HUMEX, Study on the Survivability and Adaptation of Humans to Long-Duration Exploratory Missions. ESA-SP-1264, ESA/ESTEC, Noordwijk, The Netherlands

Horneck G, Rettberg P, Rabbow E, Strauch W, Seckmeyer G, Facius R, Reitz G, Strauch K, and Schott JU (1996) Biological dosimetry of solar radiation for different simulated ozone column thicknesses. J. Photochem. Photobiol. B: Biol. 32, pp. 189-196

Hughes-Fulford M (2002) Physiological Effects of Microgravity on Osteoblast Morphology and Cell Biology. In: Cogoli A ed: "Cell Biology and Biotechnology in Space," Adv. Space Biol. Med. Vol. 8, Elsevier, Amsterdam, pp. 129-158

Kanas N and Manzey D (2003) Space Psychology and Psychiatry. Kluwer Academic Publishers (now Springer), Heidelberg

Kiefer J, Kost M, and Schenk-Meuser K (1996) Radiation Biology. In: Moore D, Bie P, and Oser H eds: "Biological and Medical Research in Space," Springer, Berlin, pp. 300-367

Kuznetsov OA, Schwuchow J, Sack FD, and Hasenstein KH (1999) Curvature induced by amyloplast magnetophoresis in protonemata of the moss Ceratodon purpureus. Plant Physiol. 119, pp. 645-50

Leach Huntoon C, Cintron N, and Whitson PA (1994) Endocrine and Biochemical Functions. In: Nicogossian AE, Leach Huntoon CE, and Pool SL eds: "Space Physiology and Medicine," Lea & Febiger – a Waverly Company, Philadelphia, pp. 334-350

Lewis M L (2002) The Cytoskeleton, Apoptosis, and Gene Expression in T Lymphocytes and Other Mammalian Cells Exposed to Altered Gravity. In: Cogoli, A ed: "Cell Biology and Biotechnology in Space," Adv. Space Biol. Med. Vol. 8, Elsevier, Amsterdam, pp. 77-128

Manzey D, Lorenz B, and Poljakov VV (1998) Mental performance in extreme environments: Results from a performance monitoring study during a 438-day space mission. Ergonomics 41, pp. 537-559

Marthy H-J (2003) ed: Developmental Biology Research in Space. Adv. Space Biol. Med. Vol. 9, Elsevier, Amsterdam

Mileikowsky C, Cucinotta F, Wilson JW, Gladman B, Horneck G, Lindegren L, Melosh J, Rickman H, Valtonen M, and Zheng JQ (2000) Natural transfer of viable microbes in space, Part 1: From Mars to Earth and Earth to Mars. Icarus 145, pp. 391-427

Mills PJ, Meck JV, Waters WW, D'Aunno D, and Ziegler MG (2001) Peripheral leukocyte subpopulations and catecholamine levels in astronauts as a function of mission duration. Psychosomatic Medicine 63, pp. 886-890

Moore D, Bie P, and Oser H (1996) eds: Biological and Medical Research in Space, Springer, Heidelberg

Newman D, and Barratt M (1997) Life support and Performance Issues for Extravehicular Activity. In: Churchill, SE ed: "Fundamentals of Space Life Sciences," Vol. 2, Krieger Publishing Comp.(now Springer), Heidelberg, pp 337-364

Nicholson WL, Munakata N, Horneck G, Melosh HJ, and Setlow P (2000) Resistance of Bacillus endospores to extreme terrestrial and extraterrestrial environments. Microb. Mol. Biol. Rev. 64, pp. 548-572

Norsk P, and Epstein M (1991) Manned space flight and the kidney. Am J. Nephrol. 11, pp. 81-97

Obe G, Johannes I, Johannes C, Hallmann K, Reitz G, and Facius R (1997) Chromosomal aberrations in blood lymphocytes of astronauts after long-term space flights. Int. J. Radiat. Biol., 72, pp. 726-734

Pinsky LS, Osborne WZ, Hoffman RA, and Bailey JV (1975) Light flashes observed by astronauts on Skylab 4. Science 188, pp. 928-930

Platts SH, Ziegler MG, Waters WW, Mitchell BM, and Meck JV (2004) Midodrine prescribed to improve recurrent post-spaceflight orthostatic hypotension. Aviat Space Envir. Med. 75, pp. 554-556

Reitz G, Beaujean R, Heckeley N, and Obe G (1993) Dosimetry in the space radiation field. Clin. Investig. 71, pp. 710-717

Reschke MF, Bloomberg JJ, Harm DL, Paloski WH, Layne C, and McDonald V (1998) Posture, locomotion, spatial orientation, and motion sickness as a function of space flight. Brain Res. Rev. 28, pp. 102-117

Rummel JD (2001) Planetary exploration in the time of astrobiology: protecting against biological contamination. Proc. Natl. Acad. Sci. 98, pp. 2128-2131

Seibert G (2001) ed: "A World Without Gravity," ESA SP-1251, ESA/ESTEC, Noordwijk, The Netherlands

Slenzka K (2002) Life support for aquatic species - past, present, future. Adv. Space Res. 30, pp. 789-95

Tash JS, and Bracho GE (1999) Microgravity alters protein phosphorylation changes during initiation of sea urchin sperm motility. FASEB J. 13, S43-S54

Titze J, Shakibaei M, Schafflhuber M, Schulze-Tanzil G, Porst M, Schwind KH, Dietsch P, and Hilgers KF (2004) Glycosaminoglycan polymerization may enable osmotically inactive Na$^+$ storage in the skin. Am. J. Physiol. 287, H203-H208

West JB, Elliott AR, Guy HJB, Prisk GK (1997) Pulmonary function in space. J. Amer. Med. Assoc. 277, pp. 1957-1961

Web References
Centrifuge: http://www.yenra.com/ centrifuge/

Any Limits?

14 Challenges and Perspectives

Overleaf image: Earth atmosphere at orbital sunset (NASA)

14 Challenges and Perspectives
by Berndt Feuerbacher and Heinz Stoewer

Space technologies and developments are changing at a rapid pace. Breathtaking discoveries and exciting applications make the news in ever shorter intervals. New visions for space exploration dominate the political debates. Industry has matured and space has penetrated into many layers of society and our day-to-day lives. Simultaneously the political and economic environment is evolving. New countries with their own objectives are entering the group of space faring nations.

In spite of these encouraging developments, there is ample room for improvement in space utilization. In the commercial domain regulations, restrictions, and complex public private relationships prevent many opportunities from developing. Open competition should be the rule to encourage commercial exploitation of potential space markets. Level playing fields are an essential prerequisite for opportunities to be tackled by the best entrepreneurs. A thorough investigation of the market situation is prerequisite for a successful business, as recent setbacks in communication constellations have demonstrated.

In the public sector, important missions have been planned in an ad hoc fashion, leading to single demonstration flights where in many cases continuity requirements ask for long term consecutive mission sequences. In areas where there is demand for large, global programs, often single nations or limited cooperations launch isolated initiatives. There are however some encouraging steps such as the Global Earth Observation System of Systems (GEOSS) initiative, which hopefully will lead to closer collaborations and unrestricted availability of data concerning our planets environment in the future.

Global access to data is a precondition for progress, including world wide availability and low cost access, with suitable long term archiving to agreed standards. This would further enable rapid integration of space data with ground based in situ results,

including the necessary assimilation efforts, and enable accelerated progress to the benefit of all nations.

14.1
Focusing on Human Needs

It is unlikely that single imperatives will be fueling space exploitation in the future. The times of the cold war technology race and the need for geopolitical dominance are gone. Tangible drivers are called for to capture and sustain public and political support for funding space ventures. Space investments, more than ever, will have to respond to fundamental priorities of society, such as human welfare, sustainability of life, economic development, security, culture and knowledge.

Space utilization contributes to *human welfare* in many respect. It has improved access to and management of primary resources like water, food, and energy. Identification and quantification of the human influence on global climate change helps to insure a livable environment for future generations. Communications and mobility are supported by space technology on a global scale, improving the efficiency of the infrastructure of industrial nations and enhancing access to knowledge and trade by countries with underdeveloped ground infrastructures. Health care on Earth takes advantage of new results and methods of space medicine.

Space technology enhances *economic development* on national and global scales in many direct and indirect ways. Communication and navigation applications are strong engines for employment and economic growth. Earth and weather observation from space change agricultural production and forest management, cartography and geoinformation systems. In the long term fields as space tourism hold promise of economic potential.

Security and peace are amongst the strongest human needs. Surveillance from space provides information for monitoring international peace agreements. Co-operation in space projects like the International Space Station (ISS) has contributed to an improved dialog and understanding between many nations independent of their political priorities, thereby reducing potential conflicts. In a similar way, the use of common global resources, such as the Internet, communications networks, and navigation services increases mutual dependence between nations. Future "security networks" will have broad capabilities to support global agreements and will contain substantial means to counteract risks. Space information technology can help to reduce antagonism between people, states, and religions and thereby decrease threats posed by terrorism. The detection and mitigation of potential hazards originating from space, such as the impact of near-earth objects (comets or asteroids) also depend upon space means.

Culture is substantially enriched by space activities. Our knowledge society largely benefits from discoveries in space. Questions of the origin and future of mankind touch every person on our planet. Space science missions make ever new inroads in answering fundamental questions on the origin of life or the origin and destination of our home planet, Earth. Space cosmology will bring us closer to understanding the creation and development of our universe and give new insight into the nature of fundamental forces, also relevant for basic knowledge and terrestrial technology applications. Exciting visions of the future of space, including the human exploration of our nearby space environment, inspire young generations. *Education and knowledge* are assets of humanity and recognized drivers for human development deserving attention and improvement in all nations. Special demand and focus exists in underdeveloped and developing countries. Space technologies can make dramatic differences in the efficiency of the needed education infrastructures.

14.2
Reaching Space - the Costly First Step

Space utilization depends on the availability and cost of transport capacity from Earth to space. Cost is the

Focusing on Human Needs

In order to respond to fundamental demands of human society, space activities have to concentrate on
- human welfare and sustainability of life
- economic development and security
- culture and knowledge

key driver, in particular for orbits that are in demand for commercial applications, i.e., low earth and geostationary orbits. Cost and technology in turn are strongly influenced by the extreme reliability requirements for space launchers. If any of some 150,000 components of a typical space rocket fail, the entire mission is endangered. Requirements for human space transport systems are even higher, since crew safety forces extra design and test provisions. Missions beyond earth orbit require new space infrastructures, including intermediate way stations. Affordability and sustainability will be the crucial concerns.

Getting into Orbit, the Principle Transportation Challenge

The strongest variables for the future development of space utilization remain cost and reliability of space transportation. The principal challenge is transport from Earth into near space, i.e., to low, intermediate, or geostationary orbits. Our present generation of expendable launch vehicles, such as Proton, Sojus, Ariane, Delta, Titan and many more, deliver remarkable space transport services, albeit at reliabilities and cost which make improvements highly desirable. The reusable space shuttle is a very successful development, but could not fulfill expectations regarding cost, reliability, and safety. According to current planning it will be retired in 2010. New launch vehicles are under development and new entrepreneurs are entering the scene, but in view of the inherent complexities of launching satellites into space, breakthroughs in cost and reliability are not expected to come quickly.

The next generation of space launch vehicles, some one or more likely two decades from now, is expected to be fully reusable. They should eventually transport goods and people into earth orbit with a safety standard similar to today's commercial aircraft. Provided they can incorporate substantial technology breakthroughs, they may also be able to reduce cost. Space tourism could then rival adventure holidays on earth. More important, such transport systems would open new utilization opportunities and markets and inspire new users to enter the space frontier.

Beyond Earth Orbit - Creating a Space Infrastructure

In the past, systems designed to transport payloads out of the gravitational well of Earth have been a matter of opportunity and short-term demand. The USA and USSR have built powerful rockets to carry their nations to the moon. While the Saturn-5 production was cancelled in 1968 and the last spacecraft of this series was launched in 1973, the competing Soviet N1 rocket never took off successfully. At present, rockets optimized to carry geostationary payloads are used, often with additional upper stages, to provide transport into escape trajectories beyond earth orbit.

New in-space propulsion concepts are needed to advance the exploration of our solar system. Today's space probes need ten or more years to reach their distant targets in the outer solar system, several decades to reach beyond. In the long term, human interplanetary travel will not really be practical without substantially shortened travel times, extremely high safety standards, and intermediate way stations.

The International Space Station represents an attempt to establish a space infrastructure for a multiplicity of utilization purposes. It is not the way station of choice for exploration of the planetary system as its inclination of 51.6 degrees requires energy consuming maneuvers to propel spacecraft into the ecliptic plane. With changing U.S. priorities, the ISS is currently only being supported to the year 2016. In the meantime, other transport options or in-orbit nodes may emerge.

In view of growing ambitions and plans to embark more strongly on missions beyond earth orbit, an early focus on a new space infrastructure is necessary. This should not only be based upon the demands of a single nation, but on the priorities of a broader international space community. We should abandon the once-only philosophy applied, e.g., in the Apollo Program, where an infrastructure scenario was developed and given up later. Rather, mission goals of increasing demands should evolve by way of a stepwise build-up.

Attempts to derive suitable scenarios have been initiated by, e.g., the International Academy of Astronautics (Huntress et al. 2004), who have started their analysis based initially upon plans for extended scientific exploration of our solar system. Similar studies taking into account the broader utilization ambitions of all potential partners in such ventures should be stimulated on a multinational basis.

A suitable space infrastructure system for solar system exploration should consider missions to our nearby planetary system, extending from the moon

Transport Challenges
Key long-term drivers for transport into earth orbit are
- cost and reliability,
- reusability,
- aircraft-like operations.

Beyond earth orbit, cooperative efforts between nations should lead to space infrastructures characterized by
- affordability,
- sustainability,
- flexibility,
possibly with intermediate way stations at e.g.
- earth orbit,
- lunar outpost,
- sun-earth libration point,
- planetary or asteroid surface.

to our neighboring planets Mars and Venus and the asteroid belt. In this region, human missions may complement robotic exploration. Missions to further distant bodies, Mercury, the giant gas planets Jupiter, Saturn and their moons, Neptune and Pluto, as well as deep space operations beyond our solar system are limited to robotic means in the foreseeable future.

A new space infrastructure consisting of transportation and communications nodes in outer space must be sustainable, affordable, and flexible. Preferably it would be built up by many interested nations cooperatively. It should have a modular architecture and be constructed in well timed sequential capability steps. Intermediate way stations, such as a LEO space station, a sun-earth L_2 libration point node, or a lunar or planetary facility, could be of advantage for long-term exploration and recurring missions. Learning from previous international cooperative ventures, e.g. the ISS, would help to avoid mistakes such as exceedingly strong mutual interdependencies between contributing nations, or the ability to adapt to changing priorities on national or global scales.

14.3
Challenges of Space Utilization

Responding to and focusing upon the basic human needs highlighted earlier, future space utilization should advance along three major avenues:

- Human welfare

with emphasis on identifying and preserving basic resources like water, food, climate, energy, and health, but also social demands like security and peace. Global tools from space offer a unique means for monitoring natural and man-made changes of our planet's resources and behavior, identifying the underlying processes, and enabling actions needed to enhance sustainability of life.

- Economic development

is fuelled by a broad spectrum of public and commercial uses of space services. Many of these profoundly change our way of life. They have enabled innovative industries, created high-tech jobs and advanced technologies which have permeated deeply into other industrial sectors. This leverage of space has multiplied investments and broadly stimulated innovation and employment.

- Culture and knowledge

are propelled by fundamental questions regarding human history and future. Space discoveries have changed mankind's perception about itself and its destiny. They have stimulated generations of scientists, engineers, the general population and in particular our children. Society's role and position in future global social interaction will depend increasingly upon its level of education and knowledge management.

14.3.1
Human Welfare - Sustainability of Life

Sustaining our biosphere for future generations is a global challenge with many local facets and calls for unprecedented global cooperation. Many promising efforts to solve partial problems are under way, initiated by individual nations or as broader cooperative ventures on various scales. Examples are the observation of the current state and development of the ozone hole, or studies concerning the understanding of the earth's complex regional water cycles. These constitute important measures with good individual progress, but solutions to the general problem of sustainability are possible only with global coordination of and cooperation in all related ground and space-based observations. There is a strong rationale for joining forces and for freely sharing all related information on a global scale. This requires unconstrained global acceptance that most of the issues are part of a "system of systems" calling for coordination of existing efforts and complementary future investments.

The challenges ahead for the 21st century are enormous and bear upon many fields, such as:

Sustaining Human Life – Food, Water, and Health

The population of our earth continues to grow. As developing nations succeed in improving their standard of living, demands of individual humans for more resources increase even faster. This has severe impacts on, e.g., the need for food, water, energy, or land use. The rate of exploitation of associated resources accelerates. To cope with these developments and enable a more consistent approach to their use and sustenance, global instruments like those offered by space technology are essential.

Earth remote sensing systems in space have demonstrated their capability to monitor land use on various spatial and time scales depending on demand. Ecological mapping is the basis for a variety of applications in sustainable land management. Satellite data enable precision farming to optimize cultivation parameters like soil characteristics, crop state, or fertilizer usage. They allow recognition of hazards like water deficiency, pest threats, severe weather conditions or land degradation, often well in time to initiate warnings and countermeasures.

Water is a resource in heavy demand. Water shortages are predicted to plague all but a few northern countries in the course of the 21st century. Satellite data help to detect large scale variations in the water balance leading to a change in vegetation or desertification. On smaller scales, droughts as well as floods can be monitored or even predicted. Oceans and the cryosphere are observed from space and monitored for subtle changes that could severely impact our climate and fresh water inventory.

Space medicine contributes to individual health care on earth. Medical research under weightless and stressful conditions has proven to offer new solutions in geriatrics, particularly valuable for solving problems for the world's aging population. New methods in telemedicine contribute to health services in remote and underdeveloped regions. Methods developed in astronaut care can lead to a new relationship between physicians and patients, by changing statistical approaches to individual systemic views.

Security and Peace – Conflicts, Peace agreements, Terrorism

"The global security environment has undergone dramatic, fundamental and profound change in recent years. The threats of today have a completely different shape, direction and pace when compared to those of the cold war era. Europe, along with the rest of the world, faces threats with greater diversity, unseen command structures and business-like financing mechanisms" (EC Space and Security Panel of Experts 2005).

Global security is unthinkable without space-based intelligence, communication, monitoring, and navigation services. U.S. defense forces have amply demonstrated the value of space systems for peace keeping, albeit also for intervention purposes. Europe above all needs better eyes and ears, and means for exchanging information to support its intervention and peacekeeping forces effectively and to live up to its international commitments and ambitions. Space-based means offer a strong capability for enabling the further integration of European defense forces and coordination with strategic partners worldwide. Space technologies could also enable the United Nations to play a more efficient role in the global management of conflicts. Today's and tomorrow's security challenges need to account for terrorism, proliferation of weapons of mass destruction, regional conflicts, and diverse threats originating from organized crime, totalitarian regimes or states with bad governance, which include drug trafficking, weapons smuggle or illegal immigration. Space means can make increasingly larger contributions to combating such global and regional hazards.

Open societies are particularly vulnerable to such threats and have limited means of protection. However, global communication instruments can reduce misunderstanding and conflict potential between societies, just as increased welfare and education can. Space technologies facilitate the search and detection of specific targets in remote locations, allow us to build a substantially improved information management system, and support ground forces to fight conflicts or terrorism. They can also make essential contributions to the monitoring of peace agreements. The needed technology is available or in development; the decisions on how to use it are the real challenges facing modern democratic societies.

Natural Disasters - Prediction and Mitigation

Disasters resulting from the dynamic character of our planet and its atmosphere have many forms. Volcanoes, seaquakes and earthquakes, hurricanes, floods and drought cost many lives, disrupt societies and have painful economic impacts. Understanding and predicting their effects has advanced well for e.g. hurricanes. For others, such as earthquakes,

Space for Sustainability of Life Conditions
Global tools are indispensable to fight the challenges of a global change and for conserving our biosphere.
 Space systems contribute to:
- observing and predicting weather and climate changes
- sustaining human life – food, water, and health
- ensuring security and peace – conflicts, peace agreements, migration, terrorism
- predicting and mitigating natural disasters

even though we have lived on this planet for millennia, we are only at the beginning of understanding the underlying forces and processes. Space based systems are key for advancing on these fronts. Global space based measurements improve our understanding of Earth's geodynamics. Radar interferometry techniques have demonstrated their capability to reveal minute local movements of the earth's crust, and hold promise for predicting earthquakes and volcanic eruptions. Space based instruments have been very effective in supporting local rescue forces, as again demonstrated in the 2004 South East Asia tsunami tragedy. Space is and will be an indispensable ingredient in present and future hurricane, tsunami and other disaster warning systems.

Even though the statistical probability of disasters originating from an Earth bombardment by comets or asteroids is low, the potential magnitude of such disasters could, however, exceed our imagination. It is hence quite appropriate to prepare for the "what if" case and to develop the means capable of avoiding or mitigating such catastrophes as long as there is time. In view of present and evolving space based systems capabilities to predict and avert such disasters, it is high time to prepare the actual tools and methods.

Living with our Atmosphere – Weather and Climate

Meteorological services from space have become very reliable, comprehensive and indispensable for our daily life. Extending weather forecast in many regions of the world beyond the current predictions of some four to five days will further encourage applications and reliance upon such services. Tailor-made predictions for specific local areas or regions already today influence agriculture, tourism, the use of water or other resources, and transportation strategies for some enterprises.

Based—among other things—on space data, it is not questioned today that mankind is changing its living environment, with measurable impact on the atmosphere and an accelerated development in the future. Preservation of the biosphere is an imperative for our generation. Global tools offered by space techniques are indispensable for this task.

Many programs on earth observation are individually pursued in various nations, each with its specific priorities. To make maximum progress it is indispensable to coordinate associated efforts on a global basis, with shared tasks and globally agreed objectives. An initiative in this direction has been started recently in the form of the Global Earth Observation System of Systems (GEOSS) program, which has been agreed by 60 nations and is supported by more than 30 scientific organizations and agencies. This initiative is aimed at coordinating existing systems and closing gaps identified in the observation of environmental parameters. With a promise to improve comprehensiveness, reliability and accessibility of environmental data it will help to increase awareness and improve associated policy making. It remains to be seen however how well such voluntary global cooperation can function and the extent to which prevailing barriers to data transfer and the free exchange of information can be overcome in reality.

14.3.2
Economic Development and Innovation

The economic potential of space systems remains huge, even though the pertinent industry is small compared to established sectors such as automotive, energy, or electronics. Nonetheless many of the industries involved derive direct or indirect benefits from space developments, be it through new technologies or processes. Space developments create employment in the high-tech and high-sci domains, jobs which are all about innovation, advanced technologies, economic progress, and ultimately the survival of industrialized nations. Key challenges before us are multiple and mirror some of those pertaining to other high-tech industries.

One of the central issues for the space industry is the speed and effectiveness of transfer of knowledge gained in scientific and technological space projects into products and industries able to leverage them into commercial markets. According to many studies performed in the past the leverage factor of space investments is high, but the time scale is often very slow. New funding mechanisms, such as public-private partnerships or various public seed investments, may prove more effective than conventional schemes. On the other hand it is clear that the degree of sophistication needed and the embedded risks for most space ventures are in a class of their own. Hence transfer of space technologies into consumer industry is a rocky path and takes technology maturity, cost honing and time.

The development of commercial space markets also takes substantial perseverance, high investments, risks, and time. With the exception of space based telecommunications applications, which effectively supplement an existing multibillion terrestrial communications industry and hence piggy back on a strong commercial base, all other space markets require strong initial public investments over a relatively long period of time to develop. This is true for the weather market, the geoinformatics and navigation markets. Other applications, such as the support of agricultural production, forest management, or wild fire detection will depend for extended times upon public funding, amongst other reasons because the primary customers for such applications are government agencies or public organizations.

Some of the major challenges in the field of economic development include the following:

The Global Digital Divide - Information Availability

Communication is a basic human need. Free communication flow within countries and across the globe is hence a natural desire of its people. The 21st century will see enormous leaps in information flow between enterprises, people, and nations. Communication is fuel for the economy, for education, and for the quality of life. With increasing demands for quality and quantity, availability of information on a global scale is only possible through space based satellites. They efficiently complement terrestrial networks and enable closed and underdeveloped societies to share in the information wealth.

Space-based telecommunication systems provide the missing nodes for the information society. They will, despite recent setbacks, continue to benefit from new opportunities as broadband, Internet, and multimedia driven demands grow. Satellites are an essential element in the quest to overcome Europe's and our globe's digital divide. They can have a particularly strong impact upon regions with an underdeveloped ground infrastructure. Should the paradigm of free information access to all of the world's population gain further momentum, it can be realized relatively easily with the help of space based means.

Mobility and Geoinformation Systems

Space-based positioning, navigation and timing information have become indispensable for our industrialized economy and many sectors of our society. New systems such as Galileo will complement and further improve GPS information for people and commerce worldwide and lead to an ever broader use of associated space- and ground-based services. Ground, sea, and air transportation for example will strongly benefit from dependable services with positioning accuracies in the centimeter regime.

The increasingly popular combination of telecommunication, positioning, navigation and timing ser-

Earth Services from Space – some long-term management challenges

The next decades of future space developments could see the advent of new global government organizations and private enterprises established to improve the management of global resources, such as:

AGRISPACE: evaluation and coordination of agricultural production and food distribution on a worldwide basis, reducing overproduction in some parts of the world and stimulating growth in others, thus ensuring a more equitable distribution of the world's food supply. A complementary organization could deal with the distribution of increasingly limited and crucial water resources.

SPACERESCUE: monitoring and integrating information on the global environment, predicting disasters and supporting their mitigation while making relevant data and information freely available without restrictions to all concerned. The current GEOSS initiative may have the potential to grow into such a role.

ENERSPACE: an organization charged with coordinating increasingly scarce global energy resources, while at the same time being given the task to develop alternative, more sustainable solutions, including harnessing solar energy in space and beaming it to energy nodes on Earth.

TELESERVICES: providing information to those regions of the Earth which do not constitute a large enough economic market to attract private investments for knowledge dissemination.

vices at high accuracy and bandwidth will create many new applications, further affecting economic development, quality of life, and personal security for large segments of the population worldwide.

Technology and Innovation

Most space projects rely upon advances in technology and processes. They further the state of knowledge and engineering in many domains and they are truly multidisciplinary. Space developments are in many ways precursors to other high-tech industrial developments which also show a tendency to higher complexity, more functionality and performance. This applies to modern consumer electronics and computing devices as much as to automobiles and household devices.

Reliability and safety requirements for space projects have pioneered robust electronic components and software developments along with test processes. The need for automation and remote access for space systems operating at distances of thousands and millions of kilometers from Earth has yielded design architectures along with test and checkout processes that have inspired many machine

building industries. Space technologies, such as solar or fuel cells or complex mathematical tools, such as advanced structural or thermal analysis software, are making their way into automobile, aircraft and other industries. New materials and robotic solutions have entered the medical field and improved quality of life for many patients and disabled persons. Advanced instruments and microtechnologies have yielded new solutions for controlling and monitoring industrial installations and production processes, sometimes over distances across half of our globe.

An area of immediate relationship between space and ground based industry is materials processing in space. While the early dreams and bold predictions of orbital factories for mass production have been qualified by solid scientific data, some promising research fields have emerged. Industry relies on computer modeling for many production processes, whose results depend on the quality of input parameters. To this effect, precision measurements of thermodynamic data under weightless conditions have extended the data base accessible on ground and have offered industry a competitive advantage. New processes and novel materials have thus been developed in space for use on Earth.

While there are many leading edge research fields in our high-tech societies, it is indisputable that space research is a strong stimulus for many direct and indirect innovative applications which are invigorating more conservative or lower-technology industries.

Global Resources Management – Food, Energy, Water

Earth resources, such as food, energy, wood, minerals, etc. are exploited locally but sold on global markets. Commercial exploitation at times threatens biodiversity and endangers the environment. Enormous efficiencies could be realized and sustainability improved, if critical natural resources were monitored, and in the long term also exploited, on a more global scale.

The next decades of space developments could hence see the advent of new global government organizations and private enterprises pursuing such objectives. The enabling space technologies are within reach. Inhibiting factors are more likely the lack of global governance instruments, political will, money, and global support, which in turn depend on the acceptance of the underlying needs and benefits.

The potential of using remote sensing techniques from space for potential practical applications and as a basis for commercial ventures is steadily growing. The challenges for business applications remain the penetration of generally conservative government dominated user sectors and associated private enterprises. The commercial potential for large scale utilization of modern space based geospatial information systems is huge, but the market still largely depends on classical terrestrial information sources. It is understood that tailor-made information and all weather, long term delivery guarantees are paramount for users to rely on space in lieu of their terrestrial data sources.

The 21st century is likely to see an explosion of space generated geoinformation applications in fields such as agriculture and global food management, intelligent prospecting and exploitation of earth resources, ecological monitoring, mapping and other services. This could be akin to the explosion in the use of space based GPS generated navigation, positioning, and timing information in the past decades.

14.3.3 Knowledge and Education

In our competitive world, knowledge and education will be increasingly dominating criteria for the success of a nation. Knowledge is the only global resource which is truly inexhaustible and may even grow worldwide. Education is a cultural task and will always rely, at least for its more ambitious objectives, on public funding. Public insight into the motivation of scientific endeavors to advance the knowledge frontier, and political support for it remains indispensable.

Developed nations with few natural resources require persistent educational efforts to survive in the competition of a global knowledge society. Developing nations have long recognized the importance of a high level of education of their population in speeding economic and technological progress.

Space is a strong stimulus for the interest of youth in science and technology, which is essential for the progress of a nation (presently interest in natural sciences studies is receding worldwide). Bold efforts like the U.S. Exploration Initiative (O'Keefe 2004), intended to investigate new worlds and eventually send humans to the moon and Mars, could offer a stimulus similar to that experienced from the Apollo program and Neil Armstrong's first steps on the moon.

Space communication technology offers global access to current information, allowing advanced and timely contents for teaching on all levels, even in the most remote locations of the world. Direct involvement of students in space activities is offered by many research institutions, industries and universities (e.g., the International Space University). Outreach programs from many space agencies are offering exciting material and possibilities to students.

Scientific space research expands the understanding of nature and human perception of the universe. Recalling the focus on human needs, exploration of our universe should concentrate upon three fundamental questions of high relevance for humankind:

- Where do we come from, what is our past?

- Are we alone, are there other life forms in the universe?

- Where are we bound for, what is our destiny?

Where do we Come from, what is our Past?

Since early history, humans have been looking up to the stars and wondering about their nature. They represent, in fact, worlds like our own, and their study is the key to our roots. Astronomy has made great progress in understanding the origin of our universe, the evolution of galaxies and stars, and the formation of planetary systems. Space astrophysics has introduced a giant leap forward in our perception in recent years, as exemplified by the Hubble deep field images and the impact they had on our comprehension of the evolution of the universe. Yet they demonstrate the dimension of our ignorance by proving that the matter in the universe known to us (the baryonic content) constitutes only 5% of the total matter. The nature of the dominant constituents, termed dark matter and dark energy, is unknown at present.

A mystery of this dimension can only be tackled by a major effort like a survey of matter in the universe across the electromagnetic spectrum, investigating the structure and distribution of galaxies and stars, dust and gas. This should be augmented by the observation of gravity waves, the joint ESA-NASA project Laser Interferometer Space Antenna (LISA) being a first step in this direction.

Star formation and the evolution of planetary systems are additional keys to understanding our origin. Planetary systems seem to be quite common in the universe, as the discovery of many extrasolar planets orbiting around other stars indicates. In our own planetary system, we have access to planets and moons for remote and in-situ investigation. An important source of information on the early history of our home system are small bodies like asteroids and comets, which are remnants of the original material from which the solar system was formed.

Space astronomy relies on collecting photons from distant sources for imaging and spectral analysis over the complete range of the electromagnetic spectrum from high energy gamma radiation to radio waves. Large collecting areas are essential for this task, and these are achieved in Great Observatories using various technologies in distinct wavelength ranges. Photon collectors with dimensions of 10 to 100 m will be used in future, compared to the Hubble mirror diameter of 2.4 m. Angular resolution requires large apertures, which can be achieved by coherent coupling of several light collecting facilities. In this way, synthetic apertures of up to 100 km can be achieved. Interferometry improves angular resolution even more by combining photons collected over distances of some 10,000 km.

Near earth orbits, like that of the Hubble Space Telescope, are a great advantage over ground based sites. They still have limitations due to their proximity to Earth, which emits disturbing radiation and obscures large areas of the sky. This can be overcome by other locations in space. Depending on the task to be performed, optimum locations could be higher orbits, the moon, or the sun-earth Lagrange point L_2 (SEL2). For special applications, formation flights or constellations might be suitable.

Are we Alone – are there Other Life Forms in the Universe?

Life may be abundant in the universe in diverse forms, or it might be limited to Earth. Astrobiology (or exobiology) contributes to answers for such fundamental questions by looking for possible ways in which life might have originated and how it might have spread in the universe. Space investigations, in situ and remote, of primitive bodies (comets, asteroids) and interplanetary dust provide crucial information on the question of how life originated on Earth. A search for traces of past or existing extraterrestrial life in our planetary system should concentrate on bodies for which present knowledge indicates the potential existence of at least primitive forms of life. Those are Mars, in particular its subsurface, the underground oceans on the Jupiter satellites Europa and possibly Ganymede or Callisto, and Saturn's moon, Titan. Such investigations have been initiated on Mars and Titan by means of robotic in-situ analysis and should be followed by sample return missions, eventually by human exploration on Mars.

Challenges of the knowledge frontier
Origin of the universe
- What happened in the early phases of the universe?
- Unravel the mystery of dark energy

Other worlds and life in the universe
- Are there Earth-like planets in other stellar systems?
- Is there or has there been life elsewhere in our planetary system?
- Are there traces of other life forms out there?

Planet Earth
- How can mankind survive in an evolving biosphere?
- Can we learn from other planets about the evolution and destiny of Earth?
- What is necessary to insure sustainability of life on Earth?

A challenge for future observations will be the detection of earth-like planets in orbit around other stars. This should be followed by spectroscopic identification of traces for biologic activity in the atmosphere or on the surface of such planets. Substantial progress in photon collection, detection efficiency, and spectroscopic techniques is required for tasks of this kind.

Where are we Bound for – what is our Destiny?

Planet Earth is unique of its kind. It is the only place in the universe where we definitively know that life exists. Life has developed in a benign biosphere and the crowning glory of creation, mankind, has expanded to such a level that its own activities endanger the future of its environment. It is our responsibility to secure sustainability of our ecosystem for the generations following us.

To understand the evolution of life on Earth it is necessary to have insight in the origin and development of all life forms and of our biosphere. The sun as our life spending star deserves attention, as well as the sun-earth relationships and interactions with the magnetic and plasma environment, including the earth's magnetosphere, protecting us from the hazardous radiation prevalent in space. The solar irradiation and the earth's atmosphere contribute to our climate and its variation. In particular the long term stability of our atmosphere and any changes of natural or anthropogenic origin must be monitored over space and time. This is complemented by observations of the status and evolution of the solid surface and of the oceans with multispectral sensors.

The long term evolution of our ecosystem can be studied by comparison to other planets. Mars shared a high degree of similarity with Earth during the first billion years of planetary development, but has evolved much faster due to its smaller size. In contrast to Earth, where plate tectonics lead to a recycling of the crust, Mars preserves its history in distinct surface features and hence allows a study of processes on geological time scales.

In the very long term it is a scientific fact that the earth will be extinguished. The sun, the star of our life, will follow the cycle of any star of its class and develop into a red giant. This will take some 5 to 7 billion years. By then the sun will have a diameter which corresponds to the current orbit of our earth. Long before, life on Earth, as we know it today, will have ceased. What are our options? The only one thinkable today is to emigrate from Earth towards some other planet more distant from the sun. So it's not a question as to whether humankind will leave our cradle Earth, as Konstatin Ziolkowsky once formulated it, but it is a question of when. Is it asking too much for us to at least accept this perspective from a philosophical point of view, even though it is a very, very long-term perspective?

14.4
Conclusions

Space investments in past decades have developed the fundamental technologies for access to space and its utilization. Space research has advanced fundamental scientific knowledge about our planet, our universe, human history and destiny. Space applications have improved living conditions on Earth and stimulated economies. The space business has matured, is alive and well, and continues to grow worldwide.

The space industry is still small compared to other industrial sectors, but its contributions are innovative and strategic. The 21st century will see many more users of space assets and witness mankind's next steps towards exploring living conditions away from Earth. We may even find other life forms elsewhere in our universe. For space activities to develop their full potential, they need to interact in the future even more strongly with the terrestrial communities, and in particular convey the services they can offer for the fulfillment of society's needs and priorities.

Some of the world's best brains work in the space field. There is no lack of fascinating new ideas for future missions and applications. The scope and pace regarding publicly financed new space ventures will in the future, even more than in the past, be governed by the extent to which such missions can continue to make genuine contributions to society's real and perceived needs.

The market opportunities for privately financed space ventures will continue to grow. Many new fields could open up in the coming decades. In some cases there may be alternative terrestrial solutions to fulfill inherent needs. Space based solutions often are not unique, but can be excellent complements to terrestrial solutions.

What is the conclusion, the long-term perspective? Space developments, when oriented towards society's needs and in complement to terrestrial developments on Earth, will continue to contribute substantially towards a better future of our planet and its people.

Any limits?

There are neither limits to space, nor to ideas and imagination on what could happen out there. Scientific exploration and commercial exploitation, defense and security, or plain human adventure, if seen and dealt with in the proper context, will continue to prosper.

Both dimensions of space, the one oriented outward, towards our moon, planets, sun and universe, and the other oriented downward to our earth, offer ample opportunity for future developments. Let's deal with them wisely!

References

EC Space and Security Panel of Experts (2005) Report of the Panel of Experts on Space and Security, European Commission Project Number 200446

Huntress W, Farquhar R, Stetson D (2004) The Next Steps in exploring Deep Space, International Academy of Astronautics, Final Report

O'Keefe S (2004) A Renewed Spirit of Discovery: The President's Vision for U.S. Space Exploration, NASA, Washington DC

Stoewer H (2000) Space in Context, Emeritus Lecture, Delft University of Technology

List of Acronyms

ACES	Atomic Clock Ensemble in Space
ACIA	Arctic Climate Impact Assessment
ACIS	Advanced CCD Imaging Spectrometer
ACS	Advanced Camera for Surveys
ADH	Antidiuretic Hormone
ADM	Atmospheric Dynamics Mission
AGHF	Advanced Gradient Heating Facility
AIRS	Atmospheric Infrared Sounder
AIS	Automatic Identification System
AMF	Automatic Mirror Furnace
AMS	Alpha Magnetic Spectrometer
AMV	Atmospheric Motion Vector
AO	Adaptive Optics
AOCS	Attitude and Orbit Control System
APV	Approach with Vertical Guidance
ARGO	Array of Profiling Floats over the Global Ocean
ARTEMIS	Advanced Relay and Technology Mission
ASIC	Application-Specific Integrated Circuit
ASQF	Application Specific Qualification Facility
ASTER	Advanced Spaceborne Thermal Emission and Reflection Radiometer
ASTROD	Astrodynamical Space Test of Relativity using Optical Devices
ASU	Atomic Sagnac Unit
ATEN	Advanced Thermal Environment
ATOVS	Advanced TIROS Operational Vertical Sounder
AU	Astronomical Unit
AVHRR	Advanced very high Resolution Radiometer
AXAF	Advanced X-ray Astrophysical Facility

BDC	Baseline Data Collection
BEST	Boundary Effects near Superfluid Transitions
BIRD	Bispectral Infrared Detector
bps	bits per second
BRTS	Bilateral Ranging Transponder System
BSS	Broadcast Satellite Service
BVOC	Biogenic Volatile Organic Compounds
CAGR	Cumulative Annual Growth Rate
cAMP	Cyclic Adenosine Monophosphate
CAT	Category
CCD	Charge Coupled Device
CDMA	Code Division Multiple Access
CEBAS	Closed Equilibrated Biological Aquatic System
CEOS	Committee on Earth Observation Satellites
CFCs	Chlorofluorocarbons
CGRO	Compton Gamma Ray Observatory
CHAMP	Challenging Mini-satellite Payload
CMB	Core Mantle Boundary
CMB	Cosmic Microwave Background
CNES	Centre National d'Etudes Spatiales
CNSA	Canadian National Space Agency
COBE	Cosmic Background Explorer
CoolCop	Undercooling and Demixing of Cu-Co alloys
CORINE	Coordination of Information on the Environment
COROT	Convection Rotation and Planetary Transits
COSPAR	Committee on Space Research
COSPAS	Cosmicheskaya Sistyema Poiska Avariynich Sudov (Space System for the Search of Vessels in Distress)

COSTAR   Corrective Optics Space Telescope Axial Replacement

CSR   Center for Space Research

DAU   Defence Acquisition University

DBS   Direct Broadcasting Satellite

DECLIC   Dispositif pour l'Etude de la Croissance et des Liquides Critiques

DEM   Digital Elevation Model

DEOS   Department of Earth Observation and Space systems

DEXTRE   Special Purpose Dextrous Manipulator

DGPS   Differential Global Positioning System

DIODE   Détermination Immédiate d'Orbite par Doris Embarqué

DIS   Data and Information System

DLR   Deutsches Zentrum für Luft- und Raumfahrt (German Aerospace Center)

DMSP   Defence Meteorological Satellite Program

DNA   Deoxyribonucleic Acid

DORIS   Doppler Orbitography and Radiopositioning Integrated by Satellite

DRS   Data Relay Service

DRTS   Data Relay and Technology Satellite

DSM   Digital Surface Model

DSN   Deep Space Network

DTG   Dry-Tuned Gyro

DTH   Direct to Home

DTM   Digital Terrain Model

DWL   Doppler Wind Lidar

EADS   European Aeronautic Defense and Space Company

EarthCARE   Earth Clouds, Aerosol and Radiation Explorer

EC   European Commission

ECMWF   European Centre for Medium range Weather Forecast

ECSS   European Cooperation for Space Standardization

EFI   Electrical Field Instrument

EGM96   Earth Geopotential Model 1996

EGNOS   European Geostationary Navigation Overlay System

EIGEN   European Improved Gravity Model of the Earth by New Techniques

ELDO   European Launcher Development Organization

EML   Electromagnetic Levitation

EMU   Extravehicular Mobility Unit

ENSO   El Niño Southern Oscillation

ENVISAT   Environmental Satellite

EPIC   European Photon Imaging Camera

EPS/Metop   EUMETSAT Polar System

ERA   Exobiology Radiation Assembly

ERS   Earth Resources Sensing

ERS   European Remote Sensing

ERTMS   European Rail Traffic Management System

ERTS   Earth Resources Technology Satellite

ESA   European Space Agency

ESL   Electrostatic Levitation

ESO   European Southern Observatory

ESOC   European Space Operations Centre

ESTEC   European Space Technology Center

ESTRACK   ESA Tracking

ETM+   Enhanced Thematic Mapper

EUMETSAT   European Organisation for the Exploitation of Meteorological Satellites

EURECA   European Retrievable Carrier

EUSO   Extreme Universe Space Observatory

EVA   Extravehicular Activity

EWAN   EGNOS Wide Area Network

FANS   Future Air Navigation System

FDMA   Frequency Division Multiple Access

FEC   Forward Error Correction

FIGI   Fir proto Galaxy Imager

FIR   Far Infrared

FOG   Fibre-Optic Gyro

FOV   Field of View

FSS   Fixed Satellite Service

FUSE   Far Ultraviolet Spectroscopy Explorer

GAIA   Global Astrometic Interferometer for Astrophysics

GALEX   Galaxy Evolution Explorer

GCR   Galactic Cosmic Radiation

GEM	Goddard Earth Model		GZK	Greisen-Zatsepin-Kuzmin Cut off
GEO	Geostationary Earth Orbit		HAP	High Altitude Platforms
GEO	Group on Earth Observation		HMI	Hazardous or Misleading Information
GEODYSSEA	GEODYnamics of South and South-East Asia		HOAPS	Hamburg Ocean Atmosphere Parameters and Fluxes from Satellite Data
GEOS	Geostationary Earth Orbit System		HOTOL	Horizontal Take-Off and Landing
GEOSS	Global Earth Observation System of Systems		HPPO	High Precision Pore Optics
GEWEX	Global Energy and Water Cycle Experiment		HRC	High Resolution Camera
			HRSC	High Resolution Stereo Camera
GFO	GeoSat Follow-on		HST	Hubble Space Telescope
GFZ	Geo-Forschungszentrum (Germany's National Research Center for Geosciences)		HYPER	Hyper-precision Atom Interferometry in Space
			HZE	High Z, High Energy (particles)
GIA	Glacial Isostatic Adjustment		IASI	Infrared Atmospheric Sounding Interferometer
GIS	Geographic Information System		IB	Inverse Barometer
GLAST	Gamma-ray Large Area Space Telescope		IBMP	Institute of Biomedical Problems
GLC	Global Land Cover		ICAO	International Civil Aviation Organization
GLONASS	Global Navigation Satellite System		ICESat	Ice, Cloud and Land Elevation Satellite
GMES	Global Monitoring for Environment and Security		IDGE	Isothermal Dendritic Growth Experiment
GMS	Geostationary Meteorological Satellite		IDS	International Doris Service
GNSS	Global Navigation Satellite System		IGBP	International Geosphere-Biosphere Program
GOCE	Gravity Field and Steady-state Ocean Circulation Explorer		IGN	Institut Géographique National
GOES	Geostationary Operational Environmental Satellite		IGRF	International Geomagnetic Reference Field
GOME	Global Ozone Monitoring Experiment		IGS	International GPS Service
GOS	Global Observing System		ILRS	International Laser Ranging Service
GP-A	Gravity Probe A		IML 2	International Microgravity Laboratory 2
GP-B	Gravity Probe B		IML	International Microgravity Laboratory
GPS	Global Positioning System		IMPRESS	Industrial Materials Processing in Relation to Earth and Space Solidification
GRACE	Gravity Recovery and Climate Experiment		INCOSE	International Council on Systems Engineering
GRGS	Groupe de Recherche en Géodésie Spatiale		INMARSAT	International Maritime Satellite Organization
GRI	Gamma Ray Imager		INS	Inertial Navigation System
GSM	Global System for Mobile Communications		InSAR	Interferometric Synthetic Aperture Radar
GSTDN	Ground Spaceflight Tracking and Data Network		Integral	International Gamma-Ray Astrophysics Laboratory
GTDS	Goddard Trajectory Determination System		IOM	International Maritime Organisation

IOR	INMARSAT III Satellite Indian Ocean Region	LMTPF	Low Temperature Microgravity Physics Facility
IR	Infrared	LNA	Low Noise Amplifier
IRAC	Infrared Array Camera	LNAV	Lateral Navigation
IRAS	Infrared Astronomical Satellite	LoD	Length of Day
IRS	Infrared Spectrograph	LPEE	Liquid Phase Electroepitaxy
ISAS	Institute of Space and Astronautical Science	LPI	Local Position Invariance
ISCCP	International Satellite Cloud Climatology Project	LPV	Lateral Precision with Vertical Guidance
		LST	Land Surface Temperature
ISO	Infrared Space Observatory	LST	Large Space Telescope
ISP	Internet Service Provider	MAGSAT	Magnetic Field Satellite
ISRO	Indian Space Research Organization	mas	milli-arcsecond
ISS	International Space Station	MCAO	Multi-Conjugate Adaptive Optics
ITCZ	Intertropical Convergence Zone	MCC	Mission Control Center
ITRF	International Terrestrial Reference Frame	MELiSSA	Micro-Ecological Life Support System Alternative
IUE	International Ultraviolet Explorer	MEMS	Micro-Electro-Mechanical Systems
IVS	International VLBI Service	MEOLUT	Medium Earth Orbit Local User Terminal
IWV	Integrated Water Vapour	MEOMS	Micro-Opto-Electro-Mechanical Systems
JAXA	Japan Aerospace Exploration Agency	MERIS	Medium-Resolution Imaging Spectrometer
JDEM	Joint Dark Energy Mission	MERIT	Monitoring of Earth Rotation and Intercomparison of the Techniques of Observation and Analysis
JPEG	Joint Photographic Experts Group		
JPL	Jet Propulsion Laboratory		
JWST	James Webb Space Telescope	METCOMP	Metallic Composites
KAO	Kuiper Airborne Observatory	MICAST	Microstructure Evolution in Cast Alloys
LAAS	Local Differential GPS System at Airport	MICROSCOPE	MicroSatellite à traînée Compensée pour l'Observation du Principe d'Equivalence
LAGEOS	Laser Geodynamics Satellite		
LAI	Leaf Area Index		
LAN	Local Area Network	MIM	Microgravity Isolation Mount
LASER	Light Amplification by Stimulated Emission of Radiation	MIPS	Million Instructions per Second
		MIPS	Multiband Imaging Photometer for Spitzer
LATOR	Laser Ranging Test of Relativity		
LBNP	Lower Body Negative Pressure	MIRI	Mid Infrared Instrument
LBS	Location Based Services	MISTE	Microgravity Scaling Theory Experiment
LCC	Launch Control Center	MODIS	Moderate Resolution Imaging Spectroradiometer
LDEF	Long Duration Exposure Facility		
LEM	Lunar Exploration Module	MOLA	Mars Observer Laser Altimeter
LEO	Low Earth Orbit	MONOPHAS	Advanced Bearing Alloys from Immiscibles with Aluminium
LIDAR	Light Detection and Ranging		
LISA	Laser Interferometer Space Antenna	MPIA	Max-Planck-Institut für Astronomie (Max-Planck-Institute of Astronomy)
LLR	Lunar Laser Ranging		

MSAS	Multi-Transport Satellite Augmentation System	PKC	Protein Kinase C
MSG	Meteosat Second Generation	ppm	part per million
MSL	Material Science Laboratory	PPS	Precise Positioning Service
MSS	Mobile Satellite Service	PRN	Pseudo Random Noise
NA	Not Applicable	PVT	Position Velocity Time
NASA	National Aeronautics and Space Administration	RADAR	Radio Detection and Ranging
		RAOB	Radio Observation
NASP	National Space Plane	RGB image	Red Green Blue image
NAVWAR	Navigational Warfare	RIMS	Ranging and Integrity Monitoring Station
NDVI	Normalized Differences Vegetation Index	RLG	Ring-Laser Gyro
		RMS	Root Mean Square
NEQUISOL	Non-Equilibrium Solidification, Modeling for Microstructure Engineering of Industrial Alloys	RO	Radio Occultation
		RTK	Real Time Kinematics
		RXTE	Rossi X-Ray Timing Explorer
NGA	National Geospatial-Intelligence Agency (U.S.)	SA	Selective Availability
		SAA	South Atlantic Anomaly
NICMOS	Near Infrared Camera and Multi Object Spectrograph	SAC-C	Satélite de Aplicaciones Científicas-C
		SAC	Strategic Air Command
NIR	Near Infrared	SAF	Satellite Application Facility
NIZEMI	Niedergeschwindigkeits-Zentrifugen-Mikroskop (Low Velocity Centrifuge Microscope)	SAR	Synthetic Aperture Radar
		SARSAT	Search and Rescue Satellite-Aided Tracking
NLES	Navigation Land Earth Station	SCIAMACHY	Scanning Imaging Absorption Spectrometer for Atmospheric Cartography
NNR NUVEL	No Net Rotation Northwestern University Velocity		
NOAA	National Oceanic and Atmospheric Administration	SCR	Solar Cosmic Radiation
		SEI	Space Exploration Initiative
NPA	Non-Precision Approach	SEM	Scanning Electron Microscope
NPP	Net Primary Productivity	SETA	Solidification along an Eutectic Path in Ternary Alloys
NRL	Naval Research Laboratory		
NUVEL	Northwestern University Velocity	SEVIRI	Spinning Enhanced Visible and Infrared Imager
NWP	Numerical Weather Prediction		
OECD	Organisation of Economic Co-operation and Development	SGG	Satellite Gravity Gradiometry
		SIM	Space Interferometry Mission
OPTIS	Optical Test of the Isotropy of Space	SIRTF	Space Infrared Telescope Facility
PACF	Performance Assessment and Check-Out Facility	SLR	Satellite Laser Ranging
		SM	Service Mission
PAR	Photosynthetic Active Radiation	SMS	Space Motion Sickness
PARCS	Primary Atomic Reference Clock	SNC	Shergotty, Nakhla, Chassigny (meteorite)
PELCOM	Pan-European Land Cover Monitoring	SOFIA	Stratospheric Observatory for Far Infra-red Astronomy
PHARAO	Projet d'Horloge Atomique par Refroidissement d'Atomes en Orbite		
		SOHO	Solar Heliospheric Observatory

SPEAR	Spectroscopy of Plasma Evolution from Astrophysical Radiation		TM	Thematic Mapper
SPOT	Satellite Probatoire d'Observation de la Terre		TMI	TRMM Microwave Imager
			TOGA	Tropical Ocean/Global Atmosphere Project
SPS	Standard Positioning Service		TOMS	Total Ozone Mapping Spectrometer
SQUID	Superconducting Quantum Interference Device		TOPEX	Topography Experiment
SRTM	Shuttle Radar Topography Mission		TPF	Terrestrial Planet Finder
SSL	Space Science Laboratory		TRMM	Tropical Rainfall Measuring Mission
SSM/I	Special Sensor Microwave Imager		TWTA	Traveling Wave Tube Amplifier
SST	Satellite-to-Satellite Tracking		UGR	Universality of Gravitational Redshift
SST	Sea Surface Temperature		UHV	Ultra High Vacuum
ST-ECF	Space Telescope European Coordination Facility		UMTS	Universal Mobile Telecommunications System
STEP	Satellite Test of Equivalence Principle		UNFCCC	United Nation Framework Convention on Climate Change
STIS	Space Telescope Imaging Spectrograph		USAF	US Air Force
STM	Space-Time Mission		USOC	User Operation Center
STS	Space Transportation System		UTC	Universal Time Coordinate
STSAT	Space and Technology Satellite 1		UV	Ultraviolet
STScI	Space Telescope Science Institute		VLBI	Very Long Baseline Interferometry
SUE	Superfluid Universality Experiment		VLT	Very Large Telescope
SUMO	Superconducting Microwave Oscillator		VNAV	Vertical Navigation
TANGO	Trans-Atlantic Network for Geodynamics and Oceanography		VSAT	Very Small Aperture Terminal
			WAAS	Wide Area Augmentation System
TAO	Tropical Atmosphere Ocean Project		WCRP	World Climate Research Programme
TDRS	Tracking and Data Relay Satellite		WEGENER	Working Group for the Establishment of Networks for Earthquake Research
TDRSS	Tracking and Data Relay Satellite System			
TEC	Total Electron Content		WEP	Weak Equivalence Principle
TEMPUS	Tiegelfreies ElektroMagnetisches Prozessieren Unter Schwerelosigkeit (Containerless Electromagnetic Processing under Weightlessness)		WESTPAC	Western Pacific
			WFPC	Wide Field Planetary Camera
			WISE	Wide-Field Infrared Survey Explorer
			WLAN	Wireless Local Area Network
TES	Transition Edge Sensor		WMAP	Wilkinson Microwave Anisotropy Probe
TEXUS	Technologische Experimente Unter Schwerelosigkeit (Technological Experiments in Microgravity)		WMAP	Microwave Background Anisotropy Probe
			WMO	World Meteorological Organization
THERMOLAB	Thermophysical Properties Laboratory		WVR	Water Vapor Radiometer
			WWW	World Weather Watch
THM	Traveling Heater Method		XEUS	X-Ray Evolving Universe Spectrometer
TIROS	Television Infrared Observation Satellite		XMM	X-Ray Multi-Mirror
TITUS	Tubular Furnace with Integrated Thermal Analysis under Space Conditions			

Keyword Index